INTRODUCTION TO PROTEIN STRUCTURE PREDICTION

WILEY SERIES ON BIOINFORMATICS: COMPUTATIONAL TECHNIQUES AND ENGINEERING

Series Editors, Yi Pan & Albert Zomaya

Knowledge Discovery in Bioinformatics: Techniques, Methods and Applications / Xiaohua Hu & Yi Pan

Grid Computing for Bioinformatics and Computational Biology / Albert Zomaya & El-Ghazali Talbi

Analysis of Biological Networks / Björn H. Junker & Falk Schreiber

Bioinformatics Algorithms: Techniques and Applications / Ion Mandoiu & Alexander Zelikovsky

Machine Learning in Bioinformatics / Yanqing Zhang & Jagath C. Rajapakse

Biomolecular Networks / Luonan Chen, Rui-Sheng Wang, & Xiang-Sun Zhang

Computational Systems Biology / Huma Lodhi

Computational Intelligence and Pattern Analysis in Biology Informatics / Ujjwal Maulik, Sanghamitra, & Jason T. Wang

Mathematics of Bioinformatics: Theory, Practice, and Applications / Matthew He

Introduction to Protein Structure Prediction: Methods and Algorithms / Huzefa Rangwala & George Karypis

INTRODUCTION TO PROTEIN STRUCTURE PREDICTION

Methods and Algorithms

Edited by

HUZEFA RANGWALA
GEORGE KARYPIS

WILEY

A JOHN WILEY & SONS, INC., PUBLICATION

Published by John Wiley & Sons, Inc., Hoboken, New Jersey
Published simultaneously in Canada

Library of Congress Cataloging-in-Publication Data:

Rangwala, Huzefa.
Introduction to protein structure prediction : methods and algorithms / Huzefa Rangwala, George Karypis.
p. cm.—(Wiley series in bioinformatics; 14)
Includes bibliographical references and index.
ISBN 978-0-470-47059-6 (hardback)
1. Proteins—Structure—Mathematical models. 2. Proteins—Structure—Computer simulation. I. Karypis, G. (George) II. Title.
QP551.R225 2010
572′.633—dc22

2010028352

Printed in Singapore

10 9 8 7 6 5 4 3 2 1

CONTENTS

PREFACE

PROTEIN STRUCTURE PREDICTION

Proteins play a crucial role in governing several life processes. Stunningly complex networks of proteins perform innumerable functions in every living cell. Knowing the function and structure of proteins is crucial for the development of better drugs, higher yield crops, and even synthetic biofuels. As such, knowledge of protein structure and function leads to crucial advances in life sciences and biology. The motivation behind the structural determination of proteins is based on the belief that structural information provides insights as to their function, which will ultimately result in a better understanding of intricate biological processes.

Breakthroughs in large-scale sequencing have led to a surge in the available protein sequence information that has far outstripped our ability to characterize the structural and functional characteristic of these proteins. Several research groups have been working on determining the three-dimensional structure of the protein using a wide variety of computational methods. The problem of unraveling the relationship between the amino acid sequence of a protein and its three-dimensional structure has been one of the grand challenges in molecular biology. The importance and the far reaching implications of being able to predict the structure of a protein from its amino acid sequence is manifested by the ongoing biennial competition on "Critical Assessment of Protein Structure Prediction" (CASP) that started more than 16 years ago. CASP is designed to assess the performance of current structure prediction methods and over the years the number of groups that have been participating in it continues to increase.

This book presents a series of chapters by authors who are involved in the task of structure determination and using modeled structures for applications involving drug discovery and protein design. The book is divided into the following themes.

BACKGROUND ON STRUCTURE PREDICTION

Chapter 1 provides an introduction to the protein structure prediction problem along with information about databases and resources that are widely used. Chapters 2 and 3 provide information regarding two very important initiatives in the field: (i) the structure prediction flagship competition (CASP), and (ii) the protein structure initiative (PSI), respectively. Since many of the approaches developed have been tested in the CASP competition, Chapter 2 lays the foundation for the need for such an evaluation, the problem definitions, significant innovations, competition format, as well as future outlook. Chapter 3 describes the protein structure initiative, which is designed to determine representative three-dimensional structures within the human genome.

PREDICTION OF STRUCTURAL ELEMENTS

Within each structural entity called a protein there lies a set of recurring substructures, and within these substructures are smaller substructures. Beyond the goal of predicting the three-dimensional structure of a protein from sequence several other problems have been defined and methods have been developed for solving the same. Chapters 4–6 provide the definitions of these recurring substructures called local alphabets or secondary structures and the computational approaches used for solving these problems. Chapter 6 specifically focuses on a class of transmembrane proteins known to be harder to crystallize. Knowing the pairs of residues within a protein that are within contact or at a closer distance provides useful distance constraints that can be used while modeling the three-dimensional structure of the protein. Chapter 7 focuses on the problem of contact map prediction and also shows the use of sophisticated machine learning methods to solve the problem. A successful solution for each of these subproblems assists in solving the overarching protein structure prediction problem.

TERTIARY STRUCTURE PREDICTION

Chapters 8–11 discuss the widely used structure prediction methods that rely on homology modeling, threading, and fragment assembly. Chapters 8–9 discuss the problems of fold recognition and remote homology detection that attempt to model the three-dimensional structure of a protein using known structures. Chapters 10 and 11 discuss a combination of threading-based approaches along with modeling the protein in parts or fragments and usually helps in modeling the structure of proteins known not to have a close homolog within the structure databases. Chapter 12 is a survey of the hybrid methods that use a combination of the computational and experimental methods to achieve high-resolution protein structures in a high-throughput manner.

Chapter 17 provides information about the challenges in modeling transmembrane proteins along with a discussion of some of the widely used methods for these sets of proteins.

Chapter 13 describes the loop prediction problem and how the technique can be used for refinement of the modeled structures. Chapters 14 and 15 assess the modeled structures and provide a notion of the quality of structures. This is extremely important from a biologist's perspective who would like to have a metric that describes the goodness of the structure before use. Chapter 19 provides insights into the different conformations that a protein may take and the approaches used to sample the different conformations.

FUNCTIONAL INSIGHTS

Certain parts of the protein structure may be conserved and interact with other biomolecules (e.g., proteins, DNA, RNA, and small molecules) and perform a particular function due to such interactions. Chapter 16 discusses the problem of ligand-binding site prediction and its role in determining the function of the proteins. The approach uses some of the homology modeling principles used for modeling the entire structure. Chapter 18 introduces a computational model that detects the differences between protein structure (modeled or experimentally-determined) and its modeled mutant. Chapter 20 describes the use of molecular dynamic-based approaches for modeling mutants.

ACKNOWLEDGEMENTS

We wish to acknowledge the many people who have helped us with this project. We firstly thank all the coauthors who spent time and energy to edit their chapters and also served as reviewers by providing critical feedback for improving other chapters. Kevin Deronne, Christopher Kauffman, and Rezwan Ahmed also assisted in reviewing several of the chapters and helped the book take a form that is complete on the topic of protein structure prediction and exciting to read. Finally, we wish to thank our families and friends.

We hope that you as a reader benefit from this book and feel as excited about this field as we are.

HUZEFA RANGWALA

GEORGE KARYPIS

CONTRIBUTORS

NIR BEN-TAL, Department of Biochemistry and Molecular Biology, Tel Aviv University, Tel Aviv, Israel

AURÉLIE BORNOT, Institut National de la Santé et de la Recherche Médicale, UMR-S 665, Dynamique des Structures et Interactions des Macromolécules Biologiques (DSIMB), Université Paris Diderot, Paris, France

ALEXANDRE G. DE BREVERN, Institut National de la Santé et de la Recherche Médicale, Université Paris Diderot, Institut National de la Transfusion Sanguine, 75015, Paris, France

JIANLIN CHENG, Computer Science Department and Informatics Institute University of Missouri, Columbia, MO 65211

FENG DING, Department of Biochemistry and Biophysics University of North Carolina—Chapel Hill, NC 27599

NICHOLAS E. DIXON, School of Chemistry, University of Wollongong, NSW 2522, Australia

NIKOLAY V. DOKHOLYAN, Department of Biochemistry and Biophysics, University of North Carolina, Chapel Hill, NC 27599

ESHEL FARAGGI, Indiana University School of Informatics, Indiana University-Purdue University Indianapolis, and Center for Computational Biology and Bioinformatics, Indiana University School of Medicine, Indianapolis, IN 46202

KRZYSZTOF FIDELIS, Protein Structure Prediction Center, Genome Center, University of California, Davis, Davis, CA

ANDRAS FISER, Department of Systems and Computational Biology and Department of Biochemistry, Albert Einstein College of Medicine, Bronx, NY 10461

NARCIS FERNANDEZ-FUENTES, Leeds Institute of Molecular Medicine, University of Leeds, Leeds, UK

Adam Godzik, Program in Bioinformatics and Systems Biology, Sanford-Burnham Medical Research Institute, La Jolla, CA 92037

Thomas Huber, The University of Queensland, School of Chemistry and Molecular Biosciences, QLD, Australia

Agnel Praveen Joseph, Institut National de la Santé et de la Recherche Médicale, UMR-S 665, Dynamique des Structures et Interactions des Macromolécules Biologiques (DSIMB), Université Paris Diderot, Paris, France

Kazuhiko Kanou, School of Pharmacy, Kitasato University, Tokyo 108-8641, Japan

George Karypis, Department of Computer Science, University of Minnesota Minneapolis, MN 55455

Chris Kauffman, Department of Computer Science, University of Minnesota, Minneapolis, MN 55455

Bostjan Kobe, The University of Queensland, School of Chemistry and Molecular Biosciences, Brisbane, Australia

Andriy Kryshtafovych, Protein Structure Prediction Center, Genome Center, University of California, Davis, Davis, CA

Alberto J.M. Martin, Complex and Adaptive Systems Lab, School of Computer Science and Informatics, UCD Dublin, Ireland

Majid Massa, Department of Bioinformatics and Computational Biology, George Mason University, Manassas, VA 20110

Liam J. McGuffin, School of Biological Sciences, The University of Reading, Reading, UK

Catherine Mooney, Shields Lab, School of Medicine and Medical Science, University College Dublin, Ireland

John Moult, Institute for Bioscience and Biotechnology Research, University of Maryland, Rockville, MD 20850

Dmitri Mouradov, The University of Queensland, School of Chemistry and Molecular Biosciences, QLD, Australia

Christine Orengo, Department of Structural and Molecular Biology, University College London, London UK

Shashi Bhushan Pandit, Center for the Study of Systems Biology, School of Biology, Georgia Institute of Technology, Atlanta, GA 30318

Gianluca Pollastri, Complex and Adaptive Systems Lab, School of Computer Science and Informatics, UCD Dublin, Ireland

Huzefa Rangwala, Department of Computer Science, George Mason University, Fairfax, VA 22030

BURKHARD ROST, Department of Biochemistry and Molecular Biophysics, Columbia University, New York, NY 10032

AMBRISH ROY, Center for Computational Medicine and Bioinformatics, University of Michigan, Ann Arbor, MI 48109

MAYA SCHUSHAN, Department of Biochemistry and Molecular Biology, Tel Aviv University, Tel Aviv, Israel

AMARDA SHEHU, Department of Computer Science, George Mason University, Fairfax, VA 22030

MAYUKO TAKEDA-SHITAKA, School of Pharmacy, Kitasato University, Tokyo 108-8641, Japan

ISTVÁN SIMON, Intsitute of Enzymology, BRC, Hungarian Academy of Sciences, Budapest, Hungary

JEFFREY SKOLNICK, Center for the Study of Systems Biology, School of Biology, Georgia Institute of Technology Atlanta, GA 30318

ALLISON N. TEGGE, Computer Science Department and Informatics Institute, University of Missouri, Columbia, MO 65211

GENKI TERASHI, School of Pharmacy, Kitasato University, Tokyo 108-8641, Japan

GÁBOR E. TUSNADY, Intsitute of Enzymology, BRC, Hungarian Academy of Sciences, Budapest, Hungary

HIDEAKI UMEYAMA, School of Pharmacy, Kitasato University, Tokyo 108-8641, Japan

IOSIF I. VAISMAN, Department of Bioinformatics and Computational Biology, George Mason University, Manassas, VA 20110

IAN WALSH, Complex and Adaptive Systems Lab, School of Computer Science and Informatics, UCD Dublin, Ireland

ZHENG WANG, Computer Science Department, University of Missouri, Columbia, MO 65211

SITAO WU, Center for Computational Medicine and Bioinformatics, University of Michigan, Ann Arbor, MI 48109

SHUANGYE YIN, Department of Biochemistry and Biophysics, University of North Carolina, Chapel Hill, NC 27599

YANG ZHANG, Center for Computational Medicine and Bioinformatics, University of Michigan, Ann Arbor, MI 48109

HONGYI ZHOU, Center for the Study of Systems Biology, School of Biology Georgia Institute of Technology, Atlanta, GA 30318

Yaoqi Zhou, Indiana University School of Informatics, Indiana University-Purdue University Indianapolis, and Center for Computational Biology and Bioinformatics, Indiana University School of Medicine, Indianapolis, IN 46202

CHAPTER 1

INTRODUCTION TO PROTEIN STRUCTURE PREDICTION

HUZEFA RANGWALA
Department of Computer Science
George Mason University
Fairfax, VA

GEORGE KARYPIS
Department of Computer Science
University of Minnesota
Minneapolis, MN

Proteins have a vast influence on the molecular machinery of life. Stunningly complex networks of proteins perform innumerable functions in every living cell. Knowing the function and structure of proteins is crucial for the development of improved drugs, better crops, and even synthetic biofuels. As such, knowledge of protein structure and function leads to crucial advances in life sciences and biology.

With recent advances in large-scale sequencing technologies, we have seen an exponential growth in protein sequence information. Protein structures are primarily determined using X-ray crystallography or nuclear magnetic resonance (NMR) spectroscopy, but these methods are time consuming, expensive, and not feasible for all proteins. The experimental approaches to determine protein function (e.g., gene knockout, targeted mutation, and inhibitions of gene expression studies) are low-throughput in nature [1,2]. As such, our ability to produce sequence information far outpaces the rate at which we can produce structural and functional information.

Consequently, researchers are increasingly reliant on computational approaches to extract useful information from experimentally determined three-dimensional (3D) structures and functions of proteins. Unraveling the

Introduction to Protein Structure Prediction: Methods and Algorithms,
Edited by Huzefa Rangwala and George Karypis

relationship between pure sequence information and 3D structure and/or function remains one of the fundamental challenges in molecular biology.

Function prediction is generally approached by using inheritance through homology [2], that is, proteins with similar sequences (common evolutionary ancestry) frequently carry out similar functions. However, several studies [2–4] have shown that a stronger correlation exists between structure conservation and function, that is, structure implies function, and a higher correlation exists between sequence conservation and structure, that is, sequence implies structure (sequence → structure → function).

1.1. INTRODUCTION TO PROTEIN STRUCTURES

In this section we introduce the basic definitions and facts about protein structure, the four different levels of protein structure, as well as provide details about protein structure databases.

1.1.1. Protein Structure Levels

Within each structural entity called a protein lies a set of recurring substructures, and within these substructures are smaller substructures still. As an example, consider hemoglobin, the oxygen-carrying molecule in human blood. Hemoglobin has four domains that come together to form its quaternary structure. Each domain assembles (i.e., folds) itself independently to form a tertiary structure. These tertiary structures are comprised of multiple secondary structure elements—in hemoglobin's case α-helices. α-Helices (and their counterpart β-sheets) have elegant repeating patterns dependent upon sequences of amino acids.

1.1.1.1. Primary Structure. Amino acids form the basic building blocks of proteins. Amino acids consists of a central carbon atom (C_α) attached by an amino (NH_2), a carboxyl (COOH) group, and a side chain (R) group. The side chain group differentiates the various amino acids. In case of proteins, there are primarily 20 different amino acids that form the building blocks. A protein is a chain of amino acids linked with peptide bonds. Pairs of amino acid form a peptide bond between the amino group of one and the carboxyl group of the other. This polypeptide chain of amino acids is known as the primary structure or the protein sequence.

1.1.1.2. Secondary Structure. A sequence of characters representing the secondary structure of a protein describes the general 3D form of local regions. These regions organize themselves independently from the rest of the protein into patterns of repeatedly occurring structural fragments. The most dominant local conformations of polypeptide chains are α-helices and β-sheets. These local structures have a certain regularity in their form, attributed to the hydrogen bond interactions between various residues. An α-helix has a coil-like structure, whereas a β-sheet consists of parallel strands of residues. In addition to regular secondary structure elements, irregular shapes form an important part of the structure and function of proteins. These elements are typically termed coil regions.

Secondary structure can be divided into several types, although usually at least three classes (α-helix, coils, and β-sheet) are used. No unique method of assigning residues to a particular secondary structure state from atomic coordinates exists, although the most widely accepted protocol is based on the Dictionary of Protein Secondary Structure (DSSP) algorithm [5]. DSSP uses the following structural classes: H (α-helix), G (3_{10}-helix), I (π-helix), E (β-strand), B (isolated β-bridge), T (turn), S (bend), and – (other). Several other secondary structure assignment algorithms use a reduction scheme that converts this eight-state assignment down to three states by assigning H and G to the helix state (H), E and B to a the strand state (E), and the rest (I, T, S, and –) to a coil state (C). This is the format generally used in structure databases.

1.1.1.3. Tertiary Structure. The tertiary structure of the protein is defined as the global 3D structure, represented by 3D coordinates for each atoms. These tertiary structures are comprised of multiple secondary structure elements, and the 3D structure is a function of the interacting side chains between the different amino acids. Hence, the linear ordering of amino acids forms secondary structure; arranging secondary structures yields tertiary structure.

1.1.1.4. Quaternary Structure. Quaternary structures represent the interaction between multiple polypeptide chains. The interaction between the various chains is due to the non-covalent interactions between the atoms of the different chains. Examples of these interactions include hydrogen bonding, van Der Walls interactions, ionic bonding, and disulfide bonding.

Research in computational structure prediction concerns itself mainly with predicting secondary and tertiary structures from known experimentally determined primary structure or sequence. This is due to the relative ease of determining primary structure and the complexity involved in quaternary structure.

1.1.2. Protein Sequence and Structure Databases

The large amount of protein sequence information, experimentally determined structure information, and structural classification information is stored in publicly available databases. In this section we review some of the databases that are used in this field, and provide their availability information in Table 1.1.

1.1.2.1. Sequence Databases. The Universal Protein Resource (UniProt) [6] is the most comprehensive warehouse containing information about protein

TABLE 1.1 Protein Sequence and Structure Databases

Database	Information	Availability Link
UniProt	Sequence	http://www.pir.uniprot.org/
UniRef	Cluster sequences	http://www.pir.uniprot.org/
NCBI nr	Nonredundant sequences	ftp://ftp.ncbi.nlm.nih.gov/blast/db/
PDB	Structure	http://www.rcsb.org/
SCOP	Structure classification	http://scop.mrc-lmb.cam.ac.uk/scop/
CATH	Structure classification	http://www.cathdb.info/
FSSP	Structure classification	http://www.ebi.ac.uk/dali/fssp/
ASTRAL	Compendium	http://astral.berkeley.edu/

The databases referred to in this table are most popular for protein structure-related information.

sequences and their annotation. It is a database of protein sequences and their function that is formed by aggregating the information present in the Swiss-Prot, TrEMBL, and Protein Information Resources (PIR) databases. The UniProtKB 13.2 version of database (released on April 8, 2008) consists of 5,939,836 protein sequence entries (Swiss-Prot providing 362,782 entries and TrEMBL providing 5,577,054 entries).

However, several proteins have high pairwise sequence identity, and as such lead to redundant information. The UniProt database [6] creates a subset of sequences such that the sequence identity between all pairs of sequences within the subset is less than a predetermined threshold. In essence, UniProt contains the UniRef100, UniRef90, and UniRef50 subsets where within each group the sequence identity between a pair of sequences is less than 100%, 90%, and 50%, respectively.

The National Center for Biotechnology Information (NCBI) also provides a nonredundant (NCBI nr) database of protein sequences using sequences from a wide variety of sources. This database will have pairs of proteins with high sequence identity, but removes all the duplicates. The NCBI nr version 2.2.18 (released on March 2, 2008) contains 6,441,864 protein sequences.

1.1.2.2. Protein Data Bank (PDB). The Research Collaboratory for Structural Bioinformatics (RSCB) PDB [7] stores experimentally determined 3D structure of biological macromolecules including nucleotides and proteins. As of April 20, 2008 this database consists of 46,287 protein structures that are determined using X-ray crystallography (90%), NMR (9%), and other methods like Cryo-electron microscopy (Cryo-EM). These experimental methods are time-consuming, expensive, and need protein to crystallize.

1.1.2.3. Structure Classification Databases. Various methods have been proposed to categorize protein structures. These methods are based on the pairwise structural similarity between the protein structures, as well as the topological and geometric arrangement of atoms and predominant secondary structure like subunits. Structural Classification of Proteins (SCOP) [8], Class, Architecture, Topology, and Homologous superfamily (CATH) [9], and Families of Structurally Similar Proteins (FSSP) [10] are three widely used structure classification databases. The classification methodology involves breaking a protein chain or complex into independent folding units called domains, and then classifying these domains into a set of hierarchical classes sharing similar structural characteristics.

SCOP Database. SCOP [8] is a manually curated database that provides a detailed and comprehensive description of the evolutionary and structural relationships between proteins whose structure is known (present in the PDB). SCOP classifies proteins structures using visual inspection as well as structural comparison using a suite of automated tools. The basic unit of classification is generally a domain. SCOP classification is based on four hierarchical levels that encompass evolutionary and structural relationships [8]. In particular, proteins with clear evolutionary relationship are classified to be within the same *family*. Generally, protein pairs within the same family have pairwise residue identities greater than 30%. Protein pairs with low sequence identity, but whose structural and functional features imply probably common evolutionary information, are classified to be within the same *superfamily*. Protein pairs with similar major secondary structure elements and topological arrangement of substructures (as well as favoring certain packing geometries) are classified to be within the same *fold*. Finally, protein pairs having a predominant set of secondary structures (e.g., all α-helices proteins) lie within the same *class*. The four hierarchical levels, that is, family, superfamily, fold, and class define the structure of the SCOP database.

The SCOP 1.73 version database (released on September 26, 2007) classifies 34,494 PDB entries (97,178 domains) into 1086 unique folds, 1777 unique superfamilies, and 3464 unique families.

CATH Database. CATH [9] database is a semi-automated protein structure classification database like the SCOP database. CATH uses a consensus of three automated classification techniques to break a chain into domains and classify them in the various structural categories [11]. Domains for proteins that are not resolved by the consensus approach are determined manually. These domains are then classified into the following hierarchical categories using both manual and automated methods in conjunction.

The first level membership, *class*, is determined based on the secondary structure composition and packing within the structure. The second level, *architecture*, clusters proteins sharing the same orientation of the secondary structure element but ignoring the connectivity between these substructural units. The third level, *topology*, groups protein pairs with a high structure alignment score as determined by the SSAP [12] algorithm, and in essence share both overall shape and connectivity of secondary structures. The fourth level, *homologous* pairs, shares a common ancestor and is identified by sequence alignment as well as the SSAP structure alignment method. Structures are further classified to be within the same *sequence families* if they share a high sequence identity.

The CATH 3.1.0 version database (released on January 19, 2007) classifies 30,028 (93,885 domains) proteins from the PDB into 40 architecture-level classes, 1084 topology-level classes, and 2091 homologous-level classes.

FSSP Database. The FSSP [10] is a structure classification database. FSSP uses an automatic classification scheme that employs exhaustive structure-to-structure alignment of proteins using the DALI [13] alignment. FSSP does not provide a hierarchical classification like the SCOP and CATH databases, but instead employs a hierarchical clustering algorithm using the pairwise structure similarity scores that can be used for the definition of fold classes—however, not very accurate.

There have been several studies [14,15] analyzing the relationship between the SCOP, CATH, and FSSP databases for representing the fold space for proteins. The major disagreement between the three databases lies in the domain identification step, rather than the domain classification step. A high percentage of agreement exists between the SCOP, CATH, and FSSP databases especially at the fold level with sequence identity greater than 25%.

ASTRAL Compendium. The A Structural Alignment Library (ASTRAL) [16–18] compendium is a set of database and tools used for analysis of protein structures and sequences. This database is partially derived from, and augments, the SCOP [8] database. ASTRAL provides accurate linkage between the biological sequence and the reported structure in PDB, and identifies the domains within the sequence using SCOP. Since the majority of domain sequences in PDB are very similar to others, ASTRAL tools reduce the redundancy by selecting high-quality representatives. Using the reduced nonredundant set of representation proteins allows for sampling of all the different structures in the PDB. This also removes bias due to overrepresented structures. Subsets provided by ASTRAL are based on SCOP domains and use high-quality structure files only. Independent subsets of representative proteins are identified using a greedy algorithm with filtering criterion based on pairwise sequence identity determined using the Basic Local Alignment Search Tool (BLAST) [19], an e-value-based threshold, or a SCOP level-based filter.

1.2. PROTEIN STRUCTURE PREDICTION METHODS

One of the biggest goals in structural bioinformatics is the prediction of the 3D structure of a protein from its one-dimensional (1D) protein sequence. The goal is to be able to determine the shape (known as a fold) that a given amino acid sequence will adopt. The problem is further divided based on whether the sequence will adopt a new fold or bear resemblance to an existing fold (template) in some protein structure database. Fold recognition is easy when the sequence in question has a high degree of sequence similarity to a sequence with known structure [20]. If the two sequences share evolutionary ancestry they are said to be homologous. For such sequence pairs we can build the structure for the query protein by choosing the structure of the known homologous sequence as template. This is known as comparative modeling.

In the case where no good template structure exists for the query, one must attempt to build the protein tertiary structure from scratch. These methods are usually called *ab initio* methods. In a third-fold prediction scenario, there may not necessarily be a good sequence similarity with a known structure, but a structural template may still exist for the given sequence. To clarify this case, if one were aware of the target structure then they could extract the template using structure–structure alignments of the target against the entire structural database. It is important to note that the target and template need not be homologous. These two cases define the fold prediction (homologous) and fold prediction (analogous) problems during the Critical Assessment of Protein Structure Prediction (CASP) competition.

1.2.1. Comparative Modeling

Comparative Modeling or homology modeling is used when there exists a clear relationship between the sequence of a query protein (unknown structure) and a sequence of a known structure. The most basic approach to structure prediction for such (query) proteins is to perform a pairwise sequence alignment against each sequence in protein sequence databases. This can be accomplished using sequence alignment algorithms such as Smith-Waterman [21] or sequence search algorithms (e.g., BLAST [19]) With a good sequence alignment in hand, the challenge in comparative modeling becomes how to best build a 3D protein structure for a query protein using the template structure.

The heart of the above process is the selection of a suitable structural template based on sequence pair similarity. This is followed by the alignment of query sequence to the template structure selected to build the backbone of the query protein. Finally the entire modeled structure is refined by loop construction and side chain modeling. Several comparative modeling methods, more commonly known as modeler programs, have been developed over the past several years [22,23] focusing on various parts of the problem.

As seen in the various years of CASP [24,25], the span of comparative modeling approaches [22,23] follows five basic steps: (i) selecting one or suitable templates, (ii) utilizing sensitive sequence template alignment algorithms, (iii) building a protein model using the sequence structure alignment as reference, (iv) evaluating the quality of the model, and (v) refining the model. These typical steps for the comparative modeling process are shown in Figure 1.1.

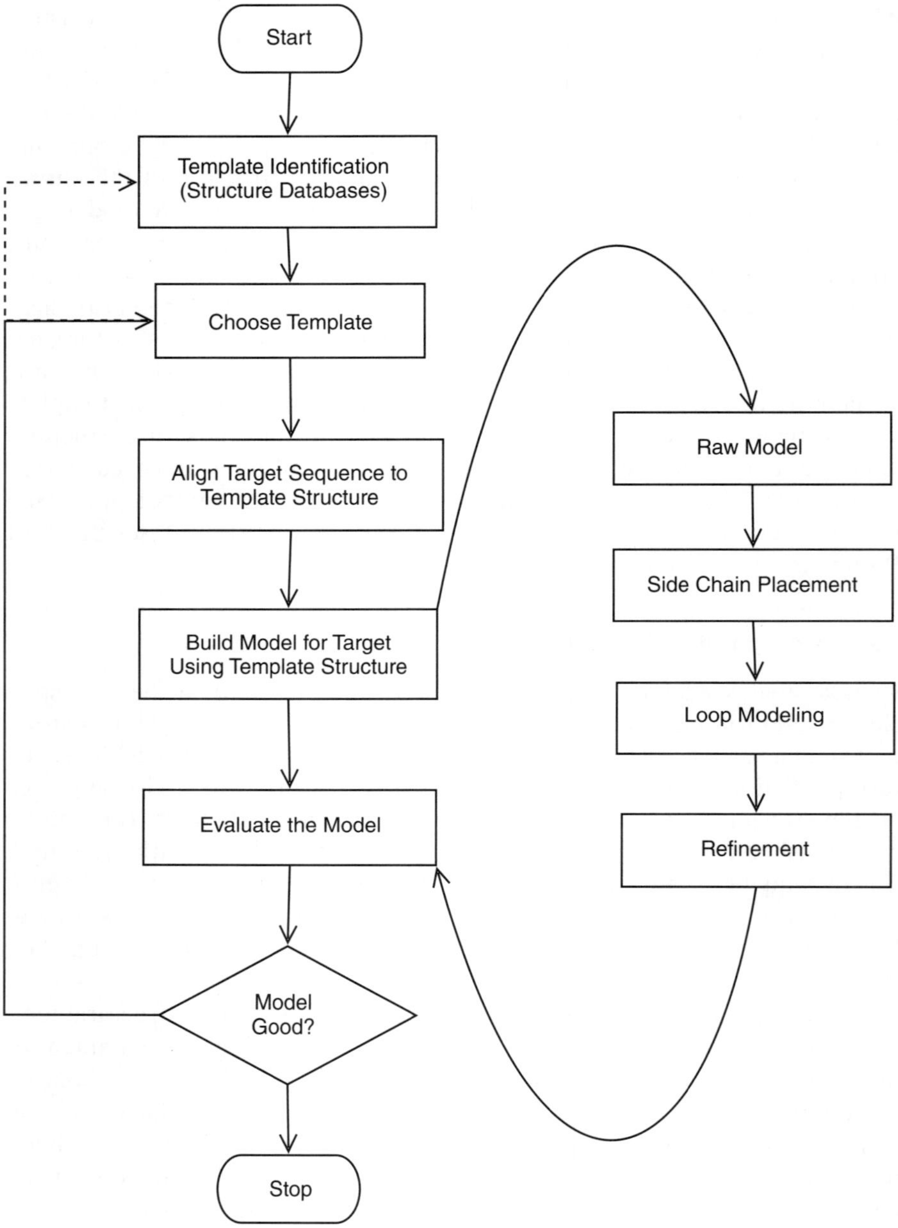

FIGURE 1.1 Flowchart for the comparative modeling process.

1.2.2. Fold Prediction (Homologous)

While satisfactory methods exist to detect homologs (proteins that share similar evolutionary ancestry) with high levels of similarity, accurately detecting homologs at low levels of sequence similarity (remote homology detection) remains a challenging problem. Some of the most popular approaches for remote homology prediction compare a protein with a collection of related proteins using methods such as Position-Specific Iterative-BLAST (PSI-BLAST) [26], protein family profiles [27], hidden Markov models (HMMs) [28,29], and Sequence Alignment and Modeling System (SAM) [30]. These schemes produce models that are generative in the sense that they build a model for a set of related proteins and then check to see how well this model explains a candidate protein.

In recent years, the performance of remote homology detection has been further improved through the use of methods that explicitly model the differences between the various protein families (classes) by building discriminative models. In particular, a number of different methods that use Support Vector Machines (SVM) [31] have been developed to produce results that are generally superior to those produced by either pairwise sequence comparisons or approaches based on generative models—provided there are sufficient training data [32–39].

1.2.3. Fold Prediction (Analogous)

Occasionally a query sequence will have a native fold similar to another known fold in a database, but the two sequences will have no detectable similarity. In many cases the two proteins will lack an evolutionary relationship as well. As the definition of this problem relies on the inability of current methods to detect sequential similarity, the set of proteins falling into this category remains in flux. As new methods continue to improve at finding sequential similarities as a result of increasing database size and better techniques, the number of proteins in question decreases. Techniques to find structures for such query sequences revolve around mounting the query sequence on a series of template structures in a process known as threading [40–42]. An objective energy function provides a score for each alignment, and the highest scoring template is chosen.

Obviously, if the correct template does not exist in the series then the method will not produce an accurate prediction. As a result of this limitation, predicting the structure of proteins in this category is as challenging as predicting protein targets that are part of the new or rare folds.

1.2.4. *Ab Initio*

Techniques to predict novel protein structure have come a long way in recent years, although a definitive solution to the problem remains elusive. Research in this area can be roughly divided into fragment assembly [43–45] and first principle-based approaches, although occasionally the two are combined [46]. The former attempt to assign a fragment with known structure to a section of the unknown query sequence. The latter start with an unfolded conformation, usually surrounded by solvent, and allow simulated physical forces to fold the protein as would normally happen *in vivo*. Usually, algorithms from either class will use reduced representations of query proteins during initial stages to reduce the overall complexity of the problem.

Even in case of these *ab initio* prediction methods, the state-of-the-art methods [46–48] determine several template structures (using the template selection methods used in comparative modeling methods). The final protein is modeled using an assembly of fragments or substructures fitted together using a highly optimized approximate energy and statistics-based potential function.

This book presents methods developed for protein structure prediction. In particular methods and problems that are prevalent in a biennial structure prediction competition (CASP) are discussed in the first half of the book. The second half of the book discusses approaches that combine experimental and computational approaches for structure prediction and also new techniques for predicting structures of transmembrane proteins. Finally, the book discusses the applications of protein structure within the context of function prediction and drug discovery.

REFERENCES

1. G. Pandey, V. Kumar, and M. Steinbach. Computational approaches for protein function prediction: A survey. Technical Report 06-23, Department of Computer Science and Engineering, University of Minnesota, 2006.
2. D. Lee, O. Redfern, and C. Orengo. Predicting protein function from sequence and structure. *Nature Reviews. Molecular Cell Biology*, 8(12):995–1005, 2007.
3. J.C. Whisstock and A.M. Lesk. Prediction of protein function from protein sequence and structure. *Quarterly Reviews of Biophysics*, 36(3):307–340, 2003.
4. D. Devos and A. Valencia. Practical limits of function prediction. *Proteins*, 41(1):98–107, 2000.
5. W. Kabsch and C. Sander. Dictionary of protein secondary structure: Pattern recognition of hydrogen-bonded and geometrical features. *Biopolymers*, 22:2577–2637, 1983.
6. UniProt Consortium. The universal protein resource (uniprot). *Nucleic Acids Research*, 36(Database issue):D190–D195, 2008.
7. H.M. Berman, T.N. Bhat, P.E. Bourne, Z. Feng, G.G.H. Weissig, and J. Westbrook. The Protein Data Bank and the challenge of structural genomics. *Nature Structural Biology*, 7:957–959, 2000.
8. A.G. Murzin, S.E. Brenner, T. Hubbard, and C. Chothia. Scop: A structural classification of proteins database for the investigation of sequences and structures. *Journal of Molecular Biology*, 247:536–540, 1995.

9. C.A. Orengo, A.D. Mitchie, S. Jones, D.T. Jones, M.B. Swindells, and J.M. Thorton. Cath- a hierarchic classification of protein domain structures. *Structure*, 5(8):1093–1108, 1997.

10. L. Holm and C. Sander. The fssp database: Fold classification based on structurestructure alignment of proteins. *Nucleic Acids Research*, 24(1):206–209, 1996.

11. S. Jones, M. Stewart, A. Michie, M.B. Swindells, C. Orengo, and J.M. Thornton. Domain assignment for protein structures using a consensus approach: Characterization and analysis. *Protein Science*, 7(2):233–242, 1998.

12. W.R. Taylor and A.C. Orengo. Protein structure alignment. *Journal of Molecular Biology*, 208(1):1–22, 1989.

13. L. Holm and C. Sander. Protein structure comparison by alignment of distance matrices. *Journal of Molecular Biology*, 233(1):123–138, 1993.

14. C. Hadley and D. Jones. A systematic comparison of protein structure classifications: Scop, cath and fssp. *Structure*, 7(9):1099–1112, 1999.

15. R. Day, D.A.C. Beck, R.S. Armen, and V. Daggett. A consensus view of fold space: Combining SCOP, CATH, and the Dali Dom ain Dictionary. *Protein Science*, 12(10):2150–2160, 2003.

16. S.E. Brenner, P. Koehl, and M. Levitt. The astral compendium for sequence and structure analysis. *Nucleic Acids Research*, 28:254–256, 2000.

17. J.-M. Chandonia, N.S. Walker, L.L. Conte, P. Koehl, M. Levitt, and S.E. Brenner. ASTRAL compendium enhancements. *Nucleic Acids Research*, 30(1):260–263, 2002.

18. J.M. Chandonia, G. Hon, N.S. Walker, L.L. Conte, P. Koehl, M. Levitt, and S.E. Brenner. The astral compendium in 2004. *Nucleic Acids Research*, 32:D189–D192, 2004.

19. S.F. Altschul, W. Gish, E.W. Miller, and D.J. Lipman. Basic local alignment search tool. *Journal of Molecular Biology*, 215:403–410, 1990.

20. P. Bourne and H. Weissig. *Structural Bioinformatics*. Hoboken, NJ: John Wiley & Sons, 2003.

21. T.F. Smith and M.S. Waterman. Identification of common molecular subsequences. *Journal of Molecular Biology*, 147:195–197, 1981.

22. P.A. Bates and M.J.E Sternberg. Model building by comparison at casp3: Using expert knowledge and computer automation. *Proteins: Structure, Functions, and Genetics*, 3:47–54, 1999.

23. A. Fiser, R.K. Do, and A. Sali. Modeling of loops in protein structures. *Protein Science*, 9:1753–1773, 2000.

24. C. Venclovas. Comparative modeling in casp5: Progress is evident, but alignment errors remain a significant hindrance. *Proteins: Structure, Function, and Genetics*, 53:380–388, 2003.

25. C. Venclovas and M. Margelevicius. Comparative modeling in casp6 using consensus approach to template selection, sequence-structure alignment, and structure assessment. *Proteins: Structure, Function, and Bioinformatics*, 7:99–105, 2005.

26. S.F. Altschul, L.T. Madden, A.A. SchÃd'ffer, J. Zhang, Z. Zhang, W. Miller, and D.J. Lipman. Gapped blast and psi-blast: A new generation of protein database search programs. *Nucleic Acids Research*, 25(17):3389–3402, 1997.

27. M. Gribskov, A.D. McLachlan, and D. Eisenberg. Profile analysis: Detection of distantly related proteins. *PNAS*, 84:4355–4358, 1987.

28. A. Krogh, M. Brown, I. Mian, K. Sjolander, and D. Haussler. Hidden Markov models in computational biology: Applications to protein modeling. *Journal of Molecular Biology*, 235:1501–1531, 1994.

29. P. Baldi, Y. Chauvin, T. Hunkapiller, and M. McClure. Hidden Markov models of biological primary sequence information. *PNAS*, 91:1053–1063, 1994.

30. K. Karplus, C. Barrett, and R. Hughey. Hidden Markov models for detecting remote protein homologies. *Bioinformatics*, 14(10):846–856, 1998.

31. V. Vapnik. *Statistical Learning Theory*. New York: John Wiley, 1998.

32. T. Jaakkola, M. Diekhans, and D. Hassler. A dscriminative framework for detecting remote protein homologies. *Journal of Computational Biology*, 7(1/2):95–114, 2000.

33. L. Liao and W.S. Noble. Combining pairwise sequence similarity and support vector machines for detecting remote protein evolutionary and structural relationships. *Proceedings of the International Conference on Research in Computational Molecular Biology*, 225–232, 2002.

34. C. Leslie, E. Eskin, and W.S. Noble. The spectrum kernel: A string kernel for svm protein classification. *Proceedings of the Pacific Symposium on Biocomputing*, 564–575, 2002.

35. C. Leslie, E. Eskin, W.S. Noble, and J. Weston. Mismatch string kernels for svm protein classification. *Advances in Neural Information Processing Systems*, 20(4):467–476, 2003.

36. Y. Hou, W. Hsu, M.L. Lee, and C. Bystroff. Efficient remote homology detection using local structure. *Bioinformatics*, 19(17):2294–2301, 2003.

37. Y. Hou, W. Hsu, M.L. Lee, and C. Bystroff. Remote homology detection using local sequence-structure correlations. *Proteins: Structure, Function, and Bioinformatics*, 57:518–530, 2004.

38. H. Saigo, J.P. Vert, N. Ueda, and T. Akutsu. Protein homology detection using string alignment kernels. *Bioinformatics*, 20(11):1682–1689, 2004.

39. R. Kuang, E. Ie, K. Wang, M. Siddiqi, Y. Freund, and C. Leslie. Profile-based string kernels for remote homology detection and motif extraction. *Journal of Bioinformatics and Computational Biology*, 3:152–160, 2004.

40. D.T. Jones, W.R. Taylor, and J.M. Thorton. A new approach to protein fold recognition. *Nature*, 358:86–89, 1992.

41. D.T. Jones. Genthreader: An efficient and reliable protein fold recognition method for genomic sequences. *Journal of Molecular Biology*, 287(4):797–815, 1999.

42. J.U. Bowie, R. Luethy, and D. Eisenberg. A method to identify protein sequences that fold into a known three-dimensional structure. *Science*, 253:797–815, 1991.

43. K.T. Simons, C. Kooperberg, E. Huang, and D. Baker. Assembly of protein tertiary structures from fragments with similar local sequences using simulated annealing and bayesian scoring functions. *Journal of Molecular Biology*, 268:209–225, 1997.

44. K. Karplus, R. Karchin, J. Draper, J. Casper, Y. Mandel-Gutfreund, M. Diekhans, and R. Hughey. Combining local-structure, fold-recognition, and new fold methods for protein structure prediction. *Proteins: Structure, Function, and Genetics*, 53:491–496, 2003.

45. J. Lee, S.-Y. Kim, K. Joo, I. Kim, and J. Lee. Prediction of protein tertiary structure using profesy, a novel method based on fragment assembly and conformational space annealing. *Proteins: Structure, Function, and Bioinformatics*, 56:704–714, 2004.
46. C.A. Rohl, C.E.M. Strauss, K.M.S. Misura, and D. Baker. Protein structure prediction using rosetta. *Methods in Enzymology*, 383:66–93, 2004.
47. Y. Zhang. I-tasser server for protein 3d structure prediction. *BMC Bioinformatics*, 9:40, 2008.
48. Y. Zhang, A.J. Arakaki, and J. Skolnick. Tasser: An automated method for the prediction of protein tertiary structures in casp6. *Proteins: Structure, Function, and Bioinformatics*, 7:91–98, 2005.

CHAPTER 2

CASP: A DRIVING FORCE IN PROTEIN STRUCTURE MODELING

ANDRIY KRYSHTAFOVYCH and KRZYSZTOF FIDELIS
Protein Structure Prediction Center
Genome Center
University of California, Davis
Davis, CA

JOHN MOULT
Center for Advanced Research in Biotechnology
University of Maryland, College Park
College Park, MD

2.1. WHY CRITICAL ASSESSMENT OF PROTEIN STRUCTURE PREDICTION (CASP) WAS NEEDED?

More than half a century has elapsed since it was shown that amino acid sequence determines the three-dimensional structure of a protein [1], but a general procedure to translate sequence into structure is still to be established. Several dozen methods for generating protein structure from sequence have been developed, providing different levels of model accuracy in different modeling circumstances. With such a variety of modeling approaches and success levels, it was important to establish an objective procedure to compare the performances of the methods and learn their advantages and weaknesses. Also, with only sparse reports on the performance of most methods it was difficult to arrive at a clear understanding of current capabilities and bottlenecks in the field. Specifically, it was not possible to address many key questions about modeling methods, in particular:

Introduction to Protein Structure Prediction: Methods and Algorithms,
Edited by Huzefa Rangwala and George Karypis

1. What are the most effective strategies for protein structure modeling?
2. What are the main factors influencing the outcome of a protein structure modeling experiment and how close can a model get to the corresponding experimental structure?
3. How can related structures on which a model can be based be identified reliably (the template identification problem)? How accurately can coordinates from the template structure be mapped to the correct positions on the target sequence (the alignment problem)? Are models produced by altering/refining templates more accurate than the models built by simply copying coordinates of the template (the refinement problem)?
4. How well can the reliability of the model in general and specific regions in particular be estimated (the quality assessment problem)?
5. How well can fully automatic modeling servers perform, compared with a combination of computing methods and human knowledge?
6. Has there been progress in the field?
7. What are the bottlenecks to further progress?
8. Where can future efforts be most productively focused?

In order to rigorously address these issues John Moult and colleagues pioneered the CASP experiment in 1994 [2]. The initiative was well accepted by the community of computational biologists, and the experiment, after eight completed rounds, continues to attract considerable attention to protein structure modelers from around the world. Two hundred thirty four predictor groups from 25 countries participated in the last completed CASP8, submitting over 80,000 predictions (see Fig. 2.1 for historical CASP participation statistics), and approximately the same number of predictor groups are participating in CASP9, which is currently (July 2010) under way.

Even though we, CASP co-organizers, continue to emphasize that CASP is primarily a scientific endeavor aimed at establishing the current state of the art in the protein structure prediction, many view it more as a "world championship" in this field of science. Thus, to a large extent, CASP owes its popularity to the twin human drives of competitiveness and curiosity. Whatever the case, a large community of structure modelers devote very considerable effort to the process, and it has now been emulated in other areas of computational biology [3–6].

2.2. CASP PRINCIPLES AND ORGANIZATION

In the pre-CASP times, protein structure modeling methods were tested using the procedure schematically shown in Figure 2.2a. Method developers selected sequences to test their own methods (usually with different research groups selecting different sets of proteins), and assessed the results by comparing models to the experimental structures already known to them at the time of

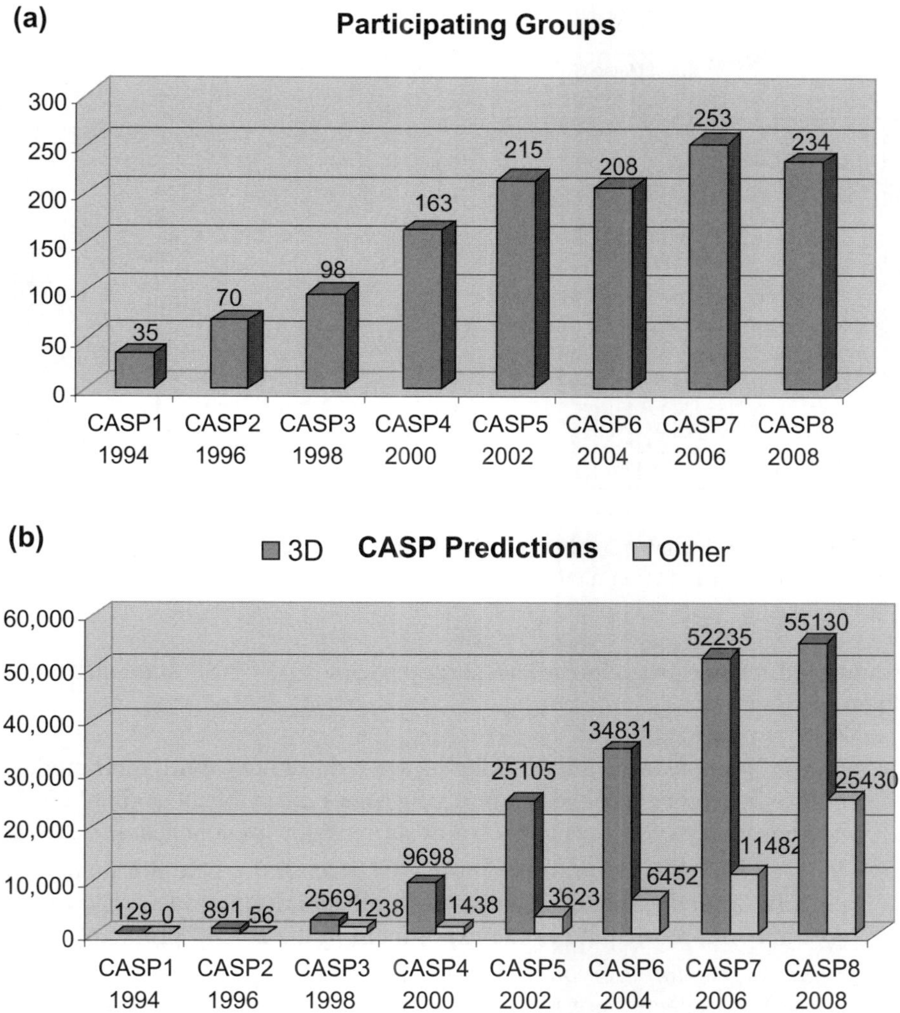

FIGURE 2.1 Statistics on (a) the number of participating groups and (b) number of submitted predictions in CASP experiments held so far. In panel (b), bars representing the number of tertiary structure predictions are shown in dark gray, while bars representing the cumulative number of predictions in other categories (secondary structure, residue-residue contacts, disorder regions, domain boundaries, function, quality assessment) are shown in light gray.

"prediction." Many apparently successful modeling results were reported in the literature but the inability of others to reproduce the results and the lack of resulting useful applications strongly suggested that this testing approach was not strict enough to ensure objective assessment of the results. In particular, many felt that the reported results were too easily influenced by the known answers. CASP was established to address the deficiencies in these

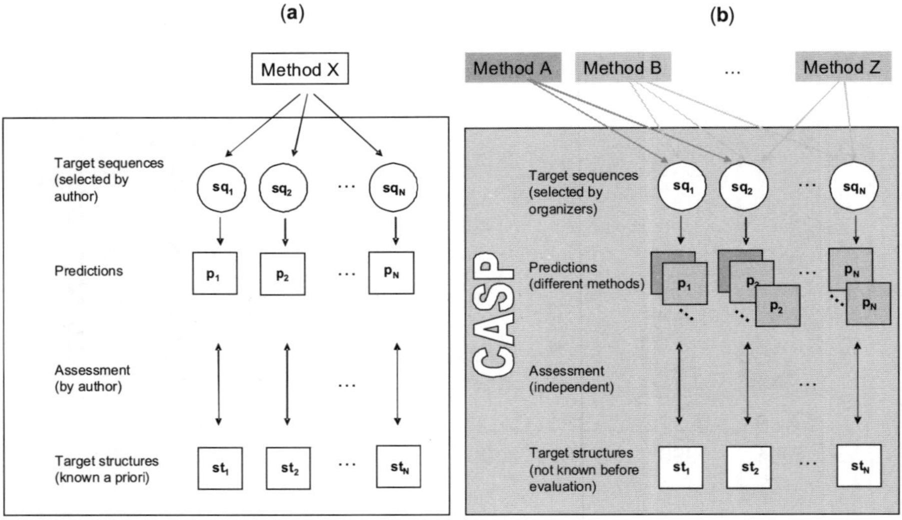

FIGURE 2.2 Schematics of (a) pre-CASP and (b) CASP testing procedures for protein structure prediction methods.

traditional testing procedures. The main principles of CASP summarized in Figure 2.2b are:

- *"Blind" prediction regime.* Predictors are required to submit their models before the answers (experimental structures) are publicly available. This is the primary CASP principle for ensuring rigorous conclusions.
- *Independent assessment of the results.* Experts in the field are invited to perform an independent assessment of all submitted models. The assessors may not participate in the experiment in the role of predictors.
- *Same targets for everyone.* Proteins for modeling ("targets" in CASP jargon) are selected not by the predictors but by the organizers who are not permitted to participate in the experiment and so have no interest in introducing any selection bias. The same set of targets is used to test all the methods, thus facilitating direct comparison of performance. Organizers strive to provide a reasonably large set of targets with a balanced range of difficulty, so that the assessment is statistically sound and shows the range of success and failure across the spectrum of structure modeling problems.
- *Anonymity of assessment.* All information that could be directly or indirectly used to identify submitting research groups are stripped off the predictions. This information is not made available to the assessors until after their analysis of the results is completed.
- *Same evaluation criteria for everyone.* All predictions are evaluated using the same set of numerical criteria.

- *Data availability for post-experiment comparisons.* All predictions and automatic evaluation results are released to the public upon completion of each CASP experiment, so as to allow others to reproduce the results, and to facilitate methods development.
- *Control of the experiment by the participants.* Those participating in CASP are involved in shaping the rules and scope of the experiment through a variety of mechanisms, particularly a discussion forum (FORCASP) and a predictors' meeting at each conference, where motions for change are considered and voted upon.

Together, these principles ensure a more objective determination of capabilities in the field of protein structure modeling than the conventional peer-review publication system. They make unjustified claims more difficult to publish, and provide a powerful mechanism for predictors to establish the strength of their methods.

The principles remain untouched from one experiment to another, but a number of changes and additions to the details have been introduced, and these are summarized in Table 2.1.

2.3. CASP PROCESS

CASP is a complicated process, requiring careful planning, data management, and security. The Protein Structure Prediction Center, established to support the experiment at the Lawrence Livermore Laboratory in 1996 and in 2005 at the University of California, Davis, provides the infrastructure for methods testing, develops method evaluation and visualization tools, and handles all data management issues [7].

Experiments are held every 2 years. The timetable of a typical CASP round is schematically shown in Figure 2.3. The experiment is open to all. The Prediction Center releases targets for prediction and collects models from registered participants for approximately 3 months. Targets for structure prediction are either structures soon-to-be solved by X-ray crystallography or nuclear magnetic resonance (NMR) spectroscopy, or structures already solved but not yet publicly accessible. Prediction methods are divided into two categories—those using a combination of computational methods and human experience, and those relying solely on computational methods. The integrity of the latter category is ensured by requiring that servers process target information and return models automatically. A window of 3 weeks is usually provided for prediction of a target by human-expert groups and 3 days by servers. Following closing of the server prediction window, the server models are posted at the Prediction Center web site. These models can then be used by human-expert predictors as starting points for further, more detailed modeling. They are also used for testing model quality assessment methods in CASP. Once all models of a target have been collected and the experimental

TABLE 2.1 Changes in the Consecutive CASPs

CASP	Prediction Categories	Main Evaluation Measures/ Packages	General
CASP1 (1994)	TS, AL, SS. Protein tertiary structure (TS—coordinates format, AL—alignment format). Secondary structure (SS).	RMSD.	Main CASP principles were established.
CASP2 (1996)	TS, AL, SS. Prediction of protein-ligand complexes introduced.		Prediction Center established to support CASP.
CASP3 (1998)	TS, AL, SS. Prediction of complexes dropped.	ProSup and DALI packages were used for structural superpositions. New evaluation software tested at the Prediction Center to replace RMSD with a measure more suitable for model-target comparison.	CAFASP experiment to evaluate fold recognition servers run as a satellite to CASP.
CASP4 (2000)	TS, AL, SS, RR. Residue–residue (RR) contact prediction introduced.	New evaluation software further developed, resulting in the LGA package [9]. The GDT_TS measure of structural similarity, and AL0 score for correctness of the model-target alignment used as basic CASP measures.	
CASP5 (2002)	TS, AL, RR, DR. SS dropped. Disordered regions (DR) prediction introduced.		

CASP6 (2004)	TS, AL, RR, DR, DP, FN Domain boundary (DP) prediction introduced. Function prediction (FN) introduced.	DAL, nonrigid body structure superposition software, used for scoring models in addition to LGA.	CASP moved to the independent of CAFASP server testing procedure. Time for server response was set to 48 hours plus 24 hours for potential format corrections. Release of server predictions to human-expert groups 72 hours after target release.
CASP7 (2006)	TS, AL, RR, DR, DP, FN, QA, TR Model quality assessment (QA) category introduced. Model refinement (TR) category introduced. Prediction of multimers introduced.	MAMMOTH structure superposition program additionally used for analysis of the results.	Structural assessment categories changed from classic division on comparative modeling/fold recognition/*ab initio* to template-based/template-free. High-accuracy modeling category separately assessed.
CASP8 (2008)	TS, AL, RR, DR, DP, FN, QA, TR Prediction of multimers dropped. FN category was narrowed to binding site prediction.	DALI structure superposition program was additionally used for analysis of the results. Prediction Center automatically calculated group rankings for comparative modeling targets according to different measures.	Limit on number of targets for human-expert groups. Division of targets into human-server and server only categories. Time for server response was set to 72 hours. Separate assessor for contacts, domains, and function predictions.
CASP9 (2010, under way)	TS, AL, RR, DR, FN, QA, TR DP prediction is dropped. Prediction of multimers is reinstated.		

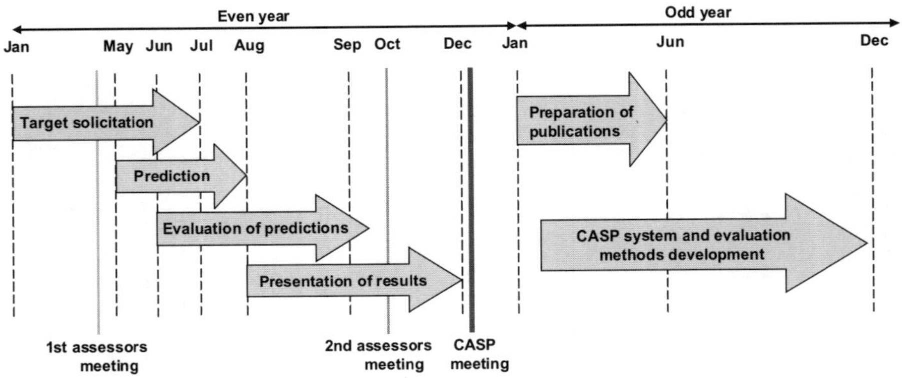

FIGURE 2.3 Timetable of the CASP experiment.

structure is available, the Prediction Center performs a standard numerical evaluation of the models, taking the experimental structure as the gold standard. A battery of tools is used for the numerical evaluation of predictions—LGA [8], ACE [9], DAL [10], MAMMOTH [11], DALI [12]. If the target consists of more than one well-defined structural domain, the evaluation is performed on each of these as well as on the complete target (the official domain boundaries are defined by the assessors). The results of automatic evaluation are made available to the independent assessors, who typically add their own analysis methods and make more subjective assessments of the merits and faults of the models. The identity of the predictors is concealed from the assessors while they conduct their analysis. Assessment outcomes are presented to the community at the predictors' meeting usually held in December of a CASP year. At that time, results of the evaluations are also made publicly available through the Prediction Center web site (http://predictioncenter.org) allowing predictors to compare their own models with those submitted by other groups. Details of all the experiments completed so far and their results are available through this web site. The web site also hosts a discussion forum, FORCASP, allowing exchange of thoughts by the predictors. The articles by the assessors, the organizers, and the most successful prediction groups are published in special issues of the journal *Proteins: Structure, Function, and Bioinformatics.* There are currently eight such issues available, one for each of the eight CASP experiments [2,13–19]. The articles in the special issues discuss in detail the methods tested in CASP, the evaluation results, and the analysis of the progress made. Below we briefly summarize the state of the art in different CASP modeling categories.

2.4. METHOD CLASSES AND PREDICTION DIFFICULTY CATEGORIES

In evaluating the ability of prediction methods, it is important to realize that difficulty of a modeling problem is determined by many factors. In theory, it is possible to calculate the structure of any protein from knowledge of its amino acid composition and environmental conditions alone, since it has long been established that these factors determine the functional conformation [1]. In practice, it is not yet possible to follow the detailed folding behavior of a system with as many atoms and degrees of freedom as a protein, nor to thoroughly search for the global free energy minimum of such a system [20–22]. Two types of methods for combating these limitations have been developed. One, by far the most effective at present, utilizes experimental structures of evolutionarily related proteins, providing templates on which to base a model. For cases where no such relationship exists, or none can be discovered, partially effective structure prediction techniques have been developed using simplified energy functions and employing approximate energy landscape search strategies. These two approaches define the main two classes of prediction methods—template-based modeling (TBM), sometimes referred to as comparative or homology modeling, and template-free modeling. Historically, template-free methods were often termed *ab initio* (or first principles), but members of the CASP community objected on the grounds that these methods often make use of knowledge-based potentials to evaluate interactions and assemblies of observed peptide fragment conformations to generate trial structures. Template-free methods are currently effective only for modeling small proteins (100 residues or less). Template-based methods can be applied wherever it is possible to identify a structurally similar protein that can be used as a template for building the model, irrespective of size. When the two approaches have been applied to the same modeling problem, template-based methods have usually proven more accurate than template-free methods. Thus, the most significant division in modeling difficulty is between cases where a model can be built based on templates derived from known experimental structures, and those where it cannot. At one extreme, high-resolution models competitive with experiments can be produced for proteins with sequences very similar to that of a known structure. At the other extreme, low resolution, very approximate models can be generated by template-free methods for proteins with no detectable sequence or structure relationship to known structures. To properly assess method successes and failures, CASP subdivides modeling into these two separate categories, each with its own challenges, and hence requiring its own evaluation procedures.

2.5. TBM

Whenever there is a detectable sequence relationship between two proteins, the corresponding structures have been found to be similar. Thus, if at least a single structure within a family of homologous proteins is determined experimentally, then template-based methods can be used to model practically all proteins in that family. The potential of this modeling is huge—by some estimates, structures are already known for a quarter of the protein single-domain families of significant size and half of all known sequences can be partially modeled due to their membership in these families (M. Levitt in [23]).

A typical template-based method consists of several consecutive steps: identifying probable templates; selecting/combining suitable templates; aligning target-template(s) sequence; copying structurally conserved regions from the selected template(s); modeling structurally variable regions; packing side chains; refining the model; and evaluating its quality. Each modeling step is prone to errors, but, as a rule, the earlier in the process the error is introduced, the costlier it is. As the template-based category covers a wide range of structure similarity, different kinds of errors are typical for different modeling difficulty subcategories.

2.5.1. High-Resolution TBM

The most reliable models can be built in cases where there is a strong sequence relationship between the target protein and a template (i.e., higher than ~40% sequence identity between target and template). In these situations target and template are expected to have very similar structures. Template selection and alignment errors are rare here, and simply copying the backbone of a suitable template may be sufficient in producing a model that may rival NMR or low-resolution X-ray structures in accuracy (~1 Å C-alpha atom root-mean-square deviations [RMSD] from the experimental structure). The main effort in this class of prediction shifts to modeling of regions of structure not present in a template (loops), proper placement of side chains, and fine adjustment of the structure (refinement).

Such high-resolution models often present a level of detail that is sufficient for detecting sites of protein–protein interactions, understanding enzyme reaction mechanisms, interpreting disease-causing mutations, molecular replacement in solving crystal structures, and occasionally even drug design.

2.5.2. Medium Difficulty Range TBM

New, more sensitive methods of detecting remote sequence relationships, especially Position-Specific Iterative-*Basic Local Alignment Search Tool* (PSI-BLAST) and profile–profile methods, have greatly extended our ability to utilize structure templates based on more remote sequence relationships. The quality of models in this category has steadily improved over the course of the CASP experiments. Models with quite accurate core (typically 2–3Å C-alpha atom RMSD from the native structure) can now often be generated. Factors still limiting progress include difficulty in recognizing best templates, combining information from several templates, aligning target sequences with template structures, adjusting for considerable shifts in conserved regions of structure, and modeling regions not represented in any of the available templates. As in high-resolution homology modeling, refinement methods play a role in improving the accuracy of final models.

Even though less accurate than high-resolution models, these models can also be used in many biological applications such as detecting of probable sites of protein–protein interactions, identifying the approximate role of disease-associated substitutions, or assessing the likely role of alternative splicing in protein function.

2.5.3. Difficult TBM

In cases where no evolutionary relationship can be detected based on sequence, it is still likely that the fold of a target protein is nevertheless similar to that of a known structure (implying a very remote evolutionary relationship or convergence of folds). Methods that check the compatibility of a target protein with the experimental structures use more sophisticated analyses (e.g., secondary structure comparison, knowledge-based structural potentials of various types) and can sometimes assist in identifying templates for modeling. As in such cases the templates have no explicit sequence relationship with the target, alignment is often not reliable and not surprisingly, the accuracy of the resulting model is often low. Nevertheless impressive models are sometimes obtained, and there has been substantial progress over the course of CASP experiments. We attribute this progress to both methodological improvements and the increased size of sequence and structure databases.

Although models for hard TBM targets may not provide accurate structural detail, they are useful for providing an overall idea of what a structure is like, recognizing approximate domain boundaries, helping choose residues for mutagenesis experiments, and providing approximate information about molecular function.

2.5.4. Progress and Challenges in the TBM

Assessment of template-based predictions over the several rounds of CASP clearly showed an indisputable progress in the area, and the accuracy of the models has grown substantially [24–28]. One measure of this is that for the majority of targets the best models for each target are now closer to the native structure than any of the available template structures. Despite this very evident progress, there are many challenges still remaining. After years of development, finding a good template and the alignment still remain the two issues with a major impact on the quality of models. The coverage of the target by the template imposes the upper limit on the fraction of residues that can be aligned between the template and the target. Figure 2.4 shows the maximum alignability together with the alignment accuracy for the best models in the latest four CASPs (see our article [28], pp. 196, 198, for the definitions). It can be observed that the trend in all CASPs is the same—both maximum alignability and alignment accuracy fall steadily and approximately linearly with increasing target difficulty. The slope of the fall off for these two measures, however, is different. For the easiest targets, predictors can routinely achieve

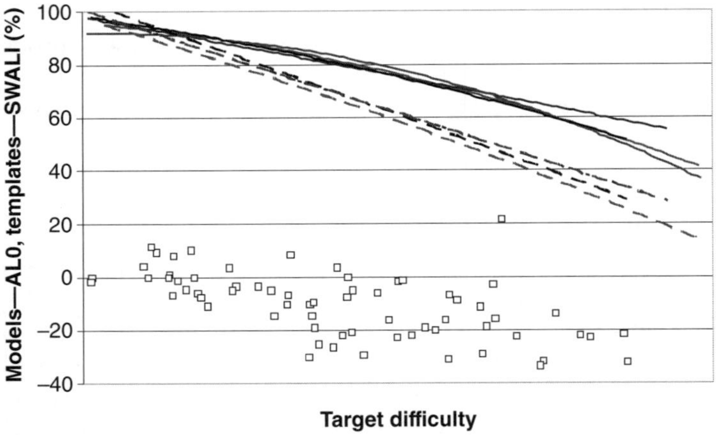

FIGURE 2.4 Maximum template-imposed alignability (SWALI, solid lines) and alignment accuracy of the best template-based models (AL0, dashed lines) from CASP5–8 as a function of target difficulty. Maximum alignability is defined as the fraction of equivalent residues in superposition of the target and best template structure; target difficulty combines coverage of the target structure by the best template and target-template sequence identity. CASP8—black lines; CASP7—blue; CASP6—brown; CASP5—red. Squares represent the difference between alignment quality and maximum alignability for CASP8 targets. Points over the 0% level represent targets where alignment accuracy was better than maximum alignability. (See color insert.)

alignment accuracy close to the maximum possible from a single template or even better; in the mid range of difficulty best alignments are typically within 20% of the optimum, but up to 40% of the structure cannot be aligned at all; for the difficult targets the gap between the maximum alignability and alignment accuracy grows to 30% with the percentage of nonaligned residues increasing to 70% [28]. Predictors often manage to achieve alignment accuracy higher than a single template maximum by using additional templates or by employing free modeling methods for the structurally nonconserved regions such as loops, insertions or deletions. It is encouraging to see an increase in the number of such cases: there are 22 targets in all CASPs where predictors superseded maximum alignability by at least 2%; out of these nine cases were from CASP8 (squares above 0% level in Fig. 2.4), eight from CASP7, four from CASP6, and one from CASP5.

Improvement in alignment over the best template shows only one side of the effectiveness of TBM methods. Analysis of the overall quality of the models (measured in terms of Global Distance Test_Total Score [GDT_TS]) shows that typically the best models are superior to the corresponding naïve models built by simply copying coordinates of aligned residues from the best possible template. This additional gain in quality can be associated with the modeling regions not present in the best template, and also with improving the quality of the model by refinement. Figure 2.5 provides comparison of

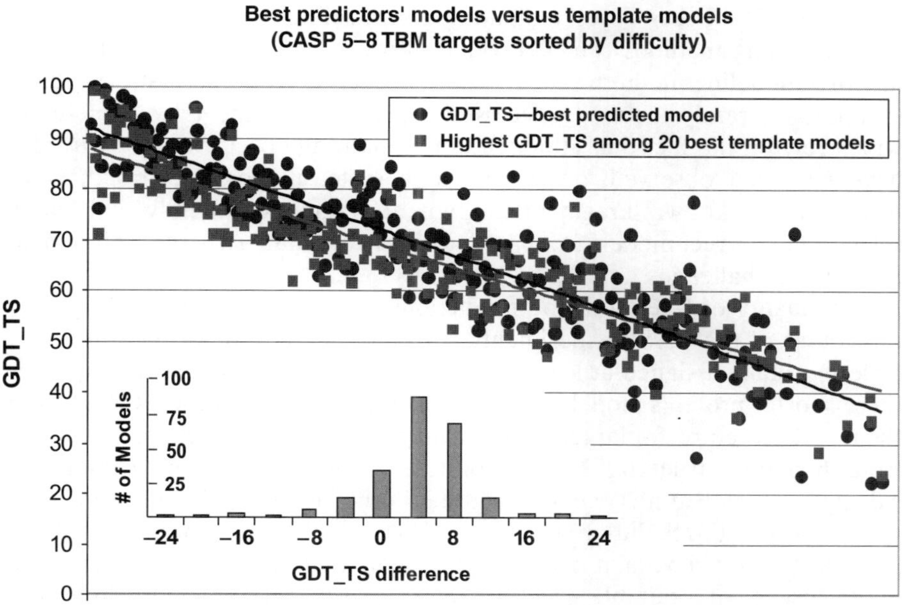

FIGURE 2.5 GDT_TS score of the best submitted model and the best naïve model built on a single template for each TBM target in CASP5–8. The darker trendline corresponds to the predicted models; the lighter one, to the naive ones. Naïve models are built on the top 20 templates according to the target coverage for each target, and the score for the naïve model with the highest GDT_TS is shown. For the easier three-fourths of the difficulty scale, best models in general outperform naïve models. The inset histogram shows number of models registering differences in GDT_TS scores between the best model and naïve model (bins stretch 4 GDT_TS units). The most representative bin is 0–4 GDT_TS difference (86 targets), followed by the 4–8 GDT_TS bin (67 targets).

quality of the best submitted models versus naïve models built on the best single template. Data trend lines indicate that in general the best submitted models are better than the corresponding naïve models, except for the targets representing the hardest one-fourth of the difficulty scale. In CASP6–8, over 70% of the best models in the template-based category have registered added value over the naïve model. The inset histogram shows that the majority of best predictions (153 out of 242) are up to eight GDT_TS units above the corresponding best naïve model. The median difference between the best model and naïve model equals 2.74 GDT_TS units (mean—2.07 GDT_TS).

2.6. FREE MODELING OF NEW FOLD PROTEINS

A quarter of the protein sequences in the contemporary databases do not appear to match any sequence pattern corresponding to an already known structure [23]. In such cases, template-free modeling methods must be used. Free modeling methods can be divided into two categories: structure-based *de-novo* modeling methods and *ab initio* (modeling from the first principles) methods. Currently, the more successful approaches are the *de novo* methods, which rely on the fact that although not all naturally occurring protein folds have yet been observed, on some length scale, all possible structures of fragments are known. Fragment assignment, fragment assembly, and finally selection of correct models from among many candidate structures all remain formidable challenges.

The quality of free modeling predictions has increased dramatically over the course of the CASP experiments, with most small proteins (100 residues or less) usually assigned at least the correct overall fold by a few groups. For these shorter proteins models are typically 4–10 Å C-alpha atom RMSD from the native structure; for larger proteins, models are usually over the 10 Å away from the native structure. This level of detail is insufficient for many biomedical applications. But it is encouraging that in the last three CASPs there were examples of high-resolution accuracy (<2 Å) models for a few small proteins [29,30]. Recently, several notable cases of high-resolution structure prediction in the absence of a suitable structural template were reported in the literature [31,32].

2.7. OTHER MODELING CATEGORIES

Besides the three-dimensional protein structure prediction, CASP evaluates several other structure-related modeling categories.

The *Secondary structure prediction* category was included in early CASP experiments. Initial substantial progress gave way to incremental improvements too small to evaluate with the amount of data collected, and so the category was dropped in 2002.

The *Disorder region prediction* category was introduced in CASP in 2002 to address growing recognition that some regions of proteins do not adopt a single three-dimensional structure but nevertheless are involved in the signaling, regulating, or controlling functions of the protein [33]. The three most recent CASPs have shown that the field has converged and that new ideas to improve the predictions are needed [34].

Prediction of intramolecular *residue–residue contacts* could in principle be helpful for predicting protein structure per se as well as for inferring mutations in the proteins or distinguishing between correct and incorrect protein docking models [35]. This type of prediction is still an area of active research, and continues to be assessed in CASP (starting in CASP4). However, there has been no detectable progress in that period, and current methods do not appear sufficiently accurate to be of any significant use.

Many proteins contain multiple domains, and identifying *domain boundaries* is important not only in modeling but also in selecting constructs for protein expression. Assessment in this area started in CASP6. In general, approximate identification of domain boundaries is straightforward when these lie between TBM regions. There are too few challenging multi-domain template-free targets in CASP to evaluate those cases. As a result, this category will be dropped from future CASPs.

One of the primary uses of a three-dimensional model is to deduce more about the protein's function. Testing of methods for *function prediction* began in CASP6 [36]. Assessment in this category faced difficulties connected with unavailability of experimental data to verify the predictions. To make the analysis more stringent, in CASP8 the category was narrowed to ligand binding site prediction.

As structure modeling has assumed a more prominent role in biology, the need to have reliable estimates of overall and detailed structure accuracy has become apparent. The necessity for an unbiased evaluation of model *quality assessment* methods led to the introduction of a separate category in CASP, starting in 2006. CASP quality assessment evaluation has demonstrated that at the moment the most accurate methods rely on the availability of multiple models for the same protein (called consensus-based or clustering methods) [37]. These methods are based on the observation that the more different modeling methods agree on structure, either overall or in particular regions, the more likely that structure is correct. The best quality assessment methods can provide ranking of overall models significantly correlated with accuracy, but are not able to consistently select the best model from the entire collection of models. It has been encouraging to see some quality assessment methods showing promising results in assessing accuracy of specific regions in a model, at times reproducing almost the exact C-alpha-C-alpha deviation along the sequence. The main challenges in the quality assessment category is developing methods that can be competitive with consensus-based methods but rely on the structural and sequential features of a single model and improving the performance of methods for determining local model accuracy.

Starting in CASP7, special attention has been given to model *tertiary structure refinement*. All-atom structure refinement is one of the challenges in protein structure prediction and the development of reliable refinement procedures would help in bringing the models up to the high-resolution standards. The assessment of predictions in this category suggests that while there are no methods that can consistently improve over the initial model, refinement can sometimes result in structures that are much closer to the target than the template (not a trivial task, especially for high accuracy targets).

2.8. SERVERS IN CASP

There are now many millions of proteins for which reasonable models could be produced, and meeting such a large-scale demand requires automatic generation of models. Even though it is apparent that not all human expertise can be encoded in automatic servers, CASP shows that the best servers are not much worse than the best human predictors. Moreover, sometimes the difference in human-server performance is just due to the fact that human experts have more time for the modeling (and quite often, base their models on initial structures obtained from servers). Under these circumstances, the importance of automatic servers for the biomedical community cannot be over-estimated.

Server performance has been continuously checked by CASP starting in CASP3 (originally with the help of the Critical Assessment of Fully Automated Structure Prediction Methods [CAFASP] [38] and since CASP6, independently). Analysis of server performance in successive CASPs [28,39] shows that the best human-expert groups in CASP still outperform the best server groups, but the gap between the best servers and the best human-expert groups is narrowing. Especially in the case of easy TBM, progress of automated servers is impressive, with the fraction of targets where at least one server model is among the best six submitted models increasing from 35% in CASP5 to 65% in CASP6, to over 90% in CASP7, and slightly decreasing to 83% in CASP8. This statistics confirms the notion that the impact of human expertise on modeling of easy comparative targets is now marginal. In general, in both CASP7 and CASP8 servers were at least at par with humans (three or more models in the best six) for about 20% of targets, and significantly worse than the best human model for only very few targets.

2.9. MODELING CHALLENGES AND CASP INITIATIVES

Despite evident progress in protein structure modeling, many challenges still remain to be addressed. In TBM, refinement of high-accuracy models and improvement of template-to-target alignments in nontrivial cases are the major limiting factors. In free modeling, the challenge is to predict larger proteins (over 100 residues) more reliably and to routinely generate models within 2–3 Å RMSD from the native structure for smaller proteins. The methods tested in CASP in this category have run out of steam and new modeling techniques seem necessary. Besides the traditional CASP categories, we have already conducted two in-between-CASP experiments to test the prediction of mutation sites and refinement of models. Assessment of model quality is currently an area of active research, and more than a dozen of papers on the subject have been published since this category was introduced in CASP7 (2006). It is also planned to conduct additional in-between-CASP experiments in modeling of membrane proteins and in selecting models from decoy sets. Within the main CASP track, in CASP9 we are reviving the prediction of quaternary structure. An initiative to continuously test free modeling methods is currently underway.

REFERENCES

1. M. Sela, F.H. Jr. White, and C.B. Anfinsen. Reductive cleavage of disulfide bridges in ribonuclease. *Science*, 125(3250):691–692, 1957.
2. J. Moult et al. A large-scale experiment to assess protein structure prediction methods. *Proteins*, 23(3):ii–v, 1995.
3. J.M Bujnicki et al. LiveBench-2: Large-scale automated evaluation of protein structure prediction servers. *Proteins*, S(5):184–191, 2001.
4. J. Janin et al. CAPRI: A critical assessment of predicted interactions. *Proteins*, 52(1):2–9, 2003.
5. V.A. Eyrich et al. EVA: Continuous automatic evaluation of protein structure prediction servers. *Bioinformatics*, 17(12):1242–1243, 2001.
6. M.G. Reese et al. Genome annotation assessment in Drosophila melanogaster. *Genome Research*, 10(4):483–501, 2000.
7. A. Kryshtafovych et al. New tools and expanded data analysis capabilities at the Protein Structure Prediction Center. *Proteins*, 69(8):19–26, 2007.
8. A. Zemla. LGA: A method for finding 3D similarities in protein structures. *Nucleic Acids Research*, 31(13):3370–3374, 2003.
9. A. Zemla et al. Processing and evaluation of predictions in CASP4. *Proteins*, (S5):13–21, 2001.
10. A. Kryshtafovych et al. CASP6 data processing and automatic evaluation at the Protein Structure Prediction Center. *Proteins*, 61(S7):19–23, 2005.
11. A.R. Ortiz, C.E. Strauss, and O. Olmea. MAMMOTH (matching molecular models obtained from theory): An automated method for model comparison. *Protein Science*, 11(11):2606–2621, 2002.
12. L. Holm et al. Searching protein structure databases with DaliLite v.3. *Bioinformatics*, 24(23):2780–2781, 2008.
13. J. Moult et al. Critical assessment of methods of protein structure prediction-Round VIII. *Proteins*, 77(S9):1–4, 2009.
14. J. Moult et al. Critical assessment of methods of protein structure prediction-Round VII. *Proteins*, 69(S8):3–9, 2007.
15. J. Moult et al. Critical assessment of methods of protein structure prediction-Round VI. *Proteins*, 61(S7):3–7, 2005.
16. J. Moult et al. Critical assessment of methods of protein structure prediction (CASP)-round V. *Proteins*, 53(6):334–339, 2003.
17. J. Moult et al. Critical assessment of methods of protein structure prediction (CASP): round IV. *Proteins*, (S5):2–7, 2001.
18. J. Moult et al. Critical assessment of methods of protein structure prediction (CASP): Round III. *Proteins*, (S3):2–6, 1999.
19. J. Moult et al. Critical assessment of methods of protein structure prediction (CASP): Round II. *Proteins*, (S1):2–6, 1997.
20. K.M. Misura and D. Baker. Progress and challenges in high-resolution refinement of protein structure models. *Proteins*, 59(1):15–29, 2005.
21. H. Lei and Y. Duan. Protein folding and unfolding by all-atom molecular dynamics simulations. *Methods in Molecular Biology*, 443:277–295, 2008.

22. Y. He et al. Exploring the parameter space of the coarse-grained UNRES force field by random search: Selecting a transferable medium-resolution force field. *Journal of Computational Chemistry*, 30(13):2127–2135, 2009.
23. T. Schwede et al. Outcome of a workshop on applications of protein models in biomedical research. *Structure*, 17(2):151–159, 2009.
24. J. Kopp et al., Assessment of CASP7 predictions for template-based modeling targets. *Proteins*, 69(S8):38–56, 2007.
25. R.J. Read and G. Chavali. Assessment of CASP7 predictions in the high accuracy template-based modeling category. *Proteins*, 69(S8):27–37, 2007.
26. D. Cozzetto et al. Evaluation of template-based models in CASP8 with standard measures. *Proteins*, 77(S9):18–28, 2009.
27. D. Keedy et al. The other 90% of the protein: Assessment beyond the C-alphas for CASP8 template-based and high-accuracy models. *Proteins*, 77(S9):29–49, 2009.
28. A. Kryshtafovych, K. Fidelis, and J. Moult. Progress from CASP6 to CASP7. *Proteins*, 69(S8):194–207, 2007.
29. P. Bradley et al. Free modeling with Rosetta in CASP6. *Proteins*, 61(S7):128–134, 2005.
30. R. Das et al. Structure prediction for CASP7 targets using extensive all-atom refinement with Rosetta@home. *Proteins*, 69(S8):118–128, 2007.
31. P. Bradley, K.M. Misura, and D. Baker, Toward high-resolution de novo structure prediction for small proteins. *Science*, 309(5742):1868–1871, 2005.
32. B. Qian et al. High-resolution structure prediction and the crystallographic phase problem. *Nature*, 450(7167):259–264, 2007.
33. P. Radivojac et al. Intrinsic disorder and functional proteomics. *Biophysical Journal*, 92(5):1439–1456, 2007.
34. L. Bordoli, F. Kiefer, and T. Schwede. Assessment of disorder predictions in CASP7. *Proteins*, 69(S8):129–136, 2007.
35. J.M. Izarzugaza et al., Assessment of intramolecular contact predictions for CASP7. *Proteins*, 69(S8):152–158, 2007.
36. S. Soro and A. Tramontano. The prediction of protein function at CASP6. *Proteins*, 61(S7):201–213, 2005.
37. D. Cozzetto et al. Assessment of predictions in the model quality assessment category. *Proteins*, 69(S8):175–183, 2007.
38. D. Fischer et al. CAFASP-1: Critical assessment of fully automated structure prediction methods. *Proteins*, (S3):209–217, 1999.
39. J.N. Battey et al. Automated server predictions in CASP7. *Proteins*, 69(S8):68–82, 2007.

CHAPTER 3

THE PROTEIN STRUCTURE INITIATIVE

ANDRAS FISER
Department of Systems and Computational Biology
Department of Biochemistry
Albert Einstein College of Medicine
Bronx, NY

ADAM GODZIK
Program in Bioinformatics and Systems Biology
Sanford-Burnham Medical Research Institute
La Jolla, CA

CHRISTINE ORENGO
Department of Structural and Molecular Biology
University College London
London, UK

BURKHARD ROST
Department of Biochemistry and Molecular Biophysics
Center for Computational Biology
Columbia University
New York, NY

3.1. BACKGROUND, RATIONALE, AND HISTORY

High-throughput sequencing projects started to pour in an unprecedented amount of genomic information in the mid 1990s. Subsequently a strong interest emerged for even more ambitious high-throughput experiments that would assign 3D shapes to all known proteins in all genomes. Three-dimensional structures of proteins are often more informative than their sequences alone because interactions take place in the 3D space and because patterns formed by residues within the same protein that are far in sequence often form a recognizable motif in space. The large-scale efforts to target and solve protein

Introduction to Protein Structure Prediction: Methods and Algorithms,
Edited by Huzefa Rangwala and George Karypis

structures were dubbed as structural genomics projects and were launched worldwide with a variety of different focuses in Europe, Japan, and the United States. The U.S. efforts were spearheaded by National Institutes of Health-National Institute of General Medical Sciences (NIH-NIGMS) and named as the Protein Structure Initiative (PSI).

The scientific rationale for structural genomics is the recognition of the fact that the many million known protein sequences seem to cluster into much fewer structural families or folds. A variety of estimates exists about the anticipated number of protein folds ranging between 1000 and 20,000 [1–4]. In addition the size distribution of protein folds is very uneven. Domain superfamilies from the 12 most highly populated folds (superfolds; [5]) cover approximately half of a typical genome, with such representatives as the Rossman, TIM, OB, or Ig fold [6]. On the other hand, many thousands of much less populated domain superfamilies compose the rest of the genomes. The rationale behind structural genomics is that a few thousand carefully selected and solved protein structures will provide a means to structurally characterize, at least in part, up to 80% of all existing sequences by using the solved structures as templates and employing comparative protein structure modeling techniques to model the rest of the proteins in each superfamily [4,7]. This rationale sheds light on the two most critical computational aspects of structural genomics efforts: target selection and structure modeling. While target selection is chiefly responsible for efficiently mapping the fold universe and properly directing efforts, structure modeling is the actual tool to provide the 3D characterization for more than 99% of all proteins.

The underlying hypothesis of PSI is that high-throughput pipelines could be developed to produce high-quality protein structure representatives of large protein families with little or no prior structural representation. This goal sets the PSI apart from traditional structural biology approaches that normally take a much more highly focused and pragmatic approach through study of one or a limited number of macromolecules via "hypothesis-driven" research.

After 1 year of preparation, a pilot phase of PSI started in year 2000, establishing nine PSI centers around the United States. These centers were charged with the initial goals to set up automated pipelines for structural genomics projects and to start producing 3D structures with an increasing efficiency that includes both an increased number of solved experimental structures and the reduction of the cost per structure. While the overall goal of PSI was to explore the fold universe and to target structurally uncharacterized families, various centers focused this global aim within more specific biologically defined frameworks, for example, targeting new folds in specific genomes of biomedical interest, such as *Thermotoga maritima or Mycobacterium tuberculosis*, or targeting human and other eukaryotic proteins, or targeting proteins involved in metabolic pathways and cancer. The pilot phase of PSI was a success, with an unprecedented number of 1300 structures solved, already making an impact on the composition of Protein Data Bank (PDB), solving far more structures than all conventional structural biology labs solved with a comparable amount of funding. Thereby, PSI-1 achieved its goals, namely to demonstrate the feasibility of efficient pipelines and to significantly reduce the cost of solving protein structures. The second, the so-called *production phase* of PSI started in 2005, funding four production centers and six specialized research centers. The four large-scale centers (Joint Center for Structural Genomics [JCSG; http://www.jcsg.org]; Midwest Center for Structural Genomics [MCSG; http://www.mcsg.anl.gov]; New York SGX Research Center for Structural Genomics [NYSGXRC; http://www.nysgrc.org]; and Northeast Structural Genomics Consortium [NESG; http://www.nesg.org]) were all selected for their proven high-throughput capability and were charged to synchronize target selection efforts providing a concerted effort to uncover fold space. Meanwhile specialized centers focused on addressing technological problems and known bottlenecks that include solving membrane proteins, proteins in higher eukaryotes (especially in human), and small protein complexes, and/or developing technology for portability, applicability, and scalability in general.

In this chapter we will review the current state of PSI efforts with focus on target selection and the impact on coverage of the fold universe.

3.2. OVERVIEW, PIPELINE, AND RESOURCES

Structural genomics centers established pipelines where all steps of the production are highly automated. The pipelines include both experimental and computational modules and typically contain the following steps: target selection (and target tracking); protein production (including cloning, expression and purification); protein characterization (e.g. solubility tests); Heteronuclear Single Quantum Coherence (HSQC) or nuclear magnetic resonance (NMR) assignment or crystallization (if the structure determination technology used is NMR or X-ray crystallography, respectively); structure determination (by X-ray crystallography or NMR spectroscopy); and modeling of related protein sequences. While the pipelines among various centers differ in their details and by the specific experimental technologies employed, the previously described major steps are common. These common steps (e.g., target selection, cloning, expression, solubility experiments, purification, biophysical characterization, crystallization, diffraction or NMR assignments, solved structure) are also reflected in the databases (Structural Genomics Target Search [TARGETDB; http://targetdb.pdb.org/] and Protein Expression Purification and Crystallization DataBase [PEPCDB; http://pepcdb.pdb.org/]) that track experimental progress at all PSI centers. Additional resources of PSI efforts include the PSI Materials Repository (http://psimr.asu.edu) that collects all of the PSI center clones and processes them for general distribution. The PSI Knowledgebase (http://kb.psi-structuralgenomics.org/) serves as the public face of PSI efforts, providing all available technologies, experimental structures and homology models to the community. A partnership with the Nature

Gateway provides exposure through highlight publication for the PSI and its products.

3.3. TARGET SELECTION AND TARGET CATEGORIES

3.3.1. Centralized Target Selection

PSI applies the structural genomics paradigm at several levels of biological investigation: (i) structural coverage of the protein universe, (ii) structural coverage of proteins targeted in collections of organisms (i.e., metagenomics), (iii) structural coverage of proteins targeted in specific organisms or organelles (e.g., *T. maritima*), (iv) structural coverage of systems of cofunctioning proteins and protein networks (e.g., protein phosphatases), (v) analysis of structure/function diversity across a large domain family, and (vi) structural analysis of membrane proteins.

Target selection for the large centers in PSI is overseen by the BioInformatics Groups (BIG4). Each center is represented by one representative in this group—the authors of this chapter. Targets in PSI can be assigned in different categories. About 70% of all targets are centrally compiled and selected among the four centers by BIG. Another 15% of targets are solved through collaborations with the research community, while the remaining 15% of targets are picked by each center individually because of their biomedical relevance. Target selection is centralized among the four production centers to increase efficiency, which is achieved by avoiding overlaps among the centers and by balancing efforts on various target lists available.

3.3.2. Modeling Families

PSI targets protein "families" because of their predicted structural "novelty." These subjective qualifiers required a more precise, operational definition. Definition of a protein family is subjective because it can range from a low-resolution definition, such as the fold of a protein, to a high-resolution definition as any structural novelty within the same fold family, such as a novel sub-domain or a longer loop insertion. In PSI the operational concept is to map the protein universe into sequence clusters, so-called Modeling Families, at a resolution, where the experimentally solved proteins can serve as suitable template for comparative protein structure modeling of similar proteins. A general guideline relies on the large-scale study that concluded that if a template-target pair shares more than 30% sequence identity a comparative model can be built, for which accuracy is expected to be within 2 Ang root-mean-square difference (RMSD) of the native structure [8]. Consequently within the PSI a practical definition of Modeling Families refers to groups of sequences, where any two sequences in a Modeling Family share more than 30% with one other (Fig. 3.1). Therefore if the structure of at least one member in a Modeling Family is solved experimentally all other family

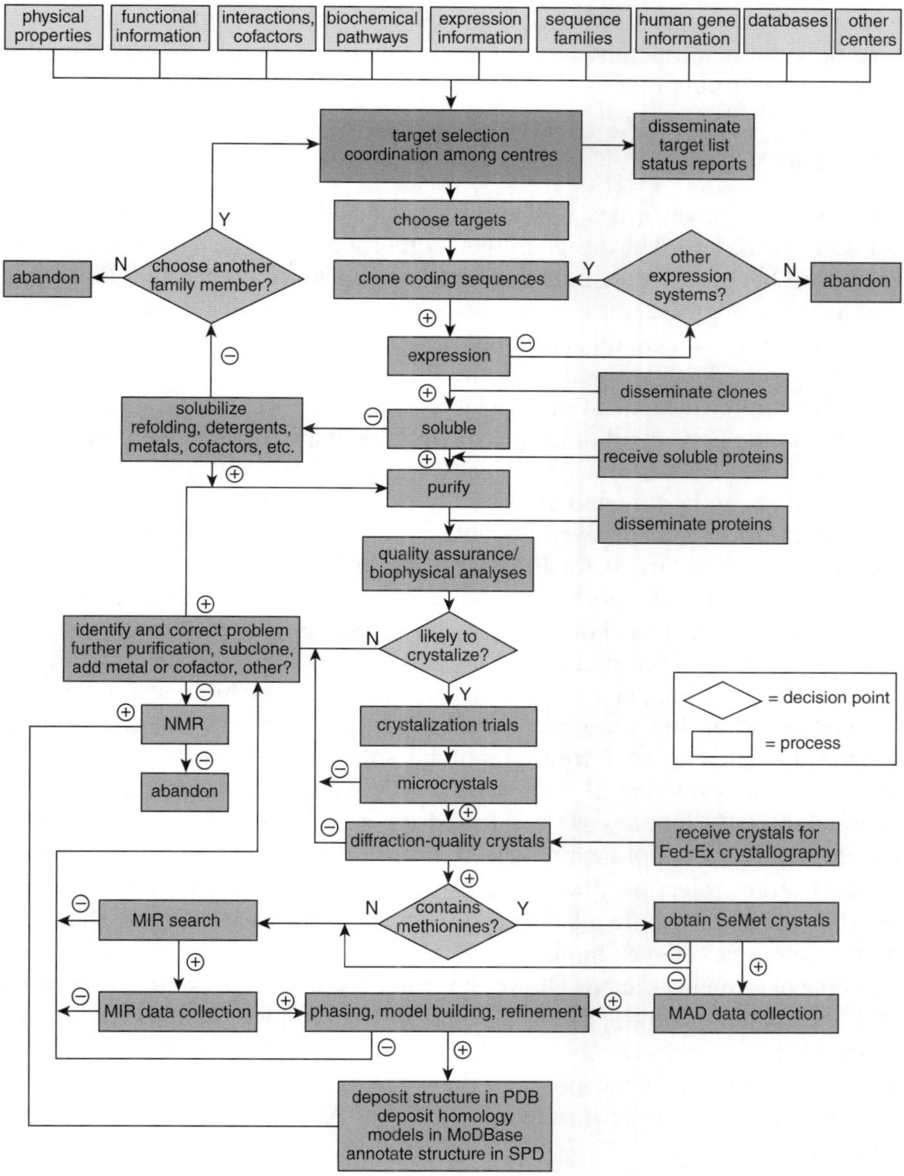

FIGURE 3.1 Flowchart of a typical PSI pipeline.

members can be accurately modeled. With the continuously improving technologies in comparative modeling, this threshold, or in general the definition of the concept of Modeling Family, can be revised, which will impact on the number of Modeling Families. In the PSI the category of biomedical targets does not necessarily follow this rule, as these proteins have proven a strong potential for immediate biomedical application that necessitates a higher resolution approach. For these targets very detailed structural information may be required, for example, to explore the docking of suitable inhibitors and therefore potentially all members of a family could be targeted by experimental structure determination.

3.3.3. BIG Families

The largest segment of the PSI effort is realized by solving targets from a centrally agreed list. The design of this target list reflects the nature of the fold universe. One part of the target list focuses on such protein families that completely lack any structural characterization (BIG families). Definition of domain families in general is a complex issue on its own. Initially PSI centers explored and selected targets from the Pfam [9] classification. Pfam is one of the best curated classification available and therefore it was selected and pursued in the first round of efforts solving the first structural representatives for almost 500 Pfams, which essentially exhausted the list of suitable Pfams families that can be targeted successfully with reasonable efforts. Meanwhile, one has to acknowledge several problematic features of Pfam to be the sole source of possible targets for PSI. All of these issues originate from the fact that the aims of PSI efforts and the rules guiding Pfam classifications are similar but not identical. For example, the sequences clustered in Pfam sometimes represent a multi-domain family rather than a single-domain family. A reverse problem happens in proteins that have been chopped into partial domains in Pfam but are never found separately and may not constitute a proper domain and therefore cannot be solved experimentally. Therefore after an initial targeting of large, structurally uncharacterized Pfam families, subsequent BIG families were selected on a consensus basis. Each center developed its own approach to select and prioritize BIG families and a centralized target selection protocol then compared these target lists to avoid overlap and to distribute a balanced mix of targets for each center. The techniques to search for such families in databases of Pfam-B or the entire known universe of sequences in NR [10] or the Universal Protein Resource (Uniprot) [11] included Position-Specific Scoring Matrix or hidden Markov profile-based [12] sequence searches employing a variety of alignment techniques including global dynamic programming and iterative heuristic approaches such as Position-Specific Iterative-Basic Local Alignment Search Tool (PSI-BLAST) [13].

3.3.4. MEGA Families

Another segment of the centralized target list (MEGA families) focuses on very large protein families that are structurally very divergent. In these very large and diverged MEGA families, analysis of the structures solved and of relatives identified in the genomes reveals up to fivefold increases in the sizes of the structures, sometimes to an extent that the fold of the relatives can be considered to have changed [14]. This structural divergence is often clearly correlated with divergence in function. These families cannot be characterized by one or a few experimentally solved structures. This is reflected quantitatively in large sequence differences among various members (more than 70%), which identify many Modeling Families within each MEGA family. In some of the largest MEGA families less than 10% of the Modeling Families have a solved structure. Technically MEGA families were identified from the Class, Architecture, Topology, and Homologous superfamily (CATH) database as the largest (top 200 most populated) CATH domain families with incomplete structural characterization. These top 200 CATH domains cover about two-thirds of sequences in a typical genome. Putative domains of MEGA families were identified in the Gene3D database (currently holds about 5 million sequences, http://gene3d.biochem.ucl.ac.uk) by scanning sequences against hidden Markov models (HMMs) derived from the CATH domain database. As of August 2008, approximately 37% of residues in protein sequences from Gene3D can be assigned to families of known structure in CATH and approximately 55% of protein sequences in Gene3D contain at least one domain that can be assigned to a family in CATH. The largest four of the very large MEGA families, also called SUPERMEGA families with more than 10,000 representative sequences each, were not allocated to any single center but instead, each center prioritized individual Modeling Families within these superfamilies, largely on the basis of features, which suited their experimental pipelines (e.g., presence of homologs in the reagent genomes used by the center). Modeling subfamilies from these four largest families were then assigned to each center using a draft pick protocol.

3.3.5. META Families

Besides BIG and MEGA families PSI also targeted the so-called META families. In recent years, metagenomics experiments focusing on specific environments such as the ocean bed or the human gut have revealed the extent of uncharacterized sequence families in the protein universe. Domain families that are overrepresented in these metagenomes also became targets in PSI, initially focusing on the human distal gut microbiome. Metagenomics can shed light on important functional roles being carried out by the bacterial communities found in specific habitats [15], for instance, many bacterial proteins in the human gut are essential for breaking down complex food substrates and synthesizing vital nutrients such as vitamins. Understanding how these communities function is likely to be important for understanding and promoting human health. In practice, these gut metagenome families constitute a subset of either BIG or MEGA families. Protein sequences found in the gut microbiome were first grouped into homologous clusters. Comparing numbers of homologs from these clusters found in the gut microbiome and in other bacterial genomes allowed the identification of clusters that are significantly overrepresented in the gut. The largest and most overrepresented clusters were considered as potential targets in PSI.

3.4. PERFORMANCE OF PSI

Several reports in the literature have detailed the success of PSI according to different criteria [4,16,17], all suggesting that PSI is successful in significantly increasing the proportion of novel distinct protein structures deposited in the PDB (http://www.pdb.org), as well as the proportion of novel structural superfamilies and novel fold groups.

Out of the currently classified 10,340 Pfams, [9] 3319 have structural representatives available. PSI-2 embarked on structural studies of an initial set of 1369 largest, structurally uncharacterized Pfam families. As of September 2008, PSI-2 was first to determine structures for 482 of these families. If we move to a higher resolution and look at the level of Modeling Families, out of the structurally characterized 6642 Modeling Families within Pfam PSI-2 provided a representative first in 1179 cases.

As we described above further attempts to improve structure coverage have focused on MEGA families that can contain thousands to hundreds of thousands of sequences. These efforts have provided first structures for another 399 subfamilies. These successes have enabled significant leverage by homology modeling such that almost 600,000 sequences can be modeled, of which around 120,000 represent completely novel leverage for which no remote templates were previously available.

A total of 1600 structures was solved by the four PSI centers in the first 3 years of the production phase (July 1, 2005 to July 1, 2008) and these amounted to 1502 distinct chains (~94%). This compares with a ratio of 61% of distinct chains to all other PDB entries (excluding PSI structures) over the same period of time.

Previous analyses of structural coverage of known proteomes suggest that up to 30–40% of protein residues, and ~50% of domain sequences, can currently be assigned a structure by modeling [6,18]. About 20–30% of proteins contain disordered or membrane-spanning regions [19], which makes them experimentally much less amenable for high-throughput structure determination and they will be suitable targets for specialized centers [20]. A significant proportion of the remaining structurally uncharacterized sequences was targeted by the BIG list. The remaining targets chosen by the four centers came from 48 MEGA families and 136 META families. A total of 193,249 targets has been selected over the 3 years since the start of PSI-2.

In raw numbers PSI centers have contributed almost 9% of all structures worldwide and all structural genomics centers have contributed almost 18% of all structures. When comparing the annual contribution from the PSI of novel structures (i.e., <30% sequence identity with any other structure in the PDB at the time of deposition) to that from all other sources, the PSI deposits now almost as many novel structures as all the other depositors in the United States combined.

A variety of different measures can be used to assess the impact of PSI. The total number of solved structures is not necessarily indicative of the actual

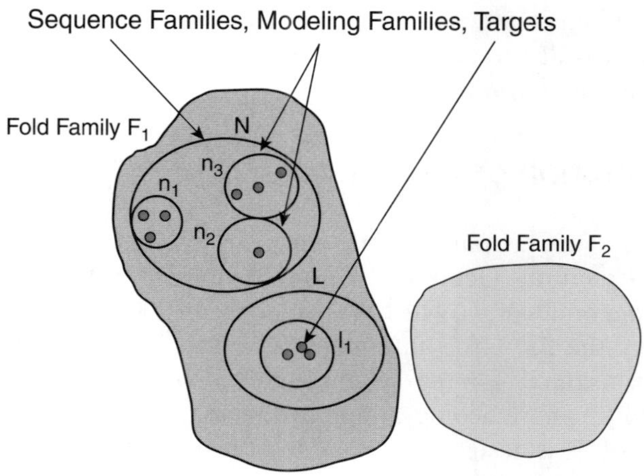

FIGURE 3.2 Definition of fold families, sequence families, modeling families, and protein targets.

impact and leverage of PSI-2, or of its success at meeting its objectives, especially given that PSI was initiated with the incentive to make the biggest possible impact from a predefined amount of resources. Therefore the fact that PSI not only met its productivity goals but clearly surpassed them is not informative by itself. A more informative question is to assess the impact of PSI on "novelty." This can be measured through the extent to which PSI structures are affecting the structural coverage of known proteomes.

Novel leverage is demonstrated most clearly when one compiles the annual increase in novel structural coverage per deposited structure. The per-structure leverage of PSI has consistently been five to eight times higher than the corresponding number for non-structural genomics structures. Given that novel leverage is an important criterion in PSI-2 target selection, PSI-generated structures also possess more novel leverage than structures from non-structural genomics groups. Worldwide structural genomics have contributed about 18% of the structures since 2005 and about 37% of all novel protein leverage (Fig. 3.2). More than three-quarters of the novel leverage since 2005 has come from PSI-2 centers.

The fraction of UniProt proteins that can be structurally modelled is now reaching 48%. This represents an increase of about 10% over the past 3 years only, with a contribution of more than 2% from PSI-2 structures. PSI-2 large-scale centers only account for around 13% of the distinct structures released in the PDB since July 1, 2005 but it contributed 23% of the increase in structural coverage of proteins in UniProt [21].

Meanwhile the impact of PSI on specific proteomes greatly depends on the type of organism. For example, in case of human proteins only 231 (i.e., 3.3% of the structural coverage increase and 0.3% of the human proteome) structures were solved by the four large-scale centers. In contrast, the structural coverage contribution is 37% for *Escherichia coli* over the same period of time, that is, 206 out of 560 proteins for which structure can now be modeled. This discrepancy is due to the fact that large-scale centers have preferentially targeted prokaryotic proteins.

3.5. DISSEMINATION OF RESULTS

The results of PSI are disseminated through various venues. A central gateway is the PSI Structural Genomics Knowledgebase (PSI SGKB; http://kb.psi-structuralgenomics.org/), which was launched in 2008. It is designed to turn the products of the PSI effort into major advances in knowledge that can be used as a key resource for the advancement of biology, biochemistry, functional genomics, pharmacology, bioinformatics, chemistry, education and clinical medicine. It is designed to provide a "marketplace of ideas" that connects protein sequence information to 3D structures and homology models, enhances functional annotations, and provides access to new experimental protocols and materials.

The initial implementation of the PSI SGKB is a gateway to a series of individual portals; these include Experimental Tracking, Material Repository, Models, Annotation, Publication, and Technology. The experimental tracking portal contains TARGETDB that tracks the status of all targets under study in the PSI, and PEPCDB that provides information about the protocols for all steps of protein production. There are links to all of the PSI centers and major annotation resources such as The Open Protein Structure Annotation Network (Topsan; http://www.topsan.org/), which follows a Wikipedia design, essentially a community-organized protein annotation system.

Nature Gateway highlights the mission, advances, and significant breakthroughs of the PSI projects each month through articles about the key research findings related to the structure and function of biological molecules as well as those about technological breakthroughs.

Lastly, all centers maintain an information-rich web site about all targets pursued and structures solved, links to bioinformatics programs, databases and annotations, and up-to-date statistics.

3.6. CONCLUSIONS AND FUTURE PERSPECTIVES

PSI-1 (2000–2005) was a feasibility study for large-scale structural genomics, mainly focusing on establishing the pipelines and techniques required by this challenge. PSI-2 (2005–2010) ramped up the performance of the centers to production level, delivering about 200 structures per year by each center, bringing the total number of PSI-solved structures beyond 4000 by the end of 2009, and making a lasting impact in the process of unraveling the protein structure universe. After several community-wide reviews and assessments, at the time of writing this chapter, a third phase of PSI has been announced by NIH-NIGMS with a new focus, which emphasizes on the importance of structure in a functional context, hence the name is PSI:Biology. In the various assessments during PSI-2 it was concluded that it appears to be useful to continue efforts along the lines of the overall subject of sequence space coverage and of the leverage of solved structures, but at a reduced level of intensity. Meanwhile emphasis will be placed on the biomedically significant regions of sequence space. The goal of the PSI:Biology initiative will be to test whether the new paradigm of high-throughput structure determination via highly organized networks of investigators can be applied to a broad range of biological problems. The focus of PSI:Biology will be divided among (1) centers defining biological themes amenable for high-throughput structure determination, (2) the actual large-scale centers solving the targets, and (3) centers focusing on membrane proteins, that are currently vastly underrepresented in PDB. The network of centers in PSI:Biology will be completed with the continued existence of the Material Repository that makes all experimental clones available for the scientific community and the Knowledgebase that is responsible for disseminating all results.

PSI efforts and other similar large-scale initiatives in sequencing, epigenomics, transcriptomics, or proteomics are impacting scientific research in several ways. First, these efforts demonstrate just how much scientific research accelerated in the last decade, when revolutionary new experimental techniques are rapidly, within years, becoming parts of resource facilities and are made available for a broad community. However, the biggest impact of genome-level experimental techniques is probably the cultural shift in the way research is already done and certainly how it will be pursued in the future. Data production—the most challenging and time-consuming part of research in previous decades—is becoming secondary to data interpretation. High-throughput techniques delivered earlier unimaginable amount of data. This presents a natural need for stronger advances in theoretical, quantitative techniques in systems biology and bioinformatics to analyze, interpret, and model all the available data.

REFERENCES

1. C. Chothia. One thousand families for the molecular biologist. *Nature*, 357:543, 1992.
2. Y.I. Wolf, N.V. Grishin, and E.V. Koonin. Estimating the number of protein folds and families from complete genome data. *J Mol Biol*, 299:897–905, 2000.
3. A.E. Todd, C.A. Orengo, and J.M. Thornton. Evolution of function in protein superfamilies, from a structural perspective. *J Mol Biol*, 307:1113, 2001.
4. M. Levitt. Growth of novel protein structural data. *Proc Natl Acad Sci U S A*, 104:3183–3188, 2007.
5. C.A. Orengo, F.M. Pearl, J.E. Bray, A.E. Todd, A.C. Martin, C.L. Lo, and J.M. Thornton. The CATH Database provides insights into protein structure/function relationships. *Nucleic Acids Res*, 27:275, 1999.
6. R.L. Marsden, T.A. Lewis, and C.A. Orengo. Towards a comprehensive structural coverage of completed genomes: A structural genomics viewpoint. *BMC Bioinformatics*, 8:86, 2007.

7. S.K. Burley, A. Joachimiak, G.T. Montelione, and I.A. Wilson. Contributions to the NIH-NIGMS Protein Structure Initiative from the PSI Production Centers. *Structure*, 16:5–11, 2008.
8. D. Vitkup, E. Melamud, J. Moult, and C. Sander. Completeness in structural genomics. *Nat Struct Biol*, 8:559, 2001.
9. A. Bateman, L. Coin, R. Durbin, R.D. Finn, V. Hollich, S. Griffiths-Jones, A. Khanna, M. Marshall, S. Moxon, E.L. Sonnhammer, D.J. Studholme, C. Yeats, and S.R. Eddy. The Pfam protein families database. *Nucleic Acids Res*, 32(Database issue):D138–141, 2004.
10. B. Boeckmann, A. Bairoch, R. Apweiler, M.C. Blatter, A. Estreicher, E. Gasteiger, M.J. Martin, K. Michoud, C. O'Donovan, I. Phan, S. Pilbout, and M. Schneider. The SWISS-PROT protein knowledgebase and its supplement TrEMBL in 2003. *Nucleic Acids Res*, 31:365, 2003.
11. C.H. Wu, R. Apweiler, A. Bairoch, D.A. Natale, W.C. Barker, B. Boeckmann, S. Ferro, E. Gasteiger, H. Huang, R. Lopez, M. Magrane, M.J. Martin, R. Mazumder, C. O'Donovan, N. Redaschi, and B. Suzek. The Universal Protein Resource (UniProt): An expanding universe of protein information. *Nucleic Acids Res*, 34:D187–191, 2006.
12. E.L. Sonnhammer, S.R. Eddy, E. Birney, A. Bateman, and R. Durbin. Pfam: Multiple sequence alignments and HMM-profiles of protein domains. *Nucleic Acids Res*, 26:320, 1998.
13. S.F. Altschul, T.L. Madden, A.A. Schaffer, J. Zhang, Z. Zhang, W. Miller, and D.J. Lipman. Gapped BLAST and PSI-BLAST: A new generation of protein database search programs. *Nucleic Acids Res*, 25:3389, 1997.
14. G.A. Reeves, T.J. Dallman, O.C. Redfern, A. Akpor, and C.A. Orengo. Structural diversity of domain superfamilies in the CATH database. *J Mol Biol*, 360:725–741, 2006.
15. C.S. Riesenfeld, P.D. Schloss, and J. Handelsman. Metagenomics: Genomic analysis of microbial communities. *Annu Rev Genet*, 38:525–552, 2004.
16. A.E. Todd, R.L. Marsden, J.M. Thornton, and C.A. Orengo. Progress of structural genomics initiatives: An analysis of solved target structures. *J Mol Biol*, 348:1235–1260, 2005.
17. J.M. Chandonia and S.E. Brenner. The impact of structural genomics: Expectations and outcomes. *Science*, 311:347–351, 2006.
18. U. Pieper, N. Eswar, H. Braberg, M.S. Madhusudhan, F.P. Davis, A.C. Stuart, N. Mirkovic, A. Rossi, M.A. Marti-Renom, A. Fiser, B. Webb, D. Greenblatt, C.C. Huang, T.E. Ferrin, and A. Sali. MODBASE, a database of annotated comparative protein structure models, and associated resources. *Nucleic Acids Res*, 32(Database issue):D217, 2004.
19. J. Liu and B. Rost. Comparing function and structure between entire proteomes. *Protein Sci*, 10:1970, 2001.
20. J.C. Norvell and J.M. Berg. Update on the protein structure initiative. *Structure*, 15:1519–1522, 2007.
21. B.H. Dessailly, R. Nair, L. Jaroszewski, J.E. Fajardo, A. Kouranov, D. Lee, A. Fiser, A. Godzik, B. Rost, and C. Orengo. PSI-2: structural genomics to cover protein domain family space. *Structure*, 17(6):869–881, 2009.

CHAPTER 4

PREDICTION OF ONE-DIMENSIONAL STRUCTURAL PROPERTIES OF PROTEINS BY INTEGRATED NEURAL NETWORKS

YAOQI ZHOU and ESHEL FARAGGI
Indiana University School of Informatics
Center for Computational Biology and Bioinformatics
Indiana University School of Medicine
Indiana University-Purdue University Indianapolis
Indianapolis, IN

4.1. INTRODUCTION

Proteins are linear polymeric chains made of various combinations of 20 amino acid residues. The exact sequence of residues for each protein is encoded in the DNA sequence. Despite their underlying chemical simplicity, proteins can perform a wide range of biological functions from molecular signaling and transportation, molecular motors, structural support to catalyzing chemical reactions as enzymes. Such a multifaceted functionality is made possible by their ability to form different three-dimensional structures/shapes for different sequences of combinations of residues.

A direct prediction of three-dimensional structures from protein sequences has proven challenging, as discussed in several chapters in this book. As a result, many scientists search for other structural properties that are easier to predict and are useful for aiding three-dimensional structure prediction as restraints and/or protein-specific scoring functions. Many of those structural properties can be classified as one-dimensional structural properties because they can be represented as a one-dimensional vector along the protein sequence. This type of one-to-one prediction has been a computationally

Introduction to Protein Structure Prediction: Methods and Algorithms,
Edited by Huzefa Rangwala and George Karypis

attractive problem for more than 50 years. The number and quality of one-dimensional structural properties predicted from protein sequences continue to increase.

Currently predicted one-dimensional structural properties can be classified as local and global structural properties. Local structural properties refer to structural features of locally connected residues (sequence neighbors) only. Examples are secondary structure and backbone torsion angles. Global structural properties, on the other hand, depend on structural neighbors that are not necessary neighbors in terms of their sequence positions. Examples are solvent accessible surface area, residue coordination (contact) number, residue contact order, and residue depth. This chapter will first review the current literature to discuss the trend and progress in prediction of these structural properties. The discussion is followed by a demonstration of two recently developed methods called Structural Properties predicted by Integrated NEural networks (SPINE) for secondary structure and solvent accessibility prediction [1] and Real-SPINE for real-value prediction of backbone torsion angles and solvent accessibility [2,3,4].

Sequence-based predictions are also made for some one-dimensional properties related to protein structures. Examples are protein domain divisions [5–7], dynamical properties of structures (residue fluctuation or temperature B-factor [8,9], conformational variable or switching regions [10–12], intrinsic disorder [13–17], and residue structural entropy [18,19]). They will not be discussed in this chapter.

One-dimensional functional properties are another focus of recent studies. These include protein-protein binding interface [20], ubiquitylation sites [21], functional residues [22], DNA-binding residues [23,24], pathological mutations [25], subcellular localization [26], prediction of coding regions [27], and catalytic residues [28,29]. More references can be found in recent reviews [30–32]. These studies belong to the category of function prediction and are out of the scope of this chapter. Most of these function predictors, however, rely on predicted one-dimensional structural properties discussed here.

4.2. LOCAL STRUCTURAL PROPERTIES

4.2.1. Secondary Structure Prediction

The most important local structural property is, perhaps, the secondary structure of proteins [33]. Secondary structure refers to the three-dimensional structural patterns of sequentially linked, local segments. They are formally defined by the patterns of hydrogen bonds between the backbone amide and carboxyl groups. Two most common regular structural patterns (motifs) are α-helices and β-strands. α-Helices have a hydrogen bond between the ith and the $i+4$th residues. β-Strands are characterized by sequential hydrogen bonding between neighboring segments rather than within a local segment as in α-helices. The most widely used method for automated assignment of secondary structure of a given protein structure is the program called Defining Secondary Structures of Proteins (DSSP) [34], which defines eight types of secondary structure based on the pattern of hydrogen bonds. Of these, G, H, and I refer to 3_{10}-helix, α-helix, and π-helix, respectively; B and E refer to β-bridge and extended sheet, respectively; T, S, and others refer to turn, high-curvature region, and others, respectively. In practice, these eight types are grouped into helix (G, H, and I), strand (E and B), and coil (all others) for secondary structure prediction. A typical secondary structure prediction method attempts to make a sequence-based prediction that matches secondary structure elements assigned by DSSP [34].

Historically, secondary structure prediction was attempted [35–37] even before the publication of the first X-ray protein structures [38,39]. Despite its long history (more than 50 years), secondary structure prediction continues to be an active field of research. Early methods for secondary structure prediction were based on statistical analysis of single residues [40–42] and their neighboring residues [43–46]. The statistics of single residues reveal the trend of certain residues as helix breaker (such as Gly and Pro), helix formers (such as Met, Ala, Leu, Glu, and Lys), and sheet formers (large aromatic and C_β branched amino acid residues) while analysis of pairs, trios, or more residues further refines the preferences. These, however, are not strong enough to produce reliable prediction as the accuracy is only around 60%. The accuracy of prediction took a leap forward by replacing single sequence input with a sequence profile from multiple sequence alignment [47]. The sequence profile is a position-specific substitution matrix based on the observed probability of an amino acid residue type at a given sequence position from a multiple sequence alignment. The accuracy of prediction gradually increased, first above 70% [48,49] then to a tenfold cross-validated 80% by SPINE [1]. As machine-learning techniques continue to improve as well as the size and quality of the database, further improvement in secondary structure prediction is expected. Because earlier work on secondary structure prediction can be found in several excellent reviews [48–51], we will mostly focus on the latest trends and attempt to address several outstanding questions.

4.2.1.1. Searching for the Best Algorithm. One common theme in recent studies of secondary structure prediction is continued experimentation with novel and/or improved machine-learning algorithms. This includes variants of Support Vector Machine models [52–59], variants of neural networks [60–69], cascaded nonlinear components analysis [70], the consensus or ensemble of multiple predictors [64,65,71–81], Bayesian or hidden semi-Markov network [82–85], multiple linear regression [86,87], k-nearest neighborhood [88], dynamic programming algorithm [89], hybrid genetic-neural system [90], sequence pre-clustering [91,92], and conditional random fields for combined prediction [77]. Some notable new proposals are Bayesian neural network rather than commonly used backpropagation neural networks [93], density-weighted quality estimates for better generalization [94], bidirectional segmented-memory recurrent neural network for nonlocal interactions [95], large-scale recursive neural network [96], and guided learning in Real-SPINE [4].

Neural networks and Support Vector Machines are the most popular machine learning tools. They are usually better than probabilistic models [85]. However, a clear winner for the best machine-learning technique is yet to emerge for secondary structure prediction. Differences in implementation and in the database for training and testing have further hampered the ability to compare different techniques. A recent review provides a detailed analysis on machine-learning techniques (up to the year 2005) for secondary structure prediction [51].

4.2.1.2. Employing Homologous Sequences Improves Prediction Accuracy. One noticeable trend is the development of methods that take advantage of homologous sequences and/or structural fragments (templates). Examples are HYPROSP [80,97], PROTEUS [98], MUpred [99], DISTILL [100], a combination of GOR V and fragment database mining [101], and a profile-profile alignment to rank fragments for secondary structure prediction [102]. Highly accurate predictions are achieved when sequence-homologous structural fragments are available.

4.2.1.3. Searching for the Next Powerful Feature. Secondary structure prediction relies on the quality of input feature vectors that characterize the input sequence. The sequence profile from multiple sequence alignment [47] indisputably is the single most important feature responsible for improving the accuracy from around 60% to 80% today. Recently introduced new features are pentapeptide statistics [103,104], conserved domain profile [105], frequent amino acid patterns [106], predicted torsion angles [107,108], predicted residue contact maps [109], predicted residue solvent accessibility (RSA) [1,110], and predicted tertiary structure [111]. An interesting new input feature is pseudo-energy parameters for helix formation [112]. Several methods also integrate experimental data such as circular dichroism and infrared spectroscopic data for secondary structure prediction [113–115]. Most of these new computational features are limited to local interactions and lead to small improvement. For example, in SPINE [1], the features that do improve secondary structure prediction beyond sequence profiles are seven representative amino acid properties identified by Meiler et al. [116]. A tenfold cross-validation over 2640 proteins indicates that these properties increase Q_3 by 0.5%. On the other hand, Adamczak et al. [213] showed that introducing predicted RSA improves Q_3 accuracy from 76.6–77.9% to 80.3–81.8% for four small datasets of 135 to 163 chains while Wood and Hirst [107] made iterative use of predicted ψ dihedral angles to increase Q_3 accuracy from 77.2% to 79.4%. However, incorporating more accurately predicted solvent accessibility and ψ angles did not improve SPINE beyond 80% accuracy [1]. Thus, there is a need to reevaluate proposed new features on the most accurate method available. It is quite possible that some features may be useful for improving a poor predictor but not a good predictor.

4.2.1.4. Chameleon Sequences Are Not the Source of Errors in Secondary Structure Prediction. Chameleon sequences are those short sequence fragments (typically less than seven or eight residues long) that have different types of secondary structure in different proteins [117]. Yoon and Welsh predicted contact-dependent secondary structure propensity [118]. Boden et al. developed a method to predict probability of secondary structure types based on solved nuclear magnetic resonance (NMR) structures [119]. Costantini et al. investigated fold-dependent secondary structures with fold-dependent propensities [120]. On the other hand, Guo et al. showed that the existence of chameleon sequences does not affect the accuracy of secondary structure prediction [121], consistent with previous conclusion [122]. The secondary structures of chameleon sequences can be predicted as accurately as other short sequences likely because a sliding window of 21 residues is much longer than the typical length of chameleon sequences (eight residues or less).

4.2.1.5. Errors Do Not Cluster at the Boundaries of Secondary Structure in SPINE. It was suspected that prediction errors are concentrated at the boundary (capping) of helices and strands. As a result, some studies have focused on analysis and prediction of capping residues. FarzadFard et al. [123] analyzed the propensity of amino acid residues in sheet initiating and capping. Wilson et al. [124] improved the prediction of N-termini of α-helix for secondary structure prediction. A separate neural network for predicting the ends of secondary structure segments yields an improved prediction of secondary structure [125]. However, a detailed analysis of the location of predicted errors of secondary structures in 2640 proteins by SPINE indicates that errors occur evenly at the boundary and inside of a secondary structure element [1]. Thus, one has to look elsewhere to further improve the accuracy of secondary structure prediction.

4.2.1.6. X-ray Resolution Has Minor Effect on Secondary Structure Prediction. The accuracy of secondary structure prediction by SPINE [1] was analyzed for proteins with X-ray resolution of 1.5 Å or less, 2 Å or less, 2.5 Å or less, and 3 Å or less. The difference was found to be less than 0.1% for Q_3, the fraction of residues with correct predicted secondary structure. Thus, one could construct a large database by relaxing the requirement on X-ray resolution. Machine learning can gain more from a larger database than lose from reduced structural precision.

4.2.1.7. Local Interactions Can Be Captured Well by SPINE. How accurate can machine-learning methods capture local interactions? This can be demonstrated by examining the prediction accuracy of fully exposed residues because fully exposed residues only interact with neighboring residues and solvent molecules. SPINE [2] can achieve 87.5% for fully exposed residues, which is very close to the theoretical limit of 88–90% [48,126]. This indicates that SPINE provides a near perfect prediction of secondary structures when local interactions dominate.

4.2.1.8. The Role of Nonlocal, Long-Range Interactions. Nonlocal, long-range interactions refer to the interactions between the amino acid residues that are not sequence neighbors. Kihara reexamined the role of long-range interactions using a database of more than 2000 proteins and a parameter of residue contact order [126]. A lower contact order indicates a higher number of nonlocal contacts. He found that there is a significant negative correlation between the prediction accuracy of the secondary structure of a residue and its contact order. This conclusion is different from an earlier study [127]. Such a difference was attributed to smaller database size and a less clear cut parameter for defining nonlocal contacts to characterize long-range interactions in the early study [126]. Information analysis on the entropy densities of primary and secondary structure sequences also implied the role of long-range interactions [128]. Indeed, an interaction-enriched bidirectional recurrent neural network is capable of improving prediction accuracy to 84.6% when native contacts are employed as a part of the input [129,130]. An attempt to capture long-range information was also made by a method called bidirectional segmented-memory recurrent neural network by gradually adding more segments onto previous segments [95]. However, these methods are yet to be applied to a large database with more comprehensive testing to confirm the findings.

4.2.1.9. The Limitation of Secondary-Structure Prediction: Assignment. The difficulty to capture nonlocal interactions is only one of the problems facing secondary structure prediction. The prediction accuracy is further limited by somewhat arbitrary definition of a few discrete states for secondary structures. While the DSSP [34] is the de facto standard, many other methods have been proposed ([131–140]; for a recent review, see Reference [141]). Because these methods are based on different combinations of hydrogen bond patterns, geometric features, and expert knowledge, their secondary structure assignments differ significantly. For example, disagreement between DSSP, P-CURVE, and DEFINE can be as large as 25% [142]. More β-sheets are assigned by XTLSSTR [137] and more π-helix by SECSTR [138] than by DSSP. The discrepancy among the different methods is caused by non-ideal configurations of helices and sheets [143–145] and lack of objective, commonly accepted standard assignments.

Recently, Zhang et al. proposed a criterion that is based on relative assignments rather than "absolute" assignments [146]. More specifically, secondary structure assignment is assessed from the similarity of the secondary structures assigned to structurally aligned non-homologous sequences. It assumes that structurally aligned residues have the same secondary structures. Using this criterion, STRIDE [132] and KAKSI [135] are 1–4% better than DSSP in consistency for assigning structurally aligned residues. A consensus method SKSP [146] made of STRIDE [132], KAKSI [135], SECSTAR [138] and P-SEA [134] provides an additional 1% improvement. Improving the consistency of secondary structure assignment will likely further improve the accuracy of secondary structure prediction. Zhang et al. noted that the consistency of assigned secondary structures between structurally aligned residues in homolog proteins (sequence identity >30%) is at 90%. This less-than-perfect consistency, caused by errors in assignment and natural structural variation in evolution, limits the accuracy for any machine-learning techniques based on sequence profiles from multiple sequence alignments of sequence homologs. This theoretical limit confirms other proposed limits according to similar reasoning [48,126].

4.2.1.10. The Limitation of Secondary-Structure Prediction: Coarse-Graining. Three-state secondary structure is only a coarse-grained representation of local backbone structures. Ideal helix and strands are rare in protein structures and coils do not have well-defined structures. This coarse-grained representation can be improved by more refined classifications (such as eight states by DSSP [34]). However, the accuracy of multistate prediction (>3) [66] is too low to be practically useful for less populated states. Another approach is to make a dedicated prediction for a particular state such as various types of turns. Examples of some recent studies can be found in References [147–150]. Obviously, a continuous description of local structure is more desirable because it will avoid the arbitrary definition of discrete states and the associated assignment problem.

4.2.2. Backbone Torsion Angle Prediction

4.2.2.1. Backbone Torsion Angles as a Replacement/Supplement for Secondary Structure. One suitable candidate for a continuous description of local structure is backbone torsion angles. Two rotation angles (torsion angles) about the C_α–N bond (ϕ) and the C_α–C bond (ψ) essentially determine the structure of a protein backbone. This is so because the polypeptide backbone is a linked sequence of rigid planar peptide groups and the rotational angle about the C–N bond, ω, is fixed at 180° for the common trans and 0° for the rare cis conformation. Indeed, various secondary structure types are clustered at different regions in the Ramachandran $\phi - \psi$ diagram [151]. For example, an ideal helix and ideal parallel β-sheet are located at ($\phi = -57°$, $\psi = -47°$) and ($\phi = -119°$, $\psi = 113°$), respectively [33]. As a result, torsion angles are often employed as a replacement of, or supplement to, secondary structures for refined local-structure classifications. This drives the development of sequence-based methods for multistate torsion-angle prediction [108,152–156]. A method for predicting cis/trans isomerization was also developed [157].

4.2.2.2. Backbone Torsion Angles as Structural Building Blocks. Torsion angle prediction is also driven by the search for efficient sampling techniques. Ramachandran et al. [151] showed that not all ϕ and ψ angles can be sampled because of internal steric constraints. As a result, sampling in torsional space is one of the most commonly used methods for efficient exploration of protein conformational space (for examples see References [158–160]). Simple models were developed based on a few torsion-angle states [161–166]. This led to the development of methods to predict torsion-angle states [161,164,167–176]. An excellent review on local protein structures can be found in Reference [141].

4.2.2.3. Backbone Torsion Angles: Real-Value Prediction. The accuracy of multistate prediction typically becomes lower as the number of states increases. Classifying ϕ and ψ angles into a few states is somewhat arbitrary and the representative center of a state is a poor approximation for those angles at the state boundary. Moreover, real-value angles allow more accurate sampling [177] of protein structures. Xue et al. introduced Real-SPINE for a real-value prediction of both ϕ and ψ torsion angles [3]. They subsequently improved its accuracy by guided learning and other techniques [4]. It was found that multistates derived from predicted real values are comparably accurate to, or more accurate than, the direct prediction of multiple states [3]. For example, a 16-state prediction leads to a mean error of 25° for ϕ and 38° for ψ [175]. The corresponding result based on real-value prediction is 22° for ϕ and 36° for ψ in Real-SPINE 3 [4]. The real-value prediction allows a more accurate characterization of coil or turn residues that control the direction of helices or strands—where nonlocal interactions play an active role. Thus, real-value prediction should be more useful than coarse-grained secondary structure prediction for tertiary structure prediction.

4.2.2.4. Methods. Most methods for multistate prediction of torsion angles are similar to the methods for secondary structure prediction. They range from early methods based on residue or fragment statistics [152,153,161,167] to probabilistic techniques, such as hidden Markov models [154,168] and Bayesian probabilistic methods [164,169,170–173], to machine-learning techniques, such as Support Vector Machines [155,174,175,178], backpropagation neural networks [155,156,176], and bidirectional recurrent neural network [108]. Newly emerged real-value prediction of torsion angles are based on simple backpropagation neural networks [3,4]. Unlike secondary structure prediction, torsion-angle prediction has a short history. Thus, many more methods will likely emerge in the future.

4.2.2.5. Application of Predicted Angles. While applications of predicted secondary structures are well-established, applications of predicted angles are only starting to emerge. Currently these applications are limited to fold recognition [154,178,179], sequence alignment [180], and secondary structure prediction [107,108]. For example, Zhang et al. [179] found that using predicted torsion angles from Real-SPINE 2 [3] leads to about 2% improvement in alignment accuracy and 7% improvement in recognizing correct structural folds. As the accuracy of real-value prediction improves, it is expected that predicted torsion angles will gradually replace and/or supplement the predicted secondary structures because it contains significantly more information than the coarse-grained secondary structure.

4.3. GLOBAL STRUCTURAL PROPERTIES

Predicted secondary structure and torsion angles of a residue provide mostly local structural information along the sequence. The global structural properties of a residue, on the other hand, should provide some information about its position in addition to its orientation relative to covalently bonded sequence neighbors. Commonly used one-dimensional global structural properties are parameters that measure the solvent exposure of a residue, including normalized solvent accessible surface area (solvent accessibility) [181], residue depth (the distance of a residue from the nearest solvent molecule) [182], residue coordination or contact number (the number of residues within a cutoff distance) [183], half-sphere exposure (orientation-dependent contact numbers) [184], and recursive convex hull class [185]. While the methods for predicting residue depth [186,187], coordination (or contact) numbers [9,188–191], half-sphere exposure [192], and recursive convex hull class [185] emerged recently, solvent accessibility prediction has a relatively long history.

Earlier methods for predicting solvent accessibility mimicked the methods for secondary structure prediction by making a two-state (buried and exposed) or three-state (buried, intermediate, and exposed) prediction [1,183,193–204]. More recent studies make real-value prediction of solvent accessibilities [2,4,199,202,205–208]. The approaches range from neural networks [2,4,199,205,206], Support Vector Machines [202], information theory [209], multiple linear regression [207], and a constrained energy optimization [208]. For real-value prediction, prediction accuracy is often measured by the correlation coefficient between predicted and actual solvent accessible surface areas. There is a steady improvement in correlation coefficient from 0.50–0.65 [199,206,207] to 0.74 by Real-SPINE [2,4]. Two or three states derived from predicted real values are as accurate or more accurate than direct two-state or three-state prediction [2,4]. This indicates that arbitrary division of a few states for predicting solvent accessibility is unnecessary.

Predicted solvent accessibility was employed initially for aiding structure prediction and more recently for function prediction. A random sample of recent applications include fold recognition [154,179,210], sequence alignment [154,211,212], secondary structure prediction [1,110,213], function effects of single amino acid residue substitutions (SNP) [214–216], interaction prediction [217,218], and functional site prediction [22].

4.4. SPINE AND REAL-SPINE

SPINE [1] and Real-SPINE [2,3,4] are described here as an illustrative example of how accurate one-dimensional structural properties of proteins can be predicted by a carefully trained backpropagation neural network over a large dataset.

4.4.1. Objective of SPINE

A literature survey [1] in 2005 revealed that the accuracy of secondary structure prediction was stagnant around 77%. Moreover, most reported accuracies are not multiply cross validated and/or result from several small datasets. This led to a simple objective for SPINE development: to build a large dataset and perform careful large-scale training so that a reliable estimate of the prediction accuracy can be obtained. The same method was applied to three-state solvent accessibility prediction.

4.4.2. Tenfold Cross-Validation and Overfit Protection

The construction of a large dataset was aided by the availability of the protein sequence culling server PISCES [219]. Two thousand six hundred forty protein chains were obtained based on criteria of sequence identity less than 25% and X-ray resolution ≤3 Å. This dataset contains a total of 591,797 residues. To make a reliable estimate of the prediction accuracy, we perform a tenfold cross-validation as illustrated in Figure 4.1. The training set of 2640 chains was divided into 10 parts (264 proteins in each), nine of which is for training and the rest for testing. The process is repeated 10 times. During training, an additional 5% randomly selected chains (132 chains) within a training set of nine folds are excluded from training and employed as the stop criteria for

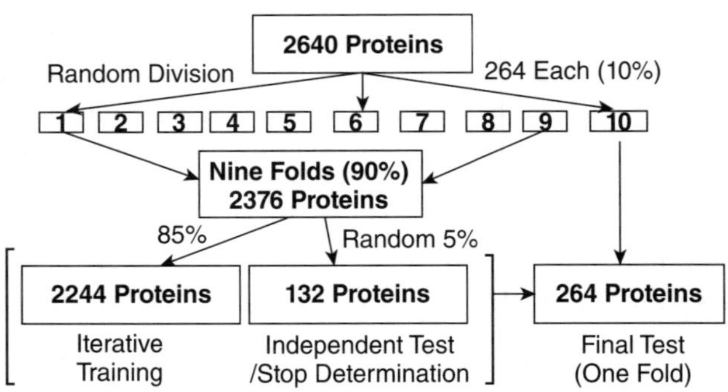

FIGURE 4.1 A schematic diagram illustrating how the data are divided for training, independent test, and final one-fold test. This is done 10 times, one for each fold (10% of the data), while the rest of the data are for training (85% data) and independent test (5% data).

weight optimization. Iterations for learning stop if there is more than a fixed number of continuous iterations that decrease (or do not increase) the prediction accuracy of the 5% data. This criterion is essential for avoiding possible over-fit as more than 200,000 weights are trained in SPINE [2].

4.4.3. Integrated Neural Networks in SPINE

Neural networks were chosen as the main machine-learning tool for SPINE [1] because of their versatility, because they are simple to implement, and since there is no evidence that other techniques are superior to them in secondary structure prediction. Moreover, it is relatively easy to modify from the neural networks for multistate classification as in secondary structure prediction to the neural networks for real-value prediction of other structural properties.

Neural networks attempt to learn the nonlinear relationship between the multivariate input features, x_i, and multivariate responses to generate desired output y_j (variables to be predicted). This is accomplished by mimicking the action of a biological neuron that accepts weighted input signals ($S_k^1 = \sum_j w_{jk}^1 \cdot x_j$) and generates a response h_k according to an activation function $f(S)\left[h_k^2 = f\left(S_k^1\right)\right]$. The responses in the hidden layers are the input signals for the neurons in the next layer. For a network with a single hidden layer, the third layer is made of the output neurons and $y_k = f\left(S_k^2\right)$ with $S_k^2 = \sum_j w_{jk}^2 \cdot h_j^2$. Here, w_{jk}^1 and w_{jk}^2 are the weights from the first (input) to second (hidden) layer and weights from the second (hidden) to the third (output) layer, respectively.

The nonlinear relationship between x_i and y_j is learned by optimizing weights to minimize the mean-squared error between the predicted y_j and actual y_j^{Expt} in a training set. $E\left(w_{jk}^1, w_{kl}^2\right) = \frac{1}{2}\sum_{m=1}^{M}\left(y_m - y_m^{\text{Expt}}\right)^2$. This error function is minimized by the steepest gradient descent method, that is, updating the weights according to $\dot{w}_{jk}^1 = -\eta \dfrac{\delta E}{\delta w_{jk}^1}$ with η being the learning rate. A similar equation for $\dot{w}_{jk}^2$ can be obtained. This method is referred to as backpropagation [220] because the weights are corrected based on the prediction error being backpropagated from the output layer toward the input layer.

SPINE adopts an architecture of neural networks established by Rost and Sander [72] and employed by many others (e.g., [62,206]). As shown in Figure 4.2, it is made of two-level neural networks A1 and A2. Both A1 and A2 employ a sliding window of 21 residues, a single hidden layer (but with different number of hidden nodes) and a three-state output. The main difference between A1 and A2 is in the input layer. The input for A1 is seven representative amino acid properties identified by Meiler et al. [116] (Table 4.1) and 20 values from the Position-Specific Scoring Matrix (PSSM) obtained from Position-Specific Iterative-Basic Local Alignment Search Tool (PSI-BLAST) [221] with three iterations of searching against nonredundant (NR) sequence database (ftp://ftp.ncbi.nih.gov/blast/db/FASTA/nr.gz). These seven parameters are a steric parameter (graph-shaped index), hydrophobicity, volume, polarizability, isoelectric point, helix probability, and sheet probability

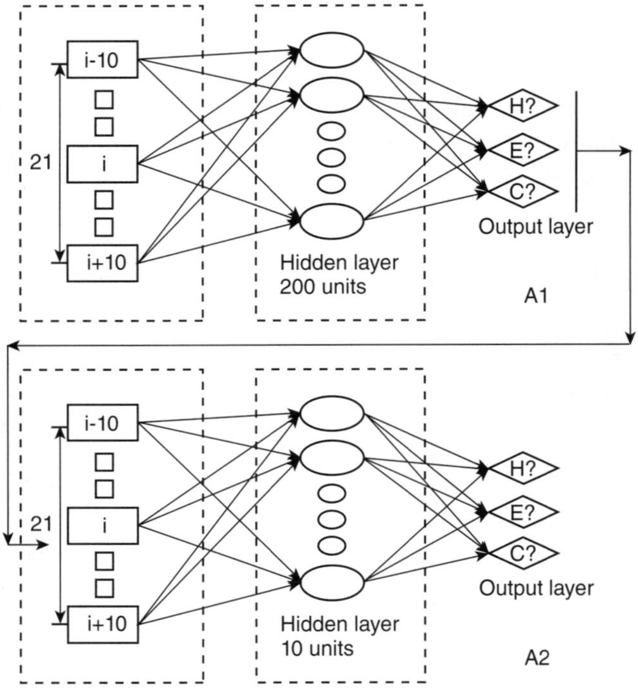

FIGURE 4.2 A schematic diagram showing how two neural networks (A1 as the first three-state classifier and A2 as the filter) are set up to make a final prediction of secondary structure in SPINE.

TABLE 4.1 The values of seven input properties (steric parameter, polarizability, volume, hydrophobicity, isoelectrical point, helix probability, and sheet probability). They are linearly normalized such that their maximum and minimum values for the 20 residue types fall at 0.9 and –0.9, respectively

Type	Steric	Polarizability	Volume	Hydrophobicity	Isoelectrical	Helix	Sheet
R	0.105	0.373	0.466	–0.900	0.900	0.528	–0.371
K	–0.088	0.066	0.163	–0.889	0.727	0.279	–0.265
D	–0.213	–0.417	–0.281	–0.767	–0.900	–0.155	–0.635
E	–0.230	–0.241	–0.058	–0.696	–0.868	0.900	–0.582
N	–0.213	–0.329	–0.243	–0.674	–0.075	–0.403	–0.529
Q	–0.230	–0.110	–0.020	–0.464	–0.276	0.528	–0.371
H	0.384	0.110	0.138	–0.271	0.195	–0.031	–0.106
Y	0.363	0.417	0.541	0.188	–0.274	–0.155	0.476
W	0.479	0.900	0.900	0.900	–0.209	0.279	0.529
S	–0.337	–0.637	–0.544	–0.364	–0.265	–0.466	–0.212
T	0.402	–0.417	–0.321	–0.199	–0.288	–0.403	0.212
G	–0.900	–0.900	–0.900	–0.342	–0.179	–0.900	–0.900
P	0.247	–0.900	–0.294	0.055	–0.010	–0.900	0.106
A	–0.350	–0.680	–0.677	–0.171	–0.170	0.900	–0.476
M	0.110	0.066	0.087	0.337	–0.262	0.652	–0.001
C	–0.140	–0.329	–0.359	0.508	–0.114	–0.652	0.476
F	0.363	0.373	0.412	0.646	–0.272	0.155	0.318
L	0.213	–0.066	–0.009	0.596	–0.186	0.714	–0.053
V	0.677	–0.285	–0.232	0.331	–0.191	–0.031	0.900
I	0.900	–0.066	–0.009	0.652	–0.186	0.155	0.688

(Table 4.1). The input for A2 is the output from A1 with a sliding widow of 21 residues. Here a sliding window of 21 residues means that the information of 20 sequence-neighboring residues are employed to predict the secondary structure of the central residue. The number of input attributes is $21 \times 27 + 1$ with one additional attribute for the bias used for refining the network.

In addition to neural networks A1 and A2 described above, SPINE further independently trained B1 and B2 by different initial random numbers for neural network weights. Different random initial weights in principle should reach the same solution if a global minimum is found during weight optimization. In reality, slightly different minima are obtained with different initial guesses for weights. In SPINE, a consensus prediction is made based on the output of A2 and B2.

4.4.4. Real-SPINE for Real-Value Prediction of Backbone Torsion Angles and Solvent Accessibility

The objective of Real-SPINE is to remove the arbitrary definition of three states and make real-value prediction of backbone torsion angles (to replace secondary structure) and solvent accessibility. While the methods for real-value prediction of solvent accessibility had achieved reasonable accuracy prior to Real-SPINE (a correlation of above 0.60 [199,207]), prediction of real-value backbone angle appeared more challenging with a correlation coefficient between predicted and measured real-value ψ angles at only 0.47 [107].

Real-SPINE [2] employs a simpler architecture than SPINE since real-value prediction requires only one output node, rather than three states in SPINE. It was further found that a filter (A2 and B2) is not useful for improving real-value prediction. That is, only A1 and B1 networks were employed for consensus prediction.

Real-SPINE 2 [3] recognized that one can take advantage of the periodicity of angles for improving prediction. This is done by shifting angles so that least populated angles are located at the edges (−180° and 180°). For example, the ψ angles were shifted by adding 100° to the angles between −100° and 180°, and adding 460° to the angles between −180° and −100°. This shift ensures that a minimum number of angles occur at the ends of the sigmoidal function. This region is inherently difficult to predict in a neural network-based machine-learning method.

Real-SPINE 3 [4] introduced a guided learning technique that gives different importance to the weights depending on their distance from the central residue. The guiding factors are designed to take into account the fact that in most cases, the contribution of a residue to the structural properties of another residue is inversely proportional to the distance between them along in the sequence. To implant this expectation in the weighting scheme, we separate each weight w_{jk}^i into a fixed guiding component g_{jk}^i (Fig. 4.3) and a variable component $c_{jk}^i \left(w_{jk}^i = g_{jk}^i c_{jk}^i\right)$. The guiding component has a preset value while the initial value for c_{jk}^i is a random number within a predetermined range. Only

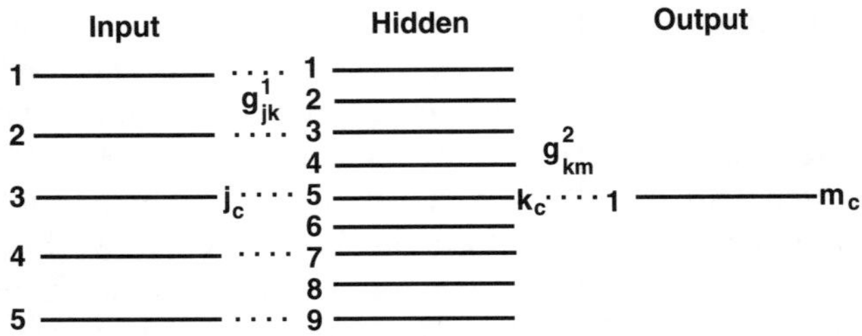

FIGURE 4.3 An illustration of guided for one hidden-layer network: A network with five input neurons, nine hidden neurons, and one output neuron. The guide factor is the highest for g_{11}^1, g_{23}^1, g_{35}^1, g_{47}^1, g_{59}^1, and g_{51}^2.

the c_{jk}^i component is updated during each optimization cycle. In Real-SPINE 3, we tested a guiding component g_{jk}^i given below:

$$g_{jk}^1 = \frac{1}{\sqrt{1+\left((k-1)\dfrac{J-1}{K-1}-(j-1)\right)^2}}, \tag{4.1}$$

$$g_{kl}^2 = \frac{1}{\sqrt{1+\left((k-k_c)\dfrac{J-1}{K-1}-(l-l_c)\dfrac{J-1}{L-1}\right)^2}}, \tag{4.2}$$

and

$$g_{lm}^3 = \frac{1}{\sqrt{1+\left((l-l_c)\dfrac{J-1}{L-1}-(m-m_c)\right)^2}}, \tag{4.3}$$

with $k_c = \dfrac{K+1}{2}$, $l_c = \dfrac{L+1}{2}$, and $m_c = \dfrac{M+1}{2}$, the central location of the two hidden and output layers, respectively. These simple weights impose a condition so that residues that are closer (in sequence distance) to a given amino acid residue will contribute more in determining predicted properties. In addition, Real-SPINE 3 employed two-layer neural network with a hyperbolic activation function, rather than a single-layer neural network with a sigmoidal activation function as in SPINE and Real-SPINE 1.0/2.0.

4.4.5. Data Preparation and Processing

Experimental values of secondary structure, ψ and ϕ angles and solvent accessible surface areas are obtained by applying the DSSP program to 2640 proteins [34]. The eight types of secondary structure are grouped into helix (G, H, and I), strand (E and B), and coil (all others) for secondary structure prediction. Solvent accessibility is solvent accessible surface area of a residue normalized by the area of its "unfolded" state [205] (in SPINE/Real-SPINE) or the maximum area in the database (in Real-SPINE 3.0). The three states of solvent accessibility (SA) are buried if SA is ≤25%, somewhat exposed if 25% < SA ≤ 75%, and fully exposed if SA > 75%. All input values for each network are normalized to be within the range from 0 to 1 with a linear normalization based on maximum and minimum (–1 to 1 in Real-SPINE 3 with hyperbolic activation). Neural network weights were initially generated by random number generators between –0.5 and 0.5. A momentum coefficient of 0.4 is used in all methods but different learning rates are employed (0.001 for SPINE/Real-SPINE, 0.0001 for Real-SPINE 2, and 0.01 for Real-SPINE 3). In Real-SPINE 3, a faster learning rate allows for more efficient learning.

4.4.6. Algorithm Optimization

The accuracy of SPINE/Real-SPINE was achieved only after extensive testing and parameter optimization. Table 4.2 shows that seven amino acid properties, the size of database, doubled hidden nodes, or a sliding window for filter each makes a contribution for improving the accuracy of SPINE (Q_3) by 0.7%, 0.5%, 0.2%, and 0.2%, respectively. Seven amino acid properties and doubled hidden nodes are also useful in improving solvent-accessibility prediction as shown in Table 4.3. Similarly, Table 4.4 shows that guided learning and the

TABLE 4.2 Six experiments in secondary structure prediction with different sizes of data base, the size of filter window, input features, and number of neural network units

#	# Chains	# Residues	Chain Size	Filter Window	Input Profile	NN Units	Q_3 %
1	1373	248299	40-2000	1	PSSM	100	77.9
2	1373	248299	40-2000	1	PSSM + PROP	100	78.6
3	2640	591797	40-2000	1	PSSM + PROP	100	79.1
4	2640	591797	40-2000	1	PSSM + PROP	200	79.3
5	2640	591797	40-2000	11..21[a]	PSSM + PROP	200	79.5
6	1952	313006	50-300	11..21[a]	PSSM + PROP	200	80.0

[a]The sizes of filter window (A_2, B_2) tested are 11, 13, 15, 17, 19, and 21. They yielded the same performance.

TABLE 4.3 Four experiments in solvent accessibility 3-state prediction

Experiment	# Chains	# Residues	Chain Size	Input Profile	NN Units	Q_3 Score %
1	2640	591797	40–2000	PSSM	100	72.2
2	2640	591797	40–2000	PSSM	200	72.4
3	2640	591797	40–2000	PSSM + PROP	100	72.8
4	2640	591797	40–2000	PSSM + PROP	200	73.0

TABLE 4.4 The tenfold cross-validated accuracy for predicting the ϕ and ψ angles, and RSA from five experiments. Standard deviations between 10 folds are also shown for Experiments III, IV, and V

Experiment	I		II		III		**IV**		V	
	(500,1,21)[a]		(500,2,21)[a]		(2479,1,21)[a]		(2479,2,21)[a]		(2479,2,41)[a]	
Guided?	No[b]	Yes[c]	No	Yes	No	Yes	No	Yes	No	Yes
ψ-Q_{10}[d]	43.5%	45.3%	48.1%	49.5%	47.0 ± 0.8%	48.4 ± 0.5%	49.8 ± 0.5%	**50.7 ± 0.5%**	48.5 ± 0.4%	50.1 ± 0.6%
ψ-$Q_{10\%}$[e]	61.1%	63.0%	65.6%	66.8%	64.6 ± 0.7%	65.8 ± 0.5%	67.3 ± 0.4%	**68.5 ± 0.5%**	65.7 ± 0.4%	67.8 ± 0.6%
ψ-PCC[f]	0.72	0.729	0.757	0.770	0.741 ± 0.007	0.746 ± 0.007	0.743 ± 0.007	**0.746 ± 0.007**	0.729 ± 0.007	0.743 ± 0.007
ψ-MAE[g]	41.5°	39.8°	39.8°	38.1°	38.3 ± 0.8°	37.3 ± 0.8°	36.8 ± 0.9°	**36.1 ± 0.8°**	38.2 ± 0.9°	36.6 ± 0.8°
ϕ-Q_{10}[d]					54.6 ± 0.5%	55.6 ± 0.5%	54.8 ± 0.5%	**56.1 ± 0.5%**	54.9 ± 0.4%	56.1 ± 0.4%
ϕ-$Q_{10\%}$[e]					81.7 ± 0.4%	82.1 ± 0.4%	82.0 ± 0.4%	**82.4 ± 0.4%**	81.2 ± 0.4%	82.2 ± 0.3%
ϕ-PCC[f]					0.653 ± 0.005	0.658 ± 0.005	0.653 ± 0.005	**0.659 ± 0.005**	0.642 ± 0.006	0.654 ± 0.006
ϕ-MAE[g]					22.8 ± 0.4°	22.3 ± 0.4°	22.6 ± 0.3°	**22.2 ± 0.4°**	22.8 ± 0.4°	22.3 ± 0.3°
RSA-Q_{10}[d]					39.0 ± 0.8%	39.7 ± 0.5%	39.2 ± 0.7%	**39.9 ± 0.4%**	38.7 ± 1.4%	39.7 ± 0.8%
RSA-$Q_{10\%}$[e]					57.0 ± 0.9%	58.0 ± 0.5%	57.4 ± 0.8%	**58.1 ± 0.3%**	56.5 ± 1.5%	57.7 ± 0.8%
RSA-PCC[f]					0.737 ± 0.004	0.744 ± 0.005	0.738 ± 0.004	**0.745 ± 0.004**	0.725 ± 0.005	0.742 ± 0.004
RSA-MAE[g]					0.114 ± 0.002	0.112 ± 0.001	0.113 ± 0.002	**0.111 ± 0.001**	0.117 ± 0.002	0.112 ± 0.001

[a]The number of proteins in the dataset, the number of hidden layers, and input window size.
[b]No guided weights.
[c]Guided weights are used.
[d]Q_{10}: Fraction of residues with correctly predicted states. Angles are divided into 10 states with 36° per bin.
[e]$Q_{10\%}$: Fraction of residues whose angles are predicted within 36° from the true value.
[f]Pearson's correlation coefficient between predicted and actual values.
[g]Mean-absolute error between predicted and actual values. Degrees are used for the ϕ and ψ angles and [0,1] normalization for the RSA.

additional layer of the neural network in Real-SPINE 3 make nearly equal contribution for improving ψ and ϕ angles, respectively. For example, introducing the guided learning improves between 0.9% and 2.2% for $Q_{10\%}$ in ψ, 0.4% and 1.3% for $Q_{10\%}$ in ϕ, and 0.7% and 1.2% for $Q_{10\%}$ in RSA while the mean absolute errors of ψ, ϕ, and RSA are reduced by 2–4%. Here $Q_{10\%}$ denotes the fraction of residues whose predicted ψ or ϕ torsion angles are within 36° of native angles. These improvements are consistent regardless of the number of hidden layers, the size of input window, the size of the database for training and cross-validation, and the parameter that measures the accuracy. These results indicate that a significant improvement in accuracy requires multiple techniques and careful selections of parameters.

4.5. CONCLUSION AND OUTLOOK

The challenging problem of protein structure prediction demands more accurate prediction of one-dimensional structural properties. While the upper limit for secondary structure prediction is estimated at around 90% [48,126], the exact contribution from nonlocal interaction is not clear. There is no doubt that the current record of 80% will be broken because guided learning can contribute an additional 1% improvement (Faraggi and Zhou, in preparation). Improving assignment consistency will likely push further the limit. Real-value prediction of solvent accessibility, on the other hand, will be more challenging to improve because nonlocal interactions play more important role in RSA than in secondary structure formation. Moreover, solvent accessibility is not as conserved as secondary structure and the correlation coefficient of solvent accessibility between homologs is only 0.77 [194], compared with 0.74 currently reached by Real-SPINE based on homolog-derived sequence profiles [3,4]. On the other hand, predicting torsion angles is just the beginning. More significant improvement is expected in the near future. More accurately predicted torsion angles will likely replace the dominant role played by predicted secondary structure in protein three-dimensional structure prediction as found in the development and application of SPINE XI [222].

REFERENCES

1. O. Dor and Y. Zhou. Achieving 80% ten-fold cross-validated accuracy for secondary structure prediction by large-scale training. *Proteins*, 66:838–845, 2007.
2. O. Dor and Y. Zhou. Real-SPINE: An integrated system of neural networks for real-value prediction of protein structural properties. *Proteins*, 68:76–81, 2007.
3. B. Xue, O. Dor, E. Faraggi, and Y. Zhou. Real-value prediction of backbone torsion angles. *Proteins*, 72:427–433, 2008.
4. E. Faraggi, B. Xue, and Y. Zhou. Improving the accuracy of predicting real-value backbone torsion angles and residue solvent accessibility by guided learning through two-layer neural networks. *Proteins*, 74:857–871, 2009.

5. J. Cheng, M.J. Sweredoski, and P. Baldi. DOMpro: Protein domain prediction using profiles, secondary structure, relative solvent accessibility, and recursive neural networks. *Data Mining and Knowledge Discovery*, 13:1–10, 2006.
6. J. E. Gewehr and R. Zimmer. SSEP-Domain: Protein domain prediction by alignment of secondary structure elements and profiles. *Bioinformatics*, 22:181–187, 2006.
7. M. Tress, J. Cheng, P. Baldi, K. Joo, J. Lee, J.-H. Seo, J. Lee, D. Baker, D. Chivian, D. Kim, and I. Ezkurdia. Assessment of predictions submitted for the casp7 domain prediction category. *Proteins*, 69(8):137–151, 2007.
8. A. Schlessinger and B. Rost. Protein flexibility and rigidity predicted from sequence. *Proteins*, 61:115–126, 2005.
9. Z. Yuan, T. L. Bailey, and R. D. Teasdale. Prediction of protein B-factor profiles. *Proteins*, 58:905–912, 2005.
10. M. Young, K. Kirshenbaum, K.A. Dill, and S. Highsmith. Predicting conformational switches in proteins. *Protein Science*, 8:1752–1764, 1999.
11. M. Gross. Proteins that convert from alpha helix to beta sheet: Implications for folding and disease. *Current Protein & Peptide Science*, 1:339–347, 2000.
12. I. B. Kuznetsov. Ordered conformational change in the protein backbone: Prediction of conformationally variable positions from sequence and low-resolution structural data. *Proteins*, 72:74–87, 2008.
13. F. Ferron, S. Longhi, B. Canard, and D. Karlin. A practical overview of protein disorder prediction methods. *Proteins*, 65:1–14, 2006.
14. Z. Dosztanyi, M. Sandor, P. Tompa, and I. Simon. Prediction of protein disorder at the domain level. *Current Protein & Peptide Science*, 8:161–171, 2007.
15. J. M. Bourhis, B. Canard, and S. Longhi. Predicting protein disorder and induced folding: From theoretical principles to practical applications. *Current Protein & Peptide Science*, 8:135–149, 2007.
16. P. Radivojac, L.M. Iakoucheva, C.J. Oldfield, Z. Obradovic, V.N. Uversky, and A. K. Dunker. Intrinsic disorder and functional proteomics. *Biophysical Journal*, 92:1439–1456, 2007.
17. H. Rangwala, C. Kauffman, and G. Karypis. A kernel framework for protein residue annotation. In T. Theeramunkong, B. Kijsirikul, N. Cercone, and T.-B. Ho (Eds.), *Advances in Knowledge Discovery and Data Mining, 13th Pacific-Asia Conference, PAKDD 2009*. Bangkok, Thailand, April 27–30, 2009, Proceedings; *Lecture Notes in Computer Science*, vol. 5476, pp. 439–451. Springer, 2009.
18. C.-H. Chan, H.-K. Liang, N.-W. Hsi, M.-T. Ko, P.-C. Lyu, and J.-K. Hwang. Relationship between local structural entropy and protein thermostabilty. *Proteins*, 57:684–691, 2004.
19. S. W. Huang and J. K. Hwang. Computation of conformational entropy from protein sequences using the machine-learning method—Application to the study of the relationship between structural conservation and local structural stability. *Proteins*, 59:802–809, 2005.
20. S. Liang, C. Zhang, S. Liu, and Y. Zhou. Protein binding site prediction with an empirical scoring function. *Nucleic Acids Research*, 34:3698–3707, 2006.
21. C.W. Tung and S.Y. Ho. Computational identification of ubiquitylation sites from protein sequences. *BMC Bioinformatics*, 9:310, 2008.

22. J.D. Fischer, C.E. Mayer, and J. Soeding. Prediction of protein functional residues from sequence by probability density estimation. *Bioinformatics*, 24:613–620, 2008.

23. S. Hwang, Z. Gou, and I.B. Kuznetsov. DP-Bind: a web server for sequence-based prediction of DNA-binding residues in DNA-binding proteins. *Bioinformatics*, 23:634–636, 2007.

24. S. Ahmad and A. Sarai. PSSM-based prediction of DNA binding sites in proteins. *BMC Bioinformatics*, 6:33, 2005.

25. C. Ferrer-Costa, M. Orozco, and X. de la Cruz. Sequence-based prediction of pathological mutations. *Proteins*, 57:811–819, 2004.

26. H. Lin, H. Ding, F.B. Guo, A.Y. Zhang, and J. Huang. Predicting subcellular localization of mycobacterial proteins by using Chou's pseudo amino acid composition. *Protein & Peptide Letters*, 15:739–744, 2008.

27. J. Liu, J. Gough, and B. Rost. Distinguishing protein-coding from non-coding RNAs through support vector machines. *PLOS Genetics*, 2:529–536, 2006.

28. N.V. Petrova and C.H. Wu. Prediction of catalytic residues using Support Vector Machine with selected protein sequence and structural properties. *BMC Bioinformatics*, 7:312, 2006.

29. T. Zhang, H. Zhang, K. Chen, S. Shen, J. Ruan, and L. Kurgan. Accurate sequence-based prediction of catalytic residues. *Bioinformatics*, 24:2329–2338, 2008.

30. A. Godzik, M. Jambon, and I. Friedberg. Computational protein function prediction: Are we making progress? *Cellular & Molecular Life Sciences*, 64:2505–2511, 2007.

31. Z.R. Yang. Biological applications of support vector machines. *Briefings in Bioinformatics*, 5:328–338, 2004.

32. G. Lopez, A. Rojas, M. Tress, and A. Valencia. Assessment of predictions submitted for the CASP7 function prediction category. *Proteins*, 69(8):165–174, 2007.

33. D. Voet and J.G. Voet. *Biochemistry*. New York: John Wiley & Sons, Inc.,1995.

34. W. Kabsch and C. Sander. Dictionary of protein structure: pattern recognition of hydrogen-bonded and geometrical features. *Biopolymers*, 22:2577–2637, 1983.

35. L. Pauling, R.B. Corey, and H.R. Branson. The structure of proteins: Two hydrogenbonded helical configurations of the polypeptide chain. *Proceedings of the National Academy of Science U S A*, 37:205–234, 1951.

36. L. Pauling and R.B. Corey. Configurations of the polypeptide chains with favored orientations around single bonds: Two new pleated sheets. *Proceedings of the National Academy of Science U S A*, 37:729–740, 1951.

37. A.G. Szent-Gyorgyi and C. Cohen. Role of proline in polypeptide chain configuration of proteins. *Science*, 126:697–698, 1957.

38. J.C. Kendrew, R.E. Dickerson, B.E. Strandberg, R.G. Hart, and D.R. Davies. Structure of Myoglobin: A Three-Dimensional Fourier Synthesis at 2 Å. Resolution. *Nature*, 185:422–427, 1960.

39. M.F. Perutz, M.G. Rossmann, A.F. Cullis, G. Muirhead, G. Will, and A.T North. Structure of Haemoglobin: A Three-Dimensional Fourier Synthesis at 5.5 Å. Resolution. *Nature*, 185:416–422, 1960.

40. H.A. Scheraga. Structural studies of ribonuclease III. A model for the secondary and tertiary structure. *Journal of the American Chemical Society*, 82:3847–3852, 1960.
41. A.V. Finkelstein and O.B. Ptitsyn. Statistical analysis of the correlation among amino acid residues in helical, β-structural and non-regular regions of globular proteins. *Journal of Molecular Biology*, 62:613–624, 1971.
42. P.Y. Chou and U.D. Fasman. Prediction of protein conformation. *Biochemistry*, 13:211–215, 1974.
43. E.A. Kabat and T.T. Wu. The influence of nearest-neighbor amino acids on the conformation of the middle amino acid in proteins: Comparison of predicted and experimental determination of β-sheets in concanavalin A. *Proceedings of the National Academy of Science U S A*, 70:1473–1477, 1973.
44. F.R. Maxfield and H.A. Scheraga. Status of empirical methods for the prediction of protein backbone topography. *Biochemistry*, 15:5138–5153, 1976.
45. H.L. Holley and M. Karplus. Protein secondary structure prediction with a neural network. *Proceedings of the National Academy of Science U S A*, 86:152–156, 1989.
46. G.E. Arnold, A.K. Dunker, S.J. Johns, and R.J. Douthart. Use of conditional probabilities for determining relationships between amino acid sequence and protein secondary structure. *Proteins*, 12:382–399, 1992.
47. M.J. Zvelebil, G.J. Barton, W.R. Taylor, and M.J E. Sternberg. Prediction of protein secondary structure and active-sites using the alignment of homologous sequences. *Journal of Molecular Biology*, 195:957–961, 1987.
48. B. Rost. Review: Protein secondary structure prediction continues to rise. *Journal of Structural Biology*, 134:204–218, 2001.
49. V.A. Simossis and J. Heringa. Integrating secondary structure prediction and multiple sequence alignment. *Current Protein and Peptide Science*, 5:1–15, 2004.
50. J. Heringa. Computational methods for protein secondary structure prediction using multiple sequence alignments. *Current Protein and Peptide Science*, 1:273–301, 2000.
51. P.D. Yoo, B.B. Zhou, and A.Y. Zomaya. Machine learning techniques for protein secondary structure prediction: An overview and evaluation. *Current Bioinformatics*, 3:74–86, 2008.
52. S. Hua and Z. Sun. A novel method of protein secondary structure prediction with high segment overlap measure: support vector machine approach. *Journal of Molecular Biology*, 308(2):397–407, 2001.
53. J.J. Ward, L.J. McGuffin, B.F. Buxton, and D.T. Jones. Secondary structure prediction with support vector machines. *Bioinformatics*, 19:1650–1655, 2003.
54. H. Kim and H. Park. Protein secondary structure prediction based on an improved support vector machines approach. *Protein Engineering*, 16:553–560, 2003.
55. H.J. Hu, Y. Pan, R. Harrison, and P.C. Tai. Improved protein secondary structure prediction using support vector machine with a new encoding scheme and an advanced tertiary classifier. *IEEE Transactions on Nanobioscience*, 3:265–271, 2004.
56. L.H. Wang, Y.F. Li, J. Liu, and H.B. Zhou. Predicting protein secondary structure by a support vector machine based on a new coding scheme. *Genome Informatics*, 15:181–190, 2004.

57. J. Guo, H. Chen, Z.R. Sun, and Y.L. Lin. A novel method for protein secondary structure prediction using dual-layer SVM and profiles. *Proteins*, 54:738–743, 2004.

58. G. Karypis. YASSPP: Better kernels and coding schemes lead to improvements in protein secondary structure prediction. *Proteins*, 64:575–586, 2006.

59. M.N. Nguyen and J.C. Rajapakse. Prediction of protein secondary structure with two-stage multi-class SVMs. *International Journal of Data Mining and Bioinformatics*, 1:248–269, 2007.

60. N. Qian and T.J. Sejnowski. Predicting the secondary structure of globular proteins using neural network models. *Journal of Molecular Biology*, 202:865–884, 1988.

61. B. Rost and C. Sander. Improved prediction of protein secondary structure by use of sequence profiles and neural networks. *Proceedings of the National Academy of Science U S A*, 90:7558–7562, 1993.

62. D.T. Jones. Protein secondary structure prediction based on position-specific scoring matrices. *Journal of Molecular Biology*, 292:195–202, 1999.

63. J.A. Cuff and G.J. Barton. Application of multiple sequence alignment profiles to improve protein secondary structure prediction. *Proteins*, 40:502–511, 2000.

64. G.M. Chandonia and M. Karplus. New methods for accurate prediction of protein secondary structure. *Proteins*, 35:293–306, 1999.

65. T.N. Petersen, C. Lundegaard, M. Nielsen, H. Boher, J. Boher, S. Brunak, G.P. Gippert, and O. Lund. Prediction of protein secondary structure at 80% accuracy. *Proteins*, 41:17–20, 2000.

66. G. Pollastri, D. Przybylski, B. Rost, and P. Baldi. Improving the prediction of protein secondary structure in three and eight classes using recurrent neural networks and profiles. *Proteins*, 47:228–235, 2002.

67. G. Pollastri and A. McLysaght. Porter: A new, accurate server for protein secondary structure prediction. *Bioinformatics*, 21:1719–1720, 2005.

68. K. Lin, V.A. Simossis, W.R. Taylor, and J. Heringa. A simple and fast secondary structure prediction method using hidden neural networks. *Bioinformatics*, 21:152–159, 2005.

69. J. Chen and N.S. Chaudhari. Cascaded bidirectional recurrent neural networks for protein secondary structure prediction. *IEEE-ACM Transactions on Computational Biology and Bioinformatics*, 4:572–582, 2007.

70. S. Botelho, G. Simas, and P. Silveira. Prediction of protein Secondary Structure using nonlinear method. In I. King and J. Wang, and L. Chan, and D.L. Wang (Eds.), *Neural Information Processing, Part 3, Proceedings; Lecture Notes in Computer Science*, vol. 4234, pp. 40–47. Berlin and Heidelberg: Springer, 2006.

71. X. Zhang, J.P. Mesirov, and D.L. Waltz. Hybrid system for protein secondary structure prediction. *Journal of Molecular Biology*, 225:1049–1063, 1992.

72. B. Rost and C. Sander. Prediction of protein secondary structure at better than 70% accuracy. *Journal of Molecular Biology*, 232:584–599, 1993.

73. J.A. Cuff, M.E. Clamp, A.S. Siddiqui, M. Finlay, and G.J. Barton. Jpred: A consensus secondary structure prediction server. *Bioinformatics*, 14:892–893, 1998.

74. R.D. King, M. Ouali, A.T. Strong, A. Aly, A. Elmaghraby, M. Kantardzic, and D. Page. Is it better to combine predictions? *Protein Engineering*, 13:15–19, 2000.

75. M. Albrecht, S.C.E. Tosatto, T. Lengauer, and G. Valle. Simple consensus procedures are effective and sufficient in secondary structure prediction. *Protein Engineering*, 16:459–462, 2003.
76. A. Ceroni, P. Frasconi, A. Passerini, and A. Vullo. A combination of support vector machines and bidirectional recurrent neural networks for protein secondary structure prediction. In A. Cappelli and F. Turini (Eds.), *AIIA 2003: Advances in Artificial Intelligence, Proceedings; Lecture Notes in Artificial Intelligence*, vol. 2829, pp. 142–153. Berlin and Heidelberg: Springer, 2003.
77. Y. Liu, J. Carbonell, J. Klein-Seetharaman, and V. Gopalakrishnan. Comparison of probabilistic combination methods for protein secondary structure prediction. *Bioinformatics*, 20:3099–3107, 2004.
78. Y. Guermeur, G. Pollastri, A. Elisseeff, D. Zelus, H. Paugam-Moisy, and P. Baldi. Combining protein secondary structure prediction models with ensemble methods of optimal complexity. *Neurocomputing*, 56:305–327, 2004.
79. M. Kazemian, B. Moshiri, H. Nikbakht, and C. Lucas. Protein secondary structure classifiers fusion using OWA. In J.L. Oliveira and V. Maojo, and F. Martin Sanchez and A.S. Pereira (Eds.), *Biological and Medical Data Analysis, Proceedings; Lecture Notes in Computer Science*, vol. 3745, pp.338–345. Berlin and Heidelberg: Springer, 2005.
80. H.N. Lin, J.M. Chang, K.P. Wu, T.Y. Sung, and W.L. Hsu. Hyprosp II—A knowledge-based hybrid method for protein secondary structure prediction based on local prediction confidence. *Bioinformatics*, 21:3227–3233, 2005.
81. K.-H. Liu, J.-F. Xia, and X. Li. Efficient ensemble schemes for protein secondary structure prediction. *Protein and Peptide Letters*, 15:488–493, 2008.
82. W. Chu, Z. Ghahramani, A. Podtelezhnikov, and D.L. Wild. Bayesian segmental models with multiple sequence alignment profiles for protein secondary structure and contact map prediction. *IEEE-ACM Transactions on Computational Biology and Bioinformatics*, 3:98–113, 2006.
83. Z. Aydin, Y. Altunbasak, and M. Borodovsky. Protein secondary structure prediction for a single-sequence using hidden semi-Markov models. *BMC Bioinformatics*, 7, 2006.
84. Z. Aydin, Y. Altunbasak, and H. Erdogan. Bayesian protein secondary structure prediction with near-optimal segmentations. *IEEE Transactions on Signal Processing*, 55:3512–3525, 2007.
85. X.-Q. Yao, H. Zhu, and Z.-S. She. A dynamic Bayesian network approach to protein secondary structure prediction. *BMC Bioinformatics*, 9, 2008.
86. X.M. Pan. Multiple linear regression for protein secondary structure prediction. *Proteins*, 43:256–259, 2001.
87. S. Qin, Y. He, and X.M. Pan. Predicting protein secondary structure and solvent accessibility with an improved multiple linear regression method. *Proteins*, 61:473–480, 2005.
88. J. Sim, S.Y. Kim, and J. Lee. Prediction of protein solvent accessibility using fuzzy k-nearest neighbor method. *Bioinformatics*, 21:2844–2849, 2005.
89. M. Sadeghi, S. Parto, S. Arab, and B. Ranjbar. Prediction of protein secondary structure based on residue pair types and conformational states using dynamic programming algorithm. *FEBS Letters*, 579:3397–3400, 2005.

90. G. Armano, G. Mancosu, L. Milanesi, A. Orro, M. Saba and E. Vargiu. A hybrid genetic-neural system for predicting protein secondary structure. *BMC Bioinformatics*, 6:S3, 2005.

91. F. Jiang. Prediction of protein secondary structure with a reliability score estimated by local sequence clustering. *Protein Engineering*, 16:651–657, 2003.

92. S.H. Doong and C.Y. Yeh. Cluster-based local modeling approach to protein secondary structure prediction. *Journal of Computational and Theoretical Nanoscience*, 2:551–560, 2005.

93. J.L. Shao, D. Xu, L.Z. Wang, and Y.F. Wang. Bayesian neural networks for prediction of protein secondary structure. In X. Li and S. Wang, and Z.Y. Dong, (Eds.), *Advanced Data Mining and Applications, Proceedings; Lecture Notes in Artificial Intelligence*, vol. 3584, pp. 544–551. Berlin and Heidelberg: Springer, 2005.

94. L. Budagyan and R. Abagyan. Weighted quality estimates in machine learning. *Bioinformatics*, 22:2597–2603, 2006.

95. J. Chen and N.S. Chaudhari. Bidirectional segmented-memory recurrent neural network for protein secondary structure prediction. *Soft Computing*, 10:315–324, 2006.

96. P. Baldi and G. Pollastri. The principled design of large-scale recursive neural network architectures-DAG-RNNs and the protein structure prediction problem. *Journal of Machine Learning Research*, 4:575–602, 2004.

97. K.P. Wu, H.N. Lin, J.M. Chang, T.Y. Sung, and W.L. Hsu. HYPROSP: A hybrid protein secondary structure prediction algorithm—a knowledge-based approach. *Nucleic Acids Research*, 32:5059–5065, 2004.

98. S. Montgomerie, S. Sundararaj, W.J. Gallin, and D.S. Wishart. Improving the accuracy of protein secondary structure prediction using structural alignment. *BMC Bioinformatics*, 7, 2006.

99. R. Bondugula and D. Xu. MUPRED: A tool for bridging the gap between template based methods and sequence profile based methods for protein secondary structure prediction. *Proteins*, 66:664–670, 2007.

100. G. Pollastri, A.J.M. Martin, C. Mooney, and A. Vullo. Accurate prediction of protein secondary structure and solvent accessibility by consensus combiners of sequence and structure information. *BMC Bioinformatics*, 8:201, 2007.

101. H. Cheng, T.Z. Sen, R.L. Jernigan, and A. Kloczkowski. Consensus data mining (CDM) protein secondary structure prediction server: Combining GOR V and fragment database mining (FDM). *Bioinformatics*, 23 2628–2630, 2007.

102. J.M. Pei and N.V. Grishin. Combining evolutionary and structural information for local protein structure prediction. *Proteins*, 56:782–794, 2004.

103. A. Figureau, M.A. Soto, and J. Toha. A pentapeptide-based method for protein secondary structure prediction. *Protein Engineering*, 16:103–107, 2003.

104. G.T. Kilosanidze, A.S. Kutsenko, N.G. Esipova, and V.G. Tumanyan. Analysis of forces that determine helix formation in alpha-proteins. *Protein Science*, 13:351–357, 2004.

105. S.K. Woo, C.B. Park, and S.W. Lee. Protein secondary structure prediction using sequence profile and conserved domain profile. In D.S. Huang and X.P. Zhang, and G.B. Huang (Eds.), *Advances in Intelligent Computing, Part 2, Proceedings*; *Lecture Notes in Computer Science*, vol. 3645, pp. 1–10. Berlin and Heidelberg: Springer, 2005.

106. F. Birzele and S. Kramer. A new representation for protein secondary structure prediction based on frequent patterns. *Bioinformatics*, 22:2628–2634, 2006.
107. M.J. Wood and J.D. Hirst. Protein secondary structure prediction with dihedral angles. *Proteins*, 59:476–481, 2005.
108. C. Mooney, A. Vullo, and G. Pollastri. Protein structural motif prediction in multidimensional phi-psi space leads to improved secondary structure prediction. *J. Computational Biology*, 13:1489–1502, 2006.
109. U. Midic, A.K. Dunker, and Z. Obradovic. Exploring alternative knowledge representations for protein secondary-structure prediction. *International Journal of Data Mining and Bioinformatics*, 1:286–313, 2007.
110. A. Momen-Roknabadi, M. Sadeghi, H. Pezeshk, and S.-A. Marashi. Impact of residue accessible surface area on the prediction of protein secondary structures. *BMC Bioinformatics*, 9, 2008.
111. J. Meiler and D. Baker. Coupled prediction of protein secondary and tertiary structure. *Proceedings of the National Academy of Science U S A*, 100:12105–12110, 2003.
112. B. Gassend, C.W. O'Donnell, W. Thies, A. Lee, M. van Dijk, and S. Devadas. Learning biophysically-motivated parameters for alpha helix prediction. *BMC Bioinformatics*, 8, 2007.
113. B.I. Baello, P. Pancoska, and T.A. Keiderling. Enhanced prediction accuracy of protein secondary structure using hydrogen exchange Fourier transform infrared spectroscopy. *Analytical Biochemistry*, 280:46–57, 2000.
114. J.A. Hering, P.R. Innocent, and P.I. Haris. Neuro-fuzzy structural classification of proteins for improved protein secondary structure prediction. *Proteomics*, 3:1464–1475, 2003.
115. J.G. Lees and R.W. Janes. Combining sequence-based prediction methods and circular dichroism and infrared spectroscopic data to improve protein secondary structure determinations. *BMC Bioinformatics*, 9, 2008.
116. J. Meiler, M. Muller, A. Zeidler, and F. Schmaschke. Generation and evaluation of dimension reduced amino acid parameter representations by artificial neural networks. *Journal of Molecular Modeling*, 7:360–369, 2001.
117. M. Mezei. Chameleon sequences in the PDB. *Protein Engineering*, 11:411–414, 1998.
118. S. Yoon and W.J. Welsh. Rapid assessment of contact-dependent secondary structure propensity: Relevance to amyloidogenic sequences. *Proteins*, 60:110–117, 2005.
119. M. Boden, Z. Yuan, and T.L. Bailey. Prediction of protein continuum secondary structure with probabilistic models based on NMR solved structures. *BMC Bioinformatics*, 7, 2006.
120. S. Costantini, G. Colonna, and A.M. Facchiano. PreSSAPro: A software for the prediction of secondary structure by amino acid properties. *Computational Biology and Chemistry*, 31:389–392, 2007.
121. J.-T. Guo, J.W. Jaromczyk, and Y. Xu. Analysis of chameleon sequences and their implications in biological processes. *Proteins*, 67:548–558, 2007.
122. I. Jacoboni, P.L. Martelli, P. Fariselli, M. Compiani, and R. Casadio. Predictions of protein segments with the same aminoacid sequence and different secondary structure: A benchmark for predictive methods. *Proteins*, 41:535–544, 2000.

123. F. FarzadFard, N. Gharaei, H. Pezeshk, and S.-A. Marashi. Beta-sheet capping: Signals that initiate and terminate beta-sheet formation. *Journal of Structural Biology*, 161:101–110, 2008.

124. C.L. Wilson, P.E. Boardman, A.J. Doig, and S.J. Hubbard. Improved prediction for N-termini of alpha-helices using empirical information. *Proteins*, 57:322–330, 2004.

125. U. Midic, K. Dunker, and Z. Obradovic. Improving protein secondary-structure prediction by predicting ends of secondary-structure segments. *Proceedings of the 2005 IEEE Symposium on Computational Intelligence in Bioinformatics and Computational Biology, CIBCB 2005*. Embassy Suites Hotel La Jolla, La Jolla, CA, November 14–15, 2005, pp. 490–497. IEEE, 2005.

126. D. Kihara. The effect of long-range interactions on the secondary structure formation of proteins. *Protein Science*, 14:1955–1963, 2005.

127. A. Fiser, Z. Dosztanyi, and I. Simon. The role of long-range interactions in defining the secondary structure of proteins is overestimated. *Computer Applications in the Biosciences*, 13:297–301, 1997.

128. G.E. Crooks and S.E. Brenner. Protein secondary structure: Entropy, correlations and prediction. *Bioinformatics*, 20:1603–1611, 2004.

129. A. Ceroni and P. Frasconi. On the role of long-range dependencies in learning protein secondary structure. *IEEE Proceedings on Neural Network*, 3:1899–1904, 2004.

130. A. Ceroni, P. Frasconi, and G. Pollastri. Learning protein secondary structure from sequential and relational data. *Neural Networks*, 18:1029–1039, 2005.

131. C.A. Andersen and B. Rost. Secondary structure assignment. *Methods of Biochemical Analysis*, 44:341–363, 2003.

132. D. Frishman and P. Argos. Knowledge-based protein secondary structure assignment. *Proteins*, 23:566–579, 1995.

133. F.M. Richards and C.E. Kundrot. Identification of structural motifs from protein coordinate data: Secondary structure and first level supersecondary structure. *Proteins*, 3:71–84, 1988.

134. G. Labesse, N. Colloc'h, J. Pothier, and J.P. Mornon. P-SEA: A new efficient assignment of secondary structure from C alpha trace of proteins. *Computer Applications in the Biosciences*, 13:291–295, 1997.

135. J. Martin, G. Letellier, A. Marin, J.F. Taly, A G. de Brevern, and G.F. Gibrat. Protein secondary structure assignment revisited: a detailed analysis of different assignment mthods. *BMC Structural Biology*, 5, 2005.

136. H. Sklenar, C. Etchebest, and R. Lavery. Describing protein structure: a general algorithm yielding complete helicoidal parameters and a unique overall axis. *Proteins*, 6:46–60, 1989.

137. S.M. King and W.C. Johnson. Assigning secondary structure from protein coordinate data. *Proteins*, 3:313–320, 1999.

138. M.N. Fodje and S. Al-Karadaghi. Occurrence, conformational features and amino acid propensities for the pi-helix. *Protein Engineering*, 15:353–358, 2002.

139. M.V. Cubellis, F. Cailliez, and S.C. Lovell. Secondary structure assignment that accurately reflects physical and evolutionary characteristics. *BMC Bioinformatics*, 6:S8, 2005.

140. F. Dupuis, J.F. Sadoc, and J.P. Mornon. Protein secondary structure assignment through Voronoi tessellation. *Proteins*, 55:519–528, 2004.

141. B. Offmann, M. Tyagi, and A.G. de Brevern. Local protein structures. *Current Bioinformatics*, 2:165–202, 2007.

142. N. Colloc'h, C. Etchebest, E. Thoreau, B. Henrissat, and J.-P. Mornon. Comparison of three algorithms for the assignment of secondary structure in proteins: The advantages of a consensus assignment. *Protein Engineering*, 6:377–382, 1993.

143. G.E. Schulz, C.D. Barry, J. Friedman, P.Y. Chou, G.D. Fasman, A.V. Finkelstein, V.I. Lim, O.B. Pititsyn, E.A. Kabat, T.T. Wu, M. Levitt, B. Robson, and K. Nagano. Comparison of predicted and experimentally determined secondary structure of adenyl kinase. *Nature*, 250:140–142, 1974.

144. B. Robson and J. Garnier. *Introduction to Proteins and Protein Engineering*. Amsterdam: Elsevier Press, 1986.

145. D.J. Barlow and J.M. Thornton. Helix geometry in proteins. *Journal of Molecular Biology*, 201:601–619, 1988.

146. W. Zhang, K. Dunker, and Y. Zhou. Assessing secondary-structure assignment of protein structures by using pairwise sequence-alignment benchmarks. *Proteins*, 71:61–67, 2008.

147. P.F.J. Fuchs and A.J.P. Alix. High accuracy prediction of beta-turns and their types using propensities and multiple alignments. *Proteins*, 59:828–839, 2005.

148. Y. Wang, Z. Xue, and J. Xu. Better prediction of the location of alpha-turns in proteins with support vector machine. *Proteins*, 65:49–54, 2006.

149. A. Kirschner and D. Frishman. Prediction of beta-turns and beta-turn types by a novel bidirectional Elman-type recurrent neural network with multiple output layers (MOLE-BRNN). *Gene*, 422:22–29, 2008.

150. H. Kaur and G.P.S. Raghava. Prediction of alpha-turns in proteins using PSI-BLAST profiles and secondary structure information. *Proteins*, 55:83–90, 2004.

151. G.N. Ramachandran and V. Sasisekharan. Conformation of polypeptides and proteins. *Advances in Protein Chemistry*, 23:283–437, 1968.

152. J.F. Gibrat, B. Robson, and J. Garnier. Influence of the local amino acid sequence upon the zones of the torsional angles phi and psi adopted by residues in proteins. *Biochemistry*, 30:1578–1586, 1991.

153. H.S. Kang, N.A. Kurochkina, and B. Lee. Estimation and use of protein backbone angle probabilities. *Journal of Molecular Biology*, 229:448–460, 1993.

154. R. Karchin, M. Cline, Y. Mandel-Gutfreund, and K. Karplus. Hidden Markov models that use predicted local structure for fold recognition: alphabets of backbone geometry. *Proteins*, 51:504–514, 2003.

155. R. Kuang, C. S. Lesliei, and A.-S. Yang. Protein backbone angle prediction with machine learning approaches. *Bioinformatics*, 20:1612–1621, 2004.

156. S. Katzman, C. Barrett, G. Thiltgen, R. Karchin, and K. Karplus. PREDICT-2ND: A tool for generalized protein local structure prediction. *Bioinformatics*, 24:2453–2459, 2008.

157. J.N. Song, K. Burrage, Z. Yuan, and T. Huber. Prediction of cis/trans isomerization in proteins using PSI-BLAST profiles and secondary structure information. *BMC Bioinformatics*, 7, 2006.

158. R.A. Abagyan, M.M. Totrov, and D.A. Kuznetsov. ICM: A new method for structure modeling and design: Applications to docking and structure prediction from the distorted native conformation. *Journal of Computational Chemistry*, 15:488–506, 1994.

159. L.M. Rice and A.T. Brunger. Torsion angle dynamics: reduced variable conformational sampling enhances crystallographic structure refinement. *Proteins*, 19:277–290, 1994.

160. J.S. Evans, A.M. Mathiowetz, S.I. Chan, and W.A. Goddard III. De novo prediction of polypeptide conformations using dihedral probability grid Monte Carlo methodology. *Protein Science*, 4:1203–1216, 1995.

161. M.J. Rooman, J.P. Kocher, and S.J. Wodak. Prediction of protein backbone conformation based on seven structure assignments. Influence of local interactions. *Journal of Molecular Biology*, 221:961–979, 1991.

162. B.H. Park and M. Levitt. The complexity and accuracy of discrete state models of protein structure. *Journal of Molecular Biology*, 249:493–507, 1995.

163. X.F. de la Cruz, M.W Mahoney, and B. Lee. Discrete representations of the protein C_α chain. *Folding & Design*, 2:223–234, 1997.

164. A.G. de Brevern, C. Etchebest, and S. Hazout. Bayesian probabilistic approach for predicting backbone structures in terms of protein blocks. *Proteins*, 41:271–287, 2000.

165. H. Gong, P.J. Fleming, and G.D. Rose. Building native protein conformation from highly approximate backbone torsion angles. *Proceedings of the National Academy of Science U S A*, 102:16227–16232, 2005.

166. Y. Yang and H. Liu. Genetic algorithms for protein conformation sampling and optimization in a discrete backbone dihedral angle space. *Journal Computational Chemistry*, 27:1593–1602, 2006.

167. C. Bystroff and D. Baker. Prediction of local structure in proteins using a library of sequence-structure motifs. *Journal of Molecular Biology*, 281:565–577, 1998.

168. C. Bystroff, V. Thorsson, and D. Baker. HMMSTR: A hidden Markov model for local sequence-structure correlations in proteins. *Journal of Molecular Biology,* 301:173–190, 2000.

169. A.G. de Brevern, C. Benros, R. Gautier, H. Valadie, S. Hazout, and C. Etchebest. Local backbone structure prediction of proteins. *In Silico Biology*, 4:31, 2004.

170. C. Etchebest, C. Benros, S. Hazout, and A.G. de Brevern. A structural alphabet for local protein structures: Improved prediction methods. *Proteins*, 59:810–827, 2005.

171. O. Sander, I. Sommer, and T. Lengauer. Local protein structure prediction using discriminative models. *BMC Bioinformatics*, 7, 2006.

172. C. Benros, A.G. de Brevern, C. Etchebest, and S. Hazout. Assessing a novel approach for predicting local 3D protein structures from sequence. *Proteins*, 62:865–880, 2006.

173. A.G. De Brevern, C. Etchebest, C. Benros, and S. Hazout. "Pinning strategy": A novel approach for predicting the backbone structure in terms of protein blocks from sequence. *Journal of Biosciences*, 32:51–70, 2007.

174. O. Zimmermann and U.H.E. Hansmann. Support vector machines for prediction of dihedral angle regions. *Bioinformatics*, 22:3009–3015, 2006.

175. O. Zimmermann and U.H.E. Hansmann. LOCUSTRA: Accurate prediction of local protein structure using a two-layer support vector machine approach. *Journal of Chemical Information and Modeling*, 48:1903–1908, 2008.

176. Q. Dong, X. Wang, L. Lin, and Y. Wang. Analysis and prediction of protein local structure based on structure alphabets. *Proteins*, 72:163–172, 2008.

177. W. Boomsma, K.V. Mardia, C.C. Taylor, J. Ferkinghoff-Borg, A. Krogh, and T. Hamelryck. A generative, probabilistic model of local protein structure. *Proceedings of the National Academy of Science U S A*, 105(26):8932–8937.

178. S. Wu and Y. Zhang. MUSTER: Improving protein sequence profile-profile alignments by using multiple sources of structure information. *Proteins*, 72:547–556, 2008.

179. W. Zhang, S. Liu, and Y. Zhou. SP^5: Improving protein fold recognition by using predicted torsion angles and profile-based gap penalty. *PLoS ONE*, 6:e2325, 2008.

180. Y.M. Huang and C. Bystroff. Improved pairwise alignments of proteins in the twilight zone using local structure predictions. *Bioinformatics*, 22:413–422, 2006.

181. T.G. Pedersen, B.W. Sigurskjold, K.V. Andersen, M. Kjaer, F.M. Poulsen, C.M. Dobson, and C. Redfield. A nuclear magnetic resonance study of the hydrogen-exchange behavior of lysozyme in crystals and solution. *Journal of Molecular Biology*, 218:413–426, 1991.

182. S. Chakravarty and R. Varadarajan. Residue depth: a novel parameter for the analysis of protein structure and stability. *Structure with Folding and Design*, 15:723–732, 1999.

183. G. Pollastri, P. Baldi, P. Fariselli, and R. Casadio. Prediction of coordination number and relative solvent accessibility in proteins. *Proteins*, 47:142–153, 2002.

184. T. Hamelryck. An amino acid has two sides: A new 2D measure provides a different view of solvent exposure. *Proteins*, 59:38–48, 2005.

185. M. Stout, J. Bacardit, J.D. Hirst, and N. Krasnogor. Prediction of recursive convex hull class assignments for protein residues. *Bioinformatics*, 24:916–923, 2008.

186. Z. Yuan and Z.X. Wang. Quantifying the relationship of protein burying depth and sequence. *Proteins*, 70:509–516, 2008.

187. H. Zhang, T. Zhang, K. Chen, S. Shen, J. Ruan, and L. Kurgan. Sequence based residue depth prediction using evolutionary information and predicted secondary structure. *BMC Bioinformatics*, 9, 2008.

188. P. Fariselli and R. Casadio. RCNPRED: Prediction of the residue co-ordination numbers in proteins. *Bioinformatics*, 17:202–203, 2001.

189. C.T. Zhang and R. Zhang. Q(9), a content-balancing accuracy index to evaluate algorithms of protein secondary structure prediction. *International Journal of Biochemistry & Cell Biology*, 35:1256–1262, 2003.

190. A.R. Kinjo and K. Nishikawa. CRNPRED: Highly accurate prediction of one-dimensional protein structures by large-scale critical random networks. *BMC Bioinformatics*, 7:401, 2006.

191. T. Ishida, S. Nakamura, and K. Shimizu. Potential for assessing quality of protein structure based on contact number prediction. *Proteins*, 64:940–947, 2006.

192. J. Song, H. Tan, K. Takemoto, and T. Akutsu. HSEpred: Predict half-sphere exposure from protein sequences. *Bioinformatics*, 24:1489–1497, 2008.

193. S.R. Hobrook, S.M. Mushal, and S.H. Kim. Predicting surface exposure of amino acids from protein sequence. *Protein Engineering*, 3:659–665, 1990.

194. B. Rost and C. Sander. Conservation and prediction of solvent accessibility in protein families. *Proteins*, 20:216–226, 1994.

195. M.H. Mucchielli-Giorgi, P. Tuffery, and S. Hazout. Prediction of solvent accessibility of amino acid residues: critical aspects. *Theoretical Chemistry Accounts*, 101:186–193, 1999.

196. S. Pascarella, R. De Persio, F. Bossa, and P. Argos. Easy method to predict solvent accessibility from multiple protein sequence alignments. *Proteins*, 32:190–199, 1999.

197. X. Li and X. Pan. New method for accurate prediction of solvent accessibility from protein sequence. *Proteins*, 42:1–5, 2001.

198. S. Ahmad and M.M. Gromiha. NETASA: Neural network based prediction of solvent accessibility. *Bioinformatics*, 18:819–824, 2002.

199. A. Garg, H. Kaur, and G.P.S. Raghava. Real value prediction of solvent accessibility in proteins using multiple sequence alignment and secondary structure. *Proteins*, 61:318–324, 2005.

200. M.F. Raih, S. Ahmad, R. Zheng, and R. Mohamed. Solvent accessibility in native and isolated domain environments: general features and implications to interface predictability. *Biophysical Chemistry*, 114:63–69, 2005.

201. Z. Yuan, K. Burrage, and J.S. Mattick. Prediction of protein solvent accessibility using support vector machines. *Proteins*, 48:566–570, 2002.

202. Z. Yuan and B. Huang. Prediction of protein accessible surface areas by support vector regression. *Proteins*, 57:558–564, 2004.

203. G. Gianese, F. Bossa, and S. Pascarella. Improvement in prediction of solvent accessibility by probability profiles. *Protein Engineering*, 16:987–992, 2003.

204. H. Kim and H. Park. Prediction of protein relative solvent accessibility with support vector machines and long-range interaction 3D local descriptor. *Proteins*, 54:557–562, 2004.

205. S. Ahmad, M.M. Gromiha, and A. Sarai. Real value prediction of solvent accessibility from amino acid sequence. *Proteins*, 50:629–635, 2003.

206. R. Adamczak, A. Porollo, and J. Meller. Accurate prediction of solvent accessibility using neural networks-based regression. *Proteins*, 56:753–767, 2004.

207. J. Wang, H. Lee, and S. Ahmad. Prediction and evolutionary information analysis of protein solvent accessibility using multiple linear regression. *Proteins*, 61:481–491, 2005.

208. Z. Xu, C. Zhang, S. Liu, and Y. Zhou. QBES: Predicting real values of solvent accessibility from sequences by efficient, constrained energy optimization. *Proteins*, 63:961–966, 2006.

209. H. Naderi-Manesh, M. Sadeghi, S. Arab, and A.A.M. Movahedi. Prediction of protein surface accessibility with information theory. *Proteins*, 42:452–459, 2001.

210. S. Liu, C. Zhang, S. Liang, and Y. Zhou. Fold recognition by concurrent use of solvent accessibility and residue depth. *Proteins*, 68:636–645, 2007.

211. H.L. Chen and H.X. Zhou. Prediction of solvent accessibility and sites of deleterious mutations from protein sequence. *Nucleic Acids Research*, 33:3193–3199, 2005.

212. J. Qiu and R. Elber. SSALN: An alignment algorithm using structure-dependent substitution matrices and gap penalties learned from structurally aligned protein pairs. *Proteins*, 62:881–891, 2006.

213. R. Adamczak, A. Porollo, and J. Meller. Combining prediction of secondary structure and solvent accessibility in proteins. *Proteins*, 59:467–475, 2005.

214. V.G. Krishnan and D.R. Westhead. A comparative study of machine-learning methods to predict the effects of single nucleotide polymorphisms on protein function. *Bioinformatics*, 19:2199–2209, 2003.

215. R.J. Dobson, P.B. Munroe, M.J. Caulfield, and M.A.S. Saqi. Predicting deleterious nsSNPs: an analysis of sequence and structural attributes. *BMC Bioinformatics*, 7, 2006.

216. Y. Bromberg, G. Yachdav, and B. Rost. SNAP predicts effect of mutations on protein function. *Bioinformatics*, 24:2397–2398, 2008.

217. Y. Ofran and B. Rost. Predicted protein-protein interaction sites from local sequence information. *FEBS Letters*, 544:236–239, 2003.

218. A. Porollo and J. Meller. Prediction-based fingerprints of protein-protein interactions. *Proteins*, 66:630–645, 2007.

219. G. Wang and R.L. Jr. Dunbrack. PISCES: a protein sequence culling server. *Bioinformatics*, 19:1589–1591, 2003.

220. D.E. Rumelhart, G.E. Hinton, and R.J. Williams. Learning representations by backpropagating errors. *Nature*, 323:533–536, 1986.

221. S.F. Altschul, T.L. Madden, A.A. Schäffer, J. Zhang, Z. Zhang, W. Miller, and D.J. Lipman. Gapped BLAST and PSI-BLAST: A new generation of protein database search programs. *Nucleic Acids Research*, 25:3389–3402, 1997.

222. E. Faraggi, Y. Yang, S. Zhang, and Y. Zhou. Predicting continuous local structure and the effect of its substitution for secondary structure in fragment-free protein structure prediction. *Structure*, 17:1515–1527, 2009.

CHAPTER 5

LOCAL STRUCTURE ALPHABETS

AGNEL PRAVEEN JOSEPH and AURÉLIE BORNOT
Institut National de la Santé et de la Recherche Médicale, UMR-S 665
Dynamique des Structures et Interactions des Macromolécules Biologiques (DSIMB)
Université Paris Diderot
Paris, France

ALEXANDRE G. DE BREVERN
Institut National de la Santé et de la Recherche Médicale
Université Paris Diderot
Institut National de la Transfusion Sanguine
Paris, France

5.1. INTRODUCTION

Proteins play a crucial key role in most cellular processes. They act as enzymes, transcription factors, mediators in cell signaling, transporters, and storage molecules, or have structural, regulatory, or protective roles. Many diseases are associated with abnormality in protein functions. Today, proteins are also the most important drug targets. The protein three-dimensional (3D) structure is directly dependent on its biological function. So a good understanding of 3D structure often gives sufficient hints in understanding protein functions, and this forms the basis of structure-based drug design [1]. Only about 1% of the total number of sequenced proteins has experimentally determined structures [2] and a considerable number of these proteins are without known functions [3]. Considering the fact that the amino acid sequence of a protein determines its 3D structure, one often tries to extract the structural information embedded in the sequence.

Even before the first protein structure was solved, Pauling and Corey had proposed two major repetitive structures that could occur within protein structures: the α-helix and the β-sheet [4,5]. Since then, these repetitive structures

Introduction to Protein Structure Prediction: Methods and Algorithms,
Edited by Huzefa Rangwala and George Karypis

are not only being used to analyze protein structures but also to predict them. Nonetheless, this description has some limitations that have led to the definition of a more complex concept of structural alphabets. Here, we will present the secondary structures and the different structural alphabets designed today.

5.2. REPEATING STRUCTURAL ELEMENTS IN PROTEINS

A number of repeating structural elements have been observed in the known protein structures. Representing proteins in terms of secondary structures like helices and strands is known to be useful for visualization, prediction, classification, and analysis of protein structures [6–8]. Several methods for assigning secondary structures and other repeating elements (discussed in the following paragraphs) have been developed. Methods like Dictionary of Protein Secondary Structure (DSSP) [9] or STRIDE [10] use the information on the hydrogen-bonding patterns to characterize these secondary structures. PROSS [11] and SEGNO [12] use torsion angle information for assignments while others [13] use the inter-Cα distances either alone or along with the information on the hydrogen-bonding pattern and dihedral angles for assigning secondary structures.

5.2.1. Classical Secondary Structures

The classical way of describing protein structures is in terms of α-helices and β-sheets, the two major repetitive local structures in proteins [14]. These repeating units are characterized by the pattern of hydrogen bonds formed by the protein backbone. α-Helices involve hydrogen bonds between ith and $i + 4$th residues while β-sheets are composed of extended strands with hydrogen bonds formed between adjacent strands. β-Sheets help to bring together parts of protein that are far apart in the sequence, while helices involve consecutive residues in a sequence. The planar arrangement of β-strands gives rise to steric constraints that cause consecutive side chains to point in opposite sides of the plane.

Analysis of sequence–structure relationships has shown over- and under-representations of certain amino acids. Richardson and Richardson and Pal et al. have made a detailed analysis and shown that short and long helices have different amino acid compositions [15,16]. The sequence specificities of β-strands have also been studied [17] as of their ends [18]. Experimental and statistical works on analysis of specificity of pairs of interacting residues in neighboring strands have given limited results and failed to present pertinent laws for their associations. The recent studies mainly focus on the crucial question of protein aggregation [19]. Analysis of helix signals in proteins highlighted the hydrophobic capping, a hydrophobic interaction that straddles the helix terminus and is always found to be associated with hydrogen-bonded capping [18,20,21].

In the 1970s, predictions of regular secondary structures have been carried out using statistical approaches [22]. The introduction of Artificial Neural Networks coupled with evolutionary information has led to an impressive increase in the prediction rate, for example, PHD methodology [23]. The secondary structure prediction rate has reached a maximum limit that is slightly better than 80%. The two most widely used programs are PSI-PRED [24] and SSPRO [25,26]. No new significant improvements have been seen during the last few years. It is considered that the secondary structure prediction is no longer a research area that can be improved

5.2.2. Other Helical and Extended Conformations

Several other repeating structural elements are also observed (see Fig. 5.1. for some examples). Apart from α-helices, other helical states like 3_{10} and π are also found, covering around 4% and 0.02% of residues, respectively. 3_{10}-Helices are characterized by inter-residue hydrogen bonds between *i*th and *i* + 3th residues. Majority of 3_{10}-helices involve only one turn [27]. They are usually found at the termini of α-helices, often linking two α-helices like what is observed in hairpins and corner motifs [28]. π-Helices involve inter-residue hydrogen bonds between *i*th and *i* + 5th residues. Dynamic transitions between

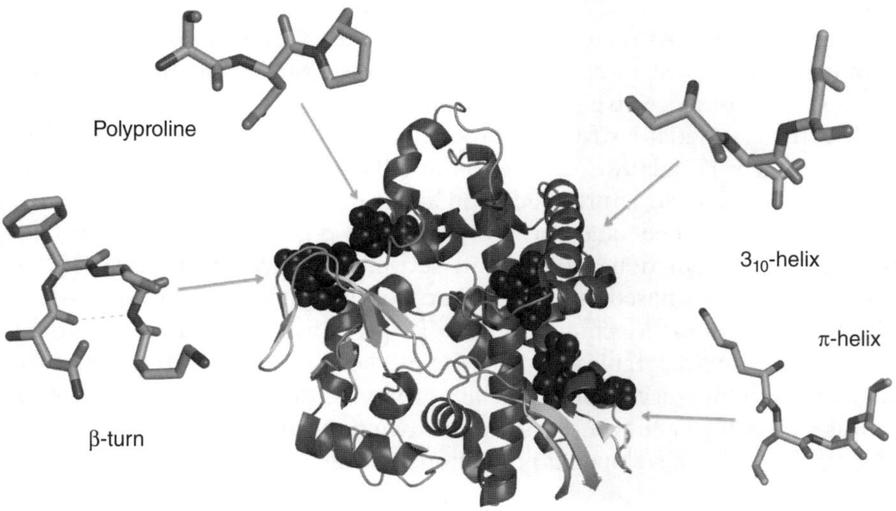

FIGURE 5.1 Some less common "secondary structures." A cytochrome P450 (PDB code 1IO7 [32]) has been assigned using DSSP and PROSS. One 3_{10}-helix has been assigned by both approaches (positions 148–150) while PROSS is the only one to have assigned one π-helix (positions 120–123). PROSS has also assigned one Polyproline II (positions 71–73). Different β-turns have also been located; the one represented encompassed the amino acid from positions 35 to 38. Visualization has been done using PyMol [33].

α- and 3_{10}- and α- and π-helices have been proposed to occur during the folding and unfolding process [29]. As shown in Figure 5.1, these different helices are short and thus difficult to assign precisely. For instance, the 3_{10}-helix shown is the only one assigned by DSSP and PROSS, each one assigned the other 3_{10}-helices of this cytochrome as coil or turn (see References [13,18,30]for more details).

Isolated extended structures that are not part of a β-sheet are also found in proteins and they are generally exposed to solvent [31]. SSPRO8 has the potentiality to predict them. However due to low occurrences, the prediction of π-helices or isolated extended structures as become difficult.

5.2.3. Turns

The first description and analysis of turns was made by Venkatachalam [34]. Turns correspond to a short return of the protein backbone. It is the third most studied secondary structure. A turn with n residues has a distance of less than 7 Å between the Cα carbons of residues i and $i + n$. Also, the central residues are not helical and at least one residue must not be extended. There are four different types of tight turns: γ-turns (three residues), β-turns (four residues), α-turns (five residues), and π-turns (six residues). Each of these is further classified into different types based on the φ/ϕ dihedral angles. γ- and β-Turns are the most widely studied types of turns. About 25% to 30% of residues correspond to β-turns. To date, seven types of β-turns have been characterized [35]. As seen in Figure 5.1, the β-turn can be easily confounded within a helical structure, for example, α-helix. Moreover, they are often multiple, that is, successive β-turns overlap.

The first secondary structure prediction method was also dedicated to predict β-turns [22]. However, due to the difficulty associated with its prediction, the secondary structure prediction had been rapidly limited to the prediction of α-helix, β-sheet, and coil. Nowadays, the prediction of β-turns is done mainly after a prediction of three-state secondary structures, as in PSI-PRED [24] or the method based on statistical approaches [36] or advanced classifiers like Support Vector Machines (SVMs) [37]. Prediction accuracy of turns is nowadays quite acceptable; however the prediction of some rarely seen turns remains low [36]. Very recently Klebe's group has done a new learning of the "turns" to define a novel classification of open and hydrogen-bond turns [38,39]. They also developed a prediction method.

5.2.4. Polyproline II (PII)

PII helices are left-handed helical structures that help in the formation of coiled coils in fibrous proteins [40]. The left handedness is characterized by specific dihedral angles and trans-isomers of peptide bonds. The φ/ϕ dihedral angles (approximately -75°C and 145°C, respectively) fall in the region that is characteristic of β-strand. These helices are often solvent exposed and also associated with high-temperature factors [41]. Nonlocal interactions suggest a prominent role for PII helices in protein-protein and protein-ligand interactions [42,43]. It must be noticed that PII can exist without any Proline. For instance, the only Polyproline observed within this cytochrome P450 contains only one Proline (see Fig. 5.1). So it has been noted that designation PII is a bit misleading, since the conformation is not just associated with Pro but can be adopted by all amino acids. In a recent and fine study, about one-third of the residues in the center of PII tripeptides are Pro; the rest include all types of amino acids. The authors proposed that the common name could be changed to a more general "polypeptide-II" conformation [44]. Only PROSS [11], XTLSSTR [45], and SEGNO [12] are capable of PII assignment; it is not the case for instance for DSSP [9], STRIDE [10], P-SEA [46], VOTAP [47], or PROSIGN [48]. To the best of our knowledge, only one group has recently developed prediction methods of PII [49].

5.2.5. Loops

Even after performing helical, strand, and turn assignments, about 50% of the residues are left out and are associated with the coil state. Thus different classification approaches have been developed to analyze the regions connecting repetitive structures. β-Hairpins are the most studied type of specific loops; thanks to their high frequency of occurrence. They connect two adjacent antiparallel β-strands. They are grouped into different classes based on their length and conformation. Other types of loops joining β-strands like the β-β corners and orthogonal β-β motifs have also been studied. Characteristic sequence patterns are often observed in the strand-loop-strand motifs and some dedicated prediction strategies based on neural networks have been developed [50–52]. Prediction rates of β-hairpins go up to 80%, leading to an overall prediction rate of 65% for the four states [52] α-α-Turns and corners have also been studied extensively [53]. Complete loop regions have also been analyzed. Most of these studies are focused on loops of length less than nine residues leading to some classifications [54]. ArchDB is an online method available to find potential compatible loops [55].

5.3. BEYOND SECONDARY STRUCTURES

Secondary structure assignments are widely used to analyze protein structures. However, it often gives a wrong representation of real protein structures. Figure 5.2 shows the idea behind the secondary structure assignment. From the atomic coordinates in the Protein Data Bank (PDB) file, (cf. Fig. 5.2a) covalent bonds can be assigned to link the atoms (cf. Fig. 5.2b) or only the protein backbone can be considered (cf. Fig. 5.2c). The secondary structure assignment as shown in Figure 5.2d is the classical way to see it, but as shown in Figure 5.2e, about half of the residues are not assigned any secondary structure.

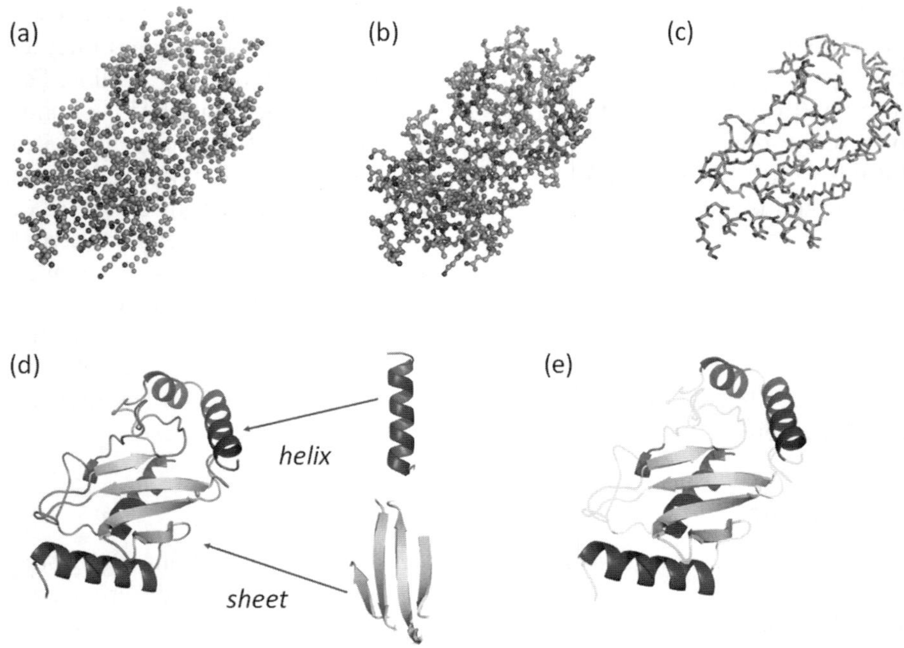

FIGURE 5.2 The different descriptions of a protein structure. (a) The atoms are presented as in the PDB file. (b) Links are done between the atoms. (c) Only the backbone is shown. (d) The secondary structures are assigned. (e) Only the regular structures are really assigned.

Moreover, it could give a wrong impression that helices and/or strands are ideal. Although helices and strands are geometrically defined as stable structural elements, local irregularities are often seen. The majority of α-helices is not linear but curved (58%) and even kinked (17%) [13,56]. Contiguous stretches of intra-helical residues exhibiting non-helical geometry have also been well defined; they are named π-bulges [57]. They are not frequently observed but are implicated in the protein function.

Like α-helices, β-strands are also found to have local stretches with non-extended conformation, called β-bulges [58,59]. An elaborate classification of β-bulges has been made by Thornton's group [60]. They are observed quite frequently.

Secondary structure assignment is often considered as a resolved problem and assignment made by DSSP is considered the true and the only possible secondary structure assignment. However, it is not the case and the huge number of different assignment methods proved it [10,11,13,18,45–48,61–65]. The most important factor is the choice of descriptors and the parameters used, for example, distances and angles. Even with similar descriptors, the assignments could be different as shown by Reference [62]. Protein flexibility also plays an important role. Comparison of different secondary structure assignment methods has shown some surprising results: difference in assignments could be seen in about one in five residues [66,67]. These different problems has led to the idea that some other descriptions of local protein structures can be useful.

5.4. LOCAL STRUCTURE LIBRARIES

The absence of secondary structure assignment for an important proportion of the residues has led some scientific teams to develop local protein structure libraries (i) that are able to approximate all (or almost all) of the local protein structures and (ii) that do not take into account the description of classical secondary structures. These libraries brought about the categorization of 3D structures without any a priori knowledge of small prototypes that are specific for local folds found in proteins. The complete set of local structure prototypes (LSPs) defines a structural alphabet [68]. A structural alphabet, being able to approximate the local structures in proteins, helps to represent the structural information in one dimension as a sequence. Such a representation also presents methods that are effective and computationally cheap for the comparison and analysis of protein structures (see Table 5.1 [69]).

5.4.1. Building Blocks

Unger et al. were the first to develop a structural alphabet using a clustering approach based on Cα root-mean-square deviation (RMSD) [70]. They had chosen hexapeptides as the smallest units that can represent unique local structural information. Using a clustering method called "of annexation" and an RMSD threshold of 1 Å for clustering, they were able to select about 100 representatives, which they called "building blocks". They were able to cover 76% of hexapeptide fragments in the dataset, with an RMSD less than 1.0 Å. [71]. They then carried out a first detailed study of those building blocks associated with extended strands.

5.4.2. Hierarchical Clustering

Rooman and Wodak extended their work on protein secondary structure prediction to the description of local protein structures [72]. For this purpose, they performed a hierarchical clustering based on Cα RMSD. They were mainly interested in prototypes of different lengths and they tested fragments of lengths ranging from four to seven residues long [73]. They selected four different prototypes for each length. This limited number was chosen based on their final purpose: perform a prediction of these local protein structures from the sequence. Using a simple statistical approach, they obtained a correct prediction rate ranging from 41% to 47% [74].

5.4.3. Cα Distances and Dihedral Angles

Pretrelski and et al. have developed a structural alphabet to support their experimental studies on trypsin-like proteins [75]. For this purpose, they used

TABLE 5.1 The Different Sets of Structural Alphabet

Research Team	Number of Proteins in Dataset	Fragment Length	Distance Measure	Learning Method	Prototype Number	Prediction
Unger et al. [70,71]	4/82	6	Cα RMSD	k-means	103	N
Rooman et al. [73,74]	75	4,5,6, 7	Cα RMSD	Hierarchical clustering	4	Y
Prestrelski et al. [75,76]	14	8	Linear Cα disance and α torsion angle	Function of Cα disance and torsion angle	113	N
Schuchhardt et al. [79]	136	9	Dihedral angles	Kohonen map	100	N
Fetrow et al. [80]	116	7	C α distance, dihedral and bond angles	Auto-ANN	6	N
Bystroff and Baker [81,83]	471	3–19	Sequence profiles, RMSD, MDA	k-means	13(later updated to 16)	Y
Camproux et al. [85,86]	100	4	Cα distance	HMM	12	N
Micheletti et al. [89]	75	4,5,6,7	Cα RMSD	Iterative clustering (Monte-Carlo like)	28,202,932,2561	N
de Brevern et al. [67,91,105,113,115,117, 118,128]	342	5	Dihedral angles	Unsupervised classifier (SOM with transition probabilities)	16	Y

Kolodony et al. [134]	145/200	4,5,6,7	Cα RMSD	Simulated annealing based on k-means	4-14,10–225, 40–300,50–250	N
Hunter and Subramaniam [90,92]	790	7	Hypercosine Cα	Hypercosine clustering	28–16336	Y
Camproux et al. [85,86]	250 * 2	4	C α distance	HMM	27	N
Etchebest et al. [114,129]	1407	5	Dihedral angles	Unsupervised classifier	16 (new evaluation)	Y
Benros et al. [109–111]	675 and 1401	11	Cα RMSD, PB-based	Hybrid protein model	120	Y
Sander et al. [93]	1999	7	Cα distance	Leader algorithm and k-means	28	Y
Tung et al. [95]	1348	5	κ and α angle	Nearest neighbor clustering	23	N
Ku and Hu [97]	18	5	Dihedral angle	SOM and k-means	18	N
Bornot et al. [112,133]	675 and 1401	11	Cα RMSD, PB-based	Hybrid protein model	120	Y
Yang [98]	268	5	Cα distances and angles	Shape object clustering	27	N

MDA, maximum deviation angle.

a combination of linear Cα distances and the Cα dihedrals to generate a set of local structural prototypes. The scoring function designed is a complex combination of Cα distances and the tangent of the dihedral angles. They could find 113 prototypes that are of five residues in length [76]. Their approach was based only on structural approximation.

5.4.4. Self-Organizing Maps (SOMs)

Schuchhardt and et al. designed a complex SOM [77,78] to generate LSPs. Their learning approach was based on protein fragments that are nine residues in length encoded as series of φ/ϕ angles, that is, 16 dihedral angles. They could characterize 100 structural prototypes [79]. Interestingly, they could also identify amino acid preferences associated with some structural prototypes that can be considered as part of protein loops.

5.4.5. Auto-Associative Neural Network

Fetrow and et al. generated a set of local protein structures using a learning method more complex than the earlier ones [80]. They used an auto-associative neural network (autoANN). This specific neural network has input and output layers with similar dimensions. The hidden layer thus does a compaction of the information. They used this hidden layer to characterize seven-residue long fragments encoded as distances, bond, and dihedral angles. They generated six structural prototypes and also performed an analysis on the amino acid composition of each prototype, underlining some specificities related to repetitive structures.

5.4.6. I-sites

Based on a library of short sequence patterns having high correlation with 3D structure, Bystroff and Baker developed an efficient method for predicting local protein structures [81]. They identified frequently occurring sequence motifs by automatic clustering and characterized their corresponding local structures. They further developed an iterative method to optimize the correspondence between sequence and structure. Sequence-based clusters were generated with the homology-derived structures of proteins (HSSP) protein families [82] and the most frequent local structure in each cluster was chosen as the structural *paradigm*. An iterative process similar to the k-means approach was then employed, by reestimating the *paradigms* obtained from clusters formed from the dataset. The clustering on the structure space was done using criteria of Cα distance and dihedral angle measure. A library of 82 sequence clusters that are 3 to 19 residues long were obtained finally. The local structural paradigms corresponding to these clusters were then structurally aligned to get 13 different sequence–structure motifs, which they called "I-sites." The library of I-sites presented new sequence–structure relationships.

In combination with the secondary structure prediction method based on profile-based neural networks, PHD, the sequence–structure relationships in the I-sites were used to develop a local structure prediction method leading to a prediction rate of ~50%. The prediction method performed well in the Critical Assessment of Protein Structure Prediction 2 (CASP2) trials and the prediction for α-spectrin SRC homology 3 (SH3) domain had good correlation with nuclear magnetic resonance (NMR) results [83].

They further generated a set of hidden Markov model (HMM)-based profiles called HMMSTR for the sequences in the I-sites library. This HMM was built using overlapping I-sites using an updated dataset [84].

5.4.7. Hidden Markov Model

The first work done by Pr. Serge Hazout (also see the PBs section) was on short protein fragments of four residues. Described as series of Cα distance, these fragments were learnt by a classical hidden Markov model [85]. Thirteen structural prototypes were obtained from the model and some of them showed specific amino acid preferences. A work dedicated for the prediction of short loops was carried out [86]. A specific work focuses on the reconstruction of protein backbone from Cα traces [87]. Another one was based on the specific learning of fragments from outer membrane proteins [88]; it has led to propose 20 structural prototypes that show some amino acid specificities. These structural models were used to discriminate CASP models.

5.4.8. Oligons

Michetelli and et al. used an iterative procedure to generate LSPs based on RMSD [89]. At the first stage the fragments were clustered based on the RMSD distribution. The representatives chosen from each cluster, named "oligons," were clustered again and this process was repeated. The optimization process is similar to the classical Monte-Carlo approach. This method helps to generate prototypes with hierarchical weights associated with them, that is, the first set of oligons is more significant than those that follow. The main aim behind this approach was to generate an increasing number of local structural prototypes. They had tested this approach on fragments of lengths varying from 3 to 10 residues. Highly satisfying results were obtained on structure reconstruction trials using oligons. The importance of the fragment length is highlighted, showing that, for longer fragments, a large number of prototypes are required for a similar 3D approximation. No specific study of amino acid specificities associated with these local protein structures was done.

5.4.9. Centroids

Using a hyper-cosine clustering method, Hunter and Subramaniam [90] clustered seven residue fragments. RMSD was used as the distance measurement. They chose a threshold to define the optimum number of clusters, which they called the centroids. Despite a detailed analysis of parameters used to select the threshold, the fragment distribution among the 28 clusters finally chosen is highly uneven. To develop a prediction method based on the set of centroids generated, they used a Bayesian predictor that gives the probability of each centroid to occur at a position in the sequence. This prediction is highly related to the prediction used for PBs (see the PBs section) [91]. An overall prediction accuracy of 40% was obtained. However, this correct prediction rate gives a wrong impression, as it is in fact highly biased. Indeed, 11 of the 28 centroids are not predicted at all, which diminishes greatly the interest of the approach [92]. Moreover, some major divergences can be noted between the two papers describing the approach.

5.4.10. k-Means

Sander and et al. have developed a novel approach based on the use of Cα distance matrix comparison [93] using a "complex" k-means. They defined 27 prototypes of eight residues comparable to those developed by Hunter and Subramaniam [92]. They also incorporated protein family information by using profiles instead of simple sequences. They have tested numerous prediction methods: C.5 classifier, SVMs, and random forest. All these approaches have led to an unbiased prediction unlike the predictions made using the Hunter and Subramaniam approach [92].

5.4.11. Kappa-Alpha Map

Tung and Yang have defined a structural alphabet dedicated to mine the PDB [94]. The main principle used in this approach is a measure based on Cα distance and a nearest neighbor clustering (NNC) algorithm. A set of 23 local prototypes were selected and used to identify similar protein structural domains and corresponding Structural Classification of Proteins (SCOP) superfamilies [95,96]. The search methodology is based on the direct use of the Basic Local Alignment Search Tool (BLAST) algorithm, similar to the work done earlier with PBs (see the PBs section), that is, Protein Block Expert (PBE) [69]. Analysis of sequence–structure relationship was not done.

5.4.12. SOMs and k-means

Recently, Ku and Hu [97] used the idea developed by Schuchhardt et al. [79] and it was used for PB design [91], namely defining the protein in terms of φ/ϕ dihedrals. Like PBs, they used five-residue long fragments to define the prototypes. The first step is a classical learning using an SOM [77,78]. After many simulations with different number of neurons, they selected a large map and analyzed it using U-matrix visualization. From these data, they clustered the results using k-means approach. Then, a substitution matrix was computed and it is optimized to detect SCOP class similarity. A FASTA methodology is used to compute the similarity score. Analysis of sequence–structure relationship was not done.

5.4.13. Protein Folding Shape Code

Recently Yang described a novel approach based on the description of protein local structures as a vector of angle and distances. He only used Cα distances and obtained 27 prototypes of length 5 [98].

5.5. PBs

5.5.1. Design of PBs

Following an earlier work, Pr. Serge Hazout developed a novel structural alphabet, with two specific goals: (i) to obtain a good local structure approximation and (ii) to predict local structures from sequence. Fragments that are five residues in length were coded in terms of the φ/ϕ dihedral angles. An RMSDA score was used to quantify the structural difference among the fragments. This idea was already used by Schuchhardt and et al. [79]. Using an unsupervised cluster analyser related to self-organized Kohonen maps [77,78], a three-step training process was carried out: (i) the learning of structural difference of fragments has been performed using only the minimal RMSDA as criterion to associate a fragment to a cluster; (ii) the transition probability (probability of transition from one fragment to another in a sequence) was also added to select the cluster associated to the protein fragment; and (iii) this last constraint was removed. The optimal number of prototypes was obtained by considering both the structural approximation and the prediction rate. A set of 16 prototypes called PBs represented as average dihedral vectors, were obtained at the end of this process [91]. Figure 5.3a shows the 16 PBs. Figure 5.4 gives an example of PB assignment.

5.5.2. Analysis of PBs

The relationship between PBs and secondary structures was analyzed. PB *m* corresponds to the central part of helices while PB *d* corresponds to strands. Some PBs are associated with the N-and C-caps of helices and strands representing subtle variations in the termini. Some PBs also represent conserved features in the coils. Specific or highly preferential transitions are observed between consecutive PBs in a sequence. The three major transitions observed correspond to about 76% of the possible transitions. The distribution of PBs, transition probabilities, and structural definitions has been evaluated and cross-checked using different datasets of proteins. These features were found to be highly consistent among the different datasets [99]. Table 5.2 shows the correspondence of all the 16 PBs and the different secondary structure

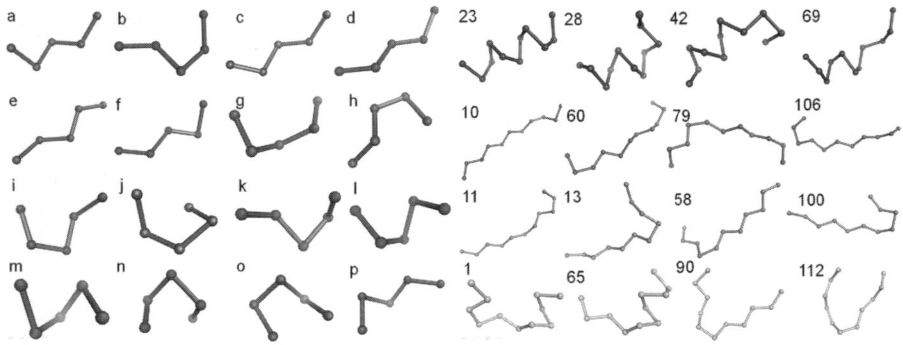

FIGURE 5.3 Protein Blocks and Local Structure Prototypes (LSPs). (left) Shows the 16 PBs (five residues in length)). (right) Shows some examples of the 120 LSPs (11 residues in length). LSPs 23, 28, 42, and 69 belong to the helical LSP, LSPs 10, 60, 79, and 106 to extended LSP, LSPs 11, 13, 58, and 100 to extended edges LSPs, and LSPs 1, 65, 90, and 112 to connection LSP.

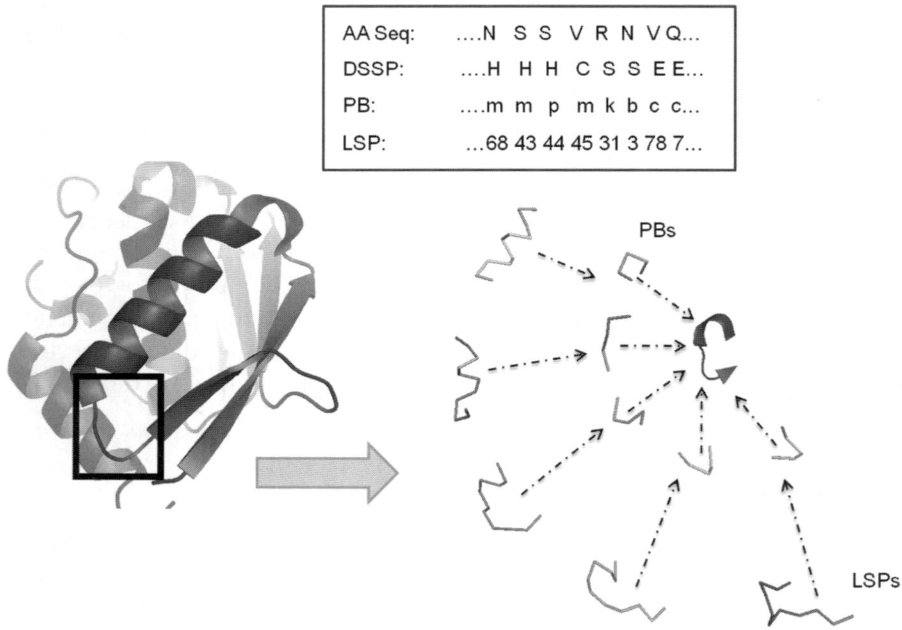

FIGURE 5.4 Example of assignment. The zinc endoprotease (PDB code 1c7k [104]) has been encoded not only in terms of secondary structures with DSSP (shown in 3D on the left), but also in terms of Protein Blocks (PBs) and Local Structure Prototypes (LSPs). The short protein fragment in the black box is detailed with the PB and the LSP sequence. The corresponding prototypes are also shown.

TABLE 5.2 S2 ←→ PBs. (a) Is Given the Relative Frequencies of Protein Blocks for Each Secondary Structure. (b) The Relative Frequencies of Secondary Structures in Each PB

(a)		Secondary Structures						Freq. Protein Block
		α-Helix	3_{10}-Helix	π-Helix	Turn	Coil	β-Strand	
Protein Blocks	*a*	*0.14*	*0.13*	*0.00*	**19.35**	**62.64**	**17.74**	3.92
	b	*0.13*	*0.10*	*0.00*	**58.38**	**25.84**	**15.54**	4.16
	c	*0.00*	*0.01*	*0.00*	**13.51**	**43.83**	**42.65**	7.93
	d	*0.00*	*0.00*	*0.00*	5.42	**21.77**	**72.81**	18.28
	e	*0.05*	*0.18*	*0.00*	9.15	**38.51**	**52.11**	2.36
	f	*0.01*	*0.01*	*0.00*	7.67	**66.36**	**25.96**	6.52
	g	*4.56*	7.83	*0.00*	**52.76**	**29.67**	5.18	1.10
	h	*0.27*	*2.54*	*0.00*	**62.35**	**16.66**	**18.17**	2.30
	i	*0.24*	*2.08*	*0.00*	**84.33**	7.63	5.72	1.79
	j	*4.55*	5.34	*0.00*	**59.58**	**21.35**	9.18	0.79
	k	**35.21**	**13.69**	*0.02*	**43.98**	6.34	*0.76*	5.41
	l	**44.90**	**17.24**	*0.02*	**31.14**	6.13	*0.57*	5.38
	m	**86.37**	*4.51*	*0.07*	6.42	*2.51*	*0.12*	31.50
	n	**64.02**	7.41	*0.14*	**24.26**	*3.49*	*0.69*	2.17
	o	**23.08**	6.45	*0.02*	**66.30**	*3.87*	*0.28*	2.86
	p	*4.05*	**12.37**	*0.00*	**62.87**	**18.91**	*1.81*	3.53

(b)		Secondary Structures					
		α-Helix	3_{10}-Helix	π-Helix	Turn	Coil	β-Strand
Protein Blocks	*a*	*0.02*	*0.12*	*0.00*	3.67	**12.57**	*3.19*
	b	*0.02*	*0.11*	*0.00*	11.74	5.49	*2.96*
	c	*0.00*	*0.02*	*0.00*	5.18	**17.78**	**15.52**
	d	*0.00*	*0.02*	*0.00*	*4.79*	**20.34**	**61.04**
	e	*0.00*	*0.11*	*0.00*	*1.04*	*4.64*	5.63
	f	*0.00*	*0.02*	*0.00*	*2.42*	**22.13**	7.77
	g	*0.15*	*2.10*	*0.00*	*2.81*	*1.67*	*0.26*
	h	*0.02*	*1.42*	*0.00*	6.93	*1.96*	*1.92*
	i	*0.01*	*0.91*	*0.00*	7.30	*0.70*	*0.47*
	j	*0.11*	*1.03*	*0.00*	*2.28*	*0.86*	*0.33*
	k	5.63	**18.01**	*4.44*	**11.50**	*1.75*	*0.19*
	l	7.14	**22.58**	*4.44*	8.10	*1.69*	*0.14*
	m	**80.42**	**34.55**	**77.78**	9.78	*4.05*	*0.18*
	n	*4.11*	*3.91*	11.11	*2.55*	*0.39*	*0.07*
	o	*1.95*	*4.49*	*2.22*	9.17	*0.57*	*0.04*
	p	*0.42*	**10.62**	*0.00*	**10.73**	*3.41*	*0.29*
freq. S2		33.83	4.11	0.03	20.68	19.56	21.80

In bold are the frequencies more than 10; in italics are frequencies less than 5%.

elements. It has been computed with a nonredundant data bank with 25% of sequence identity and a resolution better than 2.5 Å. The protein list has been taken from the PISCES web server [100] and the secondary structure assignment has been done with DSSP [9]. Table 5.2a shows the frequencies of classical secondary structures for each PB, while Table 5.2b shows the opposite. It highlights that α-helix and other helical structures are associated only to PBs *k* to *o*, while turns are found spread over all the PBs. It underlines also the non-equivalence of turns and coils that have specificities.

5.5.3. Structural Alignment

Based on PBs, a new structure comparison method (PB-ALIGN) useful for mining protein structural databases, has been developed. Using the structural homologs in the PALI database [101], encoded in terms of PBs, a dedicated PB substitution matrix was computed [69]. Using this matrix with a classical alignment approach, it is possible to find structural homologs [102] similar to what is done in the case of amino acid sequences. A recent benchmark has proven that this method is most efficient for mining the PDB to find structural homologs [103].

5.5.4. Longer Fragments

An analysis of preferential transitions of PBs of various lengths suggested that the series of five PBs (or nine residues) present interesting structural features [105]. The distribution and consistency of structural features associated with fragments representing set of five PBs were checked on different datasets and significant variation was not observed. Based on the extent to which a set of such fragments can cover a protein chain, an optimal set of 72 fragments called "Structural Words (SW)" were selected. They represented 92% of the data bank residues, nearly all the repetitive structures and 80% of the "coil." Most of these SWs were found to overlap; some had even four PBs in common. These SWs represent local structure transitions and irregularities. Quality of structural approximation was assessed, showing that a structural alphabet is meaningful even for longer fragments.

Following this idea, a novel approach was developed: the Hybrid Protein Model (HPM [106]). This specific clustering allows associating longer protein fragments to create structural prototypes with high transition between them [107–111]. From a dataset of proteins coded in the form of PB sequences, fragment sequences of PBs of varying lengths were derived. Similarity between the fragments is decided based on the propensities of PBs to occur at each position in the fragment. In this process, for a given fragment length, a hybrid protein of an optimal length that can represent the sets of preferential transitions of local structures in continuity is generated. The length of the hybrid protein and the propensities of PBs to occur at a position varied during learning. Redundant sets of PB transitions (similar propensities at the same positions). The results of an HPM approach on a dataset of fragments of length 10 residues could be effectively used for fine description of protein structures and the data were used efficiently for identifying local structural similarities between two cytochromes P450 [107]. A hybrid protein of length 233 based on 13-residue long fragments gave a better description of various local structural features [108]. Recent development has given a new hybrid protein that has been used for prediction purpose [109,112].

5.5.5. Structure Prediction Using PBs

A Bayesian probabilistic approach was utilized for the prediction of PBs from amino acid sequence. For learning the amino acid propensities associated with each PB, the set of proteins chains used in training was then encoded in terms of PBs, using the minimal RMSDA criterion. Sequence windows of length 15 residues were considered for calculating the propensities associated with each PB. For every PB, the probability of occurrence of an amino acid at each position in the sequence window was calculated and an occurrence matrix was generated for each of the 16 PBs. Bayes theorem was used to predict the structure of new sequences. A prediction rate of 34.4% was achieved [91,113].

One of the limitation of this approach is to average the sequence information associated with a PB as only one amino acid occurrence matrix corresponds to one PB. Thus, using a clustering approach related to SOM [77], amino acid occurrence matrices was split for some PBs, increasing their sequence specificities. Bayesian prediction was carried out to achieve an improved prediction rate of 40.7% [91,113]. The process of generating sequence families, including a simulated annealing approach that maximizes the prediction rate, helped to improve the overall prediction to 48.7% [113,114]. No biased or unbalanced improvements were detected among the PBs with this approach. Combining the secondary structure information with the Bayesian prediction did not result in a significant improvement of the prediction rate. A Java-based program called LocPred (see Figs. 5.5 and 5.6) is available to perform these predictions [113].

A Bayesian prediction approach (without optimization of sequence–structure relationships) similar to what was used for PB prediction was also carried out for the SWs. A 4% improvement in prediction rate could be achieved [105]. Preferable transitions were also observed between SWs occurring in a sequence and certain series of SWs were found to be highly frequent. Use of this information with an approach called "pinning strategy" helped to improve the prediction rate significantly [115]. The principle of pinning strategy is quite simple: (i) a classical Bayesian prediction is done with SWs, (ii) the positions with a high prediction confidence index are selected as "seeds," and (iii) at a seed position i, an SW (five PBs) is predicted. Therefore a selection is also done at positions $i - 1$ and $i + 1$, respectively; one SW that overlaps this SW is selected through the most probable SW. It is an iterative process, from $i - 1$ and $i + 1$, respectively; the prediction is extended through $i - n$ and $i + n$, respectively. It stops when a probability threshold is reached.

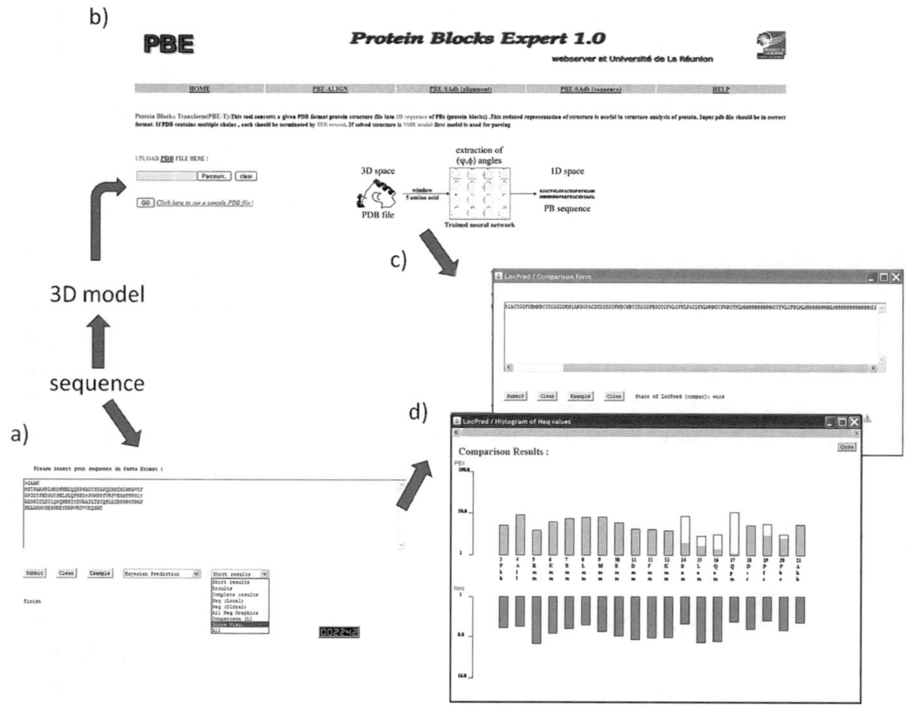

FIGURE 5.5 LocPred use with a structural model. It is possible to confront PB predictions with 3D structural model obtained by another approach. (a) The FASTA sequence is given and the prediction options are selected. (b) The structural model is encoded in terms of PBs with PBE web site (http://bioinformatics.univ-reunion.fr/PBE/). (c) The PB sequence corresponding to the structural model is placed into the comparison form. (d) The compatibility between prediction and structural model is given graphically.

A detailed analysis of PB distribution in short loop regions (6 to 10 residues) has been done [30]. The description in terms of PBs helped to understand the ambiguity associated with the assignment of the boundaries of regular secondary structures based on different assignment methods. Specific sequence–structure relationships in the short loops could be derived. A Bayesian prediction carried out based on this information gave an accuracy rate of 41.2% for the short loops and 36% for the loops in general. A recent study has shown that a specific learning of the different kinds of short loops improved greatly the prediction [116].

LocPred is useful not only in predicting protein structures in terms of PBs but also in analyzing the sequence–structure relationship of the protein of interest. The simplest output of LocPred is a list with raw prediction values with confidence indexes and different probabilities. Graphical outputs give visual representations of the probabilities associated with each predicted PB;

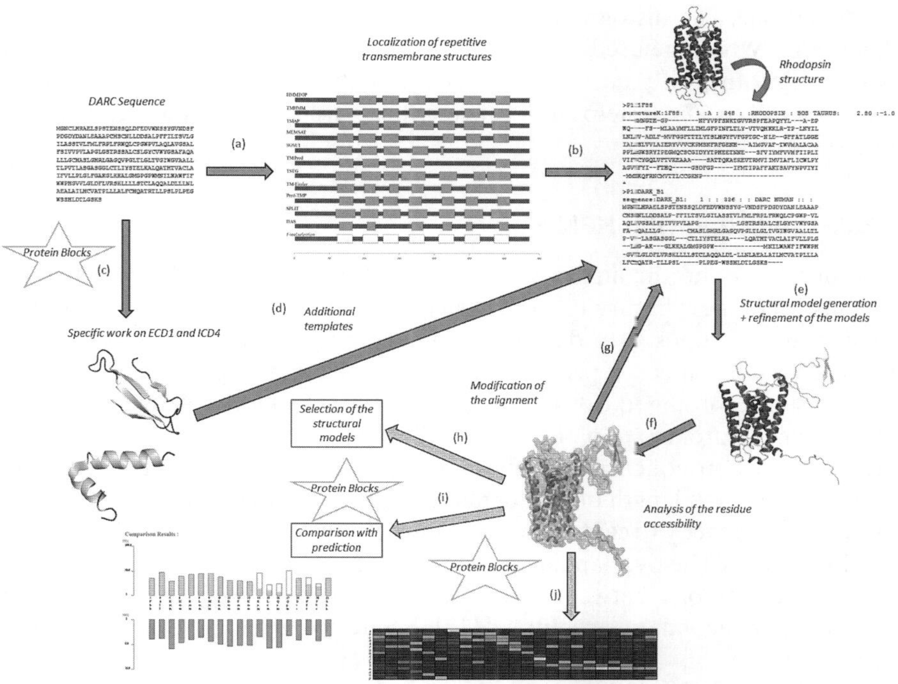

FIGURE 5.6 Building structural models of DARC. (a) Precliction of transmembrane helices. (b) Alignment of helical regions with corresponding regions of rhodopsin structure. (c) Potential structural templates for the extremities are done; thanks to Protein Blocks. (d) Addition of these results to the complete alignment for comparative modeling. (e) Structural model generation and refinement of these models. (f) Accessibility computation of amino acids and known to be exposed. (g) With regard to the results, the alignment is modified. (h) At last, some models are selected. (i) As seen in Figure 5.5, comparison between PB prediction and the PB assignment can help to locate arduous regions. (j) PBs can also be used to analyze protein molecular dynamics as in Reference [117].

it helps to have an idea of the local tendencies and the confidence index associated with each position, that is, the lower the confidence index is, the better it is. This option could be so helpful even if the user does not want to use PBs; it quantifies the sequence–structure relationship of this protein.

Figure 5.5 shows another possibility given by LocPred, that is, the comparison of a structural model and PB prediction. A prediction is performed as given in Figure 5.5a. Many different approaches, softwares, and web services allow the obtaining of structural model. Thanks to PBE web server (see Fig. 5.5b), it is simple to translate a protein structure in terms of PBs. Then, in LocPred, it is possible to compare the assigned PBs of the structural model with the PB predictions (see Figs. 5.5c and 5.6d). Figure 5.5d shows an example of such comparison. For each amino acid position is given the amino acid, the position in the sequence and the two PBs, that is, the assigned and the predicted one. The histogram corresponds to the prediction of the best predicted PBs. When the predicted and assigned PBs are the same, the histogram bar is plain; otherwise the color is smaller as in the second part of the example (positions 14 to 17). It helps to localize critical structural regions of the structural model.

5.5.6. Prediction with HPM

In order to extend the analyses of long structural fragments, HPM was used to construct a new library of local structures. One hundred twenty structural clusters were proposed to describe fragments of 11 residues in length [109]. For each class, a mean representative prototype, named LSP (see Fig. 5.3b), was chosen according to Cα RMSD criteria. These 120 LSPs enabled a satisfying average approximation of 1.6 Å for all local structures observed in known proteins. The consequences of long-range interactions are taken into account; thanks to the high length of fragments. Moreover, the major advantage of this library is its capacity to capture the continuity between the identified recurrent local structures. The overlapping properties of LSPs were used to identify very frequent transitions between them and characterize their involvement in longer super secondary structures [112]. Figure 5.4 gives an example of LSP assignment.

For each one of the 120 structural classes, high sequence relationships were observed, which led to the development of an original prediction method from single sequence and based on logistic regressions. The main purpose of local structure prediction methods is to reduce the combinatory structural possibilities for a sequence. Thus, it is worth noting that this method proposed a short list of the best structural candidates among the 120 LSPs of the library. Moreover, to identify directly regions easier or difficult to predict, each prediction is associated with a confidence index. With a geometrical assessment, a prediction rate of 51.2% was reached. This result was already very satisfying given the high length of fragments and the high number of classes [109].

Recently, an improved prediction method relying on SVMs and evolutionary information was proposed. A global prediction rate of 63.1% was achieved and corresponded to an improved prediction of 85% of proteins. A confidence index was also defined for directly assessing the relevance of the prediction at each sequence site. This method was shown to be among the most efficient cutting-edge local structure prediction strategies [112]. Taking advantage of the high length of fragments, the relationships between their structural flexibility and their predictability are now under study.

5.5.7. Solving a Biological Problem—Duffy Antigen/Receptor for Chemokines (DARC)

Local structure prediction based on PBs was used along with threading, *ab initio*, and secondary structure prediction methods to determine the fold of

DARC [117]. DARC occurs on the surface of erythrocytes and serves as a receptor for various chemokines. It was also identified as the erythrocyte receptor for *Plasmodium vivax* and *Plasmodium knowlesi* parasites. In the absence of well-defined homologs of known structure, modeling of transmembrane proteins remains a difficult task. PB predictions from the regions of low information content were highly relevant for the analysis of the models generated by energy minimization and molecular dynamics refinements. This example was a very good example of interest that helped to analyze the results of simulated annealing-based prediction with a finer description. We have recently described the use of such approaches for DARC [118] to define pertinent structural models [119]. Figure 5.6 describes the protocol used, which is based on (i) biochemical data, some residues must be accessible, (ii) transmembrane predictions, and (iii) PB approach.

5.5.8. Comparison of Predictions

As most of the structural alphabets are not available for use for the scientific community, it is very difficult to make a comparison. Comparison of prediction is not trivial, but can be done, even if they are based on unrelated methodologies. Yang and Wang developed a database of sequence profiles of nine-residue fragments, the members of each profile having similar backbone conformational state and similar sequences. These profiles are generated in a two-step process. In the first step, seed sequence profiles were generated based on φ/ϕ dihedral states defined by Reference [120] and also on the sequence similarity calculated based on structure specific amino acid substitution matrices [121]. The preliminary profiles in the form of Position-Specific Scoring Matrix (PSSMs) were then used to search for more fragments with identical backbone conformation and a good sequence profile match score. A Bayesian prediction pseudo count method was used to represent the amino acid occurrence propensities in the preliminary PSSMs. For the prediction purposes, the set of sequence profiles with a good sequence profile matching score and having at least 60% consistency with the secondary structure prediction by PSI-PRED were chosen. For each of the selected profiles, a consensus score giving an indication of the extent of backbone conformational similarity with others in the set is calculated. The one with the highest consensus score is chosen as the predicted candidate. The percent of correct predictions on a dataset were comparable to those obtained with HMMSTR. However, based on RMSD between the true and the predicted structure, this method is reported to perform better than HMMSTR. The prediction accuracy was later improved with the use of SVMs and neural networks [122]. Prediction made using HPM with linear regression [109,112] was comparable to these approaches, and the results are better with our new approaches that use SVMs with evolutionary information.

More recently, another method for predicting PBs from sequence has been developed. Li et al. propose an innovative combination of PB prediction, taking into account the information on secondary structure and solvent accessibilities [123]. Prediction rates were improved, and, interestingly, their approach was found useful for fragment threading, pseudo sequence design, and local structure predictions.

Recently, Zimmermann and Hansmann have developed a method, named *Locustra*, for predicting local structures encoded in terms of PBs from sequence [124]. The prediction was carried out using SVMs with a radial basis function kernel. For the prediction of each class of PB, a two-layer classification scheme was used. In the first step, the samples belonging to one class was considered as the positive set while those belonging to another class were considered as the negative class, that is, a pairwise coupling classifier. One hundred twenty classifiers were required. The input sequence data were enriched using the information derived from the homologs and a profile of amino acid propensities was obtained. The sequence window of 15 residues indicated a feature vector of size 315. To estimate the class probability, a cross-validation-based method was used. The probabilities at each sequence position, obtained from the 120 pairwise coupling classifiers, were used as features for the second layer. Here, a one-per-class classifier was used, where the samples belonging to one class is considered as the positive set while those belonging to all the other classes were included in the negative set. The PB having the highest number of votes in the output of the second layer was chosen as the predicted PB. The major secondary structures like helices or strands were chosen in cases of multiple predictions. The prediction accuracy reaches 61%. It was also noted that the PBs that are mispredicted were often structurally related to the true PB and these mispredictions often correspond to exposed regions of the structure.

Prediction of PBs is very simple as only a sequence in FASTA format is needed. PBs are the only structural alphabet with web service for prediction, and moreover, three different approaches are available.

5.6. CONCLUSIONS AND PERSPECTIVES

In this chapter, we have presented different facets of the protein structures at a local level, underlining some limitations of using secondary structures for describing protein structures. Global protein structures can be described by a limited set of recurring local structures [125] and in this context, the use of structural alphabets is obvious. As it is not easy to build relevant structural models directly with structural prototypes, I-sites have been added to a prediction method, namely Rosetta [126].

Recently, Dong et al. developed a set of structural alphabets with the aim of finding an optimal structural alphabet sequence from which an accurate model of the protein can be regenerated [127]. Using the standard k-means algorithm they clustered fragments that are seven residues in length, based on the Cα RMSD. The set of alphabets generated were used to reconstruct the structure of the protein such that the global RMSD is minimal. For doing so, they adopted a combination of greedy and dynamic programming algorithms. Sets of structural alphabets of sizes 4 to 100 prototypes were evaluated for both local and global structure approximations and finally a set of 28 letters were chosen. When compared with the global approximation based on PBs, this set of alphabets is reported to give slightly better results. Thus, the future of local protein structures is promising in the area of building relevant structural models.

To date, nearly all the structural alphabets are only used within the research groups that have developed them (see Table 5.1). Hence, PB structural alphabet is an exception. PB is one of the most widely used structural alphabet. Indeed, it is easy to use PBs for various applications. They have been used both to describe the 3D protein backbones [99] and to perform a local structure prediction [91,113,114,116]. The efficiency of PBs have also been proven in the description and the prediction of long fragments [67,105,107–111,115,128], to compare protein structures [69,102,103], to build globular [127] and transmembrane protein structures [117], to define a reduced amino acid alphabet dedicated to mutation design [129], to design peptides [130], or to define binding site signatures [131]. The features of this alphabet have been compared with those of eight other structural alphabets showing clearly that the PB alphabet is highly informative, with the best predictive ability of those tested [132].

Future of structural alphabets is also coupled with the taking into account more biophysical feature. One of our main axes of research is the link between local protein structure prediction and the protein flexibility [133]. For this purpose, we have studied protein dynamics from two different points of view, that is, X-ray experiments and molecular dynamics simulations. Prediction results are quite good in comparison to available methodologies.

ACKNOWLEDGMENTS

This work was supported by grants from the Ministère de la Recherche, Université Paris Diderot—Paris 7, Université de Saint-Denis de la Réunion, National Institute for Blood Transfusion (INTS) and the Institute for Health and Medical Care (INSERM). APJ has a grant from CEFIPRA number 3903-E and AB has a grant from the Ministère de la Recherche.

REFERENCES

1. O. Doppelt, F. Moriaud, F. Delfaud, and A.G. de Brevern. Analysis of HSP90 related folds with MED-SuMo classification approach. *Drug Design, Development and Therapy*, 9:3, 2009.

2. L. Slabinski, L. Jaroszewski, A.P. Rodrigues, L. Rychlewski, I.A. Wilson, S.A. Lesley, and A. Godzik. The challenge of protein structure determination—lessons from structural genomics. *Protein Science*, 16:2472–2482, 2007.
3. O. Doppelt, F. Moriaud, A. Bornot, and A.G. de Brevern. Functional annotation strategy for protein structures. *Bioinformation*, 1:357–359, 2007.
4. L. Pauling and R.B. Corey. The pleated sheet, a new layer configuration of polypeptide chains. *Proceedings of the National Academy of Sciences U S A*, 37:251–256, 1951.
5. L. Pauling, R.B. Corey, and H.R. Branson. The structure of proteins; two hydrogen-bonded helical configurations of the polypeptide chain. *Proceedings of the National Academy of Sciences U S A*, 37:205–211, 1951.
6. R.A. Sayle and E.J. Milner-White. RASMOL: Biomolecular graphics for all. *Trends in Biochemical Sciences*, 20:374–382, 1995.
7. C. Perez-Iratxeta and M.A. Andrade-Navarro. K2D2: Estimation of protein secondary structure from circular dichroism spectra. *BMC Structural Biology*, 8:25, 2008.
8. A.G. Murzin, S.E. Brenner, T. Hubbard, and C. Chothia. SCOP: A structural classification of proteins database for the investigation of sequences and structures. *Journal of Molecular Biology*, 247:536–540, 1995.
9. W. Kabsch and C. Sander. Dictionary of protein secondary structure: Pattern recognition of hydrogen-bonded and geometrical features. *Biopolymers*, 22:2577–2637, 1983.
10. D. Frishman and P. Argos. Knowledge-based protein secondary structure assignment. *Proteins*, 23:566–579, 1995.
11. R. Srinivasan and G.D. Rose. A physical basis for protein secondary structure. *Proceedings of the National Academy of Sciences U S A*, 96:14258–14263, 1999.
12. M.V. Cubellis, F. Cailliez, and S.C. Lovell. Secondary structure assignment that accurately reflects physical and evolutionary characteristics. *BMC Bioinformatics*, 6(4): S8, 2005.
13. J. Martin, G. Letellier, A. Marin, J.-F. Taly, A.G. de Brevern, and J.-F. Gibrat. Protein secondary structure assignment revisited: A detailed analysis of different assignment methods. *BMC Structural Biology*, 5:17, 2005.
14. D. Eisenberg. The discovery of the alpha-helix and beta-sheet, the principal structural features of proteins. *Proceedings of the National Academy of Sciences U S A*, 100:11207–11210, 2003.
15. J.S. Richardson and D.C. Richardson. Amino acid preferences for specific locations at the ends of alpha helices. *Science*, 240:1648–1652, 1988.
16. L. Pal, P. Chakrabarti, and G. Basu. Sequence and structure patterns in proteins from an analysis of the shortest helices: Implications for helix nucleation. *Journal of Molecular Biology*, 326:273–291, 2003.
17. L. Regan. Protein structure. Born to be beta. *Current Biology*, 4:656–658, 1994.
18. M. Tyagi, A. Bornot, B. Offmann, and A.G. de Brevern. Analysis of loop boundaries using different local structure assignment methods. *Protein Science*, 18(9):1869–1881, 2009.
19. S.D. Khare and N.V. Dokholyan. Molecular mechanisms of polypeptide aggregation in human diseases. *Current Protein & Peptide Science*, 8:573–579, 2007.

20. R. Aurora and G.D. Rose. Helix capping. *Protein Science*, 7:21–38, 1998.
21. E. Kruus, P. Thumfort, C. Tang, and N.S. Wingreen. Gibbs sampling and helix-cap motifs. *Nucleic Acids Research*, 33:5343–5353, 2005.
22. J. Garnier, D.J. Osguthorpe, and B. Robson. Analysis of the accuracy and implications of simple methods for predicting the secondary structure of globular proteins. *Journal of Molecular Biology*, 120:97–120, 1978.
23. B. Rost and C. Sander. Improved prediction of protein secondary structure by use of sequence profiles and neural networks. *Proceedings of the National Academy of Sciences U S A*, 90:7558–7562, 1993.
24. D.T. Jones. Protein secondary structure prediction based on position-specific scoring matrices. *Journal of Molecular Biology*, 292:195–202, 1999.
25. G. Pollastri and A. McLysaght. Porter: A new, accurate server for protein secondary structure prediction. *Bioinformatics*, 21:1719–1720, 2005.
26. G. Pollastri, D. Przybylski, B. Rost, and P. Baldi. Improving the prediction of protein secondary structure in three and eight classes using recurrent neural networks and profiles. *Proteins*, 47:228–235, 2002.
27. L. Pal, G. Basu, and P. Chakrabarti. Variants of 3(10)-helices in proteins. *Proteins*, 48:571–579, 2002.
28 L. Pal, B. Dasgupta, and P. Chakrabarti. 3(10)-Helix adjoining alpha-helix and beta-strand: Sequence and structural features and their conservation. *Biopolymers*, 78:147–162, 2005.
29. K.H. Lee, D.R. Benson, and K. Kuczera. Transitions from alpha to pi helix observed in molecular dynamics simulations of synthetic peptides. *Biochemistry*, 39:13737–13747, 2000.
30. L. Fourrier, C. Benros, and A.G. de Brevern. Use of a structural alphabet for analysis of short loops connecting repetitive structures. *BMC Bioinformatics*, 5:58, 2004.
31. N. Eswar, C. Ramakrishnan, and N. Srinivasan. Stranded in isolation: Structural role of isolated extended strands in proteins. *Protein Engineering*, 16:331–339, 2003.
32. S.Y. Park, K. Yamane, S. Adachi, Y. Shiro, K.E. Weiss, S.A. Maves, and S.G. Sligar. Thermophilic cytochrome P450 (CYP119) from Sulfolobus solfataricus: High resolution structure and functional properties. *Journal of Inorganic Biochemistry*, 91:491–501, 2002.
33. W.L.T. DeLano. *The PyMOL Molecular Graphics System DeLano Scientific*. San Carlos, CA, 2002. http://www.pymol.org.
34. C.M. Venkatachalam. Stereochemical criteria for polypeptides and proteins. V. Conformation of a system of three linked peptide units. *Biopolymers*, 6:1425–1436, 1968.
35. E.G. Hutchinson and J.M. Thornton. A revised set of potentials for beta-turn formation in proteins. *Protein Science*, 3:2207–2216, 1994.
36. P.F. Fuchs and A.J. Alix. High accuracy prediction of beta-turns and their types using propensities and multiple alignments. *Proteins*, 59:828–839, 2005.
37. C. Zheng and L. Kurgan. Prediction of beta-turns at over 80% accuracy based on an ensemble of predicted secondary structures and multiple alignments. *BMC Bioinformatics*, 9:430, 2008.

38. O. Koch and G. Klebe. Turns revisited: A uniform and comprehensive classification of normal, open, and reverse turn families minimizing unassigned random chain portions. *Proteins*, 74:353–367, 2009.
39. M. Meissner, O. Koch, G. Klebe, and G. Schneider. Prediction of turn types in protein structure by machine-learning classifiers. *Proteins*, 74:344–352, 2009.
40. J. Makowska, S. Rodziewicz-Motowidlo, K. Baginska, J.A. Vila, A. Liwo, L. Chmurzynski, and H.A. Scheraga. Polyproline II conformation is one of many local conformational states and is not an overall conformation of unfolded peptides and proteins. *Proceedings of the National Academy of Sciences U S A*, 103(6):1744–1749, 2006.
41. B.J. Stapley and T.P. Creamer. A survey of left-handed polyproline II helices. *Protein Science*, 8:587–595, 1999.
42. F. Eker, K. Griebenow, and R. Schweitzer-Stenner. Abeta(1–28) fragment of the amyloid peptide predominantly adopts a polyproline II conformation in an acidic solution. *Biochemistry*, 43:6893–6898, 2004.
43. J.M. Hicks and V.L. Hsu. The extended left-handed helix: A simple nucleic acid-binding motif. *Proteins*, 55:330–338, 2004.
44. S.A. Hollingsworth, D.S. Berkholz, and P.A. Karplus. On the occurrence of linear groups in proteins. *Protein Science*, 18:1321–1325, 2009.
45. S.M. King and W.C. Johnson. Assigning secondary structure from protein coordinate data. *Proteins*, 35:313–320, 1999.
46. G. Labesse, N. Colloc'h, J. Pothier, and J.P. Mornon. P-SEA: A new efficient assignment of secondary structure from C alpha trace of proteins. *Computer Application in the Biosciences*, 13:291–295, 1997.
47. F. Dupuis, J.F. Sadoc, and J.P. Mornon. Protein secondary structure assignment through Voronoi tessellation. *Proteins*, 55:519–528, 2004.
48. S. Hosseini, M. Sadeghi, H. Pezeshk, C. Eslahchi, and M. Habibi. PROSIGN: A method for protein secondary structure assignment based on three-dimensional coordinates of consecutive C(alpha) atoms. *Computational Biological Chemistry*, 32:406–4011, 2008.
49. P.K. Vlasov, A.V. Vlasova, V.G. Tumanyan, and N.G. Esipova. A tetrapeptide-based method for polyproline II-type secondary structure prediction. *Proteins*, 61:763–768, 2005.
50. M. Kuhn, J. Meiler, and D. Baker. Strand-loop-strand motifs: Prediction of hairpins and diverging turns in proteins. *Proteins*, 54:282–288, 2004.
51. M. Kumar, M. Bhasin, N.K. Natt, and G.P. Raghava. BhairPred: Prediction of beta-hairpins in a protein from multiple alignment information using ANN and SVM techniques. *Nucleic Acids Research*, 33:W154–W159, 2005.
52. X.Z. Hu and Q.Z. Li. Prediction of the beta-hairpins in proteins using support vector machine. *Protein Journal*, 27:115–122, 2008.
53. A.V. Efimov. A structural tree for alpha-helical proteins containing alpha-alpha-corners and its application to protein classification. *FEBS Letters*, 391:167–170, 1996.
54. J. Wojcik, J.P. Mornon, and J. Chomilier. New efficient statistical sequence-dependent structure prediction of short to medium-sized protein loops based on an exhaustive loop classification. *Journal of Molecular Biology*, 289:1469–1490, 1999.

55. N. Fernandez-Fuentes, E. Querol, F.X. Aviles, M.J. Sternberg, and B. Oliva. Prediction of the conformation and geometry of loops in globular proteins: Testing ArchDB, a structural classification of loops. *Proteins*, 60:746–757, 2005.

56. M. Bansal, S. Kumar, and R. Velavan. HELANAL: A program to characterize helix geometry in proteins. *Journal of Biomolecular Structure & Dynamics*, 17:811–819, 2000.

57. J.P. Cartailler and H. Luecke. Structural and functional characterization of pi bulges and other short intrahelical deformations. *Structure (Camb)*, 12:133–144, 2004.

58. E.J. Milner-White. Beta-bulges within loops as recurring features of protein structure. *Biochimica Biophysica Acta*, 911:261–265, 1987.

59. J.S. Richardson, E.D. Getzoff, and D.C. Richardson. The beta bulge: A common small unit of nonrepetitive protein structure. *Proceedings National Academy of Sciences U S A*, 75:2574–2578, 1978.

60. A.W. Chan, E.G. Hutchinson, D. Harris, and J.M. Thornton. Identification, classification, and analysis of beta-bulges in proteins. *Protein Science*, 2:1574–1590, 1993.

61. C.A. Andersen, A.G. Palmer, S. Brunak, and B. Rost. Continuum secondary structure captures protein flexibility. *Structure (Camb)*, 10:175–184, 2002.

62. M.N. Fodje and S. Al-Karadaghi. Occurrence, conformational features and amino acid propensities for the pi-helix. *Protein Engineering*, 15:353–358, 2002.

63. F.M. Richards and C.E. Kundrot. Identification of structural motifs from protein coordinate data: Secondary structure and first-level supersecondary structure. *Proteins*, 3:71–84, 1988.

64. I. Majumdar, S.S. Krishna, and N.V. Grishin. PALSSE: A program to delineate linear secondary structural elements from protein structures. *BMC Bioinformatics*, 6:202, 2005.

65. M. Parisien and F. Major. A New Catalog of Protein Beta-Sheets. *Proteins*, 61(3):545–558, 2005.

66. J. Martin, J.F. Gibrat, and F. Rodolphe. Analysis of an optimal hidden Markov model for secondary structure prediction. *BMC Structural Biology*, 6:25, 2006.

67. A.G. de Brevern, C. Benros, and S. Hazout. Structural alphabet: From a local point of view to a global description of protein 3D structures. In P.V. Yan (Ed.), *Bioinformatics: New Research*, pp. 128–187. New York: Nova Publishers, 2005.

68. B. Offmann, M. Tyagi, and A.G. de Brevern. Local Protein Structures. *Current Bioinformatics*, 3:165–202, 2007.

69. M. Tyagi, V.S. Gowri, N. Srinivasan, A.G. de Brevern, and B. Offmann. A substitution matrix for structural alphabet based on structural alignment of homologous proteins and its applications. *Proteins*, 65:32–39, 2006.

70. R. Unger, D. Harel, S. Wherland, and J.L. Sussman. A 3D building blocks approach to analyzing and predicting structure of proteins. *Proteins*, 5:355–373, 1989.

71. R. Unger, D. Harel, S. Wherland, and J.L. Sussman. Analysis of dihedral angles distribution: The doublets distribution determines polypeptides conformations. *Biopolymers*, 30:499–508, 1990.

72. M.J. Rooman and S.J. Wodak. Identification of predictive sequence motifs limited by protein structure data base size. *Nature*, 335:45–49, 1988.

73. M.J. Rooman, J. Rodriguez, and S.J. Wodak. Automatic definition of recurrent local structure motifs in proteins. *Journal of Molecular Biology*, 213:327–336, 1990.

74. M.J. Rooman, J. Rodriguez, and S.J. Wodak. Relations between protein sequence and structure and their significance. *Journal of Molecular Biology*, 213:337–350, 1990.

75. S.J. Prestrelski, D.M. Byler, and M.N. Liebman. Generation of a substructure library for the description and classification of protein secondary structure. II. Application to spectra-structure correlations in Fourier transform infrared spectroscopy. *Proteins*, 14:440–450, 1992.

76. S.J. Prestrelski, A.L. Jr. Williams, and M.N. Liebman. Generation of a substructure library for the description and classification of protein secondary structure. I. Overview of the methods and results. *Proteins*, 14:430–439, 1992.

77. T. Kohonen. Self-organized formation of topologically correct feature maps. *Biological Cybernetics*, 43:59–69, 1982.

78. T. Kohonen. *Self-Organizing Maps*, 3rd edition. Berlin: Springer, 2001.

79. J. Schuchhardt, G. Schneider, J. Reichelt, D. Schomburg, and P. Wrede. Local structural motifs of protein backbones are classified by self-organizing neural networks. *Protein Engineering*, 9:833–842, 1996.

80. J.S. Fetrow, S.R. Horner, W. Oehrl, D.L. Schaak, T.L. Boose, and R.E. Burton. Analysis of the structure and stability of omega loop a replacements in yeast iso-1-cytochrome c. *Protein Science*, 6:197–210, 1997.

81. C. Bystroff and D. Baker. Prediction of local structure in proteins using a library of sequence-structure motifs. *Journal of Molecular Biology*, 281:565–577, 1998.

82. R. Schneider, A. de Daruvar, and C. Sander. The HSSP database of protein structure-sequence alignments. *Nucleic Acids Research*, 25:226–230, 1997.

83. C. Bystroff and D. Baker. Blind predictions of local protein structure in CASP2 targets using the I-sites library. *Proteins Supplement*, 1:167–171, 1997.

84. C. Bystroff, V. Thorsson, and D. Baker. HMMSTR: A hidden Markov model for local sequence-structure correlations in proteins. *Journal of Molecular Biology*, 301:173–190, 2000.

85. A.C. Camproux, P. Tuffery, J.P. Chevrolat, J.F. Boisvieux, and S. Hazout. Hidden Markov model approach for identifying the modular framework of the protein backbone. *Protein Engineering*, 12:1063–1073, 1999.

86. A.C. Camproux, A.G. de Brevern, S. Hazout, and P. Tufféry. Exploring the use of a structural alphabet for structural prediction of protein loops. *Theoretical Chemistry Accounts*, 106:28–35, 2001.

87. J. Maupetit, R. Gautier, and P. Tuffery. SABBAC: Online structural alphabet-based protein backbone reconstruction from Alpha-Carbon trace. *Nucleic Acids Research*, 34:W147–W151, 2006.

88. J. Martin, A.G. de Brevern, and A.C. Camproux. In silico local structure approach: a case study on outer membrane proteins. *Proteins*, 71:92–109, 2008.

89. C. Micheletti, F. Seno, and A. Maritan. Recurrent oligomers in proteins: An optimal scheme reconciling accurate and concise backbone representations in automated folding and design studies. *Proteins*, 40:662–674, 2000.

90. C.G. Hunter and S. Subramaniam. Protein fragment clustering and canonical local shapes. *Proteins*, 50:580–588, 2003.

91. A.G. de Brevern, C. Etchebest, and S. Hazout. Bayesian probabilistic approach for predicting backbone structures in terms of protein blocks. *Proteins*, 41:271–287, 2000.

92. C.G. Hunter and S. Subramaniam. Protein local structure prediction from sequence. *Proteins*, 50:572–579, 2003.

93. O. Sander, I. Sommer, and T. Lengauer. Local protein structure prediction using discriminative models. *BMC Bioinformatics*, 7:14, 2006.

94. J.M. Yang and C.H. Tung. Protein structure database search and evolutionary classification. *Nucleic Acids Research*, 34:3646–3659, 2006.

95. C.H. Tung, J.W. Huang, and J.M. Yang. Kappa-alpha plot derived structural alphabet and BLOSUM-like substitution matrix for fast protein structure database search. *Genome Biology*, 8:R31, 2007.

96. C.H. Tung and J.M. Yang. fastSCOP: A fast web server for recognizing protein structural domains and SCOP superfamilies. *Nucleic Acids Research*, 35:W438–443, 2007.

97. S.Y. Ku and Y.J. Hu. Protein structure search and local structure characterization. *BMC Bioinformatics*, 9:349, 2008.

98. J. Yang. Comprehensive description of protein structures using protein folding shape code. *Proteins*, 71: 1497–1518, 2008.

99. A.G. de Brevern. New assessment of protein blocks. *In Silico Biology*, 5:283–289, 2005.

100. G. Wang and R.L. Jr. Dunbrack, PISCES: A protein sequence culling server. *Bioinformatics*, 19:1589–1591, 2003.

101. V.S. Gowri, S.B. Pandit, P.S. Karthik, N. Srinivasan, and S. Balaji. Integration of related sequences with protein three-dimensional structural families in an updated version of PALI database. *Nucleic Acids Research*, 31:486–488, 2003.

102. M. Tyagi, P. Sharma, C.S. Swamy, F. Cadet, N. Srinivasan, A.G. de Brevern, and B. Offmann. Protein Block Expert (PBE): A web-based protein structure analysis server using a structural alphabet. *Nucleic Acids Research*, 34:W119–W123, 2006.

103. M. Tyagi, A.G. de Brevern, N. Srinivasan, and B. Offmann. Protein structure mining using a structural alphabet. *Proteins*, 71:920–937, 2008.

104. G. Kurisu, Y. Kai, and S. Harada. Structure of the zinc-binding site in the crystal structure of a zinc endoprotease from Streptomyces caespitosus at 1 A resolution. *Journal of Inorganic Biochemistry*, 82:225–228, 2000.

105. A.G. de Brevern, H. Valadie, S. Hazout, and C. Etchebest. Extension of a local backbone description using a structural alphabet: A new approach to the sequence-structure relationship. *Protein Science*, 11:2871–2886, 2002.

106. A.G. de Brevern and S. Hazout. Hybrid Protein Model (HPM): A method to compact protein 3D-structures information and physicochemical properties. *IEEE—Computer Society*, S1:49–S54, 2000.

107. A.G. de Brevern and S. Hazout. Compacting local protein folds with a "hybrid protein model." *Theoretical Chemistry Accounts*, 106:36–47, 2001.

108. A.G. de Brevern and S. Hazout. "Hybrid protein model" for optimally defining 3D protein structure fragments. *Bioinformatics*, 19:345–353, 2003.

109. C. Benros, A.G. de Brevern, C. Etchebest, and S. Hazout. Assessing a novel approach for predicting local 3D protein structures from sequence. *Proteins*, 62:865–880, 2006.

110. C. Benros, S. Hazout, and A.G. de Brevern. Extension of a local backbone description using a structural alphabet. "Hybrid Protein Model": A new clustering approach for 3D local structures. In ISMIS (Ed.), *International Workshop on Bioinformatics*, pp. 36–45. Lyon, France, 2002.

111. C. Benros, A.G. de Brevern, and S. Hazout. Hybrid Protein Model (HPM): A method for building a library of overlapping local structural prototypes. sensitivity study and improvements of the training. In *IEEE Workshop on Neural Networks for Signal Processing*, pp. 53–72, 2003.

112. A. Bornot, C. Etchebest, and A.G. de Brevern. A new prediction strategy for long local protein structures using an original description. *Proteins*, 76(3):570–587, 2009.

113. A.G. de Brevern, C. Benros, R. Gautier, H. Valadie, S. Hazout, and C. Etchebest. Local backbone structure prediction of proteins. *In Silico Biology*, 4:381–386, 2004.

114. C. Etchebest, C. Benros, S. Hazout, and A.G. de Brevern. A structural alphabet for local protein structures: Improved prediction methods. *Proteins*, 62(4):865–880, 2005.

115. A.G. de Brevern, C. Etchebest, C. Benros, and S. Hazout. "Pinning strategy": A novel approach for predicting the backbone structure in terms of Protein Blocks from sequence. *Journal of Biosciences*, 32:51–72, 2007.

116. M. Tyagi, A. Bornot, B. Offmann, and A.G. de Brevern. Protein short loop prediction in terms of a structural alphabet. *Computational Biology and Chemistry*, 2009, in press.

117. A.G. de Brevern, H. Wong, C. Tournamille, Y. Colin, C. Le Van Kim, and C. Etchebest. A structural model of a seven-transmembrane helix receptor: The Duffy antigen/receptor for chemokine (DARC). *Biochimica Biophysica Acta*, 1724:288–306, 2005.

118. A.G. de Brevern, L. Autin, Y. Colin, O. Bertrand, and C. Etchebest. In silico studies on DARC. *Infectious Disorders—Drug Targets*, 9:289–303, 2009.

119. A.G. de Brevern. New opportunities to fight against infectious diseases and to identify pertinent drug targets with novel methodologies. *Infectious Disorder—Drug Targets*, 9:246–247, 2009.

120. B. Oliva, P.A. Bates, E. Querol, F.X. Aviles, and M.J. Sternberg. An automated classification of the structure of protein loops. *Journal of Molecular Biology*, 266:814–830, 1997.

121. A.S. Yang and L.Y. Wang. Local structure-based sequence profile database for local and global protein structure predictions. *Bioinformatics*, 18:1650–1657, 2002.

122. A.S. Yang and L.Y. Wang. Local structure prediction with local structure-based sequence profiles. *Bioinformatics*, 19:1267–1274, 2003.

123. Q. Li, C. Zhou, and H. Liu. Fragment-based local statistical potentials derived by combining an alphabet of protein local structures with secondary structures and solvent accessibilities. *Proteins*, 74(4):820–836, 2009.

124. O. Zimmermann and U.H. Hansmann. LOCUSTRA: Accurate prediction of local protein structure using a two-layer support vector machine approach. *Journal of Chemical Information Modeling*, 48:1903–1908, 2008.

125. N.C. Fitzkee, P.J. Fleming, H. Gong, N. Jr. Panasik, T.O. Street, and G.D. Rose. Are proteins made from a limited parts list?. *Trends in Biochemical Science*, 30:73–80, 2005.

126. R. Bonneau, C.E. Strauss, and D. Baker. Improving the performance of Rosetta using multiple sequence alignment information and global measures of hydrophobic core formation. *Proteins*, 43:1–11, 2001.

127. Q.W. Dong, X.L. Wang, and L. Lin. Methods for optimizing the structure alphabet sequences of proteins. *Computers in Biology and Medicine*, 37:1610–1616, 2007.

128. A.G. de Brevern, A.-C. Camproux, S. Hazout, C. Etchebest, and P. Tuffery. Protein structural alphabets: Beyond the secondary structure description. In S. Sangadai (Ed.), *Recent Research Developments in Protein Engineering*, pp. 319–331. Trivandrum, Kerala, India: Research Signpost, 2001.

129. C. Etchebest, C. Benros, A. Bornot, A.C. Camproux, and A.G. de Brevern. A reduced amino acid alphabet for understanding and designing protein adaptation to mutation. *European Biophysics Journal*, 36:1059–1069, 2007.

130. A. Thomas, S. Deshayes, M. Decaffmeyer, M.H. Van Eyck, B. Charloteaux, and R. Brasseur. Prediction of peptide structure: How far are we?. *Proteins*, 65:889–897, 2006.

131. M. Dudev and C. Lim. Discovering structural motifs using a structural alphabet: Application to magnesium-binding sites. *BMC Bioinformatics*, 8:212, 2007.

132. R. Karchin, M. Cline, Y. Mandel-Gutfreund, and K. Karplus. Hidden Markov models that use predicted local structure for fold recognition: alphabets of backbone geometry. *Proteins*, 51:504–514, 2003.

133. A. Bornot, B. Offmann, and A.G. de Brevern. How flexible protein structures are? New questions on the protein structure plasticity. *BIOFORUM Europe*, 11:24–25, 2007.

134. R. Kolodny, P. Koehl, L. Guibas, and M. Levitt. Small libraries of protein fragments model native protein structures accurately. *Journal of Molecular Biology*, 323(2):297–307, 2002.

CHAPTER 6

SHEDDING LIGHT ON TRANSMEMBRANE TOPOLOGY

GÁBOR E. TUSNÁDY and ISTVÁN SIMON
Intsitute of Enzymology, BRC
Hungarian Academy of Sciences
Budapest, Karolina, Hungary

6.1. INTRODUCTION

Lipid bilayers border all cells and eukaryotic cell compartments, forming a strong barrier against water-soluble materials. Therefore, they require special gates to enable the transport of such materials in a controlled way. These gates are formed by transmembrane proteins (TMPs), which play a critical role in living cells. They are involved in nutrient and metabolism transport, information flow, as well as energy production. Because of their vast functional roles, membrane proteins are important targets of pharmacological agents. According to a recent study, G-protein-coupled receptors (GPCRs), a subclass of TMPs, are the targets of approximately half of all drugs currently on the market [1]; among the 100 top-selling drugs, 25% are targeted at members of this protein family. Moreover, the various prediction tools show that about 20–25% of proteins coded in genomes sequenced so far are TMPs [2–5]. However, according to the Protein Data Bank of Transmembrane (PDBTM) database [6,7], despite the exponential growth of solved structures of TMPs, fewer than 2% of all structures of the PDB [8] form membrane-embedded TMP structures. Knowledge of three-dimentional (3D) structures of TMPs is essential for both understanding most of the cellular process and developing new drugs. Therefore, bioinformatics has emerged to help to bridge the information gap between the number of solved TMP structures and their sequences.

In this chapter, first we overview the structure of TMPs and discuss the relevant and/or current topology prediction methods from the viewpoint of

Introduction to Protein Structure Prediction: Methods and Algorithms,
Edited by Huzefa Rangwala and George Karypis

what we know about membrane protein structures and what we can learn about the various prediction methods. We also review the available structure and topology databases and training sets used by these prediction methods. In this review we focus our attention on helical TMPs, while topology prediction methods for β-barrel TMPs are reviewed only marginally. To learn more about β-barrel TMPs, readers should consult recent reviews [9,10].

6.2. STRUCTURE OF INTEGRAL MEMBRANE PROTEINS

6.2.1. Structural Elements of Integral Membrane Proteins

The lipid bilayer represents an environment from where water is expelled. Therefore, the polypeptide chain has to solve the problem of finding the best energetic solution to embed its polar amino and carbonyl groups into the bilayer, and shielding its hydrogen donor and acceptor groups. The solution is in the form of an α-helix or β-barrel structure, where all hydrogen donor and acceptor atoms are paired. The membrane-embedded parts of TMPs form an α-helical bundle nearly perpendicular or slightly tilted to the membrane plane or, rarely (mostly in bacterial porins), a β-barrel. In the central ±15 Å of the membrane (the origo is at the middle of the lipid bilayer) mostly hydrophobic amino acids can be found, and the secondary structure composition in this region is almost 100% regular element: α-helix or β-sheet [11,12].

The length of the secondary structure elements are ~20–25 and ~9–11 residues for α-helices and β-sheets, respectively, corresponding to the width of the apolar part of the membrane (~30 Å). The termini of the secondary structural elements do not necessarily coincide with the membrane water interface; sometimes these elements protrude from the membrane or turn back within the lipid bilayer by forming hairpins. As more high-resolution structures of helical membrane proteins become available, we learn that TM helices (TMHs) have a wider length distribution. If TMHs are tilted, the necessary length to cross the membrane is much larger than the length of TMHs parallel to the membrane normal. For example, very long TMHs (>35 residues) have been found in leucine transporter (LeuT), a prokaryotic homolog of mammalian neurotransmitter sodium symporters [13], where the tilt angles of TM3 and TM8 are ~50°C. Short helices usually are half helices, which are parts of the so-called reentrant loops. Reentrant loops are membrane-penetrating regions that enter and exit the membrane on the same side. According to an early analysis of these structural elements, these regions can be divided into three distinct categories based on secondary structure motifs, namely, long regions with a helix–coil–helix motif, regions of medium length with the structure helix–coil or coil–helix, and regions of short to medium length consisting entirely of irregular secondary structure [14]. By using a simple prediction method, it was shown that more than 10% of helical TMPs contain reentrant regions [14].

TMPs function as gates in the membrane, that is, they transport various solutes, ions, metabolites, and information across the membrane. To do this, their interior parts, which do not make contact with lipid side chains, are similar to the interior of globular proteins, which may contain structural elements stabilized by hydrophilic interactions. Thus, the surfaces of helices facing together sometimes contain polar side chains, prolines, or more complicated structures such as the above-mentioned membrane reentrant loops.

Proline kinks were first observed by von Heijne [15] by the investigation of the structure of the photosynthetic reaction center and of bacteriorhodopsin. It was shown that proline disrupts the TM α-helix and introduces an ~26°C kink. The possible function of prolines in surface expression, hormone binding, and cAMP induction was investigated, and it was shown that some of the membrane-embedded prolines play important roles in the function of lutropin/choriogonadotropin receptor [16]. By the analysis of the structure of "all α-fold" TMPs, the discovery of frequent non-α-helical components, such as 3^{10}-helices, π-helices, and intrahelical kinks (often due to residues other than proline), was also reported [17]. The possible evolutional history of proline kinks, and the explanation of why non-proline residues can be in proline kink, was drawn by structural analysis of GPCRs [18]. The latter analysis allows the authors to develop methods that can predict kink positions with >90% reliability.

The newly determined α-helical TMP structures revealed other structural elements mostly found in those parts of the protein that are located in the membrane–water interface region. The secondary structure of these parts of TMPs are irregular structures and interfacial helices running roughly parallel to the membrane surface, while β-strands are extremely rare [12]. In this region, hydrophobic and aromatic residues are abundant and tend to point toward the membrane, while charged/polar residues tend to point away from the membrane.

A surprising structure among β-barrels has been recently determined. The structures of β-barrels known so far are composed of an even number of β-sheets [19]. The smallest β-barrels are built up by eight sheets (e.g., OmpA, OmpW, OmpX, NspA, and PagP [20–24]), while the largest structures by 22 sheets (e.g., BtuB, FecA, FhuA, and FpvA [25–28]). However, the structure of the voltage-dependent anion channel (VDAC), which is the most abundant protein in the mitochondrial outer membrane (MOM), adopts a β-barrel architecture composed of 19 β-strands [29]. This new channel architecture is likely to be a consequence of the differences in membrane environment as well as in sorting signal and in partner proteins experienced by integral membrane proteins of the outer membrane of bacteria and those of the MOM and most likely adopted by other MOM proteins. Moreover, the finding of a 19-stranded β-barrel is in strong contrast to evolutionary theories, which predict bacterial β-barrels and related proteins to be formed by an ancient β-hairpin motif [30]. If this structure is a valid, β-barrel topology prediction methods need to be revisited.

6.2.2. Number of TM Folds

For globular-soluble proteins the total number of distinct globular folds that exist in nature is predicted to be a finite number [31], probably no more than 10,000 [32,33], regardless of the astronomical number of possible combinations of structural elements. In contrast, due to the physical constraints imposed by the lipid bilayer, most of the TMPs adhere to one principal topology, involving one or more α-helices arranged parallel to each other and oriented roughly perpendicular with respect to the membrane plane, as discussed above. The short loops between helices constrain the possible fold of TMHs, as well. Therefore, conformation space can effectively be sampled for small numbers of helices, and there are only about 30 possible folds for a TMP with three TMHs [34]. The number of combinatorially possible folds was shown to increase exponentially with the number of TMHs to 1.5 million folds for seven helices, but most probably, analogous to globular protein fold number, only a limited number of conceivable structures are actually realized in nature. The population of the distinct folds, similar to the globular proteins, is highly nonhomogeneous; the genome-wide topology predictions show that single-spanning TMPs are the most prevalent class [2]. The dependence of such fold population on the organism source was also reported [4]. We turn to the results of genome analysis later on this chapter.

6.2.3. Topology Definition and Determination

While experimental structure determination of globular protein by means of X-ray crystallography or by nuclear magnetic resonance (NMR) is becoming more routine and can be fully automated, we cannot nurse such hopes for TMPs because of the difficulties in crystallizing them in an aqueous environment and their relatively high molecular weight. Therefore, scientists have to settle for a lower resolution structure definition, the topology. Topology is defined by the number and sequential position of membrane-spanning segments and the localization of sequence segments between them relative to the membrane (cytosolic or extracytosolic). An even lower structure definition is the topography, where the number and sequential position of membrane-spanning segments are given, but the relative locations of segments between them are not defined.

Various biochemical and molecular biology techniques have been developed to get information about the topology of TMPs, including immunolocalization, molecular biology modifications of proteins, such as inserting/deleting glycosylation sites, and making fusion proteins [35]. In a recently launched database, called Topology Data Bank of Transmembrane Proteins (TOPDB) [36], a few thousands of such experiments together with topology of membrane proteins with known 3D structures have been collected for about 1500 TMPs. Genome-wide topology analysis by using C-terminal alkaline phosphatase and green fluorescent protein gene fusion combined with constrained topology prediction was reported on *Escherichia coli* inner membrane proteome [37]. With a similar technique, experimentally constrained topology models for 546 yeast proteins have been constructed [38].

6.2.4. Proteins with Ambiguous Orientation

Membrane proteins are expected to adopt only one topology in the membrane, but according to the positive inside rule [39], if there is no bias for the positively charged residues in the inside and outside loops, a TMP may be "frustrated,"that is, it cannot decide its orientation and adopts the so-called dual topology [40,41]. A dual-topology protein is defined as a single polypeptide chain that inserts into the membrane in two opposite orientations. The large-scale investigation of *E. coli* membrane proteins revealed that dual-topology proteins may exist naturally [42]. Five potential dual-topology proteins were identified with relatively few positively charged residues and little bias in the charge distribution on one side of the membrane or the other. These proteins were hypersensitive to the insertion of charged residues into the inter-membranous loops, as compared with similar proteins with strong topology determinants (e.g., with high positive charge bias). It was also shown that genes in families containing dual-topology candidates occur in genomes either as pairs or as singletons and that gene pairs encode two oppositely oriented proteins, whereas singletons encode dual-topology candidates.

The small multidrug transporter EmrE protein structure has also been reported as a dual-topology protein [43]. Although the article describing the structure of EmrE later has been retracted [44], recently the head to tail orientation in the homodimer oligomer structure has been supported by solving its 3D structure again [45]. However, in any debate about the controversial parallel or anti-parallel structure of EmrE [46–48], we should bear in mind that when solving the 3D structure of a membrane protein, removal from its native environment can have a significant impact on the structure. The presence of weak detergents, for example, can induce the formation of nonbiological oligomer structures with nonnative crystal contacts, as can be seen in the structure of rhodopsin [6,49]. Therefore, the native oligomer structure of a TMP may not be deduced from its oligomer structure found in the crystal.

Thus although there is no clear evidence that TMPs with dual topology exist, topology prediction methods should consider this possibility.

6.3. TOPOLOGY PREDICTION OF MEMBRANE PROTEINS

Topology prediction of all membrane proteins for an organism's proteome that is generating the topology of all TMPs in a genome from the nucleotide sequence may follow these steps: after gene prediction, the first task is to decide whether the given sequence codes a membrane or a water-soluble globular protein. The next step is the detection of the presence of a signal peptide. Removal of the signal sequences is followed by topography predictions, prediction of the localization of membrane-spanning segments within the amino acid sequence. Subsequently we need to decide the orientation of loops between TMHs relative to the membrane, which results in the full topology of the TMP.

In this section we overview the various algorithms and methods developed to solve these problems in the last couple of years, and we discuss how to handle the reentrant loops during the prediction and what we learned from the topology prediction methods about the TMPs. Finally, we overview the genome-wide topology prediction methods and their results on genomes sequenced so far. For discussion of older topology prediction methods we refer the reader to the reviews of von Heijne and Rost [50–53]. For those interested in structure prediction of TMPs, see Chapter 17 in this book.

6.3.1. Differentiation between Membrane and Globular Proteins

Prediction methods developed or tested for determining differences between globular and TMPs are listed in Table 6.1 (see G/TM column). It is generally considered that TM topology prediction methods can be used for discriminating TMPs and globular proteins, because the presence or absence of one or more predicted TM segments indicates the type of the protein. This is only partly true. We can tell only that the probability that a protein is a TMP increases with the number of predicted TM segments. Therefore, proteins predicted with two or more TM segments are usually regarded as TMPs, as it was applied in the preparation of global topology analysis of *E. coli* [37] and *Saccharomyces cerevisiae* [38] membrane proteome and is used in the PSORT-B algorithm to predict the subcellular location of proteins [54]. However, not all topology prediction methods were developed to differentiate between protein types. For example, the original Dense Alignment Surface (DAS) algorithm was developed to predict TMHs in prokaryotic TMPs [55], which was later upgraded to predict theTM character of the protein studied in the DAS-TMfilter method [56,57]. Another example is the algorithm of the hidden Markov model for topology prediction (HMMTOP) method [58], because the underlying statistical physics law, which is exploited in its algorithm (see below), is true only for TMPs. That is, strictly speaking, HMMTOP should be used only in the case of TMPs, as it is a topology prediction algorithm given that the query protein is a TMP. In contrast to HMMTOP, the various versions of transmembrane hidden Markov models (TMHMM) [4,59,60], Phobius [61–63], and Philius [64], which apply an HMM as a supervised machine-learning algorithm, can be used for this task.

A newly developed algorithm, called SVMtop [65,66], evaluates a hierarchical Support Vector Machine (SVM) algorithm and follows partly the same logic as described above. That is, SVM is used first to predict the sequential position of TMHs then to make a decision about the topology by a second SVM. This method classifies with a high level of the discrimination accuracy:

TABLE 6.1 Recent and/or Important Prediction Methods Used for TMP Structure Characterization

Name and Reference	Prediction						URL and Description
	G/TM	Signal	Topology	Reentrant	Homology	Genome	
DAS-TMfilter [56,57]	+					+	http://www.enzim.hu/DAS/DAS.html Dense Alignment Surface alignment method with a reversed prediction cycle
ENSEMBLE [93]	+		+		+	+	http://www.biocomp.unibo.it Ensemble of cascade-neural networks and two different hidden Markov models (HMMs)
HMMTOP2 [58,111]			+				http://www.enzim.hu/hmmtop Topology prediction methods using a five-state HMM with unsupervised learning
K4HTM [82]							Kurtosis-based hydrophobicity TM helix predictor that combines features of both hydropathy and higher order moments
MemBrain [83]							http://chou.med.harvard.edu/bioinf/MemBrain Optimized evidence-theoretic K-nearest neighbor prediction algorithm using multiple sequence information
MEMSAT3 [98]		+	+		+	+	http://bioinf.cs.ucl.ac.uk/memsat Combined neural network and model recognition approach
OCTOPUS [97]	+		+	+			http://octopus.cbr.su.se Combinations of ANN-predicted residue scores with an HMM-based global prediction algorithm. The biological language of reentrant loops is coded into the HMM architecture

TABLE 6.1 Continued

Name and Reference	Prediction						URL and Description
	G/TM	Signal	Topology	Reentrant	Homology	Genome	
Philius [64]	+	+	+		+	+	http://noble.gs.washington.edu/proj/philius Two-stage dynamic Bayesian networks decoder
Phobius [61–63]	+	+	+		+	+	http://phobius.cgb.ki.se Combined signal peptide and topology prediction method using HMM and homology information
SCAMPI [84]			+				http://topcons.cbr.su.se A simple HMM using only two parameters and the experimental scale of position-specific amino acid contributions to the free energy of membrane insertion
SPOCTOPUS [75]	+	+	+	+	+		http://octopus.cbr.su.se Combined signal peptide and topology predictor using ANN and HMM
SVMtop [65,66]			+				http://biocluster.iis.sinica.edu.tw/~bioapp/SVMtop Hierarchical SVM methods, discriminating TM, and non-TM segments, then non-TM segments into in or out by SVM
TMHMM2 [4]	+		+			+	http://www.cbs.dtu.dk/services/TMHMM Topology prediction methods using a seven-state HMM with supervised learning

TMLOOP [109]			+		http://membraneproteins.swan.ac.uk/TMLOOP Combinatorial pattern discovery approach, which used the discovered patterns as weighted predictive rules in a collective motif method
TmPro [67,68]					http://linzer.blm.cs.cmu.edu/tmpro Window-based method, which apply latent semantics analyses of amino acid properties in the sequence window, and uses only 25 free parameters
TOPMOD [14]			+	+	HMM-based method to classify the residues of a TMP sequence into four structural classes: Membrane, Reentrant, Interface, and Loop
TopPred$^{\Delta G}$ [84]	+	+		+	http://topcons.cbr.su.se Hydrophathy plot analysis using experimental scale of position-specific amino acid contributions to the free energy of membrane insertion
Zpred2 [149]		+	+		http://topcons.cbr.su.se/ Z coordinates prediction via a combined ANN/HMM method (distance of the residues from the center of the membrane)

G/TM: Globular/TMP discrimination; Signal: Signal peptide prediction; Topology: Topology prediction; Reentrant: Reentrant loop prediction; Homology: Homology sequence information used for prediction; Genome: Genome-wide prediction.

it predicts only 0.5% of globular proteins as TMP (false positive rate), and 1.2% of TMPs as globular proteins (false negative rate). A comparison of the discrimination accuracy of the current methods can also be found in the article of Lo et al. [66]. TMpro method, by using latent semantic analysis together with HMM and Artificial Neural Network (ANN), reaches much lower level of accuracy, but it uses only 25 free parameters as the results of the latent semantic analysis [67,68]. It was also shown that combination of the prediction methods increases the specificity and sensitivity of discrimination, in the case of the examination of the possible TM origin of the prion protein [69].

6.3.2. Signal Peptide Predictions

The second step in our framework to predict topologies genome-wide is the prediction of signal peptides. Signal peptides control the proper targeting of virtually all proteins to the secretory pathway. They are located at the N-terminal of the proteins and contain a hydrophobic region, which is very similar to the TMHs both in length and in amino acid composition. The signal peptides are cleaved off while the protein is translocated through the membrane; however, uncleaved signal peptides are also known. In the case of TMPs after translocating and cleaving off the signal peptides the N-terminals of the TMPs get outside the membrane into the extracytosolic space. Cleavable signal peptides can be identified by simple statistical means [70,71] or modern machine-learning approaches such as ANN, HMM, or Self-Organizing Map (SOM) with high sensitivity (95–98%) and specificity (93–98%) [72–74]. Topology prediction methods, which were developed to predict signal peptides as well, are listed in Table 6.1 (Signal column). The presence of the signal peptides on TMPs indicates the location of their N-terminus; therefore, combining signal peptide prediction and TM topology prediction reduces both false prediction of signal peptides and false prediction of TMHs [61,64,75].

Most topology predictions were reported to suffer from mis-prediction of signal peptides, due to the similar physicochemical properties of signal peptides and TMHs. The effects on the accuracy of removing of signal peptides before or after TMH prediction were also investigated [76,77]. Note, however, that signal peptide prediction and topology prediction are two different tasks. Signal peptide prediction can be regarded as the part of the sequence processing in the gene annotation project, where the aim is to predict the sequence of a mature polypeptide chain from the nucleotide sequence from the genome. After we have successfully predicted these sequences we can predict the topology of TMPs. The difference between these tasks is obvious, if we think of the different natural task of the signal peptides and TMHs, targeting proteins into the proper cellular environment and folding the polypeptide chain into the double-lipid environment, respectively. Some authors recommend that signal peptides should be removed from the amino acid sequences before the topology prediction. This was exactly the case with HMMTOP, where, as we can see later, the algorithm can be applied only for the mature sequence of TMPs [58].

6.3.3. Topography Predictions

The next step toward prediction of the topology of a TMP is to predict the sequential localization of membrane-spanning segments. While the TM segments in the case of helical TMPs are formed by hydrophobic amino acids, in case of β-barrel TMPs every second amino acid has to be hydrophobic. Moreover, the length of the membrane-spanning segments is shorter in the latter case. These two properties of β-barrels make their prediction harder than the prediction of the topography of helical TMPs. We focus on helical TM segment prediction in the following.

Earlier topography prediction methods explored the fact that membrane-spanning segment is more hydrophobic than the other parts of the protein chains. The prediction scheme contains the following steps: choose a hydrophobicity scale of the 20 amino acids (or other propensity scale), generate a plot of these values by averaging them within a sliding window over the query sequence, and identify peaks on the plot above a predefined threshold. Numerous hydrophobic scales of the amino acids have been reported to reach better prediction accuracy. For reviews of hydrophobic scales, their determination, and their application to predict topography, see References [78–80].

The so-called DAS algorithm overcomes the difficulties caused by the different hydrophobicity scales, which leads to different predictions. It was shown that in a special alignment procedure, unrelated TMPs can recognize each other without applying any hydrophobicity scales [81]. The DAS algorithm was shown to perform especially well as a topography prediction method for prokaryotic helical proteins [55]. The upgraded form of DAS, the DAS-TMfilter algorithm, is able to filter out false positive TMH predictions; therefore, it can make a distinction between proteins with and without TMHs at a reasonably low rate of false positive predictions [56,57].

Some recently published methods have also been capable of predicting topography but not topology. These methods employ new techniques and algorithms used in other areas of science, such as latent semantic analysis [67,68], higher order statistics [82], or evidence-theoretic K-nearest neighbor prediction algorithm [83].

Because the performance of the first methods based solely on the hydrophobicity plots was rather poor, novel statistical, machine-learning methods were substituted, which use hundreds of free parameters trained on the experimentally established topology information. However, after Bernsel et al. [84], the translocons responsible for membrane-protein biogenesis do not have access to statistical data but rather exploit molecular interactions to ensure that membrane proteins attain their correct topology [85,86]. Hence, those methods that are based on the same physical properties that determine translocon-mediated membrane insertion, by using properly scaled hydrophobicity values, may yield the same level of prediction accuracy as the best statistical methods. Indeed, prediction accuracy of a simple additive free-energy model derived from an experimental dataset describing the insertion of TMHs into the endoplasmic reticulum membrane in terms of free-energy contributions from individual amino acids in different positions along the membrane normal [87,88] reaches the same level as the current state-of-the-art prediction methods [84].

6.3.4. Topology Predictions

After determining the sequential localization of membrane-spanning segments, the next step for structure determination of TMPs is to orient the membrane-spanning segments from outside to inside or from the inside to outside. This is equivalent to localizing the sequence segments between membrane-spanning segments alternatively inside or outside. However, there are only a few properties of TMPs to help this task. The first and most prevalent such feature of TMPs is the so-called positive-inside rule. It was shown that positively charged amino acids close to membrane-spanning segments are more abundant in the cytosolic sequence segments than in the extracytosolic ones [39]. This rule is commonly used, from bacteria to humans [89]. Most topology predictions apply the positive-inside rule after the topography prediction to choose the more likely model from the two possible ones based on the larger differences in positively charged amino acids within intramembranous segments close to the membrane. In the first applications, where the positive inside rule is utilized [90], various models with certain and possible TM segments were generated, and the predictions were made by choosing the model in which the differences of the number of lysines and arginines was the highest between the even and odd loops. This rule is also utilized in MEMSAT method indirectly, because this method maximizes the sum of log likelihood of amino acids preferences taken from various structural parts of membrane proteins in a model recognition approach [91].

The success of MEMSAT and the various HMM-based methods [4,59–63,92–96] in topology prediction is due to the fact that the amino acid compositions of the various structural parts of TMPs are specific and the machine-learning algorithms are able to learn these compositions during supervised learning. Sequences in various compartments, such as the TM segments, the inside sequence parts close to the membrane, and the extracytosolic space, have characteristic amino acid composition. Although the novel machine-learning methods report higher and higher prediction accuracies due to the continuously growing and more reliable training sets (see later), and because of combining the various techniques (e.g., using SVM or ANN for residue prediction and HMM for segment identification [97–99]), we cannot learn from these methods about the topology-forming rules of TMPs. Moreover, to predict the topology of novel TMPs, which were not seen earlier by the machine-learning methods, these methods may need to be retrained.

One HMM method, namely HMMTOP, is based on a principle different from the other HMM-based algorithms, as described in the following. A polypeptide chain of a TMP goes through various spaces of a cell with different physicochemical properties (hydrophobic, polar, negatively charged, water-lipid interface, etc). The preference of amino acids for these spaces is different. This is why amino acid compositions of the various structural parts are characteristic and why the amino acid relative frequencies of the various structural parts of a TMP differ. However, we do not need to know and therefore do not need to learn these characteristic amino acid compositions to predict the topology of TMPs. Simply the fact that the polypeptide chain goes through various cell spaces with different properties results in the amino acid composition of the structural parts corresponding to these spaces showing maximal divergence. Therefore, topology prediction methods based on this finding should have to partition an amino acid sequence, such that the amino acid relative frequencies corresponding to the various structural spaces (inside, membrane, and outside) show maximal divergence. This partitioning can be found by HMM. Here, we have to refer to the original tutorial of Rabiner about HMM [100], where to solve the third problem of HMM (i.e., optimizing the model parameters to maximize the probability of the observation symbol sequence given the model) does not require the knowledge of emission probabilities, that is, the knowledge of the probability distribution of amino acids corresponding to the various structural parts (the distribution of balls in the urns in the tutorial). Therefore, topology prediction of a TMP is equivalent to solving the third problem of HMM, optimizing the model parameters (emission, transition state, and initial state probabilities) on every observation sequence. Therefore, HMMTOP does not require training; it predicts topology in an unsupervised manner. This is the reason why HMMTOP appears on various tests in the top [50,52,101,102], and predicts topology with high success rates for proteins, whose structures have never been seen before. It may also explain why it can increase the accuracy of the TMHMM method [4,92] (another HMM-based method developed simultaneously but independently from HMMTOP) when it was used in unsupervised mode (e.g., on each query sequence the Baum-Welch algorithms was applied) [59].

Because the prediction accuracies of the various prediction methods are not perfect (100%), measuring their reliability is an important issue. Two of the five developed reliability scores worked well on a test of Melen et al. [103], which allow the probability that the predicted topology is correct to be estimated for any proteins. Using consensus predictions of membrane protein topology not only increases the prediction accuracy by ~10% [101,104–106], but it also provides means to estimate the reliability of a predicted topology [107]. By using five topology prediction methods, it was found that the topology of nearly half of the test sequences can be predicted with high reliability (>90% correct predictions) by a simple majority-vote approach. Consensus approach was also applied to predict partial membrane topologies, that is, the part of the sequence where the majority of applied methods agree [108]. However, the sequential coverage of partial consensus topologies was quite low in the test set, 44% and 17% for prokaryotic and eukaryotic proteins, respectively. Similarly to consensus approaches, ensemble methods, that is, those that use various parallel trained methods, increase prediction accuracy by about 10% [93].

There are many measures used to calculate the prediction accuracies of methods. The most commonly used values are the "per segment" and "per protein" accuracies. Per segment accuracy measures how many membrane-spanning segments are correctly predicted relative to all TMHs. Usually, predicted membrane-spanning segments are accepted as correct if they overlap with the experimentally established TMHs to some extent. Topography of a protein is predicted as correct if all of its TMHs are predicted correctly. Topology of a protein is correctly predicted if, in addition, the in and out orientation of sequence segments between TMHs is correctly predicted. Note that the orientation of the N-terminus shows correct topology prediction if and only if the topography prediction is correct. Therefore, to measure the N-terminus orientation prediction accuracies in itself is pointless. Also, because many TMPs have large globular parts, the per residue accuracy (Q_2) is not so informative. Since the exact sequential position of TMHs in the training set is uncertain, even if the 3D structures are known, measuring the so-called helix-end accuracy is meaningless as well.

6.3.5. Reentrant Loop Predictions

Two types of reentrant loops contain medium-sized membrane helices (see above). Because these structural elements enter and exit on the same side of the membrane, and the size of the helices is comparable to the size of TMHs or is in the allowed region of the various prediction methods, TM topology prediction methods often predict them as TMHs. Not only does this result in a wrong prediction on the reentrant loop regions, but this error also influences the overall topology. Therefore, TM topology prediction methods that are not able to predict reentrant loops cannot predict per se correct topology for proteins having reentrant loops. Because more than 10% of TMPs may contain reentrant regions [14], this prediction method error greatly affects overall prediction accuracy.

There are basically three types of reentrant loop prediction methods. Lasso et al. developed an application [109] based on recognition of patterns and motifs extracted from known membrane dipping loop containing proteins in the Swiss-Prot database [110]. The characteristic amino acid composition of reentrant loops was analyzed with the TOP-MOD method [14], but its performance was not actually very high, therefore the authors of TOP-MOD developed another HMM-based prediction method, called OCTOPUS [97], by modification of the common architecture of HMM. Reentrant loop prediction methods are listed in Table 16.1 (Reentrant column).

6.3.6. Constrained Predictions

Constrained prediction methods mean that some parts of the query sequence are locked to a predefined structural part during the prediction. Thus, given a constraint that, for example, the N-terminus is inside, the constrained prediction method gives a prediction with the N-terminus being inside. This is achieved by the modification of the Baum-Welch or Viterbi algorithm of the HMM-based methods. The first such application was HMMTOP2 [111]. Later, two other HMM-based methods, TMHMM and Phobius, were also modified to include this feature [63,103,112,113]. The mathematical details of the necessary modification can be found in Reference [114]. Note that constrained prediction is not equivalent to filtering various prediction results by some conditions.

Obviously, constrained prediction increases the accuracy and reliability, as was shown in the case of the human multidrug resistance-associated protein (MRP1) [111]. The optimal placement of constraints was also investigated, and it was shown that the accuracy can be increased by 10% if the N- or C-terminal of the polypeptide chain is locked, and by 20% if one of each loop or tail residue in turn is fixed to its experimentally annotated location [115].

Constraints can be generated from different sources. Experimental results are the most commonly used constraints. In the topology analyses of *E. coli* and *S. cerevisiae*, the results of C-terminal fusion proteins were applied as constrains [37,38,115–117]. In a recently launched database, called TOPDB, more than 4500 experimental results were collected for ~1500 TMPs, and these constraints were applied for making constrained topology predictions for these ~1500 TMPs. The experimental results were applied for homologous protein sequences as well [118]. Locations of compartment-specific domains were also used as constraints. Such domains can be collected from various domain databases such as SMART [112] or Pfam [119]. Moreover, specific sequence motifs and fingerprints located conservatively on one side of TMPs were collected for this purpose into a database called TOPDOM [120]. See Table 6.2 for a list of databases.

6.3.7. Genome-Wide Topology Predictions

The number of genomic sequences is exponentially growing, generating a huge amount of data for bioinformatics. Therefore, the genome-wide identification and characterization of TMPs requires fast, efficient, and accurate prediction methods. While the early topology prediction methods based on hydropathy plot analysis were not able to predict TMPs genome-wide due to their high false positive prediction rate, several methods can now identify TMPs with better than 95% sensitivity and specificity [2–5,60,121–123]. These methods are based on statistical analysis of TMPs with experimentally established topology and on the use of machine-learning algorithms. Although the reliability of these methods measured on test sets is high, it was shown that the commonly used test sets are biased toward TMPs, whose topologies can be predicted more reliably as compared with the genomic sequences. Consequently, the prediction reliability of the genome-wide analysis should be lower than believed [103,122]. The reliability can be increased by using consensus prediction methods, and a consensus whole-genome analysis reveals that reliable

TABLE 6.2 Topology and Structure Databases of TMPs

Name	URL and Description
LocaloDom [119]	http://localodom.kobic.re.kr/LocaloDom/add_supple/LocaloDom_20.htm Localo-orientations of domains is a database that provides information about topology of Pfam domains
OPM [132,133]	http://opm.phar.umich.edu A collection of TM, monotopic, and peripherial proteins from PDB with calculated spatial arrangements in the lipid bilayer
TCDB [150]	http://www.tcdb.org Transporter classification database containing information for ~3000 hierarchical classified transporters compiled from >10,000 references
TMPDB [144]	http://bioinfo.si.hirosaki-u.ac.jp/~TMPDB Collection of experimentally characterized transmembrane topologies
TOPDB [36]	http://topdb.enzim.hu Collection of TMPs with known 3D structures and with experimentally established topology information
TOPDOM [120]	http://topdom.enzim.hu Database of domains and motifs with conservative location in transmembrane proteins

partial consensus topology can be predicted for ~70% of all membrane proteins in a typical bacterial genome and for ~55% of all membrane proteins in a typical eukaryotic genome [108].

Genome-wide prediction methods agree that about 20–30% of coded proteins are TMPs, with at least one TMH. An early study based on the analysis of 14 genomes found that larger genomes contain more TMPs than smaller ones [124], which led to the hypothesis that the complexity of multicellular organisms is proportional to the number of TMPs in its genome. The validity of this correlation should be checked with more recent and accurate methods and a much larger number of available genomic sequences, because others fail to show this observation [4,121]. By using SOSUI methods it was observed that distribution of TMHs in the predicted TMPs follows a geometric distribution [5]; that is, proteins with few TMHs are more frequent than proteins with many TMHs. Although this general trend may be valid for almost all organisms, several exceptions were shown using improved topology prediction methods. In particular proteins with seven TMHs seem to be more frequent in higher eukaryotes, mostly due to the expansion of GPCRs, whereas bacterial genomes encode large numbers of small-molecule transporters with 6 or 12 TMHs [118]. A strong bias toward an even number of TMHs with N-in, C-in topology was also observed in both bacteria and eukaryotes [37,38,118].

Although general trends on the number of TMHs can be reliably observed, as mentioned above, the overall predicted topologies by the various methods disagree strongly. Constrained Philius topology predictions were made on the same data as TMHMM and PRODIV-TMHMM, and it was reported that the Philius-predicted topology matched both of them for only 41% of proteins, whereas the constrained predictions from TMHMM and PRODIV-TMHMM match each other for 48% on the yeast TM proteome [64].

6.4. DATABASES AND BENCHMARK SETS

Collecting information from TMPs is an essential task not just because these data are useful for the molecular biologist for planning experiments and to characterize TMPs, but also because this information is used for testing topology prediction methods. Because almost all of these methods evaluate one of the numerous machine-learning algorithms, it is essential to determine what these methods learn. A learned, erroneous topology dataset may lead to false predictions. The high fluctuation of the prediction accuracies of the various methods by using different datasets can be seen from the article comparing them. In the last part of this chapter we discuss how and from where the topology data of TMPs can be collected, the various resources available on the Internet, and the benchmark sets used by the prediction methods.

6.4.1. 3D Structure Databases

The most reliable topology data of a TMP is its 3D structure. However, due to the difficulties in the crystallization of TMPs, these are highly underrepresented in the structure database; that is, fewer than 2% of solved 3D structures belong to TMPs [6,7]. Not only is the structure determination of these proteins a hard task, but their identification from the PDB [8] is problematic, because of the annotation errors and/or inconsistency of PDB entries. There are two possible ways to select TMPs from PDB. The first is by using human experts, who continuously check the deposited structures and/or the literature to collect structure of TMPs. The second one is by applying automatic software developed for this purpose, which does not use PDB's annotations, only the atomic coordinates of proteins.

There are several online available lists of TMPs with known 3D structures: the list created by H. Michel, which includes crystallization conditions and key references for the structure determinations [125]; the Molecular Probe Database (MPDB), which is an online, searchable, relational database of structural and functional information on integral, anchored, and peripheral membrane proteins and peptides [126]; the list maintained by S. White [127] containing structures determined by diffraction methods; and the list of TMPs, whose structures have been determined by NMR, collected by D. Warschawski [128]. Because these lists are created and/or maintained by human experts, they are sometimes outdated and/or biased toward specific proteins that are a focus of interest. But a more problematic issue is that the topology of TMPs cannot be gained from these lists. This is because one vital component, the membrane itself, is missing from the structure files deposited to PDB. For structure determination TMPs are taken out from the lipid bilayer, and crystallized by masking their exposed hydrophobic parts by amphiphilic detergents, so that the protein–detergent complex can be treated similarly to soluble proteins [129]. The detergent molecules are highly unstructured and are usually not visible in the X-ray picture. With the exception of a few tightly bound lipid or detergent molecules, the deposited experimental data have no direct indication that the protein is immersed into the membrane under native conditions, and do not contain information about the exact location of the lipid bilayer [130].

So far, three databases and/or methods were launched that cope with this problem, by putting the TMPs back into their native environment, into the membrane. The first such method is called TMDET [6], which uses a weighted sum of normalized lipid accessible surface area and the relative frequencies of regular secondary elements to discriminate TMPs and globular proteins and to determine the arrangement of a protein with respect to the membrane. The TMDET algorithm was applied to scan the entire PDB database to collect all TMPs with known 3D structures [7], and the algorithm is available for the public on a web server [131], to predict the membrane orientation of model structures, for example, those obtained by homology modeling. The advantage of this automatic method is that it can be used continuously week after week to update PDBTM database after the weekly update of PDB.

Another computational approach to determine the spatial arrangement of proteins in membranes has been developed by minimizing TMP's transfer energies from water to the lipid bilayer [132]. These methods were applied to representative TMPs from the PDB database, generating the OPM database [133]. The applicability and accuracy of this method was verified for a set of 24 TMPs whose orientations in membranes have been studied by spin-labeling, chemical modification, fluorescence, Attenuated Total Reflection Fourier Transform Infrared (ATR FTIR), NMR, cryo-microscopy, and neutron diffraction, and errors of ~1 Å and 2°C were reported for the calculated hydrophobic thicknesses and tilt angles of TMHs, respectively. Recently, a semi-quantitative lipid simulation-based model was adapted for simulations of membrane proteins [134], in particular to model the insertion of proteins into lipid bilayers or detergent micelles. This coarse-grain method was applied as a high-throughput approach to the prediction of membrane protein interactions with a lipid bilayer [135].

6.4.2. Databases of Experimental Established Topologies

There are numerous experimental techniques that can be used to obtain information on the topology of TMPs. Modifying the amino acid sequences, by exploring the advantage of molecular biology techniques, allowed the insertion of a given tag into a predefined position in a given sequence. By determining the accessibility of the inserted tag, the position of the insertion point relative to the membrane can be estimated. Insertion tags can be placed on N-glycolysation sites, Cys residues, antibody epitopes, proteolytic sites, or even a full reporter protein. The latter case is similar to the situation when the reporter enzyme is fused to the investigated protein in a certain position, and the activity of the reporter enzyme in the construction shows the in/out location of the insertion/fusion point. The reporters are typical molecules whose properties depend on their subcellular location. The most commonly used reporter enzymes for gene fusion studies in prokaryotes are alkaline phosphatase (encoded by the *E. coli PhoA* gene) [136], β-galactosidase (*LacZ* gene) [137], and β-lactamase (*bla* gene) [138]; while in eukaryotes HA/Suc2/His4C chimeric reporter or green fluorescent protein (GFP) is used for this purpose [38,116,139]. Readers who are interested in these methods should refer to the review of van Geest and Lolkema [35].

In the late 1990s scientists working on new prediction methods collected experimentally established topology data from the literature to check the prediction accuracy of their methods [55,91,140,141]. The first well characterized dataset of TMPs—containing 320 records—was collected by Möller et al. [142]; however, a large part of the collected data (about one-third) was based on the analysis of hydropathy plots and not on experiments. The authors faced the problem that the interpretation of individual experiments was sometimes difficult and the TM annotation was provided by human experts considering the results of the hydropathy plot analysis and experiments. The Membrane Protein Topology (MPTopo) database [143], a database of TMPs whose topologies have been verified experimentally by means of crystallography, gene fusion, and other methods, was compiled by the White lab. This database is not updated regularly; the last update took place several years ago. Another collection is the TMPDB dataset [144], containing 302 TMPs, with experimentally established topologies. This dataset includes topology data of 17 β-barrel TMPs as well. Although the references to PubMed are given for each entry, the experimental details and the method of data processing are neither included in the database nor described in the article. While the authors of TMPDB and Möller's datasets planned to maintain their collections and include updates, the Möller dataset has never been updated, while TMPDB was updated only once, in 2003, but without adding any new entries. Recently, a new database, called TOPDB [36], has been launched by collecting experimental data and 3D data of TMPs in a unified form and exactly describing the experimental details, for example, the exact position of insertion/fusion points with the activity of the reporter enzymes. To generate topology of each entry in the database, constrained predictions with the collected experimental results have been made using the HMMTOP2 algorithm [58,111].

The most common error of using these so-called low resolution datasets is that the entire topology is accepted as is; however, in many cases only a few points of the sequence are proved to be on one or on the other side, and the "experimentally established" topology for the other sequence positions are just predictions. This is why the Rost lab found the "prediction methods not significantly less accurate than low resolution experiments," because the low resolution experiments are generated mainly by predictions!

6.4.3. Benchmark Sets

Benchmark sets differ from the previous databases in that they do not contain homologous protein sequences, because their existence in the training and test set biases the prediction accuracy. The usually used sequence similarity cutoff value is 30%, but because scant reliable topology information is available, the used cutoff value can be higher. There are various commonly used benchmark sets and static servers, for example, the so-called 160 protein set [4], Möller's set [142], the nonredundant set of TMPDB [144], and the static benchmark server built in the Rost's lab [145]. Sometimes, the Swiss-Prot annotations are used (see, e.g., Reference [146]); however, as they are based on homology only, they cannot be regarded as experimentally established topology data. The benchmark sets seem to be biased in some ways. As it was shown, proteins whose topology can be predicted more reliably are more abundant in benchmark sets than they are found in the whole proteome [103]. Databases for signal peptide prediction were also collected [72,74,147].

6.5. CONCLUSIONS

It is commonly accepted that structure prediction of TMPs appears to be easier than that of water-soluble globular proteins. The available topology data and the number of known structures are scarce, and although their numbers have been growing exponentially, only 2025 TMP structures are predicted to be determined in 20 years [148]. Therefore, we do not forecast the prediction accuracies of topology prediction methods concerning TMPs to be solved in the future. It appears that the accuracies of the majority of the methods are decreasing with time. Moreover, the recently solved structures show that TMPs are more diverse than believed so far, lowering the accuracies even more.

The currently developed topology prediction methods use the same techniques as those of 10 years ago, and their accuracies are far from perfect. As techniques have changed from hydropathy plot analysis to machine-learning algorithms, significantly increasing prediction accuracies, new algorithms and/or new findings about topologies may increase the topology prediction accuracy to the level of perfection. When these new methods arrive, we should collect topology data and extend our knowledge on the structure and functions of TMPs.

REFERENCES

1. T. Klabunde and G. Hessler. Drug design strategies for targeting G-protein-coupled receptors. *Chembiochem*, 3:928–944, 2002.
2. M. Ahram, Z.I. Litou, R. Fang, and G. Al-Tawallbeh. Estimation of membrane proteins in the human proteome. *In Silico Biology*, 6:379–386, 2006.
3. D.T. Jones. Do transmembrane protein superfolds exist?. *FEBS Letters*, 423:281–285, 1998.
4. A. Krogh, B. Larsson, G. von Heijne, and E.L. Sonnhammer. Predicting transmembrane protein topology with a hidden Markov model: Application to complete genomes. *Journal of Molecular Biology*, 305:567–580, 2001.
5. S. Mitaku, M. Ono, T. Hirokawa, S. Boon-Chieng, and M. Sonoyama. Proportion of membrane proteins in proteomes of 15 single-cell organisms analyzed by the SOSUI prediction system. *Biophysical Chemistry*, 82:165–171, 1999.
6. G.E. Tusnády, Z. Dosztányi, and I. Simon. Transmembrane proteins in the protein data bank: Identification and classification. *Bioinformatics*, 20:2964–2972, 2004.
7. G.E. Tusnády, Z. Dosztányi, and I. Simon. PDB_TM: Selection and membrane localization of transmembrane proteins in the protein data bank. *Nucleic Acids Research*, 33:D275–278, 2005.
8. H.M. Berman, J. Westbrook, Z. Feng et al. The protein data bank. *Nucleic Acids Research*, 28:235–242, 2000.
9. M.M. Gromiha and M. Suwa. Current developments on beta-barrel membrane proteins: Sequence and structure analysis, discrimination and prediction. *Current Protein & Peptide Science*, 8:580–599, 2007.
10. P.G. Bagos, T.D. Liakopoulos, and S.J. Hamodrakas. Evaluation of methods for predicting the topology of beta-barrel outer membrane proteins and a consensus prediction method. *BMC Bioinformatics*, 6:7, 2005.
11. E. Wallin, T. Tsukihara, S. Yoshikawa, G. von Heijne, and A. Elofsson. Architecture of helix bundle membrane proteins: An analysis of cytochrome c oxidase from bovine mitochondria. *Protein Science*, 6:808–815, 1997.
12. E. Granseth, G. von Heijne, and A. Elofsson. A study of the membrane-water interface region of membrane proteins. *Journal of Molecular Biology*, 346:377–385, 2005.
13. S.K. Singh, A. Yamashita, and E. Gouaux. Antidepressant binding site in a bacterial homologue of neurotransmitter transporters. *Nature*, 448:952–956, 2007.
14. H. Viklund, E. Granseth, and A. Elofsson. Structural classification and prediction of reentrant regions in alpha-helical transmembrane proteins: Application to complete genomes. *Journal of Molecular Biology*, 361:591–603, 2006.
15. G. von Heijne. Proline kinks in transmembrane alpha-helices. *Journal of Molecular Biology*, 218:499–503, 1991.
16. S. Hong, K.S. Ryu, M.S. Oh, I. Ji, and T.H. Ji. Roles of transmembrane prolines and proline-induced kinks of the lutropin/choriogonadotropin receptor. *Journal of Biological Chemistry*, 272:4166–4171, 1997.
17. R.P. Riek, I. Rigoutsos, J. Novotny, and R.M. Graham. Non-alpha-helical elements modulate polytopic membrane protein architecture. *Journal of Molecular Biology*, 306:349–362, 2001.

18. S. Yohannan, S. Faham, D. Yang, J.P. Whitelegge, and J.U. Bowie. The evolution of transmembrane helix kinks and the structural diversity of G protein-coupled receptors. *Proceedings of the National Academy of Science U S A*, 101:959–963, 2004.
19. G.E. Schulz. Beta-barrel membrane proteins. *Current Opinion Structural Biology*, 10:443–447, 2000.
20. A. Pautsch and G.E. Schulz. Structure of the outer membrane protein a transmembrane domain. *Nature Structural & molecular Biology*, 5:1013–1017, 1998.
21. H. Hong, D.R. Patel, L.K. Tamm, and B. van den Berg. The outer membrane protein OmpW forms an eight-stranded beta-barrel with a hydrophobic channel. *Journal of Biological Chemistry*, 281:7568–7577, 2006.
22. J. Vogt and G.E. Schulz. The structure of the outer membrane protein OmpX from Escherichia coli reveals possible mechanisms of virulence. *Structure*, 7:1301–1309, 1999.
23. L. Vandeputte-Rutten, M.P. Bos, J. Tommassen, and P. Gros. Crystal structure of Neisserial surface protein A (NspA), a conserved outer membrane protein with vaccine potential. *Journal of Biological Chemistry*, 278:24825–24830, 2003.
24. P.M. Hwang, W. Choy, E.I. Lo et al. Solution structure and dynamics of the outer membrane enzyme PagP by NMR. *Proceedings of the National Academy of Science U S A*, 99:13560-13565, 2002.
25. D.D. Shultis, M.D. Purdy, C.N. Banchs, and M.C. Wiener. Outer membrane active transport: Structure of the BtuB:TonB complex. *Science*, 312:1396–1399. 2006.
26. A.D. Ferguson, R. Chakraborty, B.S. Smith et al. Structural basis of gating by the outer membrane transporter FecA. *Science*, 295:1715–1719, 2002.
27. A.D. Ferguson, E. Hofmann, J.W. Coulton, K. Diederichs, and W. Welte. Siderophore-mediated iron transport: Crystal structure of FhuA with bound lipopolysaccharide. *Science*, 282:2215–2220, 1998.
28. D. Cobessi, H. Celia, N. Folschweiller et al. The crystal structure of the pyoverdine outer membrane receptor FpvA from Pseudomonas aeruginosa at 3.6 angstroms resolution. *Journal of Molecular Biology*, 347:121–134, 2005.
29. M. Bayrhuber, T. Meins, M. Habeck et al. Structure of the human voltage-dependent anion channel. *Proceedings of the National Academy Science U S A*, 105:15370–15375, 2008.
30. T. Arnold, M. Poynor, S. Nussberger, A.N. Lupas, and D. Linke. Gene duplication of the eight-stranded beta-barrel OmpX produces a functional pore: A scenario for the evolution of transmembrane beta-barrels. *Journal Molecular Biology*, 366:1174–1184, 2007.
31. C. Chothia. Proteins. One thousand families for the molecular biologist. *Nature*, 357:543–544, 1992.
32. Y.I. Wolf, N.V. Grishin, and E.V. Koonin. Estimating the number of protein folds and families from complete genome data. *Journal of Molecular Biology*, 299:897–905, 2000.
33. E.V. Koonin, Y.I. Wolf, and G.P. Karev. The structure of the protein universe and genome evolution. *Nature*, 420:218–223, 2002.
34. J.U. Bowie. Helix-bundle membrane protein fold templates. *Protein Science*, 8:2711–2719, 1999.

35. M. van Geest and J.S. Lolkema. Membrane topology and insertion of membrane proteins: Search for topogenic signals. *Microbiology and Molecular Biology Reviews*, 64:13–33, 2000.

36. G.E. Tusnády, L. Kalmár, and I. Simon. TOPDB: Topology data bank of transmembrane proteins. *Nucleic Acids Research*, 36:D234–239, 2008.

37. D.O. Daley, M. Rapp, E. Granseth et al. Global topology analysis of the Escherichia coli inner membrane proteome. *Science*, 308:1321–1323, 2005.

38. H. Kim, K. Melén, M. Osterberg, and G. von Heijne. A global topology map of the Saccharomyces cerevisiae membrane proteome. *Proceedings of the National Academy of Science U S A*, 103:11142–11147, 2006.

39. G. von Heijne. The distribution of positively charged residues in bacterial inner membrane proteins correlates with the trans-membrane topology. *EMBO Journal*, 5:3021–3027, 1986.

40. I. Nilsson and G. von Heijne. Fine-tuning the topology of a polytopic membrane protein: role of positively and negatively charged amino acids. *Cell*, 62:1135–1141, 1990.

41. G. Gafvelin and G. von Heijne. Topological "frustration" in multispanning E. Coli inner membrane proteins. *Cell*, 77:401–412, 1994.

42. M. Rapp, E. Granseth, S. Seppälä, and G. von Heijne. Identification and evolution of dual-topology membrane proteins. *Nature Structural & Molecular Biology*, 13:112–116, 2006.

43. O. Pornillos, Y. Chen, A.P. Chen, and G. Chang. X-ray structure of the EmrE multidrug transporter in complex with a substrate. *Science*, 310:1950–1953, 2005.

44. G. Chang, C.B. Roth, C.L. Reyes et al. Retraction. *Science*, 314:1875, 2006.

45. Y. Chen, O. Pornillos, S. Lieu et al. X-ray structure of EmrE supports dual topology model. *Proceedings of the National Academy of Science U S A*, 104:18999–9004, 2007.

46. S. Schuldiner. When biochemistry meets structural biology: The cautionary tale of EmrE. *Trends in Biochemical Science*, 32:252–258, 2007.

47. S. Schuldiner. Controversy over EmrE structure. *Science*, 317:748–751, 2007.

48. S. Steiner-Mordoch, M. Soskine, D. Solomon et al. Parallel topology of genetically fused EmrE homodimers. *EMBO Journal*, 27:17–26, 2008.

49. K. Palczewski, T. Kumasaka, T. Hori et al. Crystal structure of rhodopsin: A G protein-coupled receptor. *Science*, 289:739–745, 2000.

50. C.P. Chen, A. Kernytsky, and B. Rost. Transmembrane helix predictions revisited. *Protein Science*, 11:2774–2791, 2002.

51. A. Elofsson and G. von Heijne. Membrane protein structure: Prediction versus reality. *Annual Review of Biochemistry*, 76:125–140, 2007.

52. M. Punta, L.R. Forrest, H. Bigelow et al. Membrane protein prediction methods. *Methods* , 41:460–474, 2007.

53. C.P. Chen and B. Rost. State-of-the-art in membrane protein prediction. *Applied Bioinformatics*, 1:21–35, 2002.

54. J.L. Gardy, C. Spencer, K. Wang et al. PSORT-B: Improving protein subcellular localization prediction for Gram-negative bacteria. *Nucleic Acids Research*, 31:3613–3617, 2003.

55. M. Cserzö, E. Wallin, I. Simon, G. von Heijne, and A. Elofsson. Prediction of transmembrane alpha-helices in prokaryotic membrane proteins: The dense alignment surface method. *Protein Engineering*, 10:673–676, 1997.
56. M. Cserzö, F. Eisenhaber, B. Eisenhaber, and I. Simon. On filtering false positive transmembrane protein predictions. *Protein Engineering*, 15:745–752, 2002.
57. M. Cserzo, F. Eisenhaber, B. Eisenhaber, and I. Simon. TM or not TM: Transmembrane protein prediction with low false positive rate using DAS-TMfilter. *Bioinformatics*, 20:136–137, 2004.
58. G.E. Tusnády and I. Simon. Principles governing amino acid composition of integral membrane proteins: Application to topology prediction. *Journal of Molecular Biology*, 283:489–506, 1998.
59. H. Viklund and A. Elofsson. Best alpha-helical transmembrane protein topology predictions are achieved using hidden Markov models and evolutionary information. *Protein Science*, 13:1908–1917, 2004.
60. R.Y. Kahsay, G. Gao, and L. Liao. An improved hidden Markov model for transmembrane protein detection and topology prediction and its applications to complete genomes. *Bioinformatics*, 21:1853–1858, 2005.
61. L. Käll, A. Krogh, and E.L.L. Sonnhammer. A combined transmembrane topology and signal peptide prediction method. *Journal of Molecular Biology*, 338:1027–1036, 2004.
62. L. Käll, A. Krogh, and E.L.L. Sonnhammer. An HMM posterior decoder for sequence feature prediction that includes homology information. *Bioinformatics*, 21(1):i251–i257, 2005.
63. L. Käll, A. Krogh, and E.L.L. Sonnhammer. Advantages of combined transmembrane topology and signal peptide prediction-the Phobius web server. *Nucleic Acids Research*, 35:W429–432, 2007.
64. S.M. Reynolds, L. Käll, M.E. Riffle, J.A. Bilmes, and W.S. Noble. Transmembrane topology and signal peptide prediction using dynamic bayesian networks. *PLoS Computational Biology*, 4:e1000213, 2008.
65. A. Lo, H. Chiu, T. Sung, and W. Hsu. Transmembrane helix and topology prediction using hierarchical SVM classifiers and an alternating geometric scoring function. *Computation System Bioinformatics Conference*, 5:31–42, 2006.
66. A. Lo, H. Chiu, T. Sung, P. Lyu, and W. Hsu. Enhanced membrane protein topology prediction using a hierarchical classification method and a new scoring function. *Journal of Proteome Research*, 7:487–496, 2008.
67. M. Ganapathiraju, C.J. Jursa, H.A. Karimi, and J. Klein-Seetharaman. TMpro web server and web service: Transmembrane helix prediction through amino acid property analysis. *Bioinformatics*, 23:2795–2796, 2007.
68. M. Ganapathiraju, N. Balakrishnan, R. Reddy, and J. Klein-Seetharaman. Transmembrane helix prediction using amino acid property features and latent semantic analysis. *BMC Bioinformatics*, 9(1):S4, 2008.
69. P. Tompa, G.E. Tusnády, M. Cserzo, and I. Simon. Prion protein: evolution caught en route. *Proceedings of the National Academy of Science U S A*, 98:4431–4436, 2001.
70. M. Gomi, F. Akazawa, and S. Mitaku. SOSUIsignal: Software system for prediction of signal peptide and membrane protein. *Genome Informatics*, 11:414–415, 2000.

71. Z. Yuan, M.J. Davis, F. Zhang, and R.D. Teasdale. Computational differentiation of N-terminal signal peptides and transmembrane helices. *Biochemical and Biophysical Research Communication*, 312:1278–1283, 2003.

72. H. Nielsen, J. Engelbrecht, S. Brunak, and G. von Heijne. Identification of prokaryotic and eukaryotic signal peptides and prediction of their cleavage sites. *Protein Engineering*, 10:1–6, 1997.

73. H. Nielsen, S. Brunak, and G. von Heijne. Machine learning approaches for the prediction of signal peptides and other protein sorting signals. *Protein Engineering*, 12:3–9, 1999.

74. J.D. Bendtsen, H. Nielsen, G. von Heijne, and S. Brunak. Improved prediction of signal peptides: SignalP 3.0. *Journal of Molecular Biology*, 340:783–795, 2004.

75. H. Viklund, A. Bernsel, M. Skwark, and A. Elofsson. SPOCTOPUS: A combined predictor of signal peptides and membrane protein topology. *Bioinformatics*, 24:2928–2929, 2008.

76. D.M. Lao, T. Okuno, and T. Shimizu. Evaluating transmembrane topology prediction methods for the effect of signal peptide in topology prediction. *In Silico Biology*, 2:485–494, 2002.

77. D.M. Lao, M. Arai, M. Ikeda, and T. Shimizu. The presence of signal peptide significantly affects transmembrane topology prediction. *Bioinformatics*, 18:1562–1566, 2002.

78. J.L. Cornette, K.B. Cease, H. Margalit et al. Hydrophobicity scales and computational techniques for detecting amphipathic structures in proteins. *Journal of Molecular Biology*, 195:659–685, 1987.

79. D.A. Phoenix, F. Harris, O.A. Daman, and J. Wallace. The prediction of amphiphilic alpha-helices. *Current Protein & Peptide Science*, 3:201–221, 2002.

80. R. Wolfenden. Experimental measures of amino acid hydrophobicity and the topology of transmembrane and globular proteins. *Journal of General Physiology*, 129:357–362, 2007.

81. M. Cserzö, J.M. Bernassau, I. Simon, and B. Maigret. New alignment strategy for transmembrane proteins. *Journal of Molecular Biology*, 243:388–396, 1994.

82. I.K. Kitsas, L.J. Hadjileontiadis, S.M. Panas. Transmembrane helix prediction in proteins using hydrophobicity properties and higher-order statistics. *Computer in Biology Medicine*, 38:867–880, 2008.

83. H. Shen and J.J. Chou. MemBrain: Improving the accuracy of predicting transmembrane helices. *PLoS ONE*, 3:e2399, 2008.

84. A. Bernsel, H. Viklund, J. Falk et al. Prediction of membrane-protein topology from first principles. *Proceedings of the National Academy of Science U S A*, 105:7177–7181, 2008.

85. N.N. Alder and A.E. Johnson. Cotranslational membrane protein biogenesis at the endoplasmic reticulum. *Journal of Biological Chemistry*, 279:22787–22790, 2004.

86. T.A. Rapoport, V. Goder, S.U. Heinrich, and K.E.S. Matlack. Membrane-protein integration and the role of the translocation channel. *Trends in Cell Biology*, 14:568–575, 2004.

87. T. Hessa, H. Kim, K. Bihlmaier et al. Recognition of transmembrane helices by the endoplasmic reticulum translocon. *Nature*, 433:377–381, 2005.

88. T. Hessa, N.M. Meindl-Beinker, A. Bernsel et al. Molecular code for transmembrane-helix recognition by the Sec61 translocon. *Nature*, 450:1026–1030, 2007.
89. J. Nilsson, B. Persson, and G. von Heijne. Comparative analysis of amino acid distributions in integral membrane proteins from 107 genomes. *Proteins*, 60:606–616, 2005.
90. G. von Heijne. Membrane protein structure prediction. Hydrophobicity analysis and the positive-inside rule. *Journal of Molecular Biology*, 225:487–494, 1992.
91. D.T. Jones, W.R. Taylor, and J.M. Thornton. A model recognition approach to the prediction of all-helical membrane protein structure and topology. *Biochemistry*, 33:3038–3049, 1994.
92. E.L. Sonnhammer, G. von Heijne, and A. Krogh. A hidden Markov model for predicting transmembrane helices in protein sequences. *Proceedings of the International Conference on Intelligent System for Molecular Biology*, 6:175–182, 1998.
93. P.L. Martelli, P. Fariselli, and R. Casadio. An ENSEMBLE machine learning approach for the prediction of all-alpha membrane proteins. *Bioinformatics*, 19(1):i205–i211, 2003.
94. H. Zhou and Y. Zhou. Predicting the topology of transmembrane helical proteins using mean burial propensity and a hidden-Markov-model-based method. *Protein Science*, 12:1547–1555, 2003.
95. W.J. Zheng, V.Z. Spassov, L. Yan, P.K. Flook, and S. Szalma. A hidden Markov model with molecular mechanics energy-scoring function for transmembrane helix prediction. *Computation Biology and Chemistry*, 28:265–274, 2004.
96. P. Fariselli, P.L. Martelli, and R. Casadio. A new decoding algorithm for hidden Markov models improves the prediction of the topology of all-beta membrane proteins. *BMC Bioinformatics*, 6(4):S12, 2005.
97. H. Viklund and A. Elofsson. OCTOPUS: Improving topology prediction by two-track ANN-based preference scores and an extended topological grammar. *Bioinformatics*, 24:1662–1668, 2008.
98. D.T. Jones. Improving the accuracy of transmembrane protein topology prediction using evolutionary information. *Bioinformatics*, 23:538–544, 2007.
99. R. Ahmed, H. Rangwala, and G. Karypis. TOPTMH: Topology predictor for transmembrane α-helices. In *Machine Learning and Knowledge Discovery in Databases*, vol. 5211, pp. 23–38. Heidelberg, Berlin: Springer, 2008.
100. L. Rabiner. A tutorial on hidden Markov models and selected applications in speech recognition. *Proceedings of the IEEE*, 77:257–286, 1989.
101. M. Ikeda, M. Arai, D.M. Lao, and T. Shimizu. Transmembrane topology prediction methods: A re-assessment and improvement by a consensus method using a dataset of experimentally-characterized transmembrane topologies. *In Silico Biology*, 2:19–33, 2002.
102. S. Möller, M.D. Croning, and R. Apweiler. Evaluation of methods for the prediction of membrane spanning regions. *Bioinformatics*, 17:646–653, 2001.
103. K. Melén, A. Krogh, and G. von Heijne. Reliability measures for membrane protein topology prediction algorithms. *Journal of Molecular Biology*, 327:735–744, 2003.

104. P.D. Taylor, T.K. Attwood, and D.R. Flower. BPROMPT: A consensus server for membrane protein prediction. *Nucleic Acids Research*, 31:3698–3700, 2003.

105. J. Xia, M. Ikeda, and T. Shimizu. ConPred_elite: A highly reliable approach to transmembrane topology predication. *Computational Biology and Chemistry*, 28:51–60, 2004.

106. M. Arai, H. Mitsuke, M. Ikeda et al. ConPred II: A consensus prediction method for obtaining transmembrane topology models with high reliability. *Nucleic Acids Research*, 32:W390–W393, 2004.

107. J. Nilsson, B. Persson, and G. von Heijne. Consensus predictions of membrane protein topology. *FEBS Letters*, 486:267–269, 2000.

108. J. Nilsson, B. Persson, and G. Von Heijne. Prediction of partial membrane protein topologies using a consensus approach. *Protein Science*, 11:2974–2980, 2002.

109. G. Lasso, J.F. Antoniw, and J.G.L. Mullins. A combinatorial pattern discovery approach for the prediction of membrane dipping (re-entrant) loops. *Bioinformatics*, 22:e290–e297, 2006.

110. B. Boeckmann, A. Bairoch, R. Apweiler et al. The SWISS-PROT protein knowledgebase and its supplement TrEMBL in 2003. *Nucleic Acids Research*, 31:365–370, 2003.

111. G.E. Tusnády and I. Simon. The HMMTOP transmembrane topology prediction server. *Bioinformatics*, 17:849–850, 2001.

112. A. Bernsel and G. Von Heijne. Improved membrane protein topology prediction by domain assignments. *Protein Science*, 14:1723–1728, 2005.

113. E.W. Xu, P. Kearney, and D.G. Brown. The use of functional domains to improve transmembrane protein topology prediction. *Journal of Bioinformatics and Computational Biology*, 4:109–123, 2006.

114. P.G. Bagos, T.D. Liakopoulos, and S.J. Hamodrakas. Algorithms for incorporating prior topological information in HMMs: Application to transmembrane proteins. *BMC Bioinformatics*, 7:189, 2006.

115. M. Rapp, D. Drew, D.O. Daley et al. Experimentally based topology models for E. Coli inner membrane proteins. *Protein Science*, 13:937–945, 2004.

116. H. Kim, K. Melén, and G. von Heijne. Topology models for 37 Saccharomyces cerevisiae membrane proteins based on C-terminal reporter fusions and predictions. *Journal of Biological Chemistry*, 278:10208–10213, 2003.

117. D. Drew, D. Sjöstrand, J. Nilsson et al. Rapid topology mapping of Escherichia coli inner-membrane proteins by prediction and PhoA/GFP fusion analysis. *Proceedings of the National Academy of Science U S A*, 99:2690–2695, 2002.

118. E. Granseth, D.O. Daley, M. Rapp, K. Melén, and G. von Heijne. Experimentally constrained topology models for 51,208 bacterial inner membrane proteins. *Journal of Molecular Biology*, 352:489–494, 2005.

119. S. Lee, B. Lee, I. Jang, S. Kim, and J. Bhak. Localizome: a server for identifying transmembrane topologies and TM helices of eukaryotic proteins utilizing domain information. *Nucleic Acids Research*, 34:W99–W103, 2006.

120. G.E. Tusnády, L. Kalmár, H. Hegyi, P. Tompa, and I. Simon. TOPDOM: Database of domains and motifs with conservative location in transmembrane proteins. *Bioinformatics*, 24:1469–1470, 2008.

121. E. Wallin and G. von Heijne. Genome-wide analysis of integral membrane proteins from eubacterial, archaean, and eukaryotic organisms. *Protein Science*, 7:1029–1038, 1998.

122. L. Käll, E.L.L. Sonnhammer. Reliability of transmembrane predictions in whole-genome data. *FEBS Letters*, 532:415–418, 2002.

123. N. Hurwitz, M. Pellegrini-Calace, and D.T. Jones. Towards genome-scale structure prediction for transmembrane proteins. *Philosophical Transaction of the Royal Society of London series B Biological Science*, 361:465–475, 2006.

124. C.G. Knight, R. Kassen, H. Hebestreit, and P.B. Rainey. Global analysis of predicted proteomes: Functional adaptation of physical properties. *Proceedings of the National Academy of Science U S A*, 101:8390–8395, 2004.

125. Membrane Proteins of Known Structure (http://www.mpibp-frankfurt.mpg.de/michel/public/memprotstruct.html).

126. P. Raman, V. Cherezov, and M. Caffrey. The Membrane Protein Data Bank. *Cellular and Molecular Life Sciences (CMLS)*, 63:36–51, 2006.

127. Membrane Proteins of Known Structure (http://blanco.biomol.uci.edu/Membrane_Proteins_xtal.html).

128. Membrane Proteins of Known Structure by NMR (http://www.drorlist.com/nmr/MPNMR.html).

129. C. Ostermeier and H. Michel. Crystallization of membrane proteins. *Current Opinion in Structural Biology*, 7:697–701, 1997.

130. A.G. Lee. Lipid-protein interactions in biological membranes: A structural perspective. *Biochim Biophys Acta*, 1612:1–40, 2003.

131. G.E. Tusnády, Z. Dosztányi, and I. Simon. TMDET: web server for detecting transmembrane regions of proteins by using their 3D coordinates. *Bioinformatics*, 21:1276–1277, 2005.

132. A.L. Lomize, I.D. Pogozheva, M.A. Lomize, and H.I. Mosberg. Positioning of proteins in membranes: a computational approach. *Protein Science*, 15:1318–1333, 2006.

133. M.A. Lomize, A.L. Lomize, I.D. Pogozheva, and H.I. Mosberg. OPM: Orientations of proteins in membranes database. *Bioinformatics*, 22:623–625, 2006.

134. P.J. Bond and M.S.P. Sansom. Insertion and assembly of membrane proteins via simulation. *Journal of American Chemical Society*, 128:2697–2704, 2006.

135. M.S.P. Sansom, K.A. Scott, and P.J. Bond. Coarse-grained simulation: a high-throughput computational approach to membrane proteins. *Biochemical Society Transaction*, 36:27–32, 2008.

136. D. Boyd, B. Traxler, and J. Beckwith. Analysis of the topology of a membrane protein by using a minimum number of alkaline phosphatase fusions. *Journal of Bacteriology*, 175:553–556, 1993.

137. S. Froshauer, G.N. Green, D. Boyd, K. McGovern, and J. Beckwith. Genetic analysis of the membrane insertion and topology of MalF, a cytoplasmic membrane protein of Escherichia coli. *Journal of Molecular Biology*, 200:501–511, 1988.

138. J.K. Broome-Smith, M. Tadayyon, and Y. Zhang. Beta-lactamase as a probe of membrane protein assembly and protein export. *Molecular Microbiology*, 4:1637–1644, 1990.

139. P.M. Deak and D.H. Wolf. Membrane topology and function of Der3/Hrd1p as a ubiquitin-protein ligase (E3) involved in endoplasmic reticulum degradation. *Journal of Biological Chemistry*, 276:10663–10669, 2001.
140. B. Rost, R. Casadio, P. Fariselli, and C. Sander. Transmembrane helices predicted at 95% accuracy. *Protein Science*, 4:521–533, 1995.
141. B. Rost, P. Fariselli, and R. Casadio. Topology prediction for helical transmembrane proteins at 86% accuracy. *Protein Science*, 5:1704–1718, 1996.
142. S. Möller, E.V. Kriventseva, and R. Apweiler. A collection of well characterised integral membrane proteins. *Bioinformatics*, 16:1159–1160, 2000.
143. S. Jayasinghe, K. Hristova, and S.H. White. MPtopo: A database of membrane protein topology. *Protein Science*, 10:455–458, 2001.
144. M. Ikeda, M. Arai, T. Okuno, and T. Shimizu. TMPDB: A database of experimentally-characterized transmembrane topologies. *Nucleic Acids Research*, 31:406–09, 2003.
145. A. Kernytsky and B. Rost. Static benchmarking of membrane helix predictions. *Nucleic Acids Research*, 31:3642–3644, 2003.
146. T.D. Liakopoulos, C. Pasquier, and S.J. Hamodrakas. A novel tool for the prediction of transmembrane protein topology based on a statistical analysis of the SwissProt database: The OrienTM algorithm. *Protein Engineering*, 14:387–390, 2001.
147. K.H. Choo, T.W. Tan, and S. Ranganathan. SPdb—a signal peptide database. *BMC Bioinformatics*, 6:249, 2005.
148. S.H. White. The progress of membrane protein structure determination. *Protein Science*, 13:1948–1949, 2004.
149. C. Papaloukas, E. Granseth, H. Viklund, and A. Elofsson. Estimating the length of transmembrane helices using Z-coordinate predictions. *Protein Science*, 17:271–278, 2008.
150. M.H. Saier, C.V. Tran, and R.D. Barabote. TCDB: The Transporter Classification Database for membrane transport protein analyses and information. *Nucleic Acids Research*, 34:D181–D186, 2006.

CHAPTER 7

CONTACT MAP PREDICTION BY MACHINE LEARNING

ALBERTO J.M. MARTIN, CATHERINE MOONEY, IAN WALSH, and GIANLUCA POLLASTRI
Complex and Adaptive Systems Lab
School of Computer Science and Informatics
UCD Dublin
Dublin, Ireland

7.1. INTRODUCTION

Proteins fold into three-dimensional (3D) structures that encode their function. Genomics and more recently metagenomics [1] projects leave us with millions of protein sequences, of which only a small fraction have their 3D structure experimentally determined. There are several structural genomics projects attempting to bridge the huge gap between sequence and structure. The high-throughput pipelines have to deal with important bottlenecks, for example, a large number of sequences are unsuitable for structural determination with current methods [2]. Therefore, computational protein structure prediction remains an irreplaceable instrument for the exploration of sequence-structure-function relationships. This is especially important for analysis at the genome or inter-genome level, where informative structural models need to be generated for thousands of gene products (or a portion of them) in reasonable amounts of time.

The more reliable and accurate procedures for structure prediction are based on the transfer of knowledge between closely related proteins deposited in sequence and structure databases—the field known as template-based modelling. Those methods typically adopt heuristics based on sequence and/or structural similarity to model the unknown target structure based on known structures that are understood to be homologous to it. Automating the

Introduction to Protein Structure Prediction: Methods and Algorithms,
Edited by Huzefa Rangwala and George Karypis

modeling process is difficult: there are several stages and critical points in the design (choice of templates, the creation of a correct sequence to structure alignment, etc.). For some of these points manual intervention leads to better predictions than fully automated methods [3]. The accuracy of template-based techniques strongly depends on the amount of detectable similarity, thus preventing the reliable application of these methods for a significant fraction of unannotated proteins. This is the realm of the so-called *ab initio* or *de novo* protein structure prediction, where models are predicted without relying on similarity to proteins with known structure. *Ab initio* techniques are obviously not as accurate as those based on templates, but the design in this case is generally more straightforward. Moreover, improvements can be obtained by relying on fragment-based algorithms [4] that use fragments of proteins of known structure to reconstruct the complete structure of the target protein. A system for the *ab initio* prediction of protein structures is generally composed of two elements: an algorithm to search the space of possible protein configurations to minimize some cost function, and the cost function itself, composed of various constraints either derived from physical laws, experimental results, or being structural features (e.g., secondary structure or solvent accessibility) predicted by machine learning or other kinds of statistical systems [5].

7.1.1. Maps Definition and Description

A two-dimensional (2D) representation of a protein structure, or more in general of a 3D object, is a map of properties of pairs of its elements, for instance the set of distances among a protein's residues. Using 2D projections of 3D objects is an attractive way of encoding geometrical information of protein structures, as these can be made scale and rotation invariant and do not depend on the coordinate frame. Therefore, 2D projections can be modeled as the output variable of learning or statistical systems trained in a supervised fashion, that is, using samples of (input, target) pairs collected from structure databases. A 2D encoding of a structure can be graphically represented as a 2D matrix. In the case of proteins, the geometrical relationship may involve fragments of the structure at different scales, using for instance amino acid [6] or secondary structure segment pairs [7], the former being much more commonly adopted than the latter.

Geometrical relationships between amino acids can be expressed as a set of distance restraints, for example, in the form $L \leq d(i, j) \leq U$, where $d(i, j)$ is the distance between residues in positions i and j and L (respectively U) is the lower (respectively upper) bound on the distance. Restraints such as the above can be experimentally determined, for example, from nuclear magnetic resonance (NMR) experiments. Indeed, algorithms for modeling protein structures from distance restraints are borrowed from the NMR literature and use, for instance, stochastic optimization methods [8,9], distance geometry [10,11], and variants of them [12–14].

FIGURE 7.1 Distance matrix in gray scale image format. White is 0 Å and black is the maximum distance in the protein.

A *distance matrix or distance map* consists of the set $\{d(i,j)\}_{i>j}$ of $N(N-1)/2$ distances between any two points in positions i and j of a protein with N amino acids. Note how the distance matrix corresponds to the above form of constraints with the lower distance bound equal to the upper one. Figure 7.1 shows a gray scale image of the distance matrix of the protein with Protein Data Bank (PDB) code 1ABV, where the distances are calculated between the C_α atoms.

There is a trade-off between the resolution of the input restraints, that is, the uncertainty with which they specify the property of the pairs, and the ability of the reconstruction algorithm to recover the correct model from these inputs. In the best case, when the complete noise-free distance matrix is available, the optimization problem can be solved analytically by finding a 3D embedding of the 2D restraints.

The distance matrix, or even detailed distance restraints, cannot be reliably determined by means of computational techniques unless experimental data are available or when there is strong homology to proteins of known structure. This is why in the past research has focused on predicting restraints derived from the distance matrix that are at the same time easier to learn than distances and are able to retain substantial structural information. The *contact map* of a protein is a labeled matrix derived by thresholding the distance

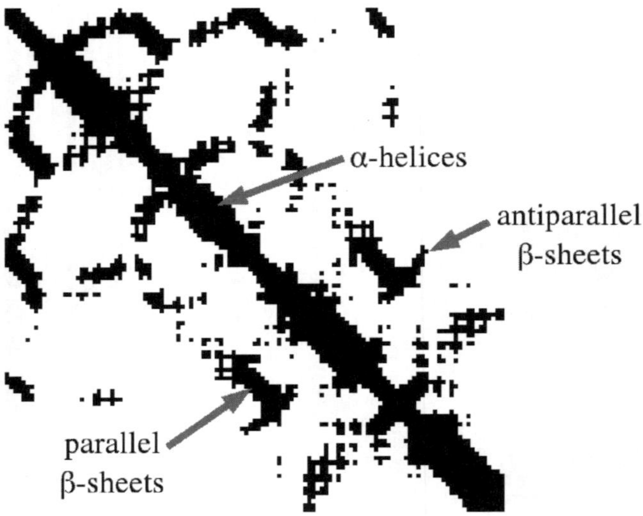

FIGURE 7.2 Different secondary structure elements like helices (thick bands along the main diagonal) and parallel—or antiparallel—β-sheets (thin bands parallel—or antiparallel—to the main diagonal) are easily detected from the contact map.

matrix and assigning labels to the resulting discrete intervals. The simplest alternative is the binary contact map, where one distance threshold t is chosen to describe pairs of residues that are in contact ($d(i, j) < t$) or not ($d(i, j) \geq t$). The binary contact map can also be seen as the adjacency matrix of the graph with N nodes corresponding to the residues. Binary contact maps are popular as noise-tolerant alternatives to the distance map, and algorithms exist to recover protein structures from these representations [8,15,16]. However, recovering high-quality models (1-2 Å resolution) from the binary map alone, even that of the native fold, is often impossible [16], unless other sources of information are taken into account.

The definition of contact among amino acids is based on a single atom (normally C_α or C_β) and depends on a geometrical threshold. This is obviously a fairly crude approximation of physical interaction, for instance when the orientation of the side chain is important. Nevertheless, it is possible to identify patterns in binary contact maps [17]—secondary structure elements can be recognized along the map diagonal as well as the interactions between them (Figure 7.2). Contact maps at 8 Å have been assessed as a special category at the Critical Assessment of Protein Structure Prediction (CASP) for several years [18].

7.1.2. Uses of Maps

It is often possible to reconstruct the 3D structure of a protein from a true contact map with a fair degree of accuracy. Reconstructed 3D structures can still be close to the native structure when a certain amount of noise is introduced into the contact map [8,16] or even when predicted contact maps are used as constraints during the reconstruction process [15]. Contact maps have been used in many other applications. These include:

- Fold recognition: Cheng and Baldi use features of predicted contact maps (with thresholds at 8 and 12 Å) as part of a set of 54 pairwise features measuring query-template similarity as input to Support Vector Machines to determine if the query and template belong to the same fold or not [19].
- Model quality assessment: In Reference [20] predicted contact maps are used to rank several different predicted 3D models for a single target.
- Building nonaligned regions for template-based models: The I-TASSER method [21,22] for 3D-model building uses templates when available, but unaligned loop regions are built by *ab initio* modeling using a potential function. The potential includes four components: (1) general knowledge-based statistics terms from the PDB (C_α/side chain correlations, H-bond, and hydrophobicity); (2) spatial restraints from threading templates; (3) sequence-based C_α contact predictions by SVMSEQ [48]; and (4) distance and contact map from segmental threading. This method ranked as the best 3D server predictor at CASP6 and CASP7 [3].
- Protein structure comparisons: Several methods have been developed that compare protein structures by making use of the fact that contact maps are rotation- and translation-invariant by searching for suboptimal solutions to the maximum contact map overlap problem [23,24]. This reduces the search from 3D to 2D, making these methods fast and accurate ways of comparing structures.
- Protein folding simulation: As contact maps represent the 3D structure of a protein they have been used to simulate folding pathways—changing a few contacts in a map may represent substantial structural changes. An initial contact map is generated and successive changes to it reflect how the structure is modified during the folding processes. These changes to the map can be directed by different energy and/or map quality functions [25,26].
- Predicting protein folding rates: In Reference [27] folding rate of a protein is found to correlate to the number of predicted long-range contacts normalized by the square of the protein length.
- Predicting intrinsically disordered regions of proteins: In Reference [28] contact map predictions by PROFcon [29] are combined with a pairwise potential to predict unstructured regions of proteins, with particular success for the case of long ones.
- Improving the prediction of protein secondary structure: High-quality predictions of contact maps and especially noise-free maps yield improved to near-perfect secondary structure predictions [30].

- Protein-protein and domain docking: While most of the methods used in docking are 3D coordinate-based, in Reference [31] protein domains are represented by their contact map and only contacts between the different domains have to be predicted. This method takes advantage of the invariance of contact maps to rotation and translation, avoiding the step of choosing the relative orientation and disposition of the domains with respect to each other.

7.1.3. Predicted Contact Map Quality Measures

In order to measure the quality of predicted contact maps several measures may be used. The most common ones take into account, correctly or incorrectly, classified contacts (true positives [TP], false positives [FP]; true negatives [TN], and false negatives [FN]) that are used to compute recall (R; Equation 7.1), precision (P; Equation 7.2) and their harmonic average, F measure (F; Equation 7.3), as follows:

$$R = \frac{TP}{(TP + FN)} \tag{7.1}$$

$$P = \frac{TP}{TP + FP} \tag{7.2}$$

$$F = \frac{2PR}{P + R} \tag{7.3}$$

Performance is usually computed for different sets of contacts based on the separation of two residues in the linear sequence. For instance at CASP [18] separations greater than {5, 11, 23} have been used. Those contacts close to the diagonal of the map (i.e., for low-sequence separations) reflect secondary structure/local interactions and are much easier to predict. Those farther away from the map diagonal reflect long-range interactions mainly between different secondary structure elements [32]. Usually, the greater the distance a map position is from the map diagonal the more difficult it is to predict as long-range contacts are less frequent, thus harder to characterize.

7.1.4. Contact Map Prediction

Although there has been improvement in contact map prediction accuracy over the last few years [18,33], the problem is still largely unsolved. Among the reasons for this are the unbalanced nature of the problem (with far fewer examples of contacts than noncontacts, although this depends on the definition of contact) and, especially, the formidable challenge of capturing long-range interactions in the maps.

In order to mitigate the intrinsic difficulty of mapping one-dimentional (1D) input sequences into 2D outputs, in virtually all existing predictive systems protein contact maps are inferred by modeling a set of independent tasks, each task being the prediction of whether two residues are in contact. For this task a variety of machine learning methods have been used: neural networks [6,29,34,35]; genetic algorithims; self-organizing maps [36]; hidden Markov models [25]; Support Vector Machines [37–39]; and Bayesian inference [20]. The best performing map predictors tend to incorporate and combine information from different sources [18]. Common inputs are conservation and co-frequency of residue pairs extracted from multiple sequences alignments (MSA); predicted secondary structure and solvent accessibility; and predicted contact density and contact order.

In Section 7.2 we describe a system for binary contact map prediction based on recursive neural networks (RNN) that we have developed. The system is capable of incorporating homology information from multiple templates from PDB, when available. Often predicted contact maps do not encode a physical 3D structure and some recent methods [40] try to eliminate nonphysical contacts to improve the physicality of the predictions. An example of this type of strategy, which we have recently developed [41], is described in Section 7.4. In Section 7.5 we introduce a new definition of multi-threshold maps along with methods we developed for their predictions. In Section 7.6 we discuss our participation to the CASP8 competition.

7.2. BINARY CONTACT MAP PREDICTION BY 2D-RNN

We split the task of predicting binary contact maps into two stages: the prediction of contact density from the primary sequence and the reconstruction of the contact map from the predicted contact density and the primary sequence. We define contact density as the principal eigenvector (PE) of a protein's residue contact map at 8 Å, multiplied by the PE. Predicting contact density from the primary sequence consists of mapping a vector into a vector. This task is less complex than mapping vectors directly into 2D matrices since the size of the problem is drastically reduced and so is the scale length of interactions that need to be learned.

Predicted contact density is incorporated into a system for contact map prediction [35,42]. Our tests show that incorporating contact density yields sizeable gains in contact map prediction accuracy, and that these gains are especially significant for long-ranged contacts, which are known to be both harder to predict and critical for accurate 3D reconstruction.

7.2.1. Methods

We define two amino acids as being in contact if the distance between their C_α is less than a given threshold. For the definition of contact density we adopt a fixed 8 Å threshold, while in the contact map prediction stage we test 8 Å and 12 Å thresholds (see Reference [6] for a more detailed explanation of contact density prediction).

The dataset used in the present simulations is extracted from the December 2003 25% pdb_select list (http://homepages.fh-giessen.de/~hg12640/pdbselect). We use the Dictionary of Protein Secondary Structure (DSSP) program [43] to assign relevant structural features and remove sequences for which DSSP does not produce an output. After processing by DSSP, the set contains 2171 proteins and 344,653 amino acids. MSA for the 2171 proteins are extracted from the NR (Non Redundant selection from SWISS-PROT plus TrEMBL) database. The alignments are generated by three runs of Position-Specific Iterative-Basic Local Alignment Search Tool (PSI-BLAST) [44] with parameters $b = 3000$, $e = 10^{-3}$, and $h = 10^{-10}$. For training and testing of the contact density predictor we split the data into a training set containing 1736 sequences and a test set of 435 (one-fifth of the total). For training and testing of the contact map predictive system the sets are first processed to remove sequences longer than 200 amino acids (for computational reasons, as in Referenc [42]), leaving 1275 proteins in the training set and 327 proteins in the test set.

7.2.1.1. Predictive Architecture. We build a system for the prediction of contact maps based on 2D-RNN, described in References [42] and [35]. This is a family of adaptive models for mapping 2D matrices of variable size into matrices of the same size. Two-dimentional-RNN-based models were among the most successful contact map predictors at the CASP5 competition [45].

In the case of prediction of contact maps, the output O of the problem is the map itself, and the input I is a set of pairwise properties of residues in the protein. For a list of input properties we have considered, see below.

Let the element of indices j and k in the output be $o_{j,k}$ (in the case of contact maps this encodes whether residues in positions j and k in the sequence are within certain distance boudaries). In a standard 2D-RNN we postulate that $o_{j,k}$ is a function of four (vectorial) quantities, representing: the input element in the corresponding position $i_{j,k}$, for example, in the case of contact maps, a vector encoding the identities of residues j and k, plus whatever individual or pairwise structural characteristics of these residues we decide to include; four contextual (vectorial) memories that encode information about different parts of the input map, $h_{j,k}^{(n)}$ for $n = 1, \ldots, 4$. Each memory vector "specializes" on a different square of the input map, for example, $h_{j,k}^{(1)}$ represents information about input elements $i_{v,t}$ for which $v \leq j$ and $t \leq k$ (the upper-left, or "northwestern" context). We do not explicitly say what to represent in a memory vector, apart from defining its functional relationship to the other memory vectors, for example, the upper-left memory $h_{j,k}^{(1)}$ has to be a function of $h_{j-1,k}^{(1)}$ (the upper-left memory in the same column and in the row above), of $h_{j,k-1}^{(1)}$ (the upper-left memory in the same row and in the column to the left), and of the input element $i_{j,k}$. The other three memory vectors in position j, k represent, respectively, the lower-left or "southwestern" context ($h_{j,k}^{(2)}$), the lower-right or "southeastern" context ($h_{j,k}^{(3)}$), and the upper-right or "northeastern" context ($h_{j,k}^{(4)}$). Overall, the set of functional relationships based on which we define a 2D-RNN is as follows:

$$
\begin{aligned}
o_{j,k} &= \mathcal{N}^{(O)}\left(i_{j,k}, h^{(1)}_{j,k}, h^{(2)}_{j,k}, h^{(3)}_{j,k}, h^{(4)}_{j,k}\right)\\
h^{(1)}_{j,k} &= \mathcal{N}^{(1)}\left(i_{j,k}, h^{(1)}_{j-1,k}, h^{(1)}_{j,k-1}\right)\\
h^{(2)}_{j,k} &= \mathcal{N}^{(2)}\left(i_{j,k}, h^{(2)}_{j+1,k}, h^{(2)}_{j,k-1}\right)\\
h^{(3)}_{j,k} &= \mathcal{N}^{(3)}\left(i_{j,k}, h^{(3)}_{j+1,k}, h^{(3)}_{j,k+1}\right)\\
h^{(4)}_{j,k} &= \mathcal{N}^{(4)}\left(i_{j,k}, h^{(4)}_{j-1,k}, h^{(4)}_{j,k+1}\right)\\
j, k &= 1, \ldots, N
\end{aligned}
$$

Let us define, for simplicity, the set of these relationships as $O = \mathcal{N}(I)$.

We assume that the five functions (the output update $\mathcal{N}^{(O)}$ and the four lateral update functions $\mathcal{N}^{(n)}$ for $n = 1, \ldots, 4$) are independent on the position on the map (the indices j, k of the input element they process), and represent them using five two-layered feed-forward neural networks, as in Reference [42]. For an instance of the problem of $N \times N$ input elements (e.g., the contact map of a protein of length N) each neural network is replicated $N \times N$ times, resulting in a single neural network parametrizing $\mathcal{N}$. This neural network may have a very large number of connections (in the order of millions in the case of proteins) but the number of free parameters is greatly restricted (to the sum of free parameter numbers of the five networks) because each parameter is shared exactly $N \times N$ times. Given that the neural network representing N does not contain cycles, it can be trained by gradient descent using the standard backpropagation algorithm.

In the latest versions of our contact map predictors use 2D-RNNs with *shortcut connections*, that is, where lateral memory connections span S-residue intervals, where $S > 1$. In this case the definition of a 2D-RNN changes into:

$$
\begin{aligned}
o_{j,k} &= \mathcal{N}^{(O)}\left(i_{j,k}, h^{(1)}_{j,k}, h^{(2)}_{j,k}, h^{(3)}_{j,k}, h^{(4)}_{j,k}\right)\\
h^{(1)}_{j,k} &= \mathcal{N}^{(1)}\left(i_{j,k}, h^{(1)}_{j-1,k} \ldots, h^{(1)}_{j-S,k}, h^{(1)}_{j,k-1}, \ldots, h^{(1)}_{j,k-S}\right)\\
h^{(2)}_{j,k} &= \mathcal{N}^{(2)}\left(i_{j,k}, h^{(2)}_{j+1,k} \ldots, h^{(2)}_{j+S,k}, h^{(2)}_{j,k-1}, \ldots h^{(2)}_{j,k-S}\right)\\
h^{(3)}_{j,k} &= \mathcal{N}^{(3)}\left(i_{j,k}, h^{(3)}_{j+1,k} \ldots, h^{(3)}_{j+S,k}, h^{(3)}_{j,k+1}, \ldots h^{(3)}_{j,k+S}\right)\\
h^{(4)}_{j,k} &= \mathcal{N}^{(4)}\left(i_{j,k}, h^{(4)}_{j-1,k} \ldots, h^{(4)}_{j-S,k}, h^{(4)}_{j,k+1}, \ldots, h^{(4)}_{j,k+S}\right)\\
j, k &= 1, \ldots, N
\end{aligned}
$$

A graphical representation of a 2D-RNN with shortcuts is depicted in Figure 7.3.

In our tests the input $i_{j,k}$ contains amino acid information, secondary structure and solvent accessibility information, and contact density information for the amino acids in positions j and k in the sequence. Amino acid information is obtained from the MSA of the protein sequence to its homologs to leverage

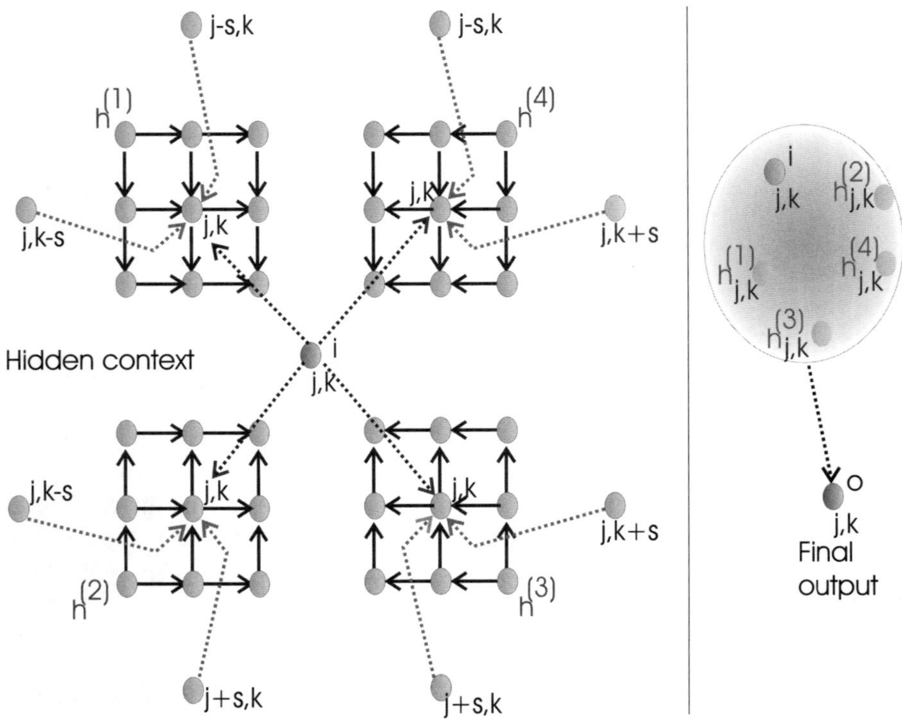

FIGURE 7.3 A graphical representation of a 2D-RNN with shortcut connections (see text for details). Nodes represent inputs element vectors ($i_{j,k}$), hidden memory vectors ($h^{(n)}_{a,b}$), and outputs ($i_{j,k}$). An edge between node A and B indicates that B is a function of A.

evolutionary information. Amino acids are coded as letters out of an alphabet of 25. Besides the 20 standard amino acids, B (aspartic acid or asparagine), U (selenocysteine), X (unknown), Z (glutamic acid or glutamine), and . (gap) are considered. The input presented to the networks is the frequency of each of the 24 non-gap symbols, plus the overall frequency of gaps in each column of the alignment, that is, if n_{jk} is the total number of occurrences of symbol j in column k, and g_k the number of gaps in the same column, the jth input to the networks in position k is:

$$\frac{n_{jk}}{\sum_{v=1}^{24} n_{vk}} \tag{7.4}$$

for $j = 1 \ldots 24$, while the 25th input is:

$$\frac{g_k}{gk + \sum_{v=1}^{24} n_{vk}} \tag{7.5}$$

This input coding scheme is richer than the simple 20-letter schemes and has proven effective in Reference [46].

The systems are trained by minimizing the cross-entropy error between the output and target probability distributions, using gradient descent with no momentum term or weight decay. We use a hybrid between online and batch training, with 580 gradient updates (roughly one every three proteins) per training epoch. The training set is also shuffled at each epoch, so that the error does not decrease monotonically. When the error does not decrease for 50 consecutive epochs, the learning rate is divided by 2. Training stops after 1000 epochs. The gradient is computed using the backpropagation through structure (BPTS) algorithm [47].

7.2.2. Results and Discussion

For comparison purposes, we encode each pair (i, j) of amino acids in the input by four different features: a 20×20 matrix corresponding to the probability distribution over all pairs of amino acids observed in the two corresponding columns of the alignment (MA); MA plus the actual discretized four-class contact density component for both residue i and j (MA_CD); MA plus the actual secondary structure (three classes) and binary thresholded (at 25%) relative solvent accessibility (MA_SS_ACC); and finally, the previous feature plus the actual four-class contact density components (MA_SS_ACC_CD). We train eight 2D-RNN ensembles, with the same architecture, one for each input feature and contact threshold. Testing takes place by encoding each pair (i, j) on input with the predicted four-class contact density. Secondary structure and solvent accessibility information input is exact during both training and testing.

Tables 7.1 and 7.2 show performance indices for all eight networks. The indices considered are P, R, and F (see Section 7.1.3). Performances are computed for three different sets of contacts, based on the separation of two residues in the linear sequence $|i - j| \geq \{6, 12, 24\}$.

The introduction of contact density predictions increases the F measure in all cases. This is true for both 8 and 12 Å maps, and for all separation thresh-

TABLE 7.1 Performance Results for Contact Map Prediction by 2D-RNN. Contact Threshold: 8 Å

	$\|i - j\| \geq 6$			$\|i - j\| \geq 12$			$\|i - j\| \geq 24$		
	P	R	F	P	R	F	P	R	F
MA	100	0	0	100	0	0	100	0	0
MA_CD	39.4	12.2	18.6	36.2	8.4	13.5	27.8	2.0	3.7
MA_SS_ACC	50.5	7.4	12.9	48.8	4.0	7.4	25.7	0.2	0.3
MA_SS_ACC_CD	43.3	11.3	17.9	38.9	7.2	12.1	25.5	2.2	4.1

TABLE 7.2 Performance Results for Contact Map Prediction by 2D-RNN. Contact Threshold: 12 Å

	$\|i-j\| \geq 6$			$\|i-j\| \geq 12$			$\|i-j\| \geq 24$		
	P	R	F	P	R	F	P	R	F
MA	60.4	10.6	18.1	55.8	0.1	0.1	38.9	0.03	0.06
MA_PE	49.5	24.5	32.8	39.4	16.8	23.6	34.5	13.6	19.5
MA_SS_ACC	61.6	19.6	29.7	48.9	7.5	13.1	40.2	2.8	5.3
MA_SS_ACC_PE	54.2	23.5	32.8	42.2	14.6	21.7	36.7	10.9	16.8

FIGURE 7.4 Examples of contact map *ab initio* predictions at 12 Å for protein 1A2P (108 amino acids). Exact map in the top-right half, predicted map in the bottom-left half. Prediction by MA_SS_ACC on the left, MA_SS_ACC_CD on the right (see text for details).

olds. An improvement is observed both in the MA_CD versus MA case and in the MA_SS_ACC_CD versus MA_SS_ACC case. In all cases the introduction of the predicted CD yields larger performance gains than the exact secondary structure and solvent accessibility combined. Interestingly, the gains become more significant for longer range contacts: for $|i - j| \geq 24$. F grows from 0.2% to 9.9% and from 3.6% to 25% in the 8 and 12 Å case, respectively. CD-based networks are more confident away from the main diagonal (an example is shown in Fig. 7.4), with a better balance between FP and FN.

7.3. INCORPORATING TEMPLATE INFORMATION

Many protein sequences show detectable identity to sequences of known structure. If this happens, homology information, when available, is provided as a further input to our predictors, similarly to Reference [48]. Almost any level of sequence identity to PDB templates (PSI-BLAST e-value as high as 10) yields more accurate maps than the *ab initio* predictor. Furthermore, in most cases the map predicted based on PDB templates is more accurate than the maps of the templates, suggesting that the combination of sequence and template information is more informative than templates alone. All the results and comparisons shown below are for 12 Å threshold maps. For each of the 1602 proteins in S1602 we search for structural templates in the PDB. We base our search on PDBFINDERII [49] as available on August 22, 2005. An obvious problem arising is that all proteins in the S1602 set are expected to be in the PDB (barring name changes), hence every protein will have a perfect template. To avoid this, we exclude from PDBFINDERII every protein that appears in S1602. We also exclude all entries shorter than 10 residues, leading to a final 66,350 chains.

To generate the actual templates for a protein, we run two rounds of PSI-BLAST against a redundancy-reduced version of the NR database with parameters $b = 3000$ (maximum number of hits), $e = 10^{-3}$ (expectation of a random hit) and $h = 10^{-10}$ (expectation of a random hit for sequences used to generate the Position-Specific Scoring Matrix [PSSM]). We then run a third round of PSI-BLAST against the PDB using the PSSM generated in the first two rounds. In this third round we deliberately use a high-expectation parameter ($e = 10$) to include hits that are beyond the usual comparative modeling scope ($e < 0.01$). We further remove from each set of hits thus found all those with sequence similarity exceeding 95% over the whole query, to exclude PDB resubmissions of the same structure at different resolution, other chains in N-mers and close homologs.

The distribution of sequence similarity of the best template and average template similarity is plotted in Figure 7.5. The average similarity for all PDB hits for each protein, not surprisingly, is generally low. This does not affect predictive performances.

For a residue at position j a structural property d_j is input to the predictors as $d_j = \frac{\sum_{p=1}^{P} \omega_p d_{pj}}{\sum_{p=1}^{P} \omega_p}$ where: P is the total number of template residues aligned; d_{pj} represents the property d of the residue in the p-th template, aligned against position j in the query. $\omega_p = id_p^3/q_p$ is the weight associated to the p-th template; id_p is the identity between template p and the query protein; and q_p is the quality of template p measured as its resolution plus R-factor divided by 20.

Table 7.3 reports the comparison between *ab initio* and template-based predictions of contact maps with a 12 Å threshold. The only decrease in performance (by 0.6%) is when there is less than 10% sequence identity between the query sequence and the best template found for that sequence. The average increase in the performance of template-based predictions over *ab initio* predictions is 6.4%.

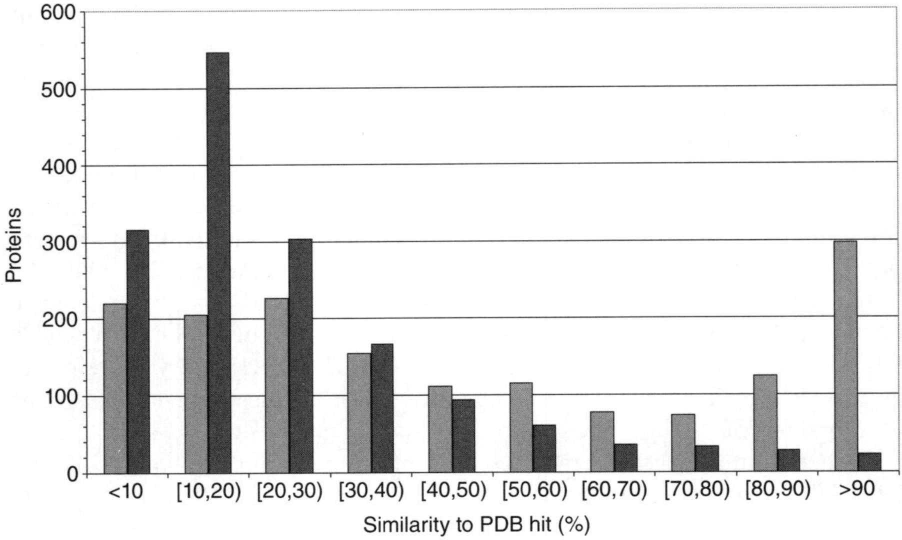

FIGURE 7.5 Distribution of best hit (light gray) and average (dark gray) sequence similarity in the PSI-BLAST templates for the S2171 set. Hits above 95% sequence similarity excluded.

TABLE 7.3 Correctly Predicted Residues (%) for the *Ab Initio* (12_{AI}) and Template-Based (12_{TE}) 12 Å Threshold Contact Map Predictor as a Function of Percentage Sequence Identity (SEQ_ID) between the Query Sequences in S1602 and the Best Template Found for Each Sequence

SEQ_ID	<10	<20	<30	<40	<50	<60	<70	<80	<90	≥90	All
12_{AI}	85.9	87.5	86.8	85.6	87.2	86.5	86.2	86.1	86.4	87.3	86.8
12_{TE}	85.3	87.8	91.3	93.6	95.7	96.0	95.8	96.4	97.0	97.3	93.2

It should be noted that template generation is an independent module in the systems. We are currently investigating whether more subtle strategies for template recognition would still benefit contact map predictions, with or without retraining the systems on the new template distributions.

7.4. FILTERING CONTACT MAPS

We implement a filtering stage (NN-Filter) to correct nonphysical contact predictions by training two different ensembles of artificial neural networks to process different regions of the contact map: the region close to the diagonal of the map, corresponding to small sequence separations, and the region further away from the diagonal. We do so because the rules governing contact probability are likely to be different for positions near the main diagonal (mainly made by backbone atoms, reflecting secondary structure contacts) and for positions away from it (where contacts mainly occur between the side chains of amino acids placed in different secondary structure elements).

The input to the NN-Filter stage is a combination of local information about the predicted map such as estimated probabilities of contact for pairs of residues or secondary structure predictions for individual residues [46] and global information, such as total number of contacts predicted for a residue or pair of residues, and a protein's residue contact order.

7.4.1. Methods

The NN-Filter is trained and tested on the same datasets and MSA that were used to train and test the binary contact map predictor (which we will refer to as XXStout), as described above in Section 7.2.1. Therefore, any improvement by NN-Filter over XXStout cannot be attributed to differences in training set size or coverage. We create a further set (T ES481) to assess the quality of filtered maps versus "raw" maps in an unbiased manner. From the July 2005 25% pdb_select list we remove all those sequences with sequence identity greater than 25% with any of the sequences found in previously used datasets and those longer than 200 residues. This set contains 481 sequences (49, 159 amino acids).

7.4.1.1. Contact Maps. As a cutoff between the filter for near-diagonal and long-range contacts we use the same distance thresholds between C_α (12 Å and 8 Å) as those defining binary contact maps that we are trying to filter. Predicted contacts are obtained from our binary contact map prediction system, XXStout, described in the previous section.

7.4.1.2. Artificial Neural Networks. NN-Filter is composed of fully connected, feed-forward neural networks trained via the backpropagation algorithm by minimizing the cross-entropy error between the output and target probability distributions. Due to the large number of examples and the long time needed to train a single network on all training instances, the training set is further divided into 20 sets (away from diagonal) and 5 sets (beside diagonal), each containing approximately the same number of examples. A different network is trained on each of these subsets and then the networks are ensembled: 20 for map positions away from the diagonal; 5 for map positions beside the diagonal.

7.4.1.3 Input Representation.

Local Inputs. The main input to the filtering neural networks is the "raw" map predicted by XXstout. We input a whole patch of 11×11 contact predictions. Thus a network that predicts the probability of contact between residues in positions i and j along the sequence is shown contact predictions (in the form of predicted probabilities of contact) for all pairs of amino acids (k, l) with $i - 5 \leq k \leq i + 5$ and $j - 5 \leq l \leq j + 5$, for a total of 121 inputs. We also input predicted secondary structure in the form of the predicted probabilities of being in one of the three standard secondary structure classes (helix, strand, coil). Sequence and evolutionary information is provided to the network in the form of identities of residues in positions i − 1, i, i + 1, and j − 1, j, j + 1 and frequency profiles for residues in positions i − 1, i, i + 1, and j − 1, j, j + 1, extracted from MSA.

Global Inputs. The groups of inputs above provide local information about the map. We also adopt a number of input features that contain global information:

Binary 8 Å and 12 Å map values for (i, j) obtained as follows: All the probabilities of contact, as predicted by XXStout, for pairs of amino acids at the same sequence separation ($|i - j|$) are added up yielding $\mathcal{L}$, then the L top-scoring pairs at the given separation are classified as contacts, where L is the integer part of $\mathcal{L}$. Notice that because of its definition, each element of these maps implicitly contains information about all those elements on the map with sequence separation $|i - j|$.

This definition of contact is also used to determine the following features.

- Number of contacts per amino acid (NC) of amino acids i and j.
- Number of amino acids in contact with both i and j (NCIJ).
- Residue Contact Order (CO): For each amino acid i its CO (CO_i) is computed using Equation 7.6, where d_{ij} is the (euclidean) distance between amino acids i and j, t is the contact threshold, L is the protein length and NC_i is the total number of contacts for i.

$$CO_i = \frac{\sum_{\forall j: d_{ij} < t} |i - j|}{L \cdot NC_i} \tag{7.6}$$

- Protein length.

7.4.2. Results and Discussion

We measure all performances on the *TES*481 dataset. The F measure for NN-Filter improves over XXStout for both 12 Å and 8 Å maps, and for all bands of sequence separation considered. For long-range contacts (sequence separation of 24 or greater) F grows from 0.286 for XXstout to 0.316 for NN-Filter for 12 Å maps and from 0.097 to 0.168 for 8 Å maps (Tables 7.4 and 7.5).

TABLE 7.4 Performances for XXstout and NN-Filter on the *TES*481 Dataset for Different Sequence Separations. 12 Å Maps

	XXStout			NN-Filter		
	F	P	R	F	P	R
$\lvert i-j\rvert \geq 6$	0.396	0.386	0.407	0.411	0.336	0.527
$\lvert i-j\rvert \geq 12$	0.323	0.313	0.334	0.353	0.278	0.484
$\lvert i-j\rvert \geq 24$	0.286	0.275	0.298	0.316	0.254	0.417
$\lvert i-j\rvert \leq 12$	0.858	0.857	0.858	0.876	0.875	0.876

TABLE 7.5 Performances for XXstout and NN-Filter on the *TES*481 Dataset for Different Sequence Separations. 8 Å Maps

	XXStout			NN-Filter		
	F	P	R	F	P	R
$\lvert i-j\rvert \geq 6$	0.213	0.354	0.153	0.271	0.277	0.265
$\lvert i-j\rvert \geq 12$	0.159	0.315	0.107	0.235	0.247	0.224
$\lvert i-j\rvert \geq 24$	0.097	0.230	0.057	0.168	0.219	0.137
$\lvert i-j\rvert \leq 8$	0.889	0.888	0.890	0.899	0.891	0.906

In all cases the improvements come from a small reduction in P, and a much greater increase in R. In the case of 8 Å maps and contacts between residues with a sequence separation of 24 or greater, P decreases from 23% to 21.9%, while R more than doubles from 5.7% to 13.7%. Thus NN-Filter is able to predict more than twice as many long-range contacts than XXstout, with only a minor loss of accuracy.

We tested both XXstout (as part of the Distill predictor) and NN-Filter at CASP7. As NN-Filter became available during the prediction season and after some of the targets has expired (i.e., no further predictions were accepted on them) we only submitted predictions for 63 of the 95 targets (68 domains). The results in Table 7.6 refer to these 63 targets. The results are for a contact threshold of 8 Å between C_α. Both XXstout and NN-Filter are compared with other CASP7 predictors of contact maps. Proteins here are split in domains, as assessed by CASP. The table broadly confirms the results on *TES*481, with NN-Filter improving over "raw" XXstout predictions on all sequence separation bands. According to the CASP assessors, NN-Filter was the top predictor at one of the quality measures (X_d) [18].

Two examples of predictions with their filtered version are reported in Figures 7.6 and 7.7.

TABLE 7.6 CASP7 Results for XXstout (Distill) and NN-Filter versus Other CASP Predictors

$\lvert i-j \rvert$	≥12			≥24		
Server	P	R	F	P	R	F
Gadjapairings	0.148	0.486	0.227	0.136	0.502	0.214
Betapro	0.269	0.150	0.193	0.230	0.120	0.158
SVMcon	0.401	0.124	0.189	0.330	0.104	0.158
NN-Filter	0.145	0.257	0.185	0.087	0.199	0.121
PROFcon	0.184	0.125	0.149	0.123	0.094	0.107
SAM_T06	0.382	0.077	0.128	0.353	0.073	0.121
Distill	0.088	0.269	0.132	0.054	0.189	0.084
Possum	0.086	0.180	0.117	0.058	0.164	0.085
GPCpred	0.064	0.119	0.083	0.034	0.131	0.054

The Results are macro-averages, that is, averaged overall proteins in the set, and sorted by descending F. Results for 68 domains (63 targets) submitted by NN-Filter. Proteins are split into domains as assessed at CASP. All other predictors tested on the same 68 domains except: Gadjapairings (48 domains); PROFcon (66 domains).

FIGURE 7.6 Example prediction for protein 1HRUA. Left side: bottom-left true map, top-right XXStout predicted 8 Å map. Right side: bottom-left true map, top-right NN-Filtered 8 Å map. In both cases the region beside the diagonal is predicted fairly well, while far from the diagonal, interactions between secondary structure elements are poorly predicted, especially interactions between strands that are not contiguous in the sequence.

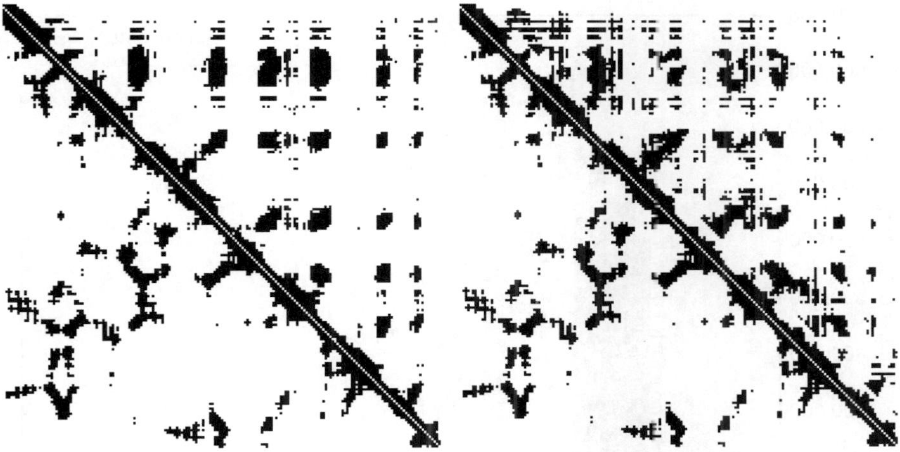

FIGURE 7.7 1HRUA. Left side: bottom-left true map, top-right XXStout predicted 12 Å map. Right side: bottom-left true map, top-right NN-Filtered 12 Å map. Similarly to the 8 Å case positions beside the diagonal are fairly well predicted, but interactions far from the diagonal are not. As for 8 Å case there is a small improvement by NN-Filter near the diagonal. Far from the diagonal, the dense patches of contacts erroneously predicted by XXStout have thinned out, although some isolated wrongly predicted contacts have appeared.

7.5. MULTI-CLASS DISTANCE MAPS

In this section we introduce a new class of distance restraints for protein structures: multi-class distance maps. We then build two predictors of four-class maps based on RNN: one *ab initio*, or relying on the sequence and on evolutionary information in the form of MSA, and one template-based, or in which homology information to known structures is provided as a further input.

We focus on a representation of the distance matrix called multi-class contact map and based on a set of categorical attributes or classes. Each class corresponds to an interval of distances that a given pair of residues may fall into. Formally, given a set of distance thresholds $\{t_k\}_{k=0\ldots T}$ (where $t_0 = 0$ and $t_T = \infty$), a multi-class contact map of a protein with N amino acids is a symmetrix $N \times N$ matrix C where the element corresponding to the amino acids in positions i and j is defined as $C_{ij} = k$ if $d(i, j) \in [t_k, t_{k+1})$. Obviously, this class of projections contains richer information than binary contact maps (so long as $T > 3$). Therefore, if thresholds are chosen carefully, native multi-class contact maps yield models of protein structures at significantly better resolutions than native binary maps [50]. Moreover, if a suitable set of distance thresholds is chosen, the number of instances in each class may be kept approximately balanced, which in turn may improve generalization

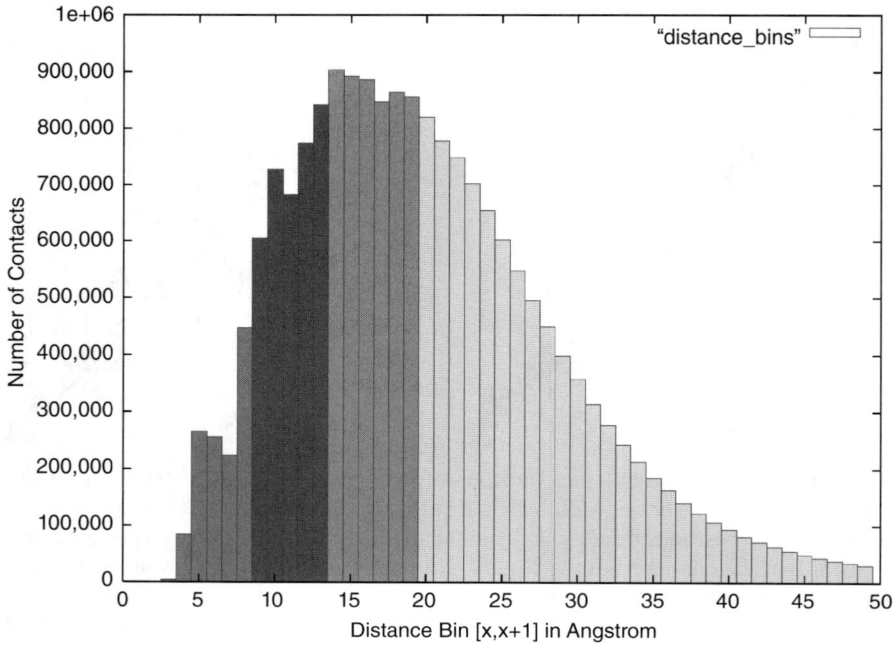

FIGURE 7.8 Distribution of contacts in [0,50] distance bins with trivial $|i - j| \leq 3$ residue contacts ignored. The classes were chosen in order to retain good distance constraints and balanced classes. Class 1 ([0,8) Å) corresponds, as a first approximation, to physical contacts.

performances of learning algorithms over the (normally unbalanced) binary prediction case.

For our experiments, we derive a set of five distance thresholds to define multi-class contact maps based on four distance intervals. As shown in Figure 7.8, the four classes are empirically chosen from the distribution of distances among amino acids in the training set, ignoring trivial pairs $|i - j| \leq 3$ and by trying to keep informative distance constraints and the classes as balanced as possibile. The resulting distance classes are [0, 8), [8, 13), [13, 19), and [19, ∞) Å. The first class roughly corresponds to physical contacts, and matches the standard contact class adopted at CASP [18]; the two middle classes are chosen to be roughly equally numerous and to span all categories of possible interaction (in Reference [16] up to 18 Å), while the fourth class represents noncontacts and is still the most numerous. Although this definition is somewhat arbitrary, our results were only minimally sensitive to small changes in the thresholds. A potential improvement beyond this choice is to automatically determine an optimal set of thresholds based on some criteria, for example, the reconstruction ability on a set of benchmarking proteins.

7.5.1. Method

7.5.1.1. Predictive Architecture. The predictive architecture for multi-class contact map prediction is essentially the same as that for binary class contact map prediction, that is, we use ensembles of 2D-RNNs trained in fivefold cross-validation as previously described in Section 7.2.1.

7.5.1.2. Datasets and Template Generation. The dataset used to train and test the predictors is extracted from the January 2007 25% pdb_select list [51]. The sequences are processed and MSA generated as described in Section 7.2.1, after which the set (S3129) contains 3129 proteins, 461,633 amino acids, and just over 100 million residue pairs. Since training is computationally very demanding we create a reduced version of S3129 from which we exclude proteins longer than 200 residues. This set contains 2452 proteins, and approximately 69 million residue pairs. All systems are trained in fivefold cross-validation. This is obtained by splitting S3129 into five approximately equal folds, then (for training purposes) removing from the folds all proteins longer than 200 residues. Testing is on the full folds, that is, including proteins longer than 200 residues. DSSP [43] assigned values for secondary structure and solvent accessibility are used during the training phase; however, predicted values are used during the testing phase. We then search the PDB for structural templates for each of the proteins as explained in Section 7.3 but against the PDB available on March 25th, 2008.

7.5.2. Results and Discussion

Table 7.3 reports the comparison between *ab initio* and template-based predictions (12_{AI} vs. 12_{TE}) for binary contact maps with a 12 Å threshold, as a function of sequence identity to the best PDB hit. The only decrease in performance for the template-based predictor compared with the *ab initio* is in the [0,10)% identity range, where the accuracy slightly decreases by 0.6%. However, the same results for multi-class maps show that template-based predictions are always more accurate on average than *ab initio* predictions (Table 7.7). Ultimately, templates improve multi-class predictions in all regions of sequence similarity (including [0,10)%), both for regions covered and regions not covered by templates.

Figures 7.9 and 7.10 show examples of a 12 Å and a multi-class map predicted for a low best hit sequence identity of 22.7%.

TABLE 7.7 Correctly Predicted Residues (%) for the *Ab Initio* (M_{AI}) and Template-Based Multi-Class Predictor (M_{TE}) as a Function of Percentage Sequence Identity (SEQ_ID) between the Query Sequences in S3129 and the Best Template Found for Each Sequence

SEQ_ID	<10	<20	<30	<40	<50	<60	<70	<80	<90	≥90	All
M_{AI}	59.3	59.4	58.4	57.3	58.3	57.4	58.3	58.5	58.2	59.9	58.8
M_{TE}	60.2	64.2	75.9	82.5	87.8	88.8	88.1	89.7	91.5	92.1	80.8

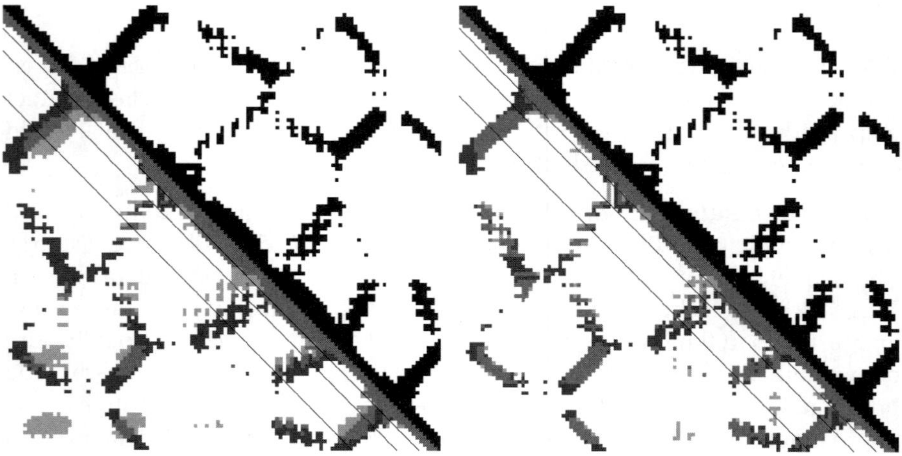

Figure 7.9 Protein 1B9LA 12 Å contact maps for *ab initio* (left) and template-based (right) predictions. The best template sequence identity is 22.7%. The top-right of each map is the true map and the bottom-left is predicted. In the predicted half white and red are true negative and positive respectively, blue and green are false negative and positive, respectively. The three black lines correspond to $|i - j| \geq 6, 12, 24$. (See color insert.)

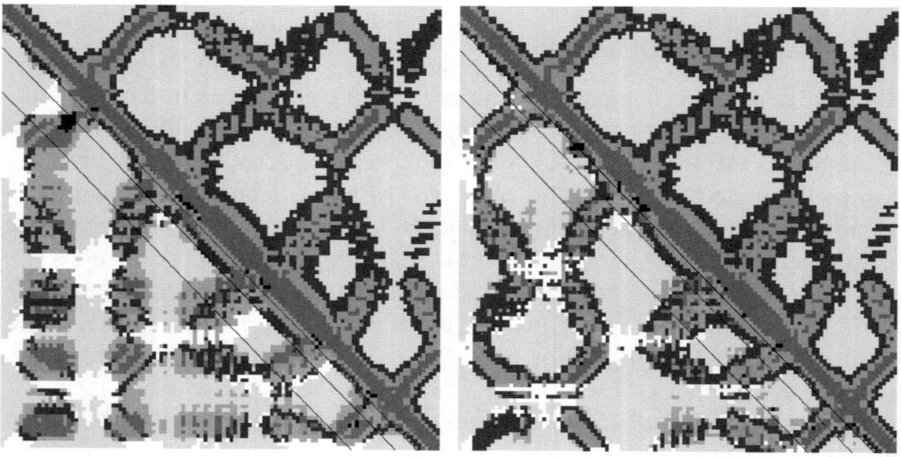

FIGURE 7.10 Protein 1B9LA multi-class contact maps for *ab initio* (left) and template-based (right) predictions. The best template sequence identity is 22.7%. The top-right of each map is the true map and the bottom left is predicted. In the predicted half red, blue, green and yellow correspond to class 0, 1, 2, and 3, respectively. The gray scale in the predicted half corresponds to falsely predicted classes. The three black lines correspond to $|i - j| \geq 6, 12, 24$. (See color insert.)

7.6. CASP8 EVALUATION

To compare our systems with other methods we participated in CASP8. We transformed multi-class maps into binary contact maps by simply recasting the second, third, and fourth distance classes into noncontacts and keeping the first one ([0,8 Å)) as contact. Table 7.8 shows the results of all the groups that participated in the map prediction category. When ranked by F measure, Distill is only outperformed by one group, SMEG-CCP. However, SMEG-CCP derives its maps by averaging 3D predictions produced by other groups and as such is not a direct predictor of contact maps. At CASP8 maps were ranked by P and only those maps with at least $\frac{L}{5}$ and $\frac{L}{10}$ were considered.

TABLE 7.8 CASP8 Evaluation. Average Values Per Target

	$\lvert i-j\rvert \geq 6$			$\lvert i-j\rvert \geq 12$			$\lvert i-j\rvert \geq 24$		
	P	R	F	P	R	F	P	R	F
SMEG-CCP (121)	0.603	0.627	0.608	0.592	0.613	0.587	0.559	0.579	0.549
Distill (122)	0.438	0.465	0.446	0.419	0.446	0.427	0.386	0.422	0.396
LEE-SERVER (79)	0.513	0.369	0.396	0.509	0.357	0.382	0.480	0.332	0.353
LEE (121)	0.470	0.362	0.372	0.465	0.343	0.356	0.433	0.308	0.321
Pairings (112)	0.616	0.191	0.289	0.599	0.210	0.308	0.508	0.192	0.281
MULTICOM-CMFR (122)	0.167	0.356	0.216	0.153	0.266	0.174	0.110	0.205	0.129
MULTICOM-RANK (122)	0.129	0.475	0.198	0.116	0.399	0.174	0.095	0.311	0.139
RR_Fang 1_(122)	0.212	0.194	0.197	0.175	0.162	0.164	0.147	0.153	0.145
MUProt (122)	0.128	0.472	0.196	0.113	0.394	0.170	0.094	0.307	0.136
FLOUDAS (37)	0.277	0.162	0.193	0.244	0.132	0.158	0.176	0.097	0.112
RR_Fang_2 (122)	0.220	0.174	0.189	0.178	0.138	0.152	0.145	0.121	0.128
LCBContacts (121)	0.350	0.136	0.189	0.324	0.099	0.149	0.291	0.065	0.104
MULTICOM (121)	0.106	0.661	0.181	0.097	0.618	0.166	0.087	0.544	0.149
Infobiotics (119)	0.094	0.445	0.152	0.082	0.365	0.130	0.072	0.289	0.110
3Dpro (118)	0.082	0.429	0.136	0.077	0.344	0.123	0.064	0.264	0.099
SAM-T08-2stage (122)	0.075	0.481	0.129	0.074	0.501	0.128	0.070	0.408	0.118
AK_RF_2 (120)	0.084	0.265	0.125	0.074	0.245	0.111	0.067	0.228	0.100
SAM-T08-server (122)	0.071	0.462	0.123	0.070	0.499	0.121	0.066	0.444	0.113
SAM-T06-server (122)	0.065	0.419	0.111	0.062	0.454	0.109	0.056	0.419	0.098
Hamilton-Torda-Huber (122)	0.171	0.105	0.124	0.154	0.073	0.097	0.139	0.050	0.070
SVMSEQ (122)	0.037	0.794	0.071	0.035	0.770	0.067	0.037	0.717	0.070
SPINE-2DA-Zhou (121)	0.146	0.018	0.032	0.146	0.022	0.037	0.146	0.030	0.049

True maps are computed from C_α instead of from C_β as in the CASP evaluation. The number in parenthesis after the predictor name is the number of targets submitted.

7.7. CONCLUSIONS

In this work we have described a machine learning pipeline for high-throughput prediction of protein contact maps. First, we introduced a system for binary map prediction and showed that the incorporation of 1D structural features, most notably contact density, into the prediction process improves prediction accuracy. Further, we described how to improve these maps by post-processing them by a filtering stage (NN-Filter) that uses first stage maps and long-range information contained in them. Then, based on the observation that protein binary contact maps are somewhat lossy representations of the structure and yield only relatively low-resolution 3D models [15], we have introduced a new representation of the contact map, the multi-class contact map.

Moreover, extending on ideas we have developed for predictors of secondary structure and solvent accessibility [48] we have presented systems for the prediction of binary and multi-class maps that use structural templates from the PDB to yield far more accurate predictions than their *ab initio* counterparts.

We have also shown that multi-class maps lead to a more balanced prediction problem than binary maps. Although it is unclear whether because of this, or because of the nature of the constraints encoded into them, the template-based systems for the prediction of multi-class maps we tested are capable of exploiting both sequence and structure information even in cases of dubious homology, significantly improving over their *ab initio* counterpart well into and below the twilight zone of sequence identity. This turns out to be only partly true, at least in our tests, for binary contact map predictors.

It is important to note how the component for homology detection in this study is basic (PSI-BLAST), and entirely modular, in that it may be substituted by any other method that finds templates without substantially altering the pipeline. Whether more subtle homology detection or fold recognition components could be substituted to PSI-BLAST, with or without retraining the underlying machine-learning systems, is the focus of our current studies. The overall pipeline, including the template-based component, is available at http://distill.ucd.ie/distill/. Protein structure predictions are based on multi-class maps and templates are automatically provided to the pipeline when available.

REFERENCES

1. S. Yooseph, G. Sutton, D.B. Rusch, A.L. Halpern, S.J. Williamson, K. Remington, J.A. Eisen, K.B. Heidelberg, G. Manning, W. Li, L. Jaroszewski, P. Cieplak, C.S. Miller, H. Li, S.T. Mashiyama, M.P. Joachimiak, C. van Belle, J.M. Chandonia, D.A. Soergel, Y. Zhai, K. Natarajan, S. Lee, B.J. Raphael, V. Bafna, R. Friedman, S.E. Brenner, A. Godzik, D. Eisenberg, J.E. Dixon, S.S. Taylor, R.L. Strausberg, M. Frazier, and J. Craig Venter. The sorcerer II global ocean sampling expedition: Expanding the universe of protein families. *PLoS Biology*, 5(3):432–466, 2007.

2. M. Adams, A. Joachimiak, R. Kim, G.T. Montelione, and J. Norvell. Meeting review: 2003 NIH protein structure initiative workshop in protein production and crystallization for structural and functional genomics. *Journal Structural Functional Genomics*, 5:1–2, 2004.

3. J. Battey, J. Kopp, L. Bordoli, R. Read, N. Clarke, and T. Schwede. Automated server predictions in CASP7. *Proteins*, 69(S8):68–82, 2007.

4. K.T. Simons, T. Kooperberg, E. Huang, and D. Baker. Assembly of protein tertiary structures from fragments with similar local sequences using simulated annealing and bayesian scoring functions. *Journal of Molecular Biology*, 268:209–225, 1997.

5. P. Larranaga, B. Calvo, R. Santana, C. Bielza, J. Galdiano, I. Inza, and J.A. Lozano. Machine learning in bioinformatics. *Briefings in bioinformatics*, 7(1):86–112, 2006.

6. A. Vullo, I. Walsh, and G. Pollastri. A two-stage approach for improved prediction of residue contact maps. *BMC Bioinformatics*, 7:180, 2006.

7. G. Pollastri, P. Baldi, A. Vullo, and P. Frasconi. Prediction of protein topologies using GIOHMMs and GRNNs. In S. Becker, S. Thrun, and K. Obermayer (Eds.), *Advances in Neural Information Processing Systems (NIPS) 15*. MIT Press, 2003.

8. M. Vendruscolo, E. Kussell, and E. Domany. Recovery of protein structure from contact maps. *Folding & Design*, 2(5):295–306, 1997.

9. D.A. Debe, M.J. Carlson, J. Sadanobu, S.I. Chan, and W.A. Goddard. Protein fold determination from sparse distance restraints: The restrained generic protein direct monte carlo method. *Journal of Physical Chemistry*, 103:3001–3008, 1999.

10. A. Aszodi, M.J. Gradwell, and W.R. Taylor. Global fold determination from a small number of distance restraints. *Journal of Molecular Biology*, 251:308–326, 1995.

11. E.S. Huang, R. Samudrala, and J.W. Ponder. Ab initio fold prediction of small helical proteins using distance geometry and knowledge-based scoring functions. *Journal of Molecular Biology*, 290:267–281, 1999.

12. J. Skolnick, A. Kolinski, and A.R. Ortiz. MONSSTER:a method for folding globular proteins with a small number of distance restraints. *Journal Molecular Biology*, 265(2):217–241, 1997.

13. P.M. Bowers, C.E. Strauss, and D. Baker. De novo protein structure determination using sparse NMR data. *Journal of Biomolecular NMR*, 18:311–318, 2000.

14. W. Li, Y. Zhang, D. Kihara, Y.J. Huang, D. Zheng, G.T. Montelione, A. Kolinski, and J. Skolnick. TOUCHSTONEX: Protein structure prediction with sparse NMR data. *Proteins: Structure, Function, and Genetics*, 53:290–306, 2003.

15. D. Baú, G. Pollastri, and A. Vullo. Distill: A machine learning approach to ab initio protein structure prediction. In S. Bandyopadhyay, U. Maulik, and J.T.L. Wang (Eds.), *Analysis of Biological Data: A Soft Computing Approach*. Singapore: World Scientific, 2007.

16. M. Vassura, L. Margara, P. Di Lena, F. Medri, P. Fariselli, and R. Casadio. FTCOMAR: Fault tolerant three-dimensional structure reconstruction from protein contact maps. *Bioinformatics*, 24(10):1313–1315, 2008.

17. J. Hu, X. Shen, Y. Shao, C. Bystroff, and M.J. Zaki. Mining protein contact maps. *BIOKDD02: Workshop on Data Mining in Bioinformatics*, Bethesda, MD, pp. 3–10, 2002.

18. J.M. Izarzugaza, O. Graña, M.L. Tress, A. Valencia, and N.D. Clarke. Assessment of intramolecular contact predictions for CASP7. *PROTEINS: Structure, Function, and Bioinformatics*, 69, Suppl 8:152–158, 2007.
19. J. Cheng and P. Baldi. A machine learning information retrieval approach to protein fold recognition. *Bioinformatics*, 22(12):1456–1463, 2006.
20. C.S. Miller and D. Eisenberg. Using inferred residue contacts to distinguish between correct and incorrect protein models. *Bioinformatics*, 24(14):1575–1582, 2008.
21. S. Wu, J. Skolnick, and Y. Zhang. Ab initio modelling of small proteins by iterative tasser simulations. *BMC Biology*, 5:17, 2007.
22. Y. Zhang. I-TASSER server for protein 3D structure prediction. *BMC Bioinformatics*, 9(40):8, 2008.
23. D.A. Pelta, J.R. Gonzal'ez, and M.M. Vega. A simple and fast heuristic for protein structure comparison. *BMC Bioinformatics*, 9(161):16, 2008.
24. M.J. Pietal, I. Tuszynska, and J.M. Bujnicki. PROTMAP2D: Visualization, comparison and analysis of 2D maps of protein structure. *Bioinformatics*, 23(11):1429–1430, 2007.
25. Y. Shao and C. Bystroff. Predicting interresidue contacts using templates and pathways. *PROTEINS: Structure, Function, and Bioinformatics*, 53:497–502, 2003.
26. M. Vendruscolo, R. Najmanovich, and E. Domany. Protein folding in contact map space. *Physical Review Letter*, 82:656–659, 1999.
27. M. Punta and B. Rost. Protein folding rates estimated from contact predictions. *Journal of Molecular Biology*, 348(3):507–12, 2005.
28. A. Schlessinger, M. Punta, and B. Rost. Natively unstructured regions in proteins identified from contact predictions. *Bioinformatics*, 23(18):2376–84, 2007.
29. M. Punta and B. Rost. PROFcon: Novel prediction of long-range contacts. *Bioinformatics*, 21:2960–2968, 2005.
30. A. Ceroni, P. Frasconi, and G. Pollastri. Learning protein secondary structure from sequential and relational data. *Neural Networks*, 18(8):1029–39, 2005.
31. S. Lise, A. Walker-Taylor, and D.T. Jones. Docking protein domains in contact space. *BMC Bioinformatics*, 7(310):14, 2006.
32. M.M. Gromiha and S. Selvaraj. Inter-residue interactions in protein folding and stability. *Progress Biophysics and Molecular Biology*, 86:235–277, 2004.
33. O. Graña, D. Baker, R.M. McCallum, J. Meiler, M. Punta, B. Rost, M.L. Tress, and A. Valencia. CASP6 assessment of contact prediction. *Proteins: Structure, Function, and Bioinformatics*, 61:214–224, 2005.
34. P. Fariselli and R. Casadio. A neural network based predictor of residue contacts in proteins. *Protein Engineering*, 12(1):15–21, 1999.
35. G. Pollastri and P. Baldi. Prediction of contact maps by recurrent neural network architectures and hidden context propagation from all four cardinal corners. *Bioinformatics*, 18, Suppl.1:S62–S70, 2002.
36. R.M. McCallum. Striped sheets and protein contact prediction. *Bioinformatics*, 20, Suppl. 1:224–231, 2004.
37. J. Cheng and P. Baldi. Improved residue contact prediction using support vector machines and a large feature set. *BMC Bioinfomatics*, 8:113, 2007.

38. S. Wu and Y. Zhang. A comprehensive assessment of sequence-based and template-based methods for protein contact prediction. *Bioinformatics*, 24(7):924–931, 2008.
39. Y. Zhao and G. Karypis. Prediction of contact maps using support vector machines. *3rd international conference on Bioinformatics and Bioengineering (BIBE)*, Bethesda, MD, pp. 26–33, 2003.
40. M. Frenkel-Morgenstern, R. Magid, E. Eyal, and S. Pietrokovski. Refining intra-protein contact prediction by graph analysis. *BMC Bioinformatics*, 8(5):5, 2007.
41. A.J.M. Martin, D. Baú, I. Walsh, A. Vullo, and G. Pollastri. Long-range information and physicality constraints improve predicted protein contact maps. *Journal of Bioinformatics and Computational Biology*, 6(5):1001–20, 2008.
42. P. Baldi and G. Pollastri. The principled design of large-scale recursive neural network architectures – DAG-RNNs and the protein structure prediction problem. *Journal of Machine Learning Research*, 4(Sep):575–602, 2003.
43. W. Kabsch and C. Sander. Dictionary of protein secondary structure: Pattern recognition of hydrogen-bonded and geometrical features. *Biopolymers*, 22(12):2577–637, 1983.
44. S.F. Altschul, T.L. Madden, and A.A. Schaffer. Gapped BLAST and PSIBLAST: A new generation of protein database search programs. *Nucleic Acids Research*, 25:3389–3402, 1997.
45. J. Moult, K. Fidelis, A. Zemla, and T. Hubbard. Critical assessment of methods of protein structure prediction (CASP)-round V. *Proteins: Structure, Function, and Genetics*, 53(6):334–339, 2003.
46. G. Pollastri and A. McLysaght. Porter: A new, accurate server for protein secondary structure prediction. *Bioinformatics*, 21(8):1719–20, 2005.
47. P. Frasconi, M. Gori, and A. Sperduti. A general framework for adaptive processing of data structures. *IEEE Transactions on Neural Networks*, 9(5):768–786, 1998.
48. G. Pollastri, A.J.M. Martin, C. Mooney, and A. Vullo. Accurate prediction of protein secondary structure and solvent accessibility by consensus combiners of sequence and structure information. *BMC Bioinformatics*, 8(201):12, 2007.
49. R.W.W. Hooft, C. Sander, and G. Vriend. The PDBFINDER database: A summary of PDB, DSSP and HSSP information with added value. *CABIOS*, 12:525–529, 1996.
50. I. Walsh, D. Baú, A.J.M. Martin, C. Mooney, A. Vullo, and G. Pollastri. Ab initio and template-based prediction of multi-class distance maps by two-dimensional recursive neural networks. *BMC Structural Biology*, 10:195, 2009.
51. U. Hobohm and C. Sander. Enlarged representative set of protein structures. *Protein Science*, 3:522–24, 1994.

CHAPTER 8

A SURVEY OF REMOTE HOMOLOGY DETECTION AND FOLD RECOGNITION METHODS

HUZEFA RANGWALA
Department of Computer Science
George Mason University
Fairfax, VA

8.1. INTRODUCTION

With recent advances in large-scale sequencing technologies, we have seen an exponential growth in protein sequence information. Currently, our ability to produce sequence information far outpaces the rate at which we can produce structural and functional information. Consequently, researchers increasingly rely on computational techniques to extract useful information from known structures contained in large databases, although such approaches remain incomplete. As such, unraveling the relationship between pure sequence information and three-dimensional (3D) structure remains one of the great fundamental problems in molecular biology.

The motivation behind the structural determination of proteins is based on the belief that structural information will ultimately result in a better understanding of intricate biological processes. Remote homology detection and fold recognition methods play a critical role in characterizing the structural and functional nature of proteins. From a comparative modeling standpoint, classifying a target protein sequence using sequence information into a protein class sharing the same fold or shape can be used as a precursor for template selection.

Formally, the *remote homology detection* problem is defined as the identification of protein pairs sharing the same evolutionary ancestry, but having less

Introduction to Protein Structure Prediction: Methods and Algorithms,
Edited by Huzefa Rangwala and George Karypis

than 30% sequence identity. *Fold recognition* is defined as the identification of protein pairs having similar structural topology and shape but no guarantee on the sequence identity. The two problems can be solved by classification of proteins into a particular class of proteins that are remote homologs or folds.

While satisfactory methods exist to detect homologs with high levels of similarity, accurately detecting homologs at low levels of sequence similarity (remote homology detection) still remains a challenging problem. Some of the most popular approaches for remote homology prediction compare a protein with a collection of related proteins using methods such as protein family profiles [1], hidden Markov models (HMMs) [2,3], Position-Specific Iterative-Basic Local Alignment Search Tool (PSI-BLAST) [4], and SAM [5]. These schemes produced models that are generative in the sense that they build a model for a set of related proteins and then check to see how well this model explains a candidate template protein.

In recent years, the performance of remote homology detection has been further improved through the use of methods that explicitly model the differences between the various protein families (classes) and build discriminative models. In particular, a number of different methods have been developed that build these discriminative models using support vector machines (SVMs) [6] and have shown, provided there is sufficient data for training, to produce results that are in general superior to those produced by either pairwise sequence comparisons or approaches based on generative models [7–14].

SVM-based approaches were designed to solve *one-versus-rest* binary classification problems, and are primarily evaluated with respect to how well each binary classifier can identify the proteins that belong to its own class (e.g., superfamily or fold). We provide a comprehensive survey of various SVM-based approaches along with other techniques used for remote homology detection and fold recognition.

SVM-based approaches usually address the design of efficient and sensitive kernel function for use in one-versus-rest classifier [15]. However, from a biologist's perspective, the problem that he or she is facing (and would like to solve) is identifying the most likely superfamily or fold (or a short list of candidates) that a particular protein belongs to. This is essentially a multi-class classification problem, in which given a set of K classes, we assign a protein sequence to one of them. We present the problem of building SVM-based multi-class classification models for remote homology prediction and fold recognition in the context of Structural Classification of Proteins (SCOP) classification scheme. We present a comprehensive study of different approaches for building such classifiers and also include hierarchical information present in the SCOP classification database [16].

8.2. LITERATURE REVIEW

Over the past few years, there have been a continual series of advances for methods to identify homologous relationships between protein pairs (both remote homologs as well as folds). More than 5000 research articles are indexed in PubMed (http://www.ncbi.nlm.nih.gov/sites/entrez) showing relevance to the term "fold recognition." Reviews by Fariselli et al. [17], Wan and Xu [18], Lindahl and Elofsson [19], and Jones and Thornton [20] describe the widely used and developed computational approaches for remote homology detection and fold recognition.

Methods to identify remote homologs and folds can be categorized into two categories, that is, prediction-based approaches and comparative approaches. A further level of categorization of these methods is dependent on the information used, that is, sequence information only, or sequence-structure information, also called threading approaches [21]. However, the most popular and successful methods (as established by the Critical Assessment of Protein Structure Prediction [CASP] experiments [22–24]) are consensus-based methods that use a combination of different techniques. Such methods are also known as *metaservers*.

In this section we review the methods developed for remote homology detection and fold recognition with a particular emphasis on kernel-based methods that use sequence information only.

8.2.1. Sequence-Based Comparative Methods

Pairwise sequence alignments are the most straightforward set of techniques to identify homologous relationships. These methods use dynamic programming-based alignments like Needleman-Wunsch [25] and Smith-Waterman algorithms [26], or heuristic-based alignments like BLAST [27], and FASTA [28] with different substitution scoring matrices [29], and sophisticated gap modeling [30] for pairwise sequence comparisons.

Over the years, the sensitivity of sequence alignment has been improved by using PSI-BLAST profiles [4] or profile HMMs generated using HMMER [31] or SAM [5] to capture evolutionary information. Sequence-to-profile scoring schemes used in methods like DIALIGN [32], FPS [33] and profile-to-profile scoring schemes in methods like HHSearch [34], FORTE [35], HMAP [21], and PICASSO [36,37] have shown far superior performance than just sequence-based methods for identification of homologous pairs with sequence identity less than 30%. The review article by Wan and Xu [18] lists all the different pairwise sequence-sequence, sequence-profile, and profile-profile alignment methods developed in the past several years.

8.2.2. Threading-Based or Sequence-Structure Methods

Since the 3D structures of proteins show better conservation during evolution in comparison to sequences, several methods use predicted protein structure information for remote homology detection and fold recognition. In fact there are several cases where protein pairs are structurally similar but share no sequence identity.

Protein threading [38–41] refers to approaches for structure prediction that "thread" a target sequence through the backbone structures of a collection of template protein structures, and compute a fitness function using the sequence-structure alignment. Such threading methods with varying scoring functions, template structure libraries, and sequence-structure alignment algorithms are used extensively for fold recognition [18].

Similar to the threading approaches are methods that build 3D structural profiles [42–45], or even use secondary structure (two-dimentional profiles) together with primary sequence [39]. These 3D profiles extend the notion of sequence profile providing spatial as well as environmental specific information (in case of FUGUE [44]) for amino acid residues. Since secondary structure of proteins is more likely to be conserved than its sequences, several methods like TOPITS [39] use predicted secondary structure for the target, and true definitions for template as additional information for the alignment. All such methods that use structure information for the template have shown to improve the performance for remote homology detection and fold recognition in comparison with pure sequence-based methods [17,18].

8.2.3. Sequence-Based Prediction Methods

There are several machine learning-based approaches that use HMMs, Neural Networks (NNs), SVMs, and Conditional Random Field (CRF). In particular, profile HMMs discussed earlier produce a probabilistic representation for profiles [31], but have shown to build accurate generative models using multiple track HMMs [5,46,47]. SAM-T04 [48] is a method that uses NNs to map the similarities between pairs of proteins, and also evaluate the alignment.

A large set of remote homology detection and fold recognition prediction approaches explicitly model the differences between the various protein families (classes) and build discriminative models. In particular, a number of different methods have been developed that build these discriminative models using SVMs [6] and have shown, provided there is sufficient data for training, to produce results that are in general superior to those produced by either pairwise sequence comparisons or approaches based on generative models [7–16]. SVMs are primarily used to solve the binary classification problem and as such these methods build one-versus-rest classification models that are evaluated with respect to how well each binary classifier can identify the proteins that belong to its own modeled class. Different SVM-based methods primarily differ in the definition of kernel function between protein sequences and are reviewed in detail in Section 8.2.3.1.

Conditional random fields-based methods [49] can also build discriminative models like SVMs, and have shown to be used for fold recognition [50].

A comparatively faster approach than using alignments and SVM-based models for remote homology detection is a recurrent NN method called the Long Short-Term Memory (LSTM) [51]. LSTM automatically determines discriminating patterns and uses correlations between these patterns for the classification models. However, this method does not achieve comparable performance to the state-of-the-art SVM-based methods.

TABLE 8.1 Kernel Methods for Remote Homology Detection and Fold Recognition

Method	Key Features
SVM-Fisher [7]	Profile HMMs for families.
SVM-pairwise [8]	Pairwise sequence similarity with sequences in training.
Spectrum kernel [9]	Exactly identical short subsequences.
Mismatch kernel [10]	Almost identical short subsequences.
eMotif kernel [54]	Common local motifs in eMotif database.
SVM-I-sites [11]	Common local motifs in I-sites library.
Cluster kernel [56]	Use of neighborhood sequences
LA kernel [13]	Direct local alignment using BLOSUM62.
SVM-HMMSTR [12]	HMM information extracted from local motifs.
Profile kernel [14]	Almost identical short subsequences scored using profiles.
SW-PSSM [15]	Profile-based direct local alignment.
AF-PSSM [15]	Direct profile-based subsequence scoring.
Oligomer kernel [57]	Similar scoring oligomers.
LSA kernel [58]	LSA-based selection of common subsequences.
Genetic kernel [55]	Genetic programming-based motif selection.

Ie et al. [52] developed schemes for combining the outputs of a set of binary SVM-based classifiers for primarily solving the remote homology prediction problem. This was in parallel to our work [16] that used the output of one-versus-rest classifiers in conjunction with hierarchical information from the SCOP [53] database for developing multi-class classification models for remote homology detection and fold recognition.

8.2.3.1. Kernel Methods. We review the methods that develop kernel function for use in an SVM-based one-versus-rest classifiers, and are also listed in Table 8.1.These kernel functions can be thought of as a measure of similarity between sequences. Different kernels correspond to different notions of similarity and can lead to discriminative functions with different performance.

There are two widely used approaches for deriving kernel functions for protein sequences. The first approach constructs them by first choosing an appropriate vector representation for the sequences, and then taking the inner product (or a function derived from them) between these representations as a kernel for the sequences [7,9,10], whereas the second approach derives a valid kernel function from an explicit protein sequence similarity measure that has been shown to be biologically relevant [8,13,15,15].

One of the early attempts with such feature-space-based approaches is the SVM-Fisher method [7], in which a profile HMM model is estimated on a set of proteins belonging to the positive class and used to extract a vector representation for each protein. Another approach is the SVM-pairwise scheme [8], which represents each sequence as a vector of pairwise similarities between all sequences in the training set. A relatively simpler feature space that contains all possible short subsequences ranging from three to eight amino acids (*k*mers) is explored in a series of papers (Spectrum kernel [9], Mismatch kernel [10], and Profile kernel [14]). All three of these methods represent a sequence X as a vector in this feature space and differ on the scheme they employ to actually determine if a particular dimension u (i.e., *k*mer) is present (i.e., has a non-zero weight) in X's vector or not. The Spectrum kernel considers u to be present if X contains u as a substring, the Mismatch kernel considers u to be present if X contains a substring that differs with u in at most a predefined number of positions (i.e., mismatches), whereas the Profile kernel considers u to be present if X contains a substring whose Position-Specific Scoring Matrix-based (PSSM-based) ungapped alignment score with u is above a user-supplied threshold. An entirely different feature space is explored by the eMotif kernel [54], SVM-I-sites [11], and SVM-HMMSTR [12] methods that take advantage of a set of local structural motifs (SVM-I-sites) and their relationships (SVM-HMMSTR). Recently, genetic programming was used to determine motif features for developing remote homology detection methods [55], referred to as the Genetic-programming kernel (GPKernel). The Cluster kernel [56] is unique that it computes the features based on the sequence membership in a cluster or neighborhood.

An alternative to measuring pairwise similarity through a dot product of vector representations is to calculate an explicit protein similarity measure. The recently developed LA kernel method [13] represents one such example of a *direct kernel function*. This scheme measures the similarity between a pair of protein sequences by taking into account all the optimal local alignment scores with gaps between all of their possible subsequences. The experiments presented in Reference [13] show that this kernel is superior to previously developed schemes that do not take into account sequence profiles and that the overall classification performance improves by taking into account all local alignments.

Another kernel, called the oligomer kernel [57], uses similar ideas to the window-based profile [15] kernels and spectrum kernels [9] to compute the similarity between the protein sequences, and another method uses latent semantic analysis [58] for selective use of *k*mer to develop kernel functions.

In our previous work [15] we developed two new classes of kernel functions derived directly from explicit similarity measures and utilize sequence profiles. The first class, referred to as window-based, determines the similarity between a pair of sequences by using different schemes to combine ungapped alignment scores of certain fixed-length subsequences. The second, referred to as local alignment-based, determines the similarity between a pair of sequences using Smith-Waterman alignments and a position independent affine gap model, optimized for the characteristics of the scoring system. Both kernel classes utilize profiles constructed automatically via PSI-BLAST and employ a profile-to-profile scoring scheme we develop by extending a recently introduced profile alignment method [37].

Experiments on two benchmarks derived from SCOP, one designed to detect remote homologs and the other designed to identify folds, show that these kernels produce results that are substantially better than those produced by all other state-of-the-art SVM-based methods. Some of the major observations from the experimental evaluation performed in the study [15] include: (i) As was the case with a number of studies on the accuracy of protein sequence alignment [37,59,60], the proper use of sequence profiles lead to dramatic improvements in the overall ability to detect remote homologs and identify proteins that share the same structural fold. (ii) Kernel functions that are constructed by directly taking into account the similarity between the various protein sequences tend to outperform schemes that are based on a feature-space representation (where each dimension of the space is constructed as one of the k-possibilities in a k-residue long subsequence or using structural motifs [I-sites] in the case of SVM-HMMSTR). (iii) Finally, time-tested methods for comparing protein sequences based on optimal local alignments (as well as global and local-global alignments), when properly optimized for the classification problem at hand, lead to kernel functions that are in general superior to those based on either short subsequences (e.g., Spectrum, Mismatch, Profile, or window-based kernel functions) or local structural motifs (e.g., SVM-HMMSTR). The fact that these widely used methods produce good results in the context of SVM-based classification is reassuring as to the validity of these approaches and their ability to capture biologically relevant information. For the rest of this chapter we base our discussion only on the profile-based local alignment kernels to train one-versus-rest classifiers, and further improve the classification accuracy by incorporating multi-class hierarchical information from the SCOP database.

8.2.4. Consensus Methods

Metaservers or methods that use a consensus of different approaches have found great success in the CASP competition [24,61]. Pcons [62–64], 3D-Jury [65], Frankenstein's monster [66], FAMS-ACE [67] and 3D-SHOTGUN [68] are examples of consensus-based methods that vary in the methods that are combined and how they are combined. Linear programming [63] as well as NN-based methods [69] has been used to select which of the input predictions are the most reliable, and how they could be combined to produce the best results.

Within the kernel formulation framework, methods have been introduced to combine heterogeneous information using semi-definite programming [70,71], second order cone programming [72], semi-infinite linear programming [73], and Bayesian framework [74]. Such multiple kernel learning approaches are not strictly consensus approaches but do intend to integrate multiple sources of information. However, such methods are not easily scalable and lead to very complex frameworks.

8.2.5. Pairwise Structure-Based Methods

Using structure-structure approaches are reliable than the sequence-structure and sequence-sequence methods described earlier [18] in detecting homologs pairs in most cases, but this does depend on the pairs in consideration [17]. The limitation of such approaches is that knowledge of 3D structure is needed for the proteins (target as well as templates).

Structure alignment methods like ACE [75], MAMMOTH [76], and MUSTANG [77] are a popular way to identify homologs. There is no clear agreement regarding the best structure alignment method for the detection problems [17]. Recently, the MAMMOTH structure alignment program was used within the SVM framework [78], on similar lines to the SVM-pairwise formulation that used sequence alignments. This use of structure information showed better performance for the classification problems.

8.3. HIERARCHICAL MULTI-CLASS CLASSIFIERS

Having designed highly accurate SVM-based binary classifiers, we studied the best way to combine the predictions of a set of SVM-based binary classifiers to solve the multi-class classification problem and assign a protein sequence to a particular superfamily or fold. We compared the multi-class classification performance between schemes that combine binary classifiers and schemes that directly build an SVM-based multi-class classification model.

In parallel research, Ie et al. [52] developed schemes for combining the outputs of a set of binary SVM-based classifiers for primarily solving the remote homology prediction problem. Specifically borrowing ideas from error-correcting output codes [79–81], they developed schemes that use a separate learning step to learn how to best scale the outputs of the binary classifiers such that when combined with a scheme that assigns a protein to the class whose corresponding scaled binary SVM prediction is the highest, it achieves the best multi-class prediction performance. In addition, for remote homology prediction in the context of the SCOP [53] hierarchical classification scheme, they also studied the extent to which the use of such hierarchical information can further improve the performance of remote homology prediction. Their experiments showed that these approaches lead to better results than the traditional schemes that use either the maximum functional output [82] or those based on fitting a sigmoid function [83].

Motivated by the positive results of Ie et al's. [52] work, we further study the problem of building SVM-based multi-class classification models for remote homology detection and fold recognition in the context of the SCOP protein classification scheme. We present a comprehensive study of different approaches for building such classifiers including (i) schemes that directly build an SVM-based multi-class model, (ii) schemes that employ a second-level learner to combine the predictions generated by a set of binary SVM-based classifiers, and (iii) schemes that build and combine binary classifiers for various levels of the SCOP hierarchy. In addition, we present and study three different approaches for combining the outputs of the binary classifiers that lead to hypothesis spaces of different complexity and expressive power, for the general K-way multi-class classification problem.

8.3.1. Methods and Algorithms

Given a set of m training examples $\{(x_1, y_1), \ldots, (x_m, y_m)\}$, where example x_i is drawn from a domain $\mathcal{X} \subseteq \Re^n$ and each of the label y_i is an integer from the set $\mathcal{Y} = \{1, \ldots, K\}$, the goal of the K-way classification problem is to learn a model that assigns the correct label from the set $\mathcal{Y}$ to an unseen test example. This can be thought of as learning a function f : $\mathcal{X} \rightarrow \mathcal{Y}$, which maps each instance x to an element y of $\mathcal{Y}$.

8.3.1.1. Direct SVM-Based K-Way Classifier Solution. One way of solving the K-way classification problem using SVMs is to use one of the many multi-class formulations for SVMs that were developed over the years [84–88]. These algorithms extend the notions of separating hyperplanes and margins and learn a model that directly separates the different classes.

In this study we evaluate the effectiveness of one of these formulations that was developed by Crammer and Singer [88], which leads to reasonably efficient optimization problems.

This formulation aims to learn a matrix W of size $K \times n$ such that the predicted class y^* for an instance x is given by

$$y^* = \arg\max_{i=1}^{K}\{\langle W_i, x\rangle\}, \tag{8.1}$$

where W_i is the ith row of W whose dimension is n.

This formulation models each class i by its own hyperplane (whose normal vector corresponds to the ith row of the matrix W) and assigns an example x to the class i that maximizes its corresponding hyperplane distance.

W itself is learned from the training data following a maximum margin with soft constraints formulation that gives rise to the following optimization problem [88]:

$$\begin{array}{ll} \min & \frac{1}{2}\beta W^2 + \sum_{i=1}^{m} \xi_i \\ \text{subject to:} & \forall i,z \;\; \langle W_y, x_i\rangle + \delta_{y_i,z} - \langle W_z, x_i\rangle \geq 1 - \xi_i \end{array} \tag{8.2}$$

where $\xi_i \geq 0$ are slack variables, $\beta > 0$ is a regularization constant, and $\delta_{y_i,z}$ is equal to 1 if $z = y_i$, and 0 otherwise.

As in the binary SVM the dual version of the optimization problem and the resulting classifier depends only on the inner products, which allows us to use any of the recently developed protein string kernels.

8.3.1.2. Merging* K *One-versus-Rest Binary Classifiers. An alternate way of solving the K-way classification problem in the context of SVM is to first build a set of K one-versus-rest binary classification models $\{f_1, f_2, \ldots, f_K\}$, use all of them to predict an instance x, and then based on the predictions of these base classifiers $\{f_1(x), f_2(x), \ldots, f_K(x)\}$ assign x to one of the K classes [79,80,83].

MaxClassifier. A common way of combining the predictions of a set of K one-versus-rest binary classifiers is to assume that the K outputs are directly comparable and assign x to the class that achieved the highest one-versus-rest prediction value, that is, the prediction y^* for an instance x is given by

$$y^* = \arg\max_{i=1}^{K}\{f_i(x)\}. \tag{8.3}$$

However, the assumption that the output scores of the different binary classifiers are directly comparable may not be valid, as different classes may be of different sizes and/or less separable from the rest of the dataset—indirectly affecting the nature of the binary model that was learned.

Cascaded SVM-Learning Approaches. A promising approach that has been explored in combining the outputs of K binary classification models is to formulate it as a cascaded learning problem in which a second-level model is trained on the outputs of the binary classifiers to correctly solve the multi-class classification problem [52,79,80].

A simple model that can be learned is the scaling model in which the final prediction for an instance x is given by

$$y^* = \arg\max_{i=1}^{K}\{w_i f_i(x)\}, \tag{8.4}$$

where w_i is a factor used to scale the functional output of the ith classifier, and the set of K w_i scaling factors make up the model that is being learned during the second-level training phase [52]. We will refer to this scheme as the *scaling* scheme.

An extension to the above scheme is to also incorporate a shift parameter s_i with each of the classes and learn a model whose prediction is given by

$$y^* = \arg\max_{i=1}^{K}\{w_i f_i(x) + s_i\}. \tag{8.5}$$

The motivation behind this model is to emulate the expressive power of the z-score approach (i.e., $w_i = 1/\sigma_i$, $s_i = -\mu_i/\sigma_i$) but learn these parameters using a maximum margin framework. We will refer to this as the *scale & shift* model.

Finally, a significantly more complex model can be learned by directly applying the Crammer–Singer multi-class formulation on the outputs of the binary classifiers. Specifically, the model corresponds to a $K \times K$ matrix W and the final prediction is given by

$$y^* = \arg\max_{i=1}^{K}\{\langle w_i, f_i(x)\rangle\}. \tag{8.6}$$

where $f(x) = (f_1(x), f_2(x), \ldots, f_K(x))$ is the vector containing the K outputs of the one-versus-rest binary classifiers. We will refer to this as the *Crammer–Singer* model.

Comparing the scaling approach to the Crammer–Singer approach we can see that the Crammer–Singer methodology is a more general version and should be able to learn a similar weight vector as the scaling approach. In the scaling approach, there is a single weight value associated with each of the classes. However, the Crammer–Singer approach has a whole weight vector of dimensions equal to the number of features per class. During the training stage, for the Crammer–Singer approach if all the weight values $w_{i,j} = 0, \forall i \neq j$ the weight vector will be equivalent to the scaling weight vector. Thus we would expect the Crammer–Singer setting to fit the dataset much better during the training stage.

8.3.1.3. Use of Hierarchical Information. One of the key characteristics of remote homology prediction and fold recognition is that the target classes are naturally organized in a hierarchical fashion. This hierarchical organization is evident in the tree-structured organization of the various known protein structures that is produced by the widely used protein structure classification schemes of SCOP [53], Class, Architecture, Topology, and Homologous superfamily (CATH) [89], and Families of Structurally Similar Proteins (FSSP) [90].

In our study we use the SCOP classification database to define the remote homology prediction and fold recognition problems. SCOP organizes the proteins into four primary levels (class, fold, superfamily, and family) based on structure and sequence similarity. Within the SCOP classification, the problem of remote homology prediction corresponds to that of predicting the superfamily of a particular protein under the constraint that the protein is not similar to any of its descendant families, whereas the problem of fold recognition corresponds to that of predicting the fold (i.e., second level of hierarchy) under the constraint that the protein is not similar to any of its descendant superfamilies. These two constraints are important because if they are violated, then we are actually solving either the family or the remote homology prediction problems, respectively.

The questions that arise are whether or not and how we can take advantage of the fact that the target classes (either superfamilies or folds) correspond to a level in a hierarchical classification scheme, so as to improve the overall classification performance.

The approach investigated in this study is primarily motivated by the different schemes presented in Section 8.3.1.2 to combine the functional outputs of multiple one-versus-rest binary classifiers. A general way of doing this is to learn a binary one-versus-rest model for each or a subset of the nodes of the hierarchical classification scheme, and then combine these models using an approach similar to the Crammer–Singer scheme described in Section 8.3.1.1.

For example, assume that we are trying to learn a fold-level multi-class model with K_f folds where K_s is the number of superfamilies that are descendants of these K_f folds, and K_c is the number of classes that are ancestors in the SCOP hierarchy. Then, we will build $K_f + K_s + K_c$ one-versus-rest binary classifiers for each one of the folds, superfamilies, and classes and use them to obtain a vector of $K_f + K_s + K_c$ predictions for a test sequence x. Then, using the Crammer–Singer approach, we can learn a second-level model W of size $K_f \times (K_f + K_s + K_c)$ and use it to predict the class of x as

$$y^* = \arg\max_{i=1}^{K}\{\langle W_i, f_i(x)\rangle\}, \tag{8.7}$$

where $f(x)$ is a vector of size $K_f + K_s + K_c$ containing the outputs of the binary classifiers.

Note that the output space of this model is still the K_f possible folds, but the model combines information both from the fold-level binary classifiers as well as the binary classifiers for superfamily- and class-level models.

In addition to Crammer–Singer-type models, the hierarchical information can also be used to build simpler models by combining selective subsets of binary classifiers. In our study we experimented with such models by focusing only on the subsets of nodes that are characteristic for each target class and are uniquely determined by it. Specifically, given a target class (i.e., superfamily or fold), the path starting from that node and moving upward toward the root of the classification hierarchy uniquely identifies a set of nodes corresponding to higher level classes containing the target class. For example, if the target class is a superfamily, this path will identify the superfamily itself, its corresponding fold, and its corresponding class in the SCOP hierarchy.

We can construct a second level classification model by combining for each target class the predictions computed by the binary classifiers corresponding to the nodes along these paths. Specifically, for the remote homology recognition problem, let K_s be the number of target superfamilies, $f_i(x)$ the prediction computed by the ith superfamily classifier, $f_{\wedge_i^f}(x)$ the prediction of the fold classifier corresponding to the ith superfamily, and $f_{\wedge_i^c}(x)$ the prediction of the class-level classifier corresponding to the ith superfamily, then we can express the prediction for instance x as

$$y^* = \arg\max_{i=1}^{K_f}\left\{w_i f_i(x) + w_{\wedge_i^f} f_{\wedge_i^f}(x) + w_{\wedge_i^c} f_{\wedge_i^c}(x)\right\}, \tag{8.8}$$

where w_i, $w_{\wedge_i^f}$ and $w_{\wedge_i^c}$ are scaling factors learned during training of the second-level model. The notation $\wedge$ denotes the predecessor or ancestral relationship operator. In particular for the superfamily i, we say that it lies under fold $\wedge_i^f$, which is under the class $\wedge_i^c$.

Note that the underlying model in Equation 8.8 is essentially an extension of the scaling model of Equation 8.4 as it linearly combines the predictions of the binary classifiers of the ancestor nodes.

In a similar fashion, we can use the scale and shift type approach for every node in the hierarchical tree. This allows for an extra shift parameter to be associated with each of the nodes being modeled. Note that similar approaches can be used to define models for fold recognition, where a weight vector is learned to combine the target fold-level node along with its specific class-level node. A model can also be learned by not considering all the levels along the paths to the root of the tree.

The generic problem of classifying within the context of a hierarchical classification system has recently been studied by the machine learning community and a number of alternative approaches have been developed [91–93].

8.3.2. Theoretical Foundations

In this section we describe how we learn the weight vectors that were introduced for integrating the binary classifiers. We learn the weight vector by a cross-validation setup on the training set using the structured SVM algorithm [91], which works on the principles of large-margin discriminative classifiers. We also introduce the notion of loss functions that are optimized for the different integration methods.

8.3.2.1. Structured Output Spaces. The various models introduced in Sections 8.3.1.2 and 8.3.1.3 can be expressed using a unified framework that was recently introduced for learning in structured output spaces [91,94–96]. This framework [91] learns a discriminant function $F: \mathcal{X} \times \mathcal{Y} \rightarrow \Re$ over input/output pairs from which it derives predictions by maximizing F over the response variable for a specific given input x. Hence, the general form of the hypothesis h is

$$h(x;\theta) = \arg\max_{y \in Y} \{F(x, y; \theta)\}, \tag{8.9}$$

where θ denotes a parameter vector. Function F is a θ-parameterized family of functions that is designed such that $F(x,y;\theta)$ achieves the maximum value for the correct output y. Among the various choices for F, if we focus on those that are linear in a combined feature representation of inputs and outputs, $\psi(x,y)$, then Equation 8.9 can be rewritten as [91]:

$$h(x;\theta) = \arg\max_{y \in Y} \{\langle \theta, \Psi(x, y) \rangle\}. \tag{8.10}$$

The specific form of Ψ depends on the nature of the problem and it is this flexibility that allows us to represent the hypothesis spaces introduced in Sections 8.3.1.2 and 8.3.1.3 in terms of Equation 8.10.

For example, consider the simple scaling scheme for the problem of fold recognition (Equation 8.4). The input space consists of the $f(x)$ vectors of the binary predictions and the output space Y consists of the set of K_f folds (labeled from 1 … K_f). Given an example x belonging to fold i (i.e., $y = i$), the function $\Psi(x, y)$ maps the (x, y) pair onto a K_f-size vector whose ith entry (i.e., the entry corresponding to x's fold) is set to $f_i(x)$ and the remaining entries are set to zero. Then, from Equation 8.10 we have

$$h(x;\theta) = \arg\max_{i=1}^{K_f} \{\langle\theta, \Psi(x, i)\rangle\} = \arg\max_{i=1}^{K_f} \theta_i f_i(x), \tag{8.11}$$

which is similar to Equation 8.4 with θ representing the scaling vector w.

Similarly, for the scale & shift approach (Equation 8.5), the $\Psi(x, y)$ function maps the (x, y) pair onto a feature space of size $2K_f$, where the first K_f dimensions are used to encode the scaling factors and the second K_f dimensions are used to encode the shift factors. Specifically, given an example x belonging to fold i, $\Psi(x, y)$ maps (x, y) onto the vector whose ith entry is $f_i(x)$, its $(2i)^{th}$ entry is one, and the remaining entries are set to zero. Then, from Equation 8.10 we have

$$\begin{aligned} h(x;\theta) &= \arg\max_{i=1}^{K_f} \{\langle\theta, \Psi(x, i)\rangle\} \\ &= \arg\max_{i=1}^{K_f} \{\theta_i f_i(x) + \theta_{2i}\}, \end{aligned} \tag{8.12}$$

which is equivalent to Equation 8.5, with the first half of θ corresponding to the scale vector w, and the second half corresponding to the shift vector s.

Finally, in the case of the Cramer-Singer approach, the $\Psi(x, y)$ function maps (x, y) onto a feature space of size $K_f \times K_f$. Specifically, given a sequence x belonging to fold i, $\Psi(x, y)$ maps (x, y) onto the vector whose K_f entries starting at $(i - 1)K_f$ are set to $f(x)$ (i.e., the fold prediction outputs) and the remaining $(K_f - 1)K_f$ entries are set to zero. Then, by rewriting Equation 8.10 in terms of the above combined input–output representation, we get

$$\begin{aligned} h(x;\theta) &= \arg\max_{i=1}^{K_f} \{\langle\theta, \Psi(x, i)\rangle\} \\ &= \arg\max_{i=1}^{K_f} \left\{\sum_{j=1}^{K_f} \theta_{(i-1)K_f+j} f_j(x)\right\}. \end{aligned} \tag{8.13}$$

This is equivalent to Equation 8.6, as θ can be viewed as the matrix W with K_f rows and K_f columns, equivalent to Equation 8.6, where θ is equal to the matrix W.

SVM-Struct, an efficient way of learning the vector θ of Equation 8.10 has been formulated as a convex optimization problem [91]. In this approach θ is learned subject to the following m nonlinear constraints

$$\forall i : \max_{y \in \mathcal{Y}} \{\langle\theta, \Psi(x_i, y)\rangle\} < \langle\theta, \Psi(x_i, y)\rangle \tag{8.14}$$

The SVM-Struct [91] algorithm is an efficient way of solving the above optimization problem in which the m nonlinear inequalities are replaced by $|\mathcal{Y}| - 1$

linear inequalities resulting in a total of $m(|\mathcal{Y}| - 1)$ linear constraints and θ is learned using the maximum-margin principle leading to the following hard-margin problem [91]:

$$\begin{aligned} &\min_{\theta} \frac{1}{2}\|\theta\|_2^2 \\ &\text{subject to } \langle \theta, \Psi(x_i,y_i) - \Psi(x_i,y) \rangle \geq 1 \\ &\quad \forall i, \forall y \in \{\mathcal{Y}/y_i\}. \end{aligned} \tag{8.15}$$

This hard-margin problem can be converted to a soft-margin equivalent to allow errors in the training set. This is done by introducing a slack variable, ξ, for every nonlinear constraint of Equation 8.14. The soft-margin problem is expressed as [91]:

$$\begin{aligned} &\min_{\theta,\xi} \frac{1}{2}\|\theta\|_2^2 + \frac{C}{n}\sum_{i=1}^{n} \xi_i, \\ &\text{subject to } \langle \theta, \Psi(x_i, y_i) - \Psi(x_i, y) \rangle \geq 1 - \xi_i \\ &\quad \forall i, \xi_i \geq 0, \forall_i, \forall y \in \{\mathcal{Y}/y_i\}. \end{aligned} \tag{8.16}$$

The results of classification depend on the value C, which is the misclassification cost that determines the trade-off between the generalization capability of the model being learned and maximizing the margin. It needs to be optimized to prevent under-fitting and over-fitting the data during the training phase.

8.3.2.2. Loss Functions. The loss function plays a key role while learning the θ parameter for the SVM-Struct optimization. Until now, our discussion focused on zero-one loss that assigns a penalty of one for a misclassification and zero for a correct prediction.

However, in cases where the class sizes vary significantly across the different folds, such a zero–one loss function may not be the most appropriate as it may lead to models where the rare class instances are often misclassified. For this reason, an alternate loss function is used, in which penalty for a misclassification is inversely proportional to the class size. This implies that the misclassification of examples belonging to smaller classes weigh higher in terms of the loss. This loss function is referred to as the balanced loss [52].

In case of the SVM-Struct formulation, the balanced loss can be optimized by reweighting the definition of separation that can be done indirectly by rescaling the slack variables ξ_i in the constraint inequalities (Equation 8.16).

While using the hierarchical information in the cascaded learning approaches (Section 8.3.1.3) we experimented with a *weighted* loss function where a larger penalty was assigned when the predicted label did not share the same ancestor compared with the case when the predicted and true class labels shared the same ancestors. This variation did not result in an improvement compared with the zero–one and balanced loss.

8.4. EXPERIMENTAL RESULTS

8.4.1. Problem Setup

We evaluated the classification performance of the profile-based kernels on a set of protein sequences obtained from the SCOP database [53]. We formulated two different classification problems. The first was designed to evaluate the performance of the algorithms for the problem of homology detection when the sequences have low sequence similarities (i.e., the remote homology detection problem), whereas the second was designed to evaluate the extent to which the profile-based kernels can be used to identify the correct fold when there are no apparent sequence similarities (i.e., the fold detection problem).

8.4.1.1. Remote Homology Detection (Superfamily Detection). Within the context of the SCOP database, remote homology detection was simulated by formulating it as a superfamily classification problem. For each family, the protein domains within the family were considered positive test examples, and protein domains within the superfamily but outside the family were considered positive training examples. Negative examples for the family were chosen from outside of the positive sequences' fold, and were randomly split into training and test sets in the same ratio as the positive examples. For example, we can visually represent the setup for the remote homology detection problem in terms of the test and training sets for a particular superfamily class (fold a.2.1) in Figure 8.1.

8.4.1.2. Fold Detection. Employing the same dataset and overall methodology as in remote homology detection, we simulated fold detection by formu-

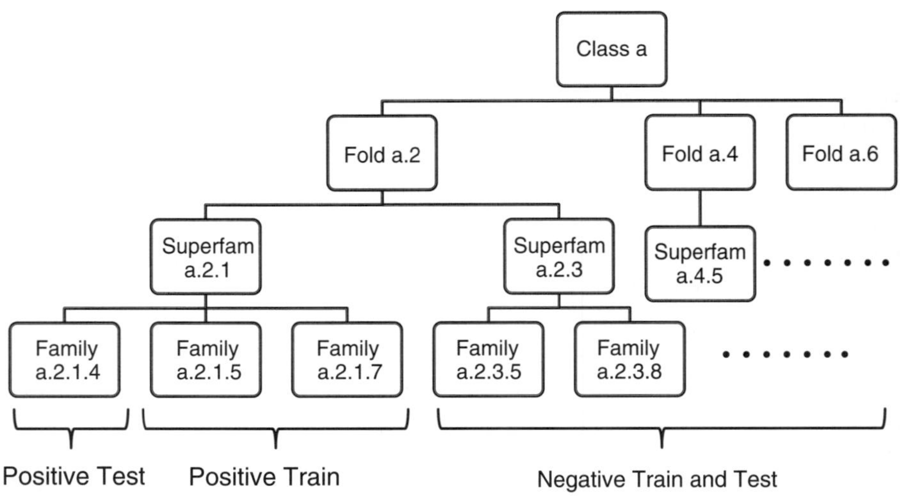

FIGURE 8.1 SCOP hierarchy tree showing the training and test instances setup for the remote homology detection problem.

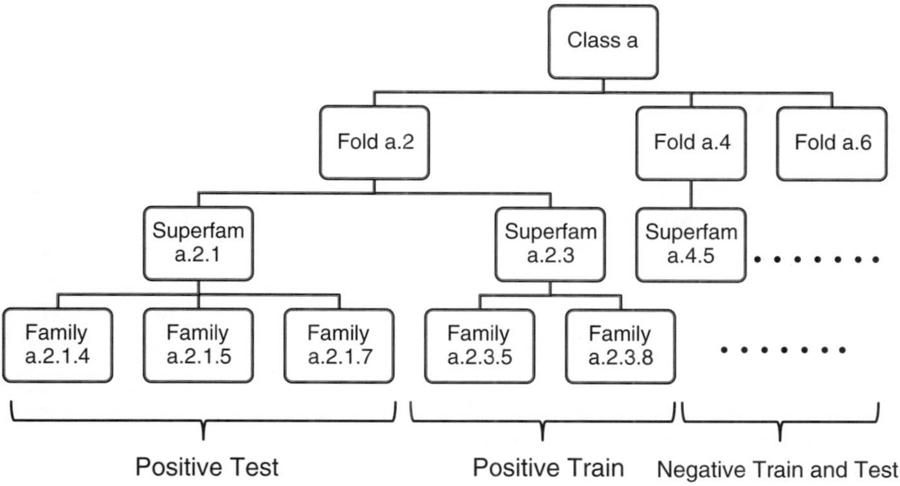

FIGURE 8.2 SCOP hierarchy tree showing the training and test instances setup for the fold recognition problem.

lating as a fold classification within the context of SCOP's hierarchical classification scheme. In this setting, protein domains within the same superfamily were considered to be positive test examples, and protein domains within the same fold but outside the superfamily were considered positive training examples.

Since the positive test and training instances were members of different superfamilies within the same fold, this new problem is significantly harder than remote homology detection, as the sequences in the different superfamilies did not have any apparent sequence similarity [53]. For example, we can visually represent the setup for the fold recognition problem in terms of the test and training sets for a particular fold class (fold a.2) in Figure 8.2.

8.4.2. Dataset Description

We assessed the performance of our multi-class classification schemes for solving the remote homology detection and fold recognition on four datasets. The first dataset, referred to as sf95 (superfamily—95%), was created by Ie et al. [52] to evaluate the performance of the multi-class classification algorithms that they developed (sf95 was designed by Ie et al. [52], whereas the other three datasets, referred to as sf40 [superfamily—40%], fd25 [fold—25%], and fd40 [fold—40%], were created for this study; sf40, fd25, and fd40 are available at the supplementary web site). The sf95 dataset was derived from SCOP 1.65, whereas the other datasets were derived from SCOP 1.67. Table 8.2 summarizes the characteristics of these datasets and presents various sequence similarity statistics.

TABLE 8.2 Dataset Statistics

Statistic	sf95	sf40	fd25	fd40
ASTRAL filtering	95%	40%	25%	40%
Number of sequences	2115	1119	1294	1651
Number of folds	25	25	25	27
Number of superfamilies	47	37	137	158
Avg. pairwise similarity	12.8%	11.5%	11.6%	11.4
Avg. Max. similarity	63.5%	33.9%	32.2%	34.3
Avg. pairwise similarity (within folds)	25.6%	17.9%	16.7%	17.4
Avg. pairwise similarity (outside folds)	10.4%	11.03%	11.2%	11.0

The percent similarity between two sequences is computed by aligning the pair of sequences using a local alignment with BLOSUM62 matrix, a gap opening of 5.0 and gap extension of 1.0. "Avg. pairwise similarity" is the average of all the pairwise percent identities; "Avg. max. similarity" is the average of the maximum pairwise percent identity for each sequence, that is, it measures the similarity to its most similar sequence. The "Avg. pairwise similarity (within folds)" and the "Avg. pairwise similarity (outside folds)" are the average of the average pairwise percent sequence similarity within the same fold and outside the fold for a given sequence.

Datasets, sf95, and sf40 are designed to evaluate the performance of remote homology detection and were derived by taking only the domains with less than 95% and 40% pairwise sequence identity according to ASTRAL [97], respectively. This set of domains was further reduced by keeping only the domains belonging to folds that (i) contained at least three superfamilies and (ii) one of these superfamilies contained multiple families. For sf95, the resulting dataset contained 2115 domains organized in 25 folds and 47 superfamilies, whereas for sf40, the resulting dataset contained 1119 domains organized in 25 folds and 37 superfamilies.

Datasets, fd25, and fd40 were designed to evaluate the performance of fold recognition and were derived by taking only the domains with less than 25% and 40% pairwise sequence identity, respectively. This set of domains was further reduced by keeping only the domains belonging to folds that (i) contained at least three superfamilies and (ii) at least three of these superfamilies contained more than three domains. For fd25, the resulting dataset contained 1294 domains organized in 25 folds and 137 superfamilies, whereas for fd40, the resulting dataset contained 1651 domains organized in 27 folds and 158 superfamilies.

8.4.3. Evaluation Methodology

The performance of the classification algorithms was assessed using the zero–one error (ZE) and the balanced error (BE) rate [52]. The ZE rate treats the various classes equally and penalizes each misclassification by one. The balanced error rate accounts for classes of varying size and assigns a lower penalty for misclassifying a test instance belonging to a larger class. The motivation behind balanced error is that larger classes are easier to predict just by chance alone and it rewards a classifier if it can also correctly predict test instances belonging to smaller classes. Following the common practice [52], we set the error of each misclassification to be inversely proportional to its true class size.

In addition, the performance of the various classifiers was evaluated using the previously established approach for evaluating fold recognition methods introduced in References [19,44] that does not penalize for certain types of misclassification. For each test instance, this scheme ranks the various classes from the most to the least likely and a test instance is considered to be correctly classified if its true class is among the highest ranked n classes (i.e., topn_n). The classes in the ranked list that are within the same next higher level SCOP ancestral class are ignored and do not count toward determining the highest ranked n classes. That is, in the case of fold recognition, the folds that are part of the same SCOP class as the test instance are ignored and they do not count in determining the n highest ranked predictions. Similarly, in case of remote homology detection, this scheme ignores the superfamilies that are part of the same SCOP fold as the test sequence. Using a small value for n that is greater than one, this measure assesses the ability of a classifier to find the correct class among its highest ranked predictions, and by penalizing only for the substantially wrong mis-predictions (i.e., different SCOP classes or folds), it can assess the severity of the misclassification of the different schemes. In our experiments we computed the error rates for $n = 1$ and $n = 3$.

8.4.3.1. Training Methodology. For each dataset we separated the proteins into test and training sets, ensuring that the test set is never used during any parts of the learning phase.

For sf95 and sf40 (fd25 and fd40), the test set is constructed by selecting from each superfamily (fold) all the sequences that are part of one family (superfamily). Thus during training, the dataset does not contain any sequences that are homologous (remote homologous) to the sequences in the test set and thus allows us to evaluate/assess remote homology prediction (fold recognition) performance. This is a standard protocol for evaluating remote homology detection and fold recognition and has been used in a number of earlier studies [7,13–15].

The models for the two-level approach can be learned in three phases by first splitting the training set into two sets, one for learning the first-level model and the other for learning the second-level model. In the first phase, the k one-versus-rest binary classifiers are trained using the training set for the first level. In the second phase, each of these k classifiers is used to predict the training set for the second level. Finally, in the third phase, the second-level model is trained using these predictions. However, due to the limited size of the available training set, we followed a different approach that does not require us to split the training set into two sets. This approach was motivated by the cross-validation methodology and is similar to that used in Reference

[52]. This approach first partitions the entire training set into 10 equal-size parts. Each part is then being predicted using the k binary classifiers that were trained on the remaining nine parts. At the end of this process, each training instance has been predicted by a set of k binary classifiers, and these prediction outputs serve as training samples for the second-level learning (using the structured SVM algorithm). Having learned the second-level model using the prediction values obtained from the first-level classifiers, we take the entire training set as a whole and retrain the first-level models.

During the evaluation stage, we compute the prediction for our untouched test dataset in two steps. In the first step, we compute the prediction values from the first-level model, which are used as features to obtain the final prediction values from the second-level model. These predictions are then evaluated using the ZE and the BE.

8.4.3.2. Model Selection. As in all learning methods, the performance of SVM methods depends on the trade-off between the variance and bias in the model. Based on the width of the margin, the model learned may be complex but will not generalize to unseen test samples. The parameter C in SVM-Struct and β for the ranking perceptron algorithm controls the complexity as well as generalizability, and hence the classification error on test samples.

As such, we perform a model selection or parameter selection step. To perform this exercise fairly, we split our test set into two equal halves of similar distributions, namely sets A and B. Using set A, we vary the controlling parameters and select the best performing model for set A. We use this selected model and compute the accuracy for set B. We repeat the above steps by switching the roles of A and B. The final accuracy results are the average of the two runs. While using the SVM-Struct program we let C take values from the set {0.0001, 0.001, 0.005, 0.01, 0.02, 0.05, 0.1, 0.2, 0.5, 1.0, 2.0, 4.0, 8.0, 10.0, 16.0, 32.0, 64.0, 128.0}. While using the perceptron algorithm we let the margin β take values in the set {0.0001, 0.005, 0.001, 0.05, 0.01, 0.02, 0.5, 0.1, 1.0, 2.0, 5.0, 10.0}.

8.4.4. Performance Results

The performance of various schemes in terms of ZE and BE is summarized in Tables 8.3 and 8.4 for remote homology detection and fold recognition, respectively. The results in Tables 8.3 and 8.4 are obtained by optimizing the balanced loss function. We use four datasets—sf95 and sf40 for remote homology detection, and fd25 and fd40 for fold recognition. We use the standard ZE and BE rates for performance assessment (described in the methods section). The schemes that are included in these tables are the following: (i) the MaxClassifier, (ii) the direct K-way classifier, (iii) the two-level learning approaches based on either the superfamily- or the fold-level binary classifiers, and (iv) the two-level learning approaches that also incorporate hierarchical information. For all two-level learning approaches (with and without hierarchical information)

TABLE 8.3 Zero–One and Balanced Error Rates for the Remote Homology Detection Problem Optimized for the Balanced Loss Function

		sf95		sf40	
		ZE	BE	ZE	BE
MaxClassifier		14.7	30.0	21.0	29.7
Direct *K*-Way Classifiers		11.5	23.1	10.9	13.0
Two-Level Approaches without Hierarchy Information					
SVM-Struct	Scaling	9.0	15.9	11.8	15.7
	Scale & shift	10.7	19.9	12.1	15.1
	Crammer–Singer	11.6	19.4	13.0	16.3
Two-Level Approaches with Fold-Level Nodes					
SVM-Struct	Scaling	11.2	19.6	14.7	21.4
	Scale & Shift	10.1	19.3	12.1	16.9
	Crammer–Singer	14.7	26.0	13.0	18.2
Two-Level Approaches with Class-Level and Fold-Level Nodes					
SVM-Struct	Scaling	11.2	20.2	13.0	18.8
	Scale & Shift	13.5	24.7	12.1	16.8
	Crammer–Singer	14.7	26.1	13.0	17.5

ZE and BE denote the zero–one error and balanced error percent rates, respectively. The results were obtained by optimizing the balanced loss function.

these tables show the results obtained by using the scaling, scale & shift, and Crammer–Singer schemes to construct the second-level classifiers.

These tables also show the performance achieved by incorporating different types of hierarchical information in the two-level learning framework. For remote homology prediction they present results that combine information from the ancestor nodes (fold and fold + class), whereas for fold recognition they present results that combine information from ancestor nodes (class), descendant nodes (superfamily), and their combination (superfamily + class).

8.4.4.1. Performance of Direct K-Way Classifier. Comparing the direct *K*-way classifiers against the MaxClassifier approach we see that the error rates achieved by the direct approach are smaller for both the remote homology detection and fold recognition problems. In many cases these improvements are substantial. For example, the direct *K*-way classifier achieves a 10.9% ZE rate for sf40 compared with a corresponding error rate of 21.0% achieved by MaxClassifier. In addition, unlike the common belief that learning SVM-based direct multi-class classifiers is computationally very expensive, we found that the Crammer–Singer formulation that we used required time that

TABLE 8.4 Zero–One and Balanced Error Rates for the Fold Recognition Problem Optimized for the Balanced Loss Function

		fd25		fd40	
		ZE	BE	ZE	BE
MaxClassifier		42.0	60.3	44.4	64.6
Direct *K*-Way Classifiers		38.4	52.3	40.4	56.9
Two-Level Approaches without Hierarchy Information					
SVM-Struct	Scaling	39.9	52.7	30.8	46.6
	Scale & Shift	39.9	52.5	28.1	42.8
	Crammer–Singer	41.3	50.5	31.1	43.3
Two-Level Approaches with Class-Level Nodes					
SVM-Struct	Scaling	39.2	52.4	29.9	45.0
	Scale & Shift	38.1	51.6	29.0	41.7
	Crammer-Singer	41.7	50.9	29.9	41.7
Two-Level Approaches with Superfamily-Level Nodes					
SVM-Struct	Scaling	40.2	52.6	30.5	44.5
	Scale & Shift	40.6	52.7	29.3	42.8
	Crammer–Singer	38.8	48.8	31.0	44.9
Two-Level Approaches with Superfamily-Level and Class-Level Nodes					
SVM-Struct	Scaling	41.0	50.9	33.7	44.6
	Scale & Shift	39.5	51.5	29.3	42.3
	Crammer–Singer	40.2	51.9	30.2	42.4

ZE and BE denote the zero–one error and balanced error percent rates, respectively. The results were obtained by optimizing the balanced loss function.

is comparable to that required for building the various binary classifiers used by the MaxClassifier approach.

8.4.4.2. Non-Hierarchical Two-Level Learning Approaches. Analyzing the performance of the various two-level classifiers that do not use hierarchical information we see that the scaling and scale & shift schemes achieve better error rates than those achieved by the Crammer–Singer scheme. Since the hypothesis space of the Crammer–Singer scheme is a superset of the hypothesis spaces of the scaling and scale & shift schemes, we found this result to be surprising at first. However, in analyzing the characteristics of the models that were learned we noticed that the reason for this performance difference is the fact that the Crammer–Singer scheme tended to over-fit the data. This was evident by the fact that the Crammer–Singer scheme had lower error rates on the training set than either the scaling or scale & shift schemes (results not reported here). Since Crammer–Singer's linear model has more parameters than the other two schemes, due to the fact that the size of the training set for all three of them is the same and rather limited, such over-fitting can easily occur. Note that these observations regarding these three approaches hold for the two-level approaches that use hierarchical information as well.

Comparing the performance of the S and SS schemes against that of the direct K-way classifier we see that the two-level schemes are somewhat worse for sf40 and fd25 and considerably better for sf95 and fd40. In addition, they are consistently and substantially better than the MaxClassifer approach across all four datasets.

8.4.4.3. Hierarchical Two-Level Learning Approaches. The results for remote homology detection show that the use of hierarchical information does not improve the overall error rates. The situation is different for fold recognition in which the use of hierarchical information leads to some improvements for fd40, especially in terms of balanced error (Table 8.4). Also, these results show that adding information from ancestor nodes is in general better than adding information from descendant nodes, and combining both types of information can sometimes improve the classification performance.

Even though the use of hierarchical information does not improve the overall classification accuracy, as the results in Tables 8.5 and 8.6 show, it does

TABLE 8.5 Error Rates ($topn_1$, $topn_3$) for the Remote Homology Detection Problem

		sf95		sf40	
		$topn_1$	$topn_3$	$topn_1$	$topn_3$
Two-Level Approaches without Hierarchy Information					
SVM-Struct	Scaling	7.5	2.6	10.1	3.8
	Scale & Shift	9.0	2.0	10.1	3.4
	Crammer–Singer	8.1	1.7	9.2	2.5
Two-Level Approaches with Fold-Level Nodes					
SVM-Struct	Scaling	4.6	0.9	6.3	1.7
	Scale & Shift	4.0	0.9	5.0	1.7
	Crammer–Singer	6.6	2.6	5.0	1.7
Two-Level Approaches with Fold-Level and Class-Level Nodes					
SVM-Struct	Scaling	5.2	1.7	5.5	1.7
	Scale & Shift	5.8	2.3	4.2	2.1
	Crammer–Singer	6.6	2.0	5.0	1.7

The results shown in the table are optimized for the balanced loss function.

TABLE 8.6 Error Rates ($topn_1$, $topn_3$) for the Fold Recognition Problem

		fd25		fd40	
		$topn_1$	$topn_3$	$topn_1$	$topn_3$
	Two-Level Approaches without Hierarchy Information				
SVM-Struct	Scaling	38.5	24.5	25.6	15.4
	Scale & Shift	37.4	24.8	24.7	15.1
	Crammer–Singer	36.3	22.7	25.0	13.4
	Two-Level Approaches with Class-Level Nodes				
SVM-Struct	Scaling	36.7	21.9	20.6	11.9
	Scale & Shift	36.3	21.6	21.2	12.2
	Crammer–Singer	37.1	22.3	25.3	13.4
	Two-Level Approaches with Superfamily-Level Nodes				
SVM-Struct	Scaling	39.9	24.5	27.9	19.5
	Scale & Shift	39.6	23.4	25.3	16.0
	Crammer–Singer	40.6	27.3	26.7	15.1
	Two-Level Approaches with Superfamily-Level and Class-Level Nodes				
SVM-Struct	Scaling	39.2	25.2	20.6	13.7
	Scale & Shift	38.5	23.0	20.9	12.2
	Crammer–Singer	37.1	23.7	24.1	12.5

The results shown in the table are optimized for the balanced loss function.

reduce the severity of the misclassifications. Comparing the $topn_1$ and $topn_3$ error rates for the two sets of schemes, we see that by incorporating hierarchical information they achieve consistently lower error rates. For remote homology detection, there is more than 50% reduction in the error rate due to the addition of fold- and class-level information, whereas somewhat smaller gains (4–20%) are obtained for fold recognition by incorporating class-level information. It is also interesting to note that there is no reduction in error rates by addition of descendant node information, that is, superfamily-level in case of fold recognition problem.

8.5. CONCLUSIONS

We have presented a comprehensive survey on the different methods for remote homology detection and fold recognition. We placed particular emphasis on reviewing kernel-based methods as a potential approach to solve the problem. We briefly summarized kernel functions derived using direct profile-to-profile sequence similarity measures. To aid the biologists we presented methods to integrate the outputs of the predictions obtained from our first-level binary classification models to solve the general K-way classification model in the context of remote homology detection and fold recognition.

Our results show that direct K-way SVM-based formulations and algorithms based on the two-level learning paradigm are quite effective in solving these problems and achieve better results than those obtained by using a set of binary one-versus-rest SVM-based classifiers. Moreover, our results and analysis showed that the two-level schemes that incorporate predictions from binary models constructed for ancestral categories within the SCOP hierarchy tend to not only lead to lower error rates but also reduce the number of errors in which a superfamily is assigned to an entirely different fold and a fold is predicted as being from a different SCOP class.

These classification methods lay the foundation for several other similar prediction problems in computation biology-protein subcellular localization, and function prediction. We used these SVM-based classifiers for selecting the best fold and searching within the predicted fold for a suitable template for a given target during the CASP7 competition.

REFERENCES

1. M. Gribskov, A.D. McLachlan, and D. Eisenberg. Profile analysis: Detection of distantly related proteins. *Proceedings of the National Academy of Sciences U S A*, 84:4355–4358, 1987.
2. A. Krogh, M. Brown, I. Mian, K. Sjolander, and D. Haussler. Hidden Markov models in computational biology: Applications to protein modeling. *Journal of Molecular Biology*, 235:1501–1531, 1994.
3. P. Baldi, Y. Chauvin, T. Hunkapiller, and M. McClure. Hidden Markov models of biological primary sequence information. *Proceedings of the National Academy of Sciences U S A*, 91:1053–1063, 1994.
4. S.F. Altschul, L.T. Madden, A.A. SchÃd'ffer, J. Zhang, Z. Zhang, W. Miller, and D.J. Lipman. Gapped blast and psi-blast: A new generation of protein database search programs. *Nucleic Acids Research*, 25(17):3389–402, 1997.
5. K. Karplus, C. Barrett, and R. Hughey. Hidden Markov models for detecting remote protein homologies. *Bioinformatics*, 14(10):846–856, 1998.
6. V. Vapnik. *Statistical Learning Theory*. New York: John Wiley, 1998.
7. T. Jaakkola, M. Diekhans, and D. Hassler. A dscriminative framework for detecting remote protein homologies. *Journal of Computational Biology*, 7(1/2):95–114, 2000.
8. L. Liao and W.S. Noble. Combining pairwise sequence similarity and support vector machines for detecting remote protein evolutionary and structural relationships. In *Proceedings of the International Conference on Research in Computational Molecular Biology*, Washington, DC: ACM, 2002, pp. 225–232.
9. C. Leslie, E. Eskin, and W.S. Noble. The spectrum kernel: A string kernel for svm protein classification. In *Proceedings of the Pacific Symposium on Biocomputing*, Stanford, CA: World Scientific Press, 2002, pp. 564–575.

10. C. Leslie, E. Eskin, W.S. Noble, and J. Weston. Mismatch string kernels for svm protein classification. *Advances in Neural Information Processing Systems*, 20(4):467–476, 2003.
11. Y. Hou, W. Hsu, M. L. Lee, and C. Bystroff. Efficient remote homology detection using local structure. *Bioinformatics*, 19(17):2294–2301, 2003.
12. Y. Hou, W. Hsu, M. L. Lee, and C. Bystroff. Remote homology detection using local sequence-structure correlations. *Proteins: Structure, Function, and Bioinformatics*, 57:518–530, 2004.
13. H. Saigo, J.P. Vert, N. Ueda, and T. Akutsu. Protein homology detection using string alignment kernels. *Bioinformatics*, 20(11):1682–1689, 2004.
14. R. Kuang, E. Ie, K. Wang, K. Wang, M. Siddiqi, Y. Freund, and C. Leslie. Profile-based string kernels for remote homology detection and motif extraction. *Computational Systems Bioinformatics*, 3:152–160, 2004.
15. H. Rangwala and G. Karypis. Profile-based direct kernels for remote homology detection and fold recognition. *Bioinformatics*, 21(23):4239–4247, 2005.
16. H. Rangwala and G. Karypis. Building multiclass classifiers for remote homology detection and fold recognition. *BMC Bioinformatics*, 7:455, 2006.
17. P. Fariselli, I. Rossi, E. Capriotti, and R. Casadio. The wwwh of remote homolog detection: The state of the art. *Briefings in Bioinformatics*, 8(2):78–87, 2007.
18. X.-F. Wan and D. Xu. Computational methods for remote homolog identification. *Current Protein and Peptide Science*, 6(6):527–546, 2005.
19. E. Lindahl and A. Elofsson. Identification of related proteins on family, superfamily and protein level. *Journal of Molecular Biology*, 295:613–625, 2000.
20. D. Jones and J. Thornton. Protein fold recognition. *Journal of Computer Aided Molecular Design*, 7(4):439–456, 1993.
21. C.L. Tang, L. Xie, I.Y.Y. Koh, S. Posy, E. Alexov, and B. Honig. On the role of structural information in remote homology detection and sequence alignment: New methods using hybrid sequence profiles. *Journal of Molecular Biology*, 334(5):1043–1062, 2003.
22. L.N. Kinch, J.O. Wrabl, S.S. Krishna, I. Majumdar, R.I. Sadreyev, Y. Qi, J. Pei, H. Cheng, and N.V. Grishin. Casp5 assessment of fold recognition target predictions. *Proteins*, 53(6):395–409, 2003.
23. C. Venclovas and M. Margelevicius. Comparative modeling in casp6 using consensus approach to template selection, sequence-structure alignment, and structure assessment. *Proteins: Structure, Function, and Bioinformatics*, 7:99–105, 2005.
24. G. Wang, Y. Jin, and R.L. Dunbrack. Assessment of fold recognition predictions in casp6. *Proteins*, 61(7):46–66, 2005.
25. S.B. Needleman and C.D. Wunsch. A general method applicable to the search for similarities in the amino acid sequence of two proteins. *Journal of Molecular Biology*, 48:443–453, 1970.
26. T.F. Smith and M.S. Waterman. Identification of common molecular subsequences. *Journal of Molecular Biology*, 147:195–197, 1981.
27. S.F. Altschul, W. Gish, E.W. Miller, and D.J. Lipman. Basic local alignment search tool. *Journal of Molecular Biology*, 215:403–410, 1990.
28. W. Pearson. Rapid and sensitive sequence comparisons with fastp and fasta. *Methods in Enzymology*, 183:63–98, 1990.

29. H. Blockeel and L. De Raedt. Top-down induction of first-order logical decision trees. *Artificial Intelligence*, 101(1–2):285–297, 1998.
30. D. Gusfield. *Algorithms on Strings, Trees, and Sequences: Computer Science and Computational Biology*. New York: Cambridge University Press, 1997.
31. S. Eddy. Profile hidden Markov models. *Bioinformatics*, 14(9):755–763, 1998.
32. B. Morgenstern, K. French, A. Dress, and T. Werner. Dialign: Finding local similarities by multiple sequence alignment. *Bioinformatics*, 14:290–294, 1998.
33. W.N. Grundy and T.L. Bailey. Family pairwise search with embedded motif models. *Bioinformatics*, 15(6):463–470, 1999.
34. J. Soding, A. Biegert, and A.N. Lupas. The hhpred interactive server for protein homology detection and structure prediction. *Nucleic Acids Research*, 33(2):W244–W248, 2005.
35. K. Tomii and Y. Akiyama. Forte: A profile-profile comparison tool for protein fold recognition. *Bioinformatics*, 20(4):594–595, 2004.
36. A. Heger and L. Holm. Picasso: Generating a covering set of protein family profiles. *Bioinformatics*, 17(3):272–279, 2001.
37. D. Mittelman, R. Sadreyev, and N. Grishin. Probabilistic scoring measures for profile-profile comparison yield more accurate short seed alignments. *Bioinformatics*, 19(12):1531–1539, 2003.
38. D.T. Jones, W.R. Taylor, and J.M. Thorton. A new approach to protein fold recognition. *Nature*, 358:86–89, 1992.
39. B. Rost, R. Schneider, and C. Sander. Protein fold recognition by prediction-based threading. *Journal of Molecular Biology*, 270(3):471–480, 1997.
40. J. Xu. Fold recognition by predicted alignment accuracy. *IEEE/ACM Transaction on Computational Biology and Bioinformatics*, 2(2):157–165, 2005.
41. D.T. Jones. Genthreader: An efficient and reliable protein fold recognition method for genomic sequences. *Journal of Molecular Biology*, 287(4):797–815, 1999.
42. J.U. Bowie, R. Luethy, and D. Eisenberg. A method to identify protein sequences that fold into a known three-dimensional structure. *Science*, 253:797–815, 1991.
43. D. Rice and D. Eisenberg. A 3d-1d substitution matrix for protein fold recognition that includes predicted secondary structure of the sequence. *Journal of Molecular Biology*, 267:1026–1038, 1997.
44. J. Shi, T.L. Blundell, and K. Mizuguchi. Fugue: Sequence-structure homology recognition using environment-specific substitution tables and structure-dependent gap penalties. *Journal of Molecular Biology*, 310:243–257, 2001.
45. L. Kelley, R. MacCallum, and M. Sternberg. Enhanced genome annotation using structural profiles in the program 3d-pssm. *Journal of Molecular Biology*, 299:523–544, 2000.
46. K. Karplus, R. Karchin, J. Draper, J. Casper, Y. Mandel-Gutfreund, M. Diekhans, and R. Hughey. Combining local-structure, fold-recognition, and new fold methods for protein structure prediction. *Proteins: Structure, Function, and Genetics*, 53:491–496, 2003.
47. K. Karplus, R. Karchin, G. Shackelford, and R. Hughey. Calibrating e-values for hidden markov models using reverse-sequence null models. *Bioinformatics*, 21(22):4107–4115, 2005.

48. K. Karplus, S. Katzman, G. Shackleford, M. Koeva, J. Draper, B. Barnes, M. Soriano, and R. Hughey. Sam-t04: What is new in protein-structure prediction for casp6. *Proteins*, 61(7):135–142, 2005.

49. J.D. Lafferty, A. McCallum, and F.C.N. Pereira. Conditional random fields: Probabilistic models for segmenting and labeling sequence data. *ICML '01: Proceedings of the Eighteenth International Conference on Machine Learning*, pp. 282–289. San Francisco, CA: Morgan Kaufmann Publishers Inc, 2001.

50. Y. Liu, J. Carbonell, P. Weigele, and V. Gopalakrishnan. Protein fold recognition using segmentation conditional random fields (scrfs). *Journal of Computational Biology*, 13(2):394–406, 2006.

51. S. Hochreiter, M. Heusel, and K. Obermayer. Fast model-based protein homology detection without alignment. *Bioinformatics*, 23(14):1728–1736, 2007.

52. E. Ie, J. Weston, W.S. Noble, and C. Leslie. Multi-class protein fold recognition using adaptive codes. *Proceedings of the 2005 International Conference on Machine Learning*, pp. 329–336. Bonn, Germany: ACM, 2005.

53. A.G. Murzin, S.E. Brenner, T. Hubbard, and C. Chothia. Scop: A structural classification of proteins database for the investigation of sequences and structures. *Journal of Molecular Biology*, 247:536–540, 1995.

54. Asa Ben-Hur and Douglas Brutlag. Remote homology detection: A motif based approach. *Bioinformatics*, 19(1):i26–i33, 2003.

55. T. HÃĕndstad, A. J.H. Hestnes, and P. Saetrom. Motif kernel generated by genetic programming improves remote homology and fold detection. *BMC Bioinformatics*, 8:23, 2007.

56. J. Weston, A. Elisseeff, D. Zhou, C. Leslie, and W. S. Noble. Protein ranking: from local to global structure in protein similarity network. *Proceedings of the National Academy of Sciences U S A*, 101:6559–6563, 2004.

57. T. Lingner and P. Meinicke. Remote homology detection based on oligomer distances. *Bioinformatics*, 22(18):2224–2231, 2006.

58. Q.-W. Dong, X.-L. Wang, and L. Lin. Application of latent semantic analysis to protein remote homology detection. *Bioinformatics*, 22(3):285–290, 2006.

59. G. Wang and R. L. Dunbrack JR. Scoring profile-to-profile sequence alignments. *Protein Science*, 13:1612–1626, 2004.

60. M. Marti-Renom, M. Madhusudhan, and A. Sali. Alignment of protein sequences by their profiles. *Protein Science*, 13:1071–1087, 2004.

61. J. Cheng and P. Baldi. A machine learning information retrieval approach to protein fold recognition. *Bioinformatics*, 22(12):1456–1463, Jun 2006.

62. B. Wallner and A. Elofsson. Pcons5: Combining consensus, structural evaluation and fold recognition scores. *Bioinformatics*, 21(23):4248–4254, 2005.

63. B. Wallner, H. Fang, and A. Elofsson. Automatic consensus-based fold recognition using pcons, proq, and pmodeller. *Proteins*, 53(6):534–541, 2003.

64. B. Wallner, P. Larsson, and A. Elofsson. Pcons.net: Protein structure prediction meta server. *Nucleic Acids Research*, 35(Web server issue):W369–W374, 2007.

65. K. Ginalski, A. Elofsson, D. Fischer, and L. Rychlewski. 3d-jury: A simple approach to improve protein structure predictions. *Bioinformatics*, 19(8):1015–1018, 2003.

66. J. Kosinski, I.A. Cymerman, M. Feder, M.A. Kurowski, J.M. Sasin, and J.M. Bujnicki. A "frankenstein's monster" approach to comparative modeling: Merging

the finest fragments of fold-recognition models and iterative model refinement aided by 3d structure evaluation. *Proteins*, 53(6):369–379, 2003.

67. G. Terashi, M. Takeda-Shitaka, K. Kanou, M. Iwadate, D. Takaya, A. Hosoi, K. Ohta, and H. Umeyama. Famsace: A combined method to select the best model after remodeling all server models. *Proteins*, 69(8):98–107, 2007.
68. D. Fischer. 3d-shotgun: A novel, cooperative, fold-recognition meta-predictor. *Proteins*, 51(3):434–441, 2003.
69. J. Lundström, L. Rychlewski, J. Bujnicki, and A. Elofsson. Pcons: A neural-network-based consensus predictor that improves fold recognition. *Protein Science*, 10(11):2354–2362, 2001.
70. G.R.G. Lanckriet, T.D. Bie, N. Cristianini, M.I. Jordan, and W.S. Noble. A statistical framework for genomic data fusion. *Bioinformatics*, 20(16):2626–2635, 2004.
71. G.R.G. Lanckriet, N. Cristianini, P. Bartlett, L.E. Ghaoui, and M.I. Jordan. Learning the kernel matrix with semi-definite programming. *Journal of Machine Learning Research*, 5:27–72, 2004.
72. F. Bach, G.R.G. Lanckriet, and M.I. Jordan. Multiple kernel learning, conic duality, and the smo algorithm. *Proceedings of the 2004 International Conference on Machine Learning*. Alberta, Canada: ACM, 2004.
73. G. Ratsch, S. Sonnenburg, and C. Schafer. Learning interpretable svms for biological sequence classification. *BMC Bioinformatics*, 7(S9), 2006.
74. T. Damoulas and M.A. Girolami. Probabilistic multi-class multi-kernel learning: On protein fold recognition and remote homology detection. *Bioinformatics*, 24(10):1264–1270, 2008.
75. I. Shindyalov and P. E. Bourne. Protein structure alignment by incremental combinatorial extension (ce) of the optimal path. *Protein Engineering*, 11:739–747, 1998.
76. D. Lupyan, A. Leo-Macias, and A.R. Ortiz. A new progressive-iterative algorithm for multiple structure alignment. *Bioinformatics*, 21(15):3255–3263, 2005.
77. A. S. Konagurthu, J. C. Whisstock, P. J. Stuckey, and A. M. Lesk. Mustang: A multiple structural alignment algorithm. *Proteins: Structure, Function, and Bioinformatics*, 64(3):559–574, 2006.
78. J. Qiu, M. Hue, A. Ben-Hur, J.-P. Vert, and W.S. Noble. A structural alignment kernel for protein structures. *Bioinformatics*, 23(9):1090–1098, 2007.
79. T.G. Dietterich and G. Bakiri. Solving multiclass learning problems via error-correcting output codes. *Journal of Artificial Intelligence Research*, 2:263–286, 1995.
80. E.L. Allwein, R.E. Schapire, and Y. Singer. Reducing multiclass to binary: A unifying approach for margin classifiers. *Proceedings of the 2000 International Conference on Machine Learning*, pp. 9–16. Stanford, CA: Morgan Kauffman, 2000.
81. K. Crammer and Y. Singer. On the learnability and design of output codes for multiclass problems. *Proceedings of the Thirteenth Annual Conference on Computational Learning Theory*, pp. 35–46, Palo Alto, CA: Morgan Kaufmann, 2000.
82. R. Rifkin and A. Klautau. In defense of one-vs-all classification. *Journal of Machine Learning Research*, 5:101–141, 2004.
83. A. J. Smola, P. Bartlett, B. Scholkopf, and D. Shuurmans, editors. Probabilistic outputs for support vector machines and comparison of regularized likelihood

methods. In P.J. Bartlett, B. Schölkopf, D. Schuurmans, and A.J. Smola (Eds.), *Advances in Large Margin Classifiers*, pp. 61–74. Cambridge, MA: MIT Press, 2000.

84. Y. Guermeur. A simple unifying theory of multi-class support vector machines. Technical Report RR-4669, INRIA, 2002.
85. Y. Guermeur, A. Elisseeff, and D. Zelus. A comparative study of multi-class support vector machines in the unifying framework of large margin classifiers. *Applied Stochastic Models in Business and Industry*, 21:199–214, 2005.
86. J. Weston and C. Watkins. Mulit-class support vector machines. Technical Report CSD-TR-89-04, Department of Computer Science, Royal Holloway, University of London, 1998.
87. F. Aiolli and A. Sperduti. Multiclass classification with multi-prototype support vector machines. *Journal of Machine Learning Research*, 6:817–850, 2005.
88. K. Crammer and Y. Singer. On the algorithmic implementation of multiclass kernel-based vector machines. *Journal of Machine Learning Research*, 2:265–292, 2001.
89. C.A. Orengo, A.D. Mitchie, S. Jones, D.T. Jones, M.B. Swindells, and J.M. Thorton. Cath—A hierarchic classification of protein domain structures. *Structure*, 5(8):1093–1108, 1997.
90. L. Holm and C. Sander. The fssp database: Fold classification based on structure-structure alignment of proteins. *Nucleic Acids Research*, 24(1):206–209, 1996.
91. I. Tsochantaridis, T. Homann, T. Joachims, and Y. Altun. Support vector machine learning for interdependent and structured output spaces. *Proceedings of the 2004 International Conference on Machine Learning*, p. 104. Banff, Canada, 2004.
92. A. Sun and E. Lim. Hierarchial text classification and evaluation. *Proceedings of the 2001 IEEE International Conference on Data Mining*, pp. 521–528. Washington, DC: IEEE Computer Society, 2001.
93. J. Rousu, C. Saunders, S. Szedmak, and J.S. Taylor. Learning hierarchial multi-category text classification methods. *Proceedings of the 22nd International Conference on Machine Learning*. Bonn, Germany, 2005.
94. M. Collins and N. Duffy. New ranking algorithms for parsing and tagging: Kernels over discrete structures and the voted perceptron. *Proceedings of the 40th Annual Meeting of the Association of Computational Linguistics*, pp. 263–270. Morristown, NJ: Association for Computational Linguistics, 2002.
95. M. Collins. Parameter estimation for statistical parsing models: Theory and practice of distribution-free methods. In H. Bunt, J. Caroll, G. Satta (Eds.), *New Developments in Parsing Technology*, pp. 1–38. Dordrecht: Kluwer, 2001.
96. B. Taskar, D. Klein, M. Collins, D. Koller, and C. Manning. Max-margin parsing. *Proceedings of the Conference on Empirical Methods in Natural Language Processing (EMNLP)*, pp. 1–8. Barcelona, Spain, 2004.
97. S.E. Brenner, P. Koehl, and M. Levitt. The astral compendium for sequence and structure analysis. *Nucleic Acids Research*, 28:254–256, 2000.

CHAPTER 9

INTEGRATIVE PROTEIN FOLD RECOGNITION BY ALIGNMENTS AND MACHINE LEARNING

ALLISON N. TEGGE, ZHENG WANG, and JIANLIN CHENG
Computer Science Department and Informatics Institute
University of Missouri
Columbia, MO

9.1. INTRODUCTION

The amino acid sequence of a protein dictates its structure, which determines its function [1–7]. In the post-genomic era, with the application of high-throughput DNA and protein sequencing technologies, the number of available protein sequences has increased exponentially, whereas the experimental determination of protein structure using X-ray crystallography [8,9] and nuclear magnetic resonance spectroscopy [10,11] is still costly and labor-intensive. Currently, only about 1.5% of protein sequences (about 40,000 out of 2.5 million) have had their structures solved. The gap between proteins with known structures and those with unknown structures is still increasing. Predicting protein structure from sequence has become one of the most fundamental problems of structural bioinformatics. Protein structure prediction software is becoming a vital tool in understanding phenomena in modern molecular and cellular biology [12] and has important applications in medical sciences, such as drug design [13].

Currently, the most successful and practical method for three-dimensional (3D) structure prediction is the template-based approach, including comparative modeling (or homology modeling) and threading [14–28]. This approach is based on the observation that nature tends to reuse existing structures/folds in order to accommodate new protein sequences and functions

Introduction to Protein Structure Prediction: Methods and Algorithms,
Edited by Huzefa Rangwala and George Karypis

during evolution [29]. Thus, despite the fact that the protein sequence space (the number of protein sequences) is very large, the protein structure space (the number of unique protein folds) is relatively small and expected to be limited [30]. Millions of protein sequences have been collected, but the number of unique structures (folds) is only about 1000 (out of about 30,000 protein structures) in the protein classification databases such as the Structural Classification of Proteins(SCOP) [31] and the Class, Architecture, Topology, and Homologous superfamily (CATH) [32]. Moreover, among the newly determined protein structures in the structural genomics projects, the novel folds account for only a small portion (about 10%), and this portion continues to decrease during the past 15 years [33].

Most protein sequences, particularly similar protein sequences within the same family and superfamily, evolving from common ancestors, have a similar structure with other proteins. Therefore, given a query protein with unknown structure, template-based prediction first tries to identify a template protein—if one exists—that has a solved, similar structure with the query protein. Next, template-based prediction uses the structure of the template protein to model the structure of the query protein based on the alignment between the query sequence and the template structure [14–16,35–38]. Figure 9.1 shows the basic procedure of the template-based protein structure prediction.

Among two kinds of template-based methods, comparative modeling traditionally refers to easy modeling cases when the query protein has high sequence similarity (≥40% identity) with the template protein, which can be identified by Position-Specific Iterative-Basic Local Alignment Search Tool

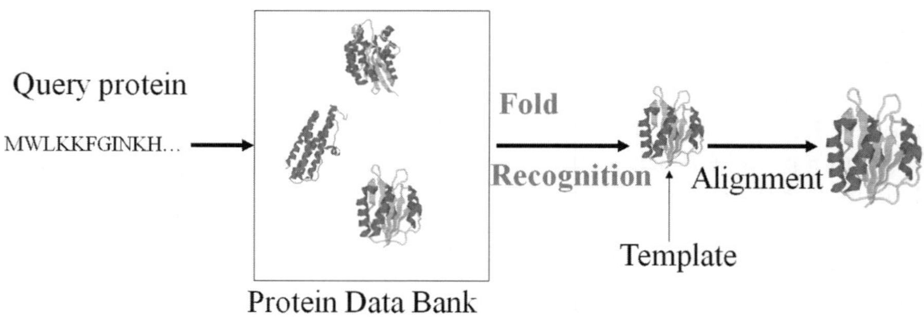

FIGURE 9.1 The basic procedure of the template-based protein structure prediction. Given a query protein without a solved structure, template-based methods try to identify a template protein structure in the Protein Data Bank (PDB) [34], which is assumed to have the similar structure with the query protein. A query-template alignment is then generated by alignment methods. Finally, based on the query-template alignment, the structure of the template protein is transferred to generate the structure model of the query protein. The simplest approach is to copy the coordinates of the residues in the template structure to their aligned counterparts of the query protein. In this procedure, the important step of identifying a template structure for the query protein is called *fold recognition*. (See color insert.)

(PSI-BLAST) [39]; fold recognition refers to hard modeling cases when the query protein has lower sequence similarity with the template protein and the structural relevance between query and template proteins is harder to recognize. However, with the development of more sensitive fold recognition methods such as *sequence-profile* alignment [24,39–48] and *profile-profile* alignment [49–60], the separation between comparative modeling and fold recognition is blurred [61]. Lately, template-based methods are used to refer to both of them and the term "fold recognition" is used to identify a template protein for a query protein no matter how similar their sequences are.

Fold recognition has been one of the most actively researched topics of protein structure prediction since it was first introduced in early 1990s [17,18]. It is still one of the most important problems in protein structure prediction because the sensitivity of identifying similar protein structures without sequence identity (i.e., analogous fold recognition) is still below 50%, and the sensitivity of identifying similar protein structures with low sequence similarity is just about 70% [28,62]. As more and more template structures are solved by structural genomics projects, more of the protein structure space has been discovered. As a result, discovering new structure folds has become a harder task. In fact, Skolnick et al. showed that the current Protein Data Bank (PDB) database is almost complete for modeling single domain proteins [63], which means any single protein in a large benchmark can be folded into at least a low-resolution structure using templates found in the PDB. However, current fold recognition methods are not sensitive enough to recall many of these templates, particularly the similar template structures without sequence identity (analogous folds). Thus, it is imperative to develop more sensitive fold recognition methods.

In the following sections, we discuss traditional alignment-based methods for fold recognition and how to integrate these methods with knowledge-based machine learning methods in order to improve protein fold recognition. We particularly focus on the integrative machine learning approach, which has the most potential to integrate protein sequence and structural features, to significantly improve analogous fold recognition.

9.2. ALIGNMENT FOLD RECOGNITION METHODS

9.2.1. Sequence-Sequence

Sequence-sequence alignment was one of the original methods for fold recognition. This method aligns a query amino acid sequence for a protein of unknown structure with that of a protein with known structure. The alignment score measures their sequence similarity. If the sequence similarity is significant, the known structural features are then inferred onto the query sequence based on their alignment. Two primary alignment algorithms have driven sequence-sequence fold recognition: global sequence alignment and local sequence alignment. These two algorithms have been incorporated into several tools that are publicly available.

Needleman and Wunsch created the global alignment algorithm that uses dynamic programming to calculate the optimal alignment of two sequences. This algorithm is often implemented as an iterative matrix method of calculations to find the best alignment of positions, while accounting for all possible deletions and insertions. Given two amino acid sequences X and Y of length m and n, respectively, the global alignment algorithm creates an $(n + 1) \times (m + 1)$ matrix. Following various pathways from the top left to the bottom right through the matrix can score all possible alignment combinations for the two amino acid sequences. The cells connected by a solid line show the path of the final sequence alignment with the highest alignment score (Fig. 9.2a).

Smith and Waterman (1981) developed a local alignment that finds the subsequences with maximum homology. This local alignment algorithm uses a similar dynamic programming method for global sequence alignment, but focuses on identifying subsequences with the greatest similarity—the highest alignment score (Fig. 9.2b). The local alignment algorithm scores an alignment similarly to the global alignment algorithm; however, the score of the alignment cannot be negative. Since many proteins only have local sequence and structure similarity, local alignment methods are more sensitive for fold recognition, when compared with global alignment methods.

Sequence-sequence alignments are scored using an e-value as the significance of sequence alignment score. The optimal sequence alignment is the alignment with the highest alignment score, or the lowest e-value. A comparison between a global and local alignment is shown in Figure 9.2, where it shows that a global and local alignment do not need to yield the same results. A local alignment actually could yield one or more alignments with the same highest alignment score. Overall, sequence-sequence alignments perform best for proteins that belong to the same protein family (having >40% sequence identity).

Global and local protein sequence alignments depend on scoring parameters for match, mismatch, and gap penalty. The two most widely used scoring matrices are blocks substitution matrices (BLOSUM) [64] and point accepted mutation (PAM) matrices [65]. The two most widely used local sequence alignment tools for fold recognition are the BLAST [66] and FASTA [67]. Both use heuristics to speed up the sequence comparison instead of implementing the full dynamic programming algorithm.

9.2.2. Sequence-Profile

Fold recognition using sequence-profile alignment methods align a query protein sequence with a profile of protein sequences with known structures, or vice versa. Protein profiles are a statistical representation of a group of protein sequences in the same protein family or superfamily. Typical profiles

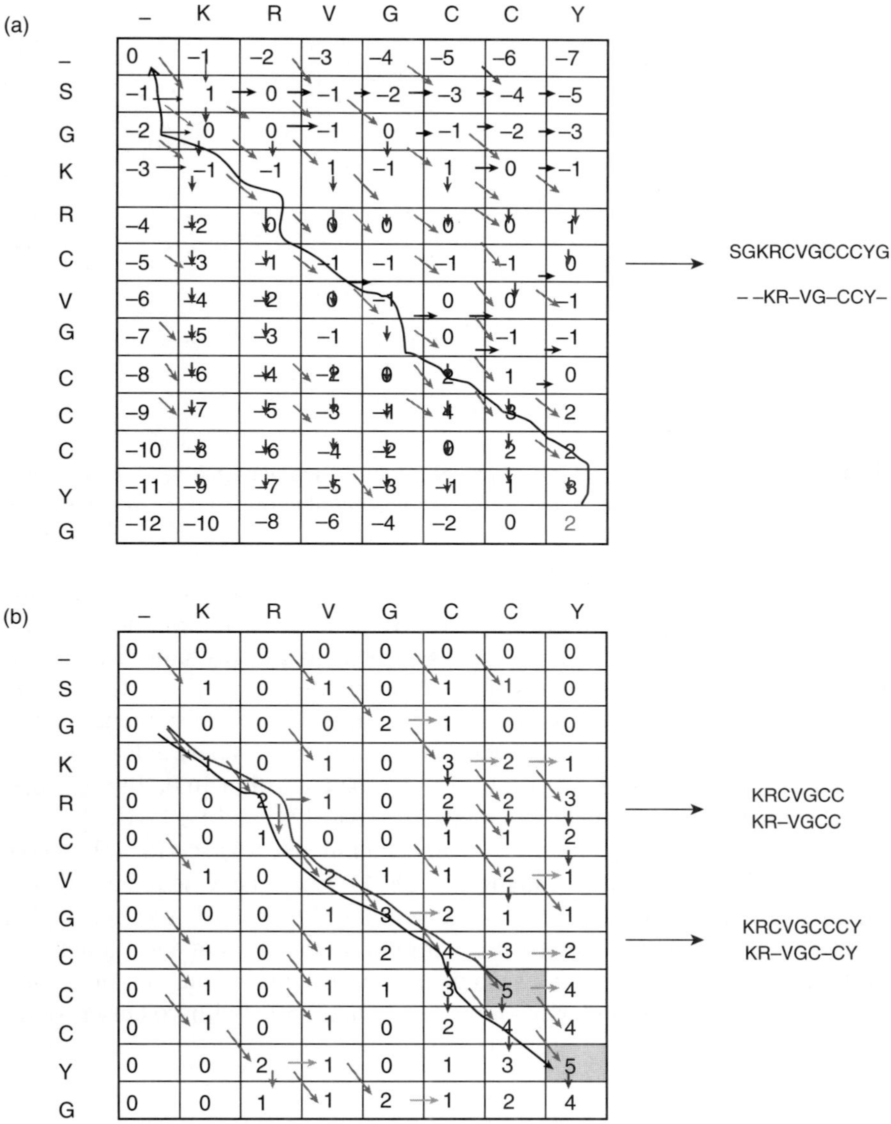

FIGURE 9.2 Sequence-sequence alignment matrices for fold recognition. (a) global alignment algorithm. (b) local alignment algorithm. The examples use a simple score function: +1 for match, -1 for mismatch, and -1 for gap. The paths denote the optimal alignments.

include multiple sequence alignments, Position-Specific Scoring Matrices (PSSM) [39], and hidden Markov models (HMMs) [41,42]. An HMM is designed containing a set of states as in Figure 9.3, where *Begin* is the starting state, *End* is the ending state, circles are deleting states, diamonds are insertion states, and squares are the main states. Both the insertion and main states

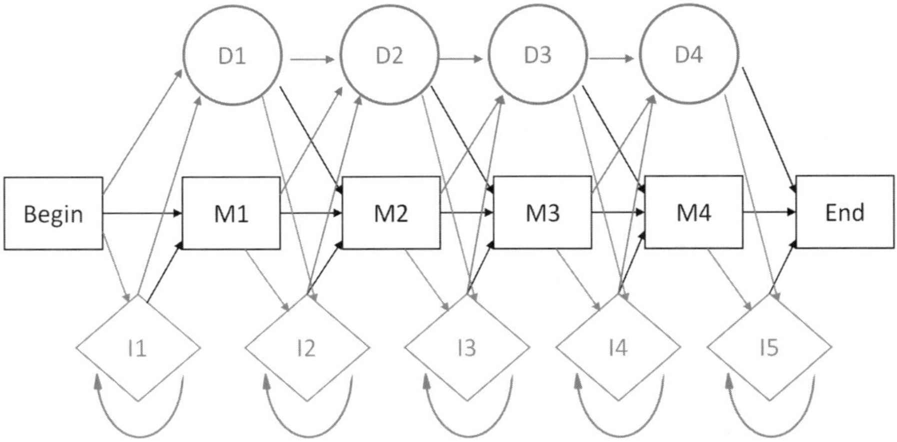

FIGURE 9.3 HMM model for sequence-profile alignment for fold recognition. Squares represent matching states, circles represent deleting states, and diamonds represent insertion states.

contain emission probabilities for each possible amino acid. The emission probabilities for the main states are the frequencies of the amino acids for each position in the protein profile, and the emission probabilities for the insertion states are uniformly distributed among the amino acids. The parameters of an HMM are usually trained on a family of protein sequences using the Baum-Welch algorithm [68], and the optimal alignment between a sequence and an HMM is determined using the Viterbi algorithm [69]. Two popular HMM sequence-profile alignment tools are SAM [24] and HMMER [44].

Another very popular sequence–profile alignment tool used for fold recognition is PSI-BLAST, which aligns the PSSM of a query sequence with a template protein sequence for fold recognition [39]. PSI-BLAST has higher sensitivity than the regular BLAST program. Profile-sequence alignment methods such as PSI-BLAST can detect remote homologous template proteins having sequence similarity as low as ~25% identity with a query protein.

9.2.3. Profile-Profile

Profile-profile alignment is to align the profiles of two proteins. Profile-profile alignments can be obtained by aligning two profile HMMs as in COACH [56,70] and HHSearch [60], by aligning two multiple sequence alignments as in CLUSTALW and T-Coffee [49,51], or by aligning two PSSM as in comparison of multiple protein alignments with assessment of statistical significance (COMPASS) [48]. Profile-profile alignment is currently the most sensitive alignment method for fold recognition, which can detect remote homologous proteins with low sequence similarity (~15% sequence identity).

CLUSTALW is a global multiple sequence alignment method that weights sequences differently, uses more than one substitution matrix, and offers local dependencies on the gap penalties [49]. By having varying weights for each sequence, those sequences that are more similar to other sequences will not bias the final scoring matrix. CLUSTALW employs various substitution matrices depending on the stage of the alignment. Sequences closely related will have matrices different from those that are highly divergent. Lastly, CLUSTALW has several scoring schemes for the gaps based on the location of the gap, as well as the physiochemical properties of the sequence around the gap. The varying gap score schemes correspond to varying places in the protein structure where gaps are acceptable or not acceptable. T-Coffee is another multiple sequence alignment method that is more sensitive but slower than CLUSTALW. It is important to note that CLUSTALW can be used in other alignment stages such as sequence-sequence alignment, in addition to profile-profile alignments.

In contrast to the multiple sequence alignment approach, COACH creates a profile HMM for a multiple sequence alignment, and then globally aligns the other alignment to that profile HMM [70]. COACH employs its own multiple sequence alignment algorithm, SATCHMO, which performs comparably to the alignments produced by CLUSTALW [56]. Overall, COACH is an iterative step that first calculates a profile HMM for one alignment. Once the initial profile HMM is determined, each sequence in the other alignment is compared individually to the profile HMM to determine a result. The combined alignment of all these individual sequences is then considered a result if it is trivially similar to the original second multiple alignment.

In addition to COACH, HHSearch is also used for the local pairwise alignment for two profile HMMs [60]. The pairwise alignment of two HMMs aligns match states and insertion states with each other, the same as in the alignment of a sequence with a profile HMM. The final pairwise alignment of the two HMMs is the sequence that is co-emitted from both HMMs.

COMPASS is a method that determines optimal profile-profile local alignments and estimates e-values [48]. The e-values are calculated using a generalized version of PSI-BLAST that has been adapted for profile-profile alignments.

Profile-profile alignment methods are sensitive to very distant homologs, when compared with sequence-sequence and sequence-profile alignment methods, and usually require >15% identity [53,55,59,71]. Identifying these distant homologs, however, is dependent on the calculation of the initial profiles being compared [50]. There are several considerations required when using profile-profile alignment methods for fold recognition. Some things to consider include selecting which sequences to use to build the profiles and how to weight similar sequences within the profile [58]. Moreover, there are different ways to account for gap penalties in a profile–profile alignment, and the varying gap penalty schemes will yield different alignment results. In addition to gap penalties, calculating the scores between positions can be done in several ways, such as a dot product, a probabilistic scoring function, or an information theoretical measure [57]. In a thorough study of similarity scores between positions, it has been shown that the probabilistic scoring functions had an advantage over other methods, as they had the best fold recognition performance [54,57].

9.2.4. Sequence–Structure

Sequence-structure alignment methods, also known as threading, are useful for recognizing similar folds between sequences with no evolutionary relationship, but with similar structures [17–19]. It tries to align a query sequence with a template structure to calculate a fitness score (e.g., pseudo energy) that measures how well the sequence fits the structure. The fitness score is calculated based on the structural environment (e.g., solvent accessibility, secondary structure [SS], neighboring residues) in which a residue fits. This alignment method takes a database of known protein structures, and then searches a sequence against all the known structures to determine if there is a significant alignment of the sequence and structure. This fold recognition requires an exhaustive search of the entire database to calculate the standardized z-scores of absolute fitness scores and to select the optimal alignment and template. Figure 9.4 shows how a query sequence is fitted into a template structure from a known protein. Each of these fittings yields a fitness score based on structural environment fitness and contact potentials, and the final structures are ranked by z-scores. One popular threading tool is mGenThreader [72,73].

Sequence-structure alignment methods were shown to be less accurate than profile-profile alignment methods at detecting remote homologous template structures. However, structural information is still useful for fold recognition. For instance, the sequence-derived predicted SS had been incorporated into sequence alignment methods to improve the sensitivity of fold recognition [23,26,55,73,74].

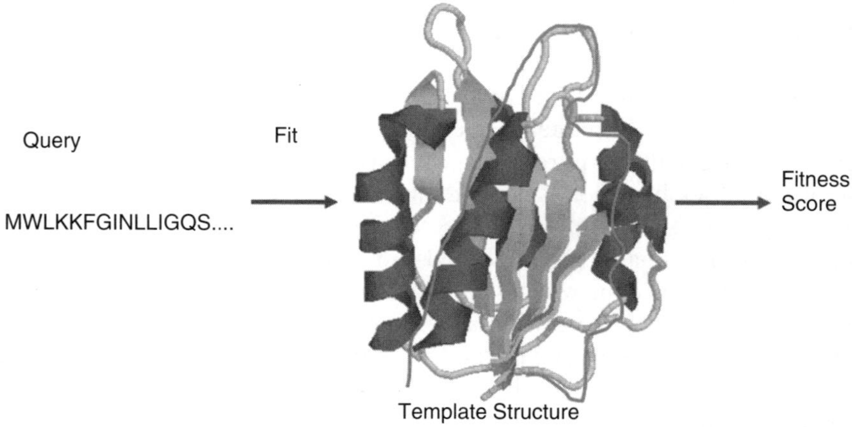

FIGURE 9.4 Sequence-structure alignment (threading) of a query protein against a known structure database. (See color insert.)

9.3. MACHINE LEARNING FOLD RECOGNITION METHODS

In fold recognition, different alignment tools are often used independently of search protein databases for similar structures. Previous research [75–79] has shown that these alignment methods are complementary and can find different correct templates. But combining these methods is difficult [76]. Meta, or jury approaches [55,71,76,80,81], collect the predicted folds or models from external fold recognition predictors and derive consensus predictions based on a small set of returned candidates (Fig. 9.5). This popular, hierarchical approach increases the breadth of fold recognition. However, it relies on the availability of external predictors and cannot recover true positive templates discarded prematurely by individual predictors.

Machine learning methods provide powerful means for integrating disparate features in pattern recognition. So far, however, most machine learning methods have been used in this area primarily for coarse homology detection, such as protein structure and fold classification [82–84] [85,86] (Fig. 9.6). Classifying proteins into a few categories or even dozens of families,

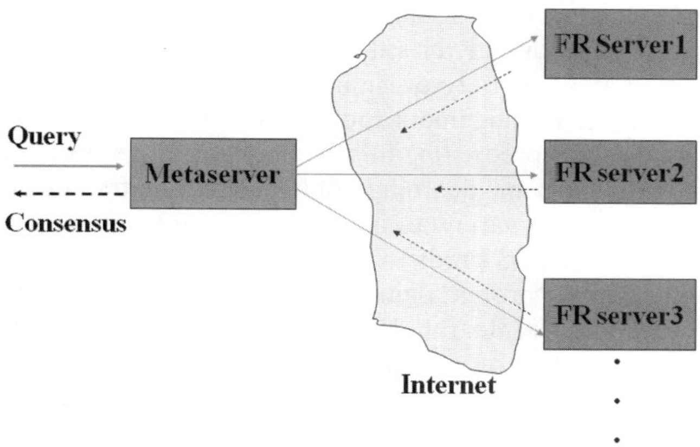

FIGURE 9.5 Meta approach for fold recognition.

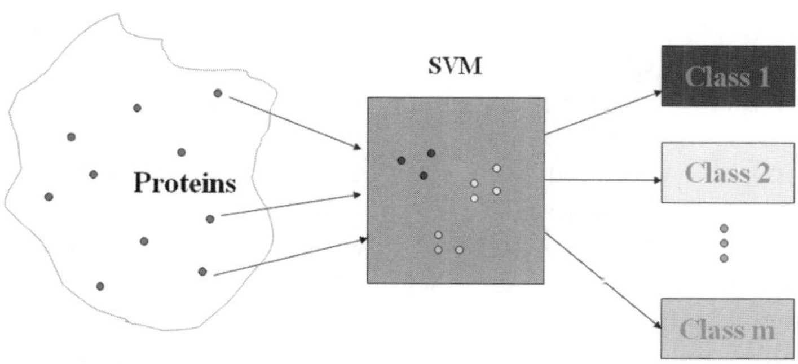

FIGURE 9.6 Protein fold classification by machine learning methods such as support vector machines (SVM).

superfamilies, and folds is very useful for function and structure annotations; however, it does not provide the specific templates required for template-based structure modeling. Furthermore, current classification methods are not likely to scale up to the thousands of families, superfamilies, and folds already present in current protein classification databases, such as SCOP [31], CATH [32], and Families of Structurally Similar Proteins (FSSP) [87].

Fold recognition is different from protein classification, in which it is fundamentally a retrieval problem, very much like finding a document or a web page [88,89]. Given a query protein, the objective of fold recognition is rather to rank all possible templates according to their structural relevance, as Google and other search engines rank web pages associated with a user's query. Machine learning methods (such as binary classifiers) have been used in threading approaches [73,74] to combine multiple scores produced by threading into a single score in order to rank the templates. In Reference [28], Cheng and Baldi generalized the idea [73,74] and derived a broad machine learning framework for the fold recognition and retrieval problem. The framework integrates a variety of similarity features and feature extraction tools, including standard alignment tools. The novelty of the machine learning approach is to pair query and template proteins and then to evaluate the structural relevance of the pairs, instead of directly classifying query proteins into categories. The framework integrates a variety of similarity features and feature extraction tools, including standard alignment tools. However, unlike meta approaches, it does not require any preexisting fold recognition programs or servers.

Specifically, the machine learning fold recognition framework has four steps (Fig. 9.7). (i) Use a variety of alignment methods and protein structure feature predictors (e.g., SS predictor) to extract pairwise similarity features for query and template protein pairs. (ii) Select a set of informative features for fold recognition. (iii) Integrate features to predict the structural similarity (e.g., having the same fold or not) between query and template proteins using machine learning methods. (iv) Use the predicted scores to rank template proteins with respect to a query protein.

The machine learning fold recognition method has some key differences from the protein classification methods [82–86]. (i) It focuses on extracting similarity features for query-template pairs, instead of extracting features of a query protein for classification. (ii) It evaluates the structural similarity level of query-template protein pairs, without trying to classify the query protein into a specific structural category. Thus, the method works around the currently impossible task of directly classifying proteins into a thousand folds. (iii) It uses the similarity scores to rank template proteins for a query protein. The method generates a template ranking list instead of a class label for the query protein. Unlike protein classification methods, the method provides the specific template for 3D structure modeling.

One major advantage of the machine learning framework is to seamlessly integrate sequence features such as sequence alignments scores with structural features such as predicted SS, solvent accessibility, and contact map. This is

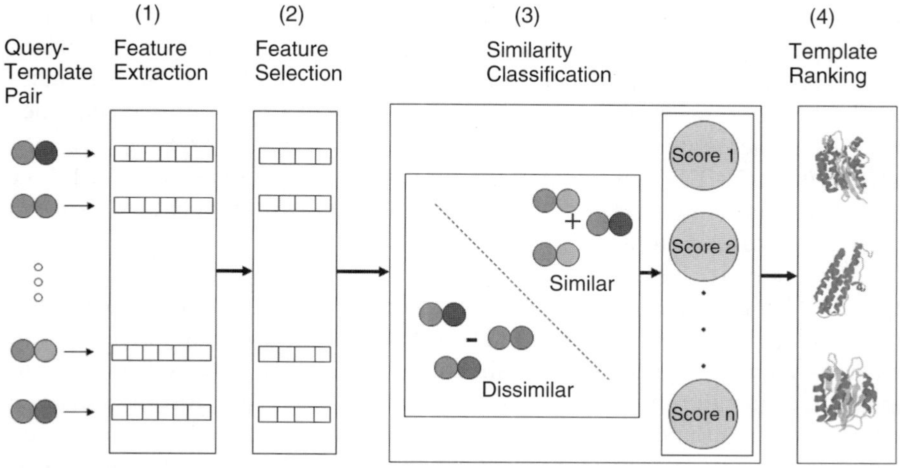

FIGURE 9.7 The machine learning framework for protein fold recognition. Step 1. Pair a query protein with template proteins and generate pairwise similarity features for each query-template pair. Step 2. Select a set of informative features. Step 3. Classify structure similarity into two categories (similar or dissimilar) using the features with machine learning methods. Step 4. Rank template proteins using classification scores with respect to the same query. (See color insert.)

particularly important for analogous fold recognition where templates cannot be identified through sequence similarity [63]. Thus, the machine learning fold recognition framework has the potential to significantly improve analogous fold recognition, one of the major challenges in protein structure prediction.

In the following sections, we describe each step of the machine learning fold recognition framework using a standard implementation of the approach [28] and discuss its advantages.

9.3.1. Feature Extraction

Extracting a set of informative features for query-template pairs is the key step of the machine learning approach. In Reference [28], five categories of pairwise similarity features for each query-template pair associated with sequence or family information, sequence alignment, sequence-profile (or profile-sequence) alignment, profile-profile alignment, and structure were extracted.

9.3.1.1. Sequence/Family Information Features. To compare the sequences of a query and template proteins, their single amino acid (monomer) and ordered pair of amino acids (dimer) compositions were computed. The composition vectors x and y of the query and template were compared and transformed into six similarity scores using the cosine, correlation, and Gaussian kernel functions, respectively. The same techniques were applied to the monomer and dimer residue composition vectors of the family of sequences associated with the query and the template to extract another set of six similarity measures, to measure the family composition similarity. The sequences for both query and template families were derived from multiple sequence alignments generated by searching the National Center for Biotechnology Information (NCBI) nonredundant sequence database [90] using PSI-BLAST [39]. Thus the sequence/family information feature subset includes 12 (6 + 6) pairwise features in total.

9.3.1.2. Sequence-Sequence Alignment Features. Two sequence alignment tools, PALIGN [57] and CLUSTALW [49], were used to extract pairwise features associated with sequence alignment scores. PALIGN used local alignment methods and produced a score and an e-value. The score was divided by the length of the query to remove any length bias. CLUSTALW generated a global sequence alignment score between the query and the template. This score was also normalized by the length of the query sequence. Thus the sequence alignment feature subset includes three pairwise features.

9.3.1.3. Sequence-Profile (or Profile-Sequence) Alignment Features. Three different profile sequence alignment tools (PSI-BLAST, HMMER-HMMSearch [44], and IMPALA [44]) were used to extract profile sequence alignment features between the query profile and the template sequence. The profiles (or multiple alignments) for queries were generated by searching the nonredundant database using PSI-BLAST, as described above. The multiple alignments were used by all profile alignment tools directly, or as the basis for building customized profiles. The alignment score normalized by the query length, the logarithm of the e-value, and the alignment length normalized by the query length from the most significant PSI-BLAST and IMPALA local alignment were used as features. The alignment scores, normalized by the length of the query sequence, and the logarithm of the e-value produced by HMMSearch alignments were also used as features. Thus the profile-sequence alignment tools generated eight pairwise features. For sequence-profile alignments, RPS-BLAST in the PSI-BLAST package and hmmpfam in the HMMER package were used to align the query sequence with the template profiles. The template profiles were generated in the same way as the query profiles. In this way, RPS-BLAST generated three features similar to PSI-BLAST. The logarithm of the e-value produced by hmmpfam was also used as one feature. Thus the subset of profile-sequence (or sequence-profile) alignment features includes 12 (8 + 4) pairwise features in total.

9.3.1.4. Profile-Profile Alignment Features. Five profile-profile alignment tools including CLUSTALW, COACH of LOBSTER [70], COMPASS [48], HHSearch [60], and PRC [53] were used to align query and template profiles. The global alignments produced by CLUSTALW and LOBSTER, and the most significant local alignments produced by COMPASS, PRC, and

HHSearch, were used to extract the pairwise features. The alignment scores produced by CLUSTALW and HHSearch were normalized by query length and used as pairwise features. PRC, an HMM profile-profile alignment tool, was used with two different kinds of profiles: HMM models built by HMMER and checkpoint profiles (.chk file) built by PSI-BLAST. In each case, PRC produced three scores (co-emission, simple, and reverse), which were normalized by query length. COMPASS was used to align query multiple alignments with template multiple alignments. The Smith-Waterman local alignment score normalized by query length and the logarithm of the e-value from the COMPASS alignments were used also as pairwise features. Thus the subset of profile-profile alignment features includes 10 pairwise features in total.

9.3.1.4. Structural Features. Based on the global profile-profile alignment between query and template obtained with LOBSTER, predicted 1D and 2D structural features including SS (3-class: alpha, beta, loop), relative solvent accessibility (RSA; 2-class: exposed or buried at 25% threshold), contact probability map at 8 Å and 12 Å, and beta-sheet residue pairing probability map were used to evaluate the compatibility between query and template structures. These structural features for query proteins were predicted using the SCRATCH suite [91–93]. The predicted SS and RSA of the query residues were compared with the nearly exact SS and RSA of the aligned residues in the template structure. The fraction of correct matches for both SS (as in Reference [73]) and RSA were used as two pairwise features. The SS and RSA composition (helix, strand, coil, exposed, and buried) were transformed into four similarity scores by cosine, correlation, Gaussian kernel, and dot product, respectively. So this 1D structural feature subset has six features in total.

For the aligned residues of the template that have sequence separation >5 and are in contact at 8 Å threshold (respectively 12 Å), the average contact probability of their counterparts in the predicted 8 Å (respectively 12 Å) contact probability map of the query were computed. The underlying assumption is that the counterparts of the contact residues in the template should have high contact probability in the query contact map if the query and template share similar structure. Similarly, for each paired beta-strand residue in the template structures, the average pairing probability of their beta-strand counterparts in the predicted beta-strand pairing probability map of the query was calculated, assuming that two proteins will share similar beta-sheet topology if they belong to the same fold.

Moreover, the contact order (the sum of sequence separation of contacts) and contact number (the number of contacts) for each aligned residue in both query and templates [28] were computed, using the predicted contact map of the query and the known structure of the templates, respectively. The contact order and contact number vectors of the aligned residues were not used directly as features. Instead, they were compared and transformed into pairwise similarity scores using the cosine and correlation functions. For both the 8 Å and 12 Å contact maps, eight pairwise features of contact order and

TABLE 9.1 54 Pairwise Features Used for Machine Learning Fold Recognition

Category	Feature	Method	Number
Seq and family information	Seq monomer compo	cos/corr/Gauss	3
	Seq dimmer compo	cos/corr/Gauss	3
	Fam monomer compo	cos/corr/Gauss	3
	Fam dimmer compo	cos/corr/Gauss	3
Sequence-sequence alignment	Local alignment	PALIGN	2
	Global alignment	CLUSTALW	1
Sequence-profile alignment	Prof versus seq	PSI-BLAST	3
	Prof versus seq	IMPALA	3
	Prof versus seq	HHMER	2
	Seq versus prof	RPS-BLAST	3
	Seq versus prof	HMMER	1
Profile-profile alignment	Multiple alignment	CLUSTALW	1
	PSSM	COMPASS	2
	HMMprof	PRC	6
	HMMprof	HHSearch	1
Structural information	SS and RSA match	ratio	2
	SS and RSA compo	cos/corr/Gauss	4
	Contact probability	average	2
	Residue contact order	cos/corr	4
	Residue contact num	cos/corr	4
	Beta-sheet pair prob.	average	1
Total	—	—	54

SS and RSA represent secondary structure and relative solvent accessibility, respectively.

contact number were extracted. Thus the 2D structural feature subset has 11 features in total, and the entire 1D and 2D structural feature subset has 17 features in total. The entire feature set contains 54 pairwise features measuring query-template similarity (Table 9.1). It is worth pointing out that structural features are very useful for fold recognition, particularly for recognizing analogous folds (similar protein folds without an evolutionary relationship) [63]. An integration of advanced machine learning techniques and the 1D and 2D structural features will improve the state of the art of analogous fold recognition.

9.3.2. Feature Selection

Selecting informative features is useful to improve classification performance. Information gain [94–96] is widely employed for feature selection in machine learning. In general, the bigger information gain a feature has, the more informative the feature. In Reference [28], a widely used machine learning toolbox, Weka [97], was employed to compute the information gain for the initial 84 features and to select the top 54 features ad hoc. The simple feature selection improved the performance by a few percentage points. Information gain can be combined with the other two feature selection techniques (forward selection and backward selection) to thoroughly evaluate the contributions of the features and to select a subset of informative features to further improve fold recognition. Forward selection is a greedy feature selection method. It starts from an empty feature subset. It adds a feature into the subset one at a time according to the information gains in decreasing order. It trains the classification methods using the subset of features and computes the accuracy. If the accuracy increases, it keeps the newly introduced feature. It repeats the process until no features are left, or the accuracy does not increase. Backward selection is the reverse process of forward selection. It first uses all features to train the system and to compute accuracy. Then it starts to remove a feature one at a time according to the information gains in increasing order, that is, remove the least informative features first. It stops removing features as soon as the accuracy does not increase anymore. The features left are the selected features. Both forward and backward selection methods are not guaranteed to find the optimal set of features, but they usually can select a very good set of features. In addition to information gain, alternative feature-ranking methods such as mutual information [48] can be used.

9.3.3. Similarity Classification

The features extracted for query-template pairs can be used by machine learning classification methods to predict their structural similarity [28]. To train support vector machines (SVM) to classify the similarity of query-template proteins into two categories (similar vs. dissimilar) (Fig. 9.8), the large benchmark dataset [76] derived from the SCOP [31] database was used. The Lindahl's dataset includes 976 proteins. The pairwise sequence identity is ≤40%. A feature vector for all 976 × 975 distinct pairs was extracted. In this dataset, 555 sequences have at least one match at the family level, 434 sequences have at least one match at the superfamily level, and 321 sequences have at least one match at the fold level. The pairs in the same family, superfamily, or fold are positive pairs. All others are negative pairs. There are only about 7500 positive pairs.

All protein pairs were evenly split into 10 subsets for tenfold cross-validation purposes (Fig. 9.9). The entire set of query-template pairs associated with the same query protein were put into the same subset. Nine subsets were used for training and the remaining subset was used for validation. The pairs that are in the training dataset , which also used queries in the test data set as templates, were removed. The procedure was repeated 10 times and the sensitivity/specificity results were computed across the 10 experiments. Training takes about 3 days for a single data split on a single node with dual Pentium processors, hence 3days for the entire tenfoldfold cross-validation experiment using 10 nodes in a cluster.

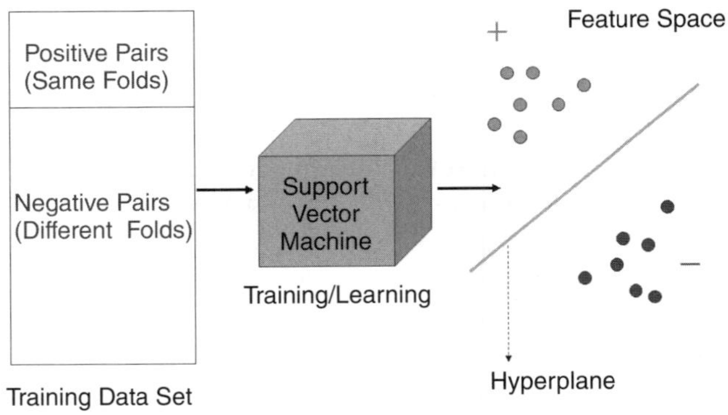

FIGURE 9.8 Protein similarity classification using support vector machines (SVMs). Query-template pairs are classified into two categories (positive: in the same fold—structurally similar; negative: not in the same fold). A classification score produced by SVM measures how likely the query-template pair shares the same fold. The larger the score, the more likely the query and template are in the same fold.

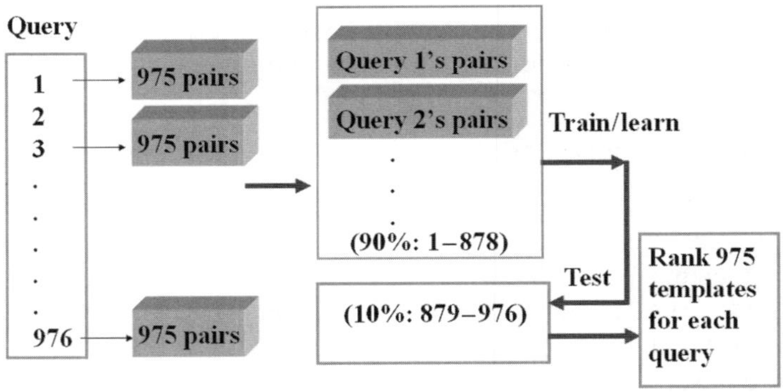

FIGURE 9.9 Tenfold cross-validation of machine learning fold recognition. Each query has 975 query-template pairs. These pairs are split into 10 folds according to queries. Nine folds were used to train support vector machines, whereas one remaining fold was used for testing. The classification scores on the test data were used to rank templates associated with the same query.

9.3.4. Template Ranking, Evaluation, and Results

The last step of machine learning fold recognition approach is to rank template proteins according to its structural similarity score with a query protein produced by machine learning classification. The templates associated with the same query can be ranked based on the similarity scores.

According to the same evaluation procedure in References [26], [76], and [98], the sensitivity of the machine learning method was evaluated by taking the top 1 or the top 5 templates in the ranking associated with each test query.

TABLE 9.2 The Sensitivity of 11-Fold Recognition Methods at Different Similarity Levels (Family, Superfamily, and Fold). It is Easiest to Recognize Similarity Structures at the Family Level and Hardest at the Fold Level. FOLDpro is the Machine Learning Fold Recognition Method [28]

Method	Family	Superfamily	Fold
PSI-BLAST [39]	72.3	27.9	4.7
HMMER [44]	73.5	31.3	14.6
SAM-T98 [24]	75.4	38.9	18.7
BLASTLINK [99]	78.9	4.06	16.5
SSEARCH [100]	75.5	32.5	15.6
SSHMM [101]	71.7	31.6	24
THREADER [102]	58.9	24.7	37.7
FUGUE [98]	85.8	53.2	26.8
RAPTOR [74]	77.8	50	45.1
SPARKS3 [103]	86.8	67.7	47.4
FOLDpro	89.9	70.0	48.3

The sensitivity at each similarity level (fold, superfamily, and family) is defined as the percentage of the query proteins having a correct template ranked as top 1 or within top 5, respectively. Table 9.2 lists the sensitivity of 11-fold recognition methods for top five ranked templates on the Lindahl's dataset. The results show that the machine learning fold recognition method (FOLDpro) has achieved the state-of-the-art performance.

9.3.5. Advantages of Machine Learning Methods

The machine learning fold recognition approach has several advantages in terms of integration, scalability, simplicity, reliability, and performance. First, the approach readily integrates complementary streams of information, from alignment to structure, and additional features can easily be added. It is worth pointing out this integrative approach is slower than some individual alignment methods such as PSI-BLAST. However, it can scan a fold library with about 10,000 templates in a few hours, for an average-size query protein. Second, most features can readily be derived using publicly downloadable sequence/profile alignment tools. This is simpler than trying to develop a new, specialized, fold recognition/threading method, which usually requires a lot of expertise. Third, machine learning approaches can be included in a metaserver, but, unlike a metaserver, they can be self-contained and do not necessarily need to rely on external fold-recognition servers. Also, unlike metaservers, this approach produces a full ranking of all the templates and does not discard any templates early on during the recognition process. Finally, the approach delivers state-of-the-art performance on current benchmarking datasets.

9.4. CONCLUSIONS

Fold recognition is a fundamental problem in protein structure prediction. We have discussed a variety of alignment methods used for fold recognition. In addition to further improving individual alignment methods, the other important direction is to integrate alignment methods and structural features to improve fold recognition, particularly analogous fold recognition. The machine learning fold recognition approach is a useful and general framework that can integrate different sources of information to increase the sensitivity of protein fold recognition. More and more advanced machine learning methods and useful features are expected to be developed to address the fold recognition problem.

ACKNOWLEDGMENTS

The work was partially supported by a UM research board grant and an MU research council grant to JC and a NLM fellowship to ANT. The authors thank the *Bioinformatics* journal for the permission to reuse and reproduce some materials originally published in Reference [28].

REFERENCES

1. L. Pauling and R. Corey. The pleated sheet: A new layer configuration of polypeptide chains. *Proceedings of the National Academy of Sciences U S A*, 37(5):251–256, 1951.
2. L. Pauling, R. Corey, and H. Branson. The structure of proteins: Two hydrogen-bonded helical configurations of the polypeptide chain. *Proceedings of the National Academy of Sciences U S A*, 37(4):205–211, 1951.
3. F. Sanger and E. Thompson. The amino-acid sequence in the glycyl chain of insulin. 1. The identification of lower peptides from partial hydrolysates. *Biochemical Journal*, 53(3):353, 1953.
4. F. Sanger and H. Tuppy. The amino-acid sequence in the phenylalanyl chain of insulin. 2. The investigation of peptides from enzymic hydrolysates. *Biochemical Journal*, 49(4):481, 1951.
5. J. Kendrew, R.E. Dickerson, B.E. Strandberg, R.G. Hart, D.R. Davies, D.C. Phillips, and V.C. Shore. Structure of myoglobin: A three-dimensional Fourier synthesis at 2 Å resolution. *Nature*, 185(4711):422–427, 1960.
6. M. Perutz, M.G. Rossmann, A.F. Cullis, H. Muirhead, G. Will, and A.C.T. North. Structure of haemoglobin: A three-dimensional Fourier synthesis at 5.5 Å resolution, obtained by X-ray analysis. *Nature*, 185(4711):416–422, 1960.
7. C. Anfinsen. Principles that govern the folding of protein chains. *Science*, 181(4096):223–230, 1973.
8. L. Bragg. The development of X-ray analysis. *Contemporary Physics*, 17(1):103–104, 1976.

9. T. Blundell and L. Johnson. *Protein Crystallography*. New York: Academic Press, 1976.

10. K. Wuthrich. NMR of proteins and nucleic acids. The George Fisher Baker non-resident lectureship in Chemistry at Cornell Unversity (USA), 1986.

11. E. Baldwin, I.T. Weber, R.S. Charles, J.C. Xuan, E. Appella, M. Yamada, K. Matsushima, B.F. Edwards, G.M. Clore, and A.M. Gronenborn. Crystal structure of interleukin 8: Symbiosis of NMR and crystallography. *Proceedings of the National Academy of Sciences U S A*, 88(2):502–506, 1991.

12. D. Petrey and B. Honig. Protein structure prediction: Inroads to biology. *Molecular Cell*, 20(6):811–819, 2005.

13. M. Jacobson and A. Sali. Comparative protein structure modeling and its applications to drug discovery. *Annual Report in Medicinal Chemistry*, 39:259–276, 2004.

14. W.J. Browne, A.C. North, D.C. Phillips, K. Brew, T.C. Vanaman, and R.L. Hill. A possible three dimensional structure of bovine alpha-lactalbumin based on that of hen's egg-white lysozyme. *Journal of Molecular Biology*, 42:65–86, 1969.

15. A. Sali. Comparative protein modeling by satisfaction of spatial restraints. *Molecular Medicine Today*, 1(6):270–277, 1995.

16. J. Greer. Comparative modeling methods: Application to the family of the mammalian serine proteases. *Proteins*, 7(4):317–334, 1990.

17. J. Bowie, R. Luthy, and D. Eisenberg. A method to identify protein sequences that fold into a known three-dimensional structure. *Science*, 253(5016):164–170, 1991.

18. D. Jones, W. Taylort, and J. Thornton. A new approach to protein fold recognition. *Nature*, 358(6381):86–89, 1992.

19. A. Godzik and J. Skolnick. Sequence-structure matching in globular proteins: Application to supersecondary and tertiary structure determination. *Proceedings of the National Academy of Sciences U S A*, 89(24):12098–12102, 1992.

20. M. Levitt. Accurate modeling of protein conformation by automatic segment matching. *Journal of Molecular Biology*, 226:507–533, 1992.

21. S. Bryant and C. Lawrence. An empirical energy function for threading protein sequence through the folding motif. *Proteins*, 16(1):92–112, 1993.

22. D. Fischer and D. Eisenberg. Protein fold recognition using sequence-derived predictions. *Protein Science*, 5(5):947, 1996.

23. B. Rost, R. Schneider, and C. Sander. Protein fold recognition by prediction-based threading. *Journal of Molecular Biology*, 270(3):471–480, 1997.

24. K. Karplus, C. Barrett, and R. Hughey. Hidden Markov models for detecting remote protein homologies. *Bioinformatics*, 14(10):846–856, 1998.

25. Y. Xu, D. Xu, and E. Uberbacher. An efficient computational method for globally optimal threading 1. *Journal of Computational Biology*, 5(3):597–614, 1998.

26. H. Zhou and Y. Zhou. Quantifying the effect of burial of amino acid residues on protein stability. *Proteins: Structure, Function, and Bioinformatics*, 54(2):315–322, 2003.

27. Y. Zhang and J. Skolnick. Automated structure prediction of weakly homologous proteins on a genomic scale. *Proceedings of the National Academy of Sciences U S A*, 101(20):7594–7599, 2004.

28. J. Cheng and P. Baldi. A machine learning information retrieval approach to protein fold recognition. *Bioinformatics*, 22(12):1456–1463, 2006.
29. C. Chothia. One thousand folds for the molecular biologist. *Nature*, 357:543–544, 1994.
30. A. Grant, D Lee, and C. Orengo. Progress towards mapping the universe of protein folds. *Genome Biology*, 5(5):107, 2004.
31. A.G. Murzin, E. Brenner, T. Hubbard, and C. Chothia. SCOP: A structural classification of proteins database for the investigation of sequences and structures. *Journal of Molecular Biology*, 247(4):536–540, 1995.
32. C. Orengo, J.E. Bray, D.W. Buchan, A. Harrison, D. Lee, F.M. Pearl, I. Sillitoe, A.E. Todd, and J.M. Thornton. The CATH protein family database: A resource for structural and functional annotation of genomes. *Proteomics*, 2(1):11–21, 2002.
33. J. Chandonia and S. Brenner. The impact of structural genomics: Expectations and outcomes. *Science*, 311(5759):347–351, 2006.
34. H. Berman, T. Battistuz, T.N. Bhat, W.F. Bluhm, P.E. Bourne, K. Burkhardt, Z. Feng, G.L. Gilliland, L. Iype, S. Jain, P. Fagan, J. Marvin, D. Padilla, V. Ravichandran, B. Schneider, N. Thanki, H. Weissig, J.D. Westbrook, and C. Zardecki. The protein data bank. *Nucleic Acids Research*, 28:235–242, 2000.
35. T. Blundell, B.L. Sibanda, M.J.E. Sternberg, and J.M. Thornton. Knowledge-based prediction of protein structures and the design of novel molecules. *Nature*, 326:347–352, 1987.
36. P. Koehl and M. Delarue. Application of a self-consistent mean field theory to predict protein side-chains conformation and estimate their conformational entropy. *Journal of Molecular Biology*, 239:249–275, 1994.
37. P. Bates, L.A. Kelley, R.M. MacCallum, and M.J. Sternberg. Enhancement of protein modeling by human intervention in applying the automatic programs 3D-JIGSAW and 3D-PSSM. *Proteins*, 45(Suppl. 5):39–46, 2001.
38. T. Schwede, J. Kopp, N. Guex, and M.C. Peitsch. SWISS-MODEL: An automated protein homology-modeling server. *Nucleic Acids Research*, 31(13):3381, 2003.
39. S. Altschul, T.L. Madden, A.A. Schaffer, J. Zhang, Z. Zhang, W. Miller, and D.J. Lipman, Gapped BLAST and PSI-BLAST: A new generation of protein database search programs. *Nucleic Acids Research*, 25(17):3389–3402, 1997.
40. T. Bailey and M. Gribskov. Score distributions for simultaneous matching to multiple motifs. *Journal of Computational Biology*, 4(1):45–59, 1997.
41. P. Baldi, Y. Chauvin, T. Hunkapiller, and M.A. McClure. Hidden Markov models of biological primary sequence information. *Proceedings of the National Academy of Sciences U S A*, 91(3):1059–1063, 1994.
42. A. Krogh, M. Brown, I.S. Mian, K. Sjolander, and D. Haussler. Hidden Markov models in computational biology. Applications to protein modeling. *Journal of Molecular Biology*, 235(5):1501–1531, 1994.
43. R. Hughe and A . Krogh. Hidden Markov models for sequence analysis: Extension and analysis of the basic method. *Bioinformatics*, 12(2):95–107, 1996.
44. S. Eddy. Profile hidden Markov models. *Bioinformatics*, 14(9):755–763, 1998.
45. J. Park, K. Karplus, C. Barrett, R. Hughey, D. Haussler, T. Hubbard, and C. Chothia. Sequence comparisons using multiple sequences detect three times

as many remote homologues as pairwise methods. *Journal of Molecular Biology*, 284(4):1201–1210, 1998.

46. K. Koretke, R. Russell, and A. Lupas. Fold recognition from sequence comparisons. *Proteins*, 45:68–75, 2001.
47. J. Gough, K. Karplus, R. Hughey, and C. Chothia. Assignment of homology to genome sequences using a library of hidden Markov models that represent all proteins of known structure. *Journal of Molecular Biology*, 313(4):903–919, 2001.
48. R. Sadreyev and N. Grishin. COMPASS: A tool for comparison of multiple protein alignments with assessment of statistical significance. *Journal of Molecular Biology*, 326(1):317–336, 2003.
49. J. Thompson, D. Higgins, and T. Gibson, CLUSTAL W: Improving the sensitivity of progressive multiple sequence alignment through sequence weighting, position-specific gap penalties and weight matrix choice. *Nucleic Acids Research*, 22(22):4673–4680, 1994.
50. L. Rychlewski, L. Jaroszewski, W. Li, and A. Godzik. Comparison of sequence profiles. Strategies for structural predictions using sequence information. *PRS*, 9(02):232–241, 2000.
51. C. Notredame, D. Higgins, and J . Heringa. T-Coffee: A novel method for fast and accurate multiple sequence alignment. *Journal of Molecular Biology*, 302(1):205–217, 2000.
52. G. Yona and M. Levitt. Within the twilight zone: A sensitive profile-profile comparison tool based on information theory. *Journal of Molecular Biology*, 315(5):1257–1275, 2002.
53. M. Madera. PRC, the profile comparer. Cited 2009 June 18, 2009. http://supfam.org/PRC/.
54. D. Mittelman, R. Sadreyev, and N. Grishin. Probabilistic scoring measures for profile-profile comparison yield more accurate short seed alignments. *Bioinformatics*, 19:1531–1539, 2003.
55. K. Ginalski, J. Pas, L.S. Wyrwicz, G. Von, J.M. Bujnicki, and L. Rychlewski. ORFeus: Detection of distant homology using sequence profiles and predicted secondary structure. *Nucleic Acids Research*, 31(13):3804, 2003.
56. R. Edgar and K. Sjolander. SATCHMO: Sequence alignment and tree construction using hidden Markov models. *Bioinformatics*, 19:1404–1411, 2003.
57. T. Ohlson, B. Wallner, and A. Elofsson. Profile-profile methods provide improved fold-recognition: A study of different profile-profile alignment methods. *Proteins*, 57(1):188–197, 2004.
58. G. Wang and D. Jr. Dunbrack. Scoring profile-to-profile sequence alignments. *Protein Science: A Publication of the Protein Society*, 13(6):1612, 2004.
59. M. Marti-Renom, A. Stuart, A. Fiser, R. Sanchez, F. Melo, and A. Sali. Comparative protein structure modeling of genes and genomes. *Annual Review of Biophysics and Biomolecular Structure*, 29:291–325, 2000.
60. J. Soding. Protein homology detection by HMM-HMM comparison. *Bioinformatics*, 21(7):951–960, 2005.
61. K. Ginalski. Comparative modeling for protein structure prediction. *Current Opinion in Structural Biology*, 16(2):172–177, 2006.

62. Y. Zhang. Progress and challenges in protein structure prediction. *Current Opinion in Structural Biology*, 18(3):342–348, 2008.
63. Y. Zhang, I.A. Hubner, A.K. Arakaki, E. Shakhnoich, and J. Skolnick. On the origin and highly likely completeness of single-domain protein structures. *Proceedings of the National Academy of Sciences U S A*, 103(8):2605–2610, 2006.
64. S. Henikoff and J. Henikoff. Amino acid substitution matrices from protein blocks. *Proceedings of the National Academy of Sciences U S A*, 89(22):10915–10919, 1992.
65. M. Dayhoff, W. Barke, and L. Hunt. Establishing homologies in protein sequences. *Methods in Enzymology*, 91:524, 1983.
66. S. Altschul, W. Gish, W. Miller, E.W. Myers, and D.J. Lipman. Basic local alignment search tool. *Journal of Molecular Biology*, 215(3):403–410, 1990.
67. W. Pearson and D. Lipman. Improved tools for biological sequence comparison. *Proceedings of the National Academy of Sciences U S A*, 85(8):2444–2448, 1988.
68. L. Baum, T. Petrie, G. Soules, and N. Weiss. A maximization technique occurring in the statistical analysis of probabilistic functions of Markov chains. *The Annals of Mathematical Statistics*, 41(1):164–171, 1970.
69. G. Jr. Forney. The viterbi algorithm. *Proceedings of the IEEE*, 61(3):268–278, 1973.
70. R. Edgar. MUSCLE: Multiple sequence alignment with high accuracy and high throughput. *Nucleic Acids Research*, 32(5):1792, 2004.
71. K. Ginalski, A. Elofsson, D. Fischer, and L. Rychlewski. 3D-Jury: A simple approach to improve protein structure predictions. *Bioinformatics*, 19:1015–1018, 2003.
72. L. McGuffin and D. Jones. Improvement of the GenTHREADER method for genomic fold recognition. *Bioinformatics*, 19:874–881, 2003.
73. D. Jones. GenTHREADER: An efficient and reliable protein fold recognition method for genomic sequences. *Journal of Molecular Biology*, 287(4):797–815, 1999.
74. J. Xu, M. Li, D. Kim, and Y. Xu. RAPTOR: Optimal protein threading by linear programming. *Journal of Bioinformatics and Computational Biology*, 1(1):95–117, 2003.
75. L. Jaroszewski, L. Rychlewski, Z. Li, W. Li, and A. Godzik. FFAS03: A server for profile-profile sequence alignments. *Nucleic Acids Research*, 33(Web server issue):W284, 2005.
76. E. Lindahl and A. Elofsson. Identification of related proteins on family, superfamily and fold level. *Journal of Molecular Biology*, 295(3):613–625, 2000.
77. Y. Shan, G. Wang, and H. Zhou. Fold recognition and accurate query-template alignment by a combination of PSI-BLAST and threading. *Proteins*, 42(1):23–37, 2001.
78. N. Vonohsen, I. Sommer, and R. Zimmer. Profile-profile alignment: A powerful tool for protein structure prediction. *Biocomputing* 2002:252, 2003.
79. B. Wallner, H. Fang, T. Ohlson, J. Frey-skott, and A. Elofsson. Using evolutionary information for the query and target improves fold recognition. *Proteins*, 54(2):342–350, 2004.

80. D. Juan, O. Grana, F. Pazos, P. Fariselli, R. Casadio, and A. Valencia. A neural network approach to evaluate fold recognition results. *Proteins*, 50(4):600–608, 2003.

81. D. Fischer. 3D-SHOTGUN: A novel, cooperative, fold-recognition meta-predictor. *Proteins*, 51(3):434–441, 2003.

82. T. Jaakkola, M. Diekhans, and D. Haussler. A discriminative framework for detecting remote protein homologies. *Journal of Computational Biology*, 7(1–2):95–114, 2000.

83. C. Ding and I. Dubchak. Multi-class protein fold recognition using support vector machines and neural networks. *Bioinformatics*, 17(4):349–358, 2001.

84. C. Leslie, E. Eskin, and W. Noble. The spectrum kernel: A string kernel for SVM protein classification. In *Proceedings of the Pacific Symposium on Biocomputing 2002*, pp. 564–575. World Scientific.

85. H. Saigo, J.P. Vert, N. Ueda, and T. Akutsu. Protein homology detection using string alignment kernels. *Bioinformatics*, 20(11):1682–1689, 2004.

86. H. Rangwala and G. Karypis. Profile-based direct kernels for remote homology detection and fold recognition. *Bioinformatics*, 21(23):4239–4247, 2005.

87. L. Holm and C. Sander. The FSSP database of structurally aligned protein fold families. *Nucleic Acids Research*, 22(17):3600, 1994.

88. J. Rocchio. Document retrieval systems: optimization and evaluation. PhD thesis, Harvard University, 1966.

89. L. Page, S. Brin, R. Motwani, and T. Winograd. *The Page Rank Citation Ranking: Bringing Order to the Web*. Stanford University, 1998. Technical report.

90. K.D. Pruitt, T. Tatusova, and D. Maglott. NCBI Reference Sequence (RefSeq): A curated non-redundant sequence database of genomes, transcripts and proteins. *Nucleic Acids Research*, 33:D501–D504, 2006.

91. G. Pollastri and P. Baldi. Prediction of contact maps by GIOHMMs and recurrent neural networks using lateral propagation from all four cardinal corners. *Bioinformatics*, 18(1):S62–S70, 2002.

92. J. Cheng, Z.R. Randall, J. Sweredoski, and P. Baldi. SCRATCH: A protein structure and structural feature prediction server. *Nucleic Acids Research*, 33:W72, 2005.

93. J. Cheng and P. Baldi. Three-stage prediction of protein [beta]-sheets by neural networks, alignments and graph algorithms. *Bioinformatics*, 21(1):i75–i84, 2005.

94. J. Quinlan. Induction of decision trees. *Machine Learning*, 1(1):81–106, 1986.

95. T. Mitchell. *Machine Learning*. Tom M. Mitchell and McGraw Hill, 1996.

96. Y. Yang and J. Pedersen. A comparative study on feature selection in text categorization. In *Proceedings of the 14th International Conference on Machine Learning*, pp. 412–420. Morgan Kaufmann, 1997.

97. I. Witten and E. Frank. Data mining: Practical machine learning tools and techniques with Java implementations. *ACM SIGMOD Record*, 31(1):76–77, 2002.

98. J. Shi, T. Blundell, and K. Mizuguchi. FUGUE: Sequence-structure homology recognition using environment-specific substitution tables and structure-dependent gap penalties. *Journal of Molecular Biology*, 310(1):243–257, 2001.

99. D. Wheeler, T. Barrett, D.A. Benson, S.H. Bryant, K. Canese, V. Chetvernin, D.M. Chruch, M. DiCuccio, R. Edgar, S. Federhen, L.Y. Geer, W. Helmberg, Y. Kapustin, D.L. Kenton, O. Khovayko, D.J. Lipman, T.L. Madden, D.R. Maglott, J. Ostell, K.D. Pruitt, G.D. Schuler, L.M. Schriml, E. Segueira, S.T. Sherry, K. Sirotkin, A. Souvoro, G. Starchenko, T.O. Suzek, R. Tatusov, T.A. Tatusova, L. Wagner, and E. Yaschenko, Database resources of the national center for biotechnology information. *Nucleic acids research*, 35(Database issue):D5, 2007.

100. T. Smith and M. Waterman. Identification of common molecular subsequences. *Journal of Molecular Biology*, 147:195–197, 1981.

101. J. Hargbo and A. Elofsson. Hidden Markov models that use predicted secondary structures for fold recognition. *Proteins*, 36(1):68–76, 1999.

102. D. Jones. THREADER: Protein sequence threading by double dynamic programming. *New Comprehensive Biochemistry*, 32:285–311, 1998.

103. H. Zhou and Y. Zhou. Single-body residue-level knowledge-based energy score combined with sequence-profile and secondary structure information for fold recognition. *Proteins*, 55(4):1005–1013, 2004.

CHAPTER 10

TASSER-BASED PROTEIN STRUCTURE PREDICTION

SHASHI BHUSHAN PANDIT, HONGYI ZHOU, and JEFFREY SKOLNICK
Center for the Study of Systems Biology
School of Biology
Georgia Institute of Technology
Atlanta, GA

10.1. INTRODUCTION

Over the past decade, genome sequencing projects have generated the complete genomic sequences for hundreds of organisms [1–4]. The availability of the complete repertoire of genes of an organism could be used to systematically understand various biological processes. An important logical step toward this goal is to get insight into the functions of gene products encoded in the genome [5]. The functional annotation of proteins involves laborious experimental studies that can be enhanced with the use of various computational methods. Sequence-based methods such as Position-Specific Iterative-Basic Local Alignment Search Tool (PSI-BLAST) [6,7], EFICAz [8], and Pfam [9] can provide clues into the functions of proteins for about 40–60% of gene products in a given genome [7,10]. However, these methods begin to fail as the sequence becomes more distant from proteins of known function [10,11]. The function prediction of these very remotely related proteins is an important challenge. The biochemical function of a protein is determined by the three-dimensional (3D) structure of the biologically active conformation of the protein. Hence, knowledge of tertiary structures of a protein with unknown function or its relationship to a protein with a known tertiary structure can assist in function annotation [12–16]. However, the experimental determination of protein tertiary structure is time-consuming and expensive; hence, the

Introduction to Protein Structure Prediction: Methods and Algorithms,
Edited by Huzefa Rangwala and George Karypis

ability to predict the native conformation of a protein has become increasingly important [12,15].

Protein structure prediction methods can be logically divided into three categories: comparative modeling (CM) [17–20], threading [21–27], and template-free (TF) [28–32], or *ab initio*, methods. CM and threading both attempt to identify a set of structurally related template proteins (with known 3D structure) to the target sequence of interest [23,33]. In the case of CM, the template proteins have a clear evolutionary relationship to the target sequence, whereas threading identifies templates with a similar fold as the target sequence; hence, homologous as well as analogous folds could in principle be found [23,25,34,35]. Once the related template is identified, the target sequence is aligned to the template structure. Over the past several years, threading and CM methods have converged; this has given rise to the term template-based, TB, modeling [36,37]. In TF, [37] or *ab initio* methods, one does not use any global template structural information as an input to the structure prediction process. Thus, the possibility of assembling a novel fold exists [38]. Among the various realizations of TF methods, some employ purely physics-based empirical potentials based on quantum mechanics [39,40], while others use predicted secondary structure and/or side chain contacts and knowledge-based potentials derived from a statistical analysis of protein structural databases [41–45]. Since these approaches do not start from a template structure, this is the hardest category with the lowest prediction accuracy.

There are two main issues involved in CM/threading methods. First, a necessary precondition for their success is the completeness of the library of solved structures in the Protein Data Bank (PDB) [46]. Several recent studies have demonstrated that the current PDB library is most likely complete; it can provide templates for all compact single domain proteins on which low-to-moderate resolution structures can be built [47–49]. Thus, in principle, the protein structure prediction problem could be solved by CM/threading methodologies for single domain proteins. However, an effective fold recognition algorithm must be developed to identify these correct template proteins and alignments.

Second, having a threading template with gapped alignments and average coverage, it is nontrivial to build a complete, full-length model of the protein. In practice, most successful structure predictions rely on the evolutionary relationship between the target and the template proteins [50]. For proteins having >50% sequence identity to the templates, CM methods provide models with a root-mean-square deviation (RMSD) from the native structure for their backbone atoms that is below 1 Å [12,33,50–52]. For proteins with 30–50% sequence identity to their templates, the models often have ~85% of their core region within an RMSD of 3.5 Å from the native structure of their backbone, with the errors mainly in loops [50–52]. When the sequence identity drops below 30%, the so-called "twilight" zone, (about one-third of known protein sequences), modeling accuracy sharply decreases because of the lack of the significant threading hits and substantial alignment errors [12,50]. In addition, the ability to accurately predict the conformation of the intervening loops between the aligned regions is rather limited. Therefore, it is essential to develop an effective automated technology that can deal with proteins in the twilight zone of sequence identity. It should build models that are closer to the native structure than their initial templates with reasonably accurate loop conformations. Furthermore, although the PDB is growing faster than ever, the gap between the numbers of sequences and solved protein structures remains large. This necessitates the need for a robust automated prediction method that is applicable to entire proteomes.

Over the past few years, our laboratory has developed the protein structure prediction methodology called Threading ASSembly Refinement (TASSER) for automated tertiary structure prediction that generates full-length models by rearranging the continuous protein structural fragments identified by threading [53–57]. In the following sections, we will describe TASSER along with its benchmarking on a large-scale, representative set of PDB structures. In addition, we will discuss its performance in various rounds of Critical Assessment of Protein Structure Prediction (CASP) [37] as well as some applications of the TASSER approach.

10.2. METHODOLOGY

The TASSER methodology consists of template identification, structure assembly, and model selection; an overview is presented in Figure 10.1.

10.2.1. Threading

The structure templates for a target sequence were originally selected from the PDB library by our iterative threading program PROSPECTOR_3 [23,58], designed to identify analogous as well as homologous templates. The scoring function of PROSPECTOR_3 includes close and distant sequence profiles, secondary structure prediction from PSIPRED [59], and chain contact pair potentials extracted from the alignments in previous threading iteration.

While we have mostly used PROSPECTOR_3 to identify the structure templates, in principle, any reliable CM/threading method could be used for template selection/alignment generation. In CASP7 [60] and CASP8, we used the 3D-jury approach [61] to select templates from PROSPECTOR_3, SP^3 [27], and SPARKS2 [24] for their subsequent use in tertiary structure prediction [60]. In our implementation of the 3D-jury approach, the 10 top-scoring templates from each threading method are compared using the structural alignment algorithm, TM-align [62], with the TM-score [63] as the similarity measure. The 3D-jury score is the sum of pairwise TM-scores for each template and is used to rank the templates. The obtained consensus templates provide the aligned fragments, tertiary distance restraints, and contact restraints used in the full-length structure assembly. Targets are classified as

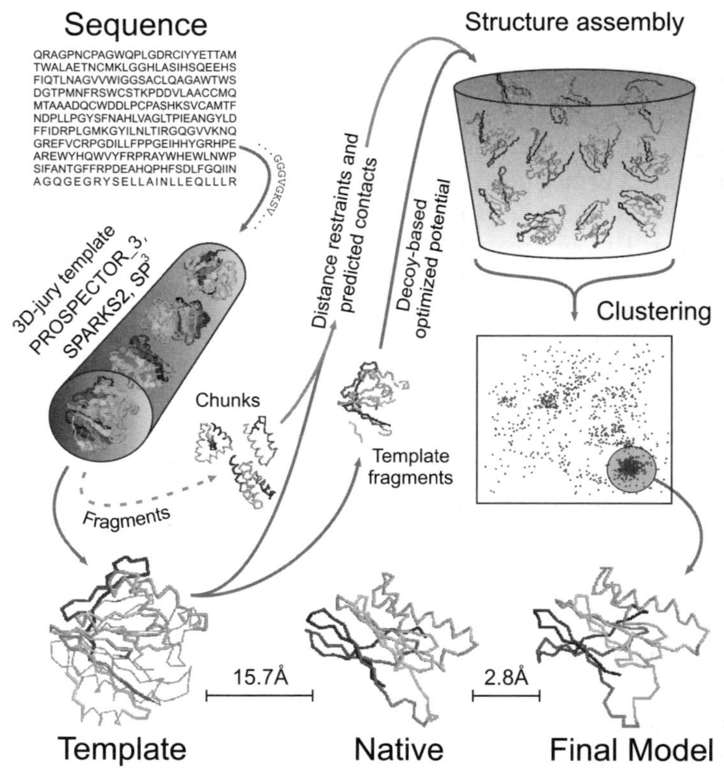

FIGURE 10.1 Overview of the TASSER/chunk-TASSER methodology. (See color insert.)

Easy, Medium, and Hard according to the template similarities of the top models from each threading method. When the top models have a TM-score >0.5 with each other, the target is defined as Easy and is likely to have a structurally related template with a good alignment; when they have a TM-score <0.4 with each other, the target is defined as Hard and the template is likely to be incorrect; all other cases are defined as Medium. The classification procedure is empirical and could be further optimized in the future, but it provides a basis of the expected accuracy of the prediction.

10.2.2. TASSER Assembly and Refinement

10.2.2.1. On-and-Off Lattice C-Alpha/Side Chain-Based (CAS) Model/Force-Field. A protein conformation is represented by its C_α atoms and side chain centers of mass (SG), the *CAS* model [55]. Based on the threading alignment, the chain is divided into continuous aligned regions (≥five residues) whose local conformations are unchanged during structure assembly and gapped regions whose conformation needs to be predicted by *ab initio* methods. For computational efficiency, the residues unaligned to the template lie on an underlying cubic lattice of grid size of 0.87 Å, whereas the aligned residues are off-lattice for maximum accuracy. The SGs are also always off-lattice [55].

The TASSER force field consists of a variety of terms based on or derived from the regularities seen in protein structures [55,64]. The force field consists of three classes of terms: (i) statistical potentials extracted from the PDB including long-range SG-pair interactions, local C_α correlations, a generic hydrogen bond potential, environmental profiles, and a residue based solvent accessibility; (ii) propensities for predicted secondary structures from PSIPRED [59]; and (iii) protein specific SG-pair potentials and tertiary distance and contact restraints extracted from the threading templates. There are also energy terms that provide a generic bias to protein-like conformations and biases toward predicted contact order and contact number, which speed up the simulation. A more detailed description of these terms has been described elsewhere [55,64].

The different energy terms in the TASSER force field serve various roles. The most important factors for overall folding in our force field are the secondary structure prediction propensities, hydrogen bonding, and the predicted tertiary contact restraints derived from threading. The first two types of terms provide a basic structural framework for the conformational space that is explored. However, the contact restraints are of critical importance in modifying the energy landscape to guide the simulations to near native states, especially for large proteins, where the general protein-like potential cannot distinguish the native state among a huge number of possible topologies. The other energy terms such as short-range correlations, environment profiles, and long-range pairwise interactions are helpful for refining the packing of side chains and local fragments.

The combination of all the energy terms was optimized by maximizing the correlation between the CAS energy and the RMSD of the decoy structures to native for a training set of 100 proteins each with 60,000 structural decoys. Following optimization on a non-homologous set of training proteins, an average correlation coefficient between RMSD and energy of 0.69 on a non-homologous set of testing proteins was obtained [55].

10.2.2.2. Template Assembly and Refinement. For a given template, an initial full-length model is built by connecting the continuous aligned regions through a random walk of C_α–C_α bond vectors where only excluded volume and geometric restraints are applied. The template fragments are connected by a random walk. If there are two few unaligned residues in the target sequence to span the gap in the template structure, a long C_α–C_α bond remains at the end of the random walk and a spring-like force is applied in Monte Carlo simulations until a physically reasonable C_α–C_α bond length is achieved. The SG positions are determined by a two-rotamer approximation that depends on whether the local backbone configuration is extended or compact.

These initial full-length models are submitted to parallel hyperbolic Monte-Carlo sampling (PHS) for assembly/refinement [65]. We consider an ensemble of 40 replicas, each at a different temperature. Two kinds of conformational updates are employed: (i) Off-lattice moves involve rigid fragment translations and rotations of its three Euler angles. The fragment length normalizes the movement amplitude so that the acceptance rate is approximately constant for different size fragments. (ii) Lattice confined residues are subjected to two to six bond movements and multi-bond sequence shifts [64].

10.2.2.3. Structure Selection. Structures generated in the 14 lowest temperature replicas are clustered using SPICKER [66]. The final models are selected on the basis of the cluster density. Usually, the five models from the five most populated clusters are used for further analysis.

10.3. BENCHMARKING TASSER FOR STRUCTURE PREDICTION

The TASSER approach has been tested on comprehensive large-scale PDB benchmark sets with templates generated by both threading and structural alignments [48,55,56]. In this section, we summarize the benchmarking results from TASSER and its variants.

10.3.1. Folding Results for Weakly Homologous Proteins

TASSER was originally benchmarked on a set of 1489 non-homologous single-domain proteins (maximum of 35% pairwise sequence identity to each other) that covered the PDB, whose lengths were below 200 residues [55]. For the purpose of benchmark evaluation, all templates whose sequence identity to their target is greater than 30% are excluded. In 927 cases, PROSPECTOR_3 identified templates having an RMSD from native <6.5 Å with ~80% alignment coverage. Following structure assembly and refinement, the number of cases with an RMSD <6.5 Å increased to 1172 proteins. This shows that for a number of proteins with no acceptable templates (with a global RMSD > 6.5 Å), TASSER could generate models with a significant RMSD to native. In Figure 10.2a, for the set of residues that are aligned to the template, we show the comparison of the best model's (among the top five clusters selected by SPICKER) RMSD to the native structure with the RMSD of the best initial threading alignment to the native structure. As is evident from the figure, there is an obvious improvement over the original template alignments. In Figure 10.2b, we show the comparison of the initial templates and optimized final models from a widely used CM tool, MODELLER [50,67]. The structure given by MODELLER is obtained by optimally satisfying tertiary restraints from templates; here, template quality mainly dictates the final result. These results show that TASSER can move the models closer to the native tertiary structure than that found in the input template alignment. Considering the

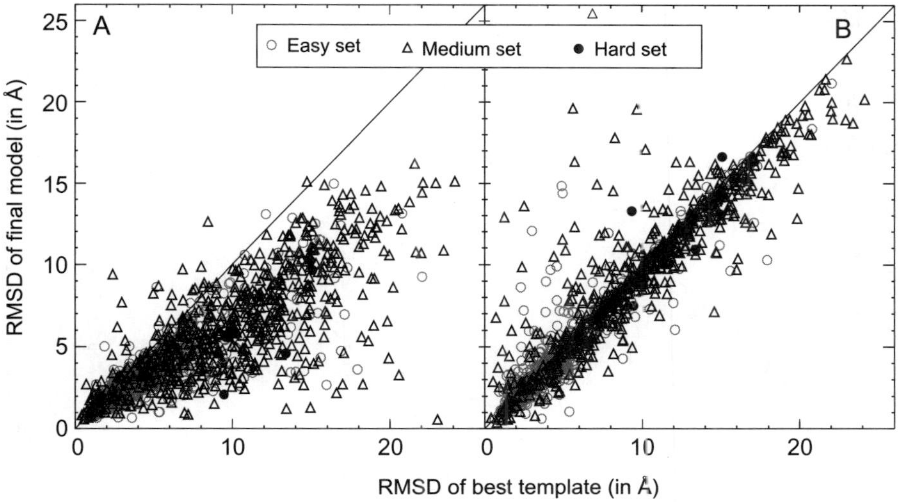

FIGURE 10.2 (A) Scatter plot of the RMSD to native for the best of five TASSER models versus the RMSD to native for the initial templates alignments from PROSPECTOR_3. The same aligned region is used in both RMSD calculations. (B) Similar data as in (A), but the models are from MODELLER.

RMSD of full-length models to the native, TASSER can fold ~2/3 (990 proteins) of proteins that are weakly homologous to their templates.

Next, we benchmarked TASSER on a nonredundant set of 745 larger proteins from various secondary structure classes with lengths between 201 and 300 residues [56]. In 365 cases (49%), PROSPECTOR_3 identified templates with an RMSD to native <6.5 Å and >70% alignment coverage. Following TASSER assembly, for about 55% of the targets, the best of the top five models has an RMSD < 6.5 Å. In Figures 10.3a,b, for Easy and Medium/Hard target, respectively, the improvement of the best of the top five models over the initial threading alignment is shown. Again, for most cases, TASSER improves the models from their initial template alignment. The lower success rate for larger proteins with respect to the set of smaller (≤200 residue), single-domain proteins is mainly due to the fact that these proteins have a higher percentage of multi-domain proteins, where TASSER fails to predict the inter-domain orientation if it differs significantly from the template structure.

10.3.2. CM

We next explored TASSER's template refinement ability for proteins in the CM regime (when the sequence identity ≥35% between the template and target sequences). Here, the initial alignments are in general quite good. TASSER was evaluated on a representative benchmark set of 901

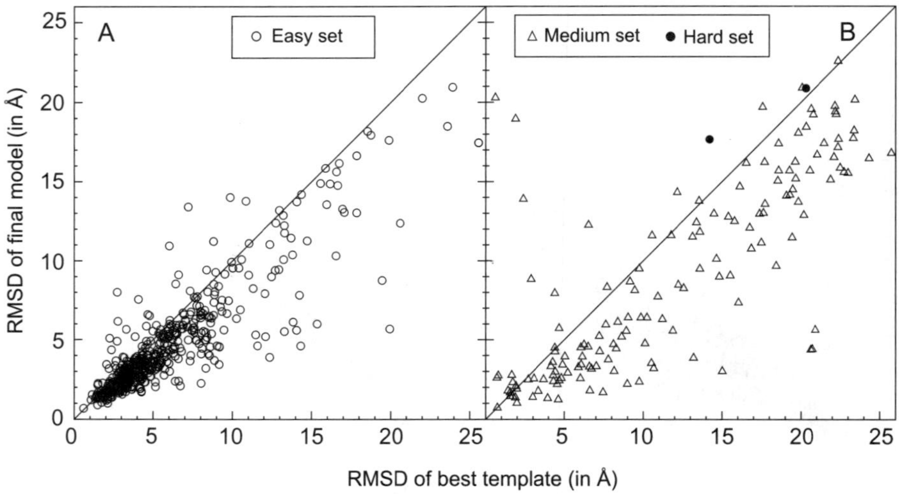

FIGURE 10.3 RMSD to native of the best of top five TASSER models versus the RMSD to native of the best initial template by PROSPECTOR_3; both RMSDs are calculated over the same aligned regions. (A) Easy set targets; (B) Medium/Hard set targets.

single-domain proteins, 41–200 residues in length, with sequence identities between 35% and 90% to the template. Using the recently developed TASSER-Lite program (which employs TASSER parameters optimized for the CM limit to give rapid results) [54], in the aligned region an average improvement of ~10% in RMSD to native over the initial template is found [54]. As shown in Figure 10.4a, for the same aligned region (provided by the threading algorithm PROSPECTOR_3), the RMSD of the best model (among the top five) to native is shown. For many cases, there is clear improvement, with a number of models refined from an RMSD > 2.5 Å to structures close to 1 Å. Figure 10.4b shows the results from the CM program MODELLER [50,67], which basically recapitulates the template alignment. For targets with an initial template RMSD to the native structure in the range of 0-1 Å, the ability of TASSER-Lite to refine over the initial template is rather limited. This is simply due to the inherent resolution of the TASSER force field, which is ~1.2 Å. Overall, for proteins in the CM regime, TASSER can generates refined models that are closer to the native structure than their input initial template alignment.

10.3.3. TF Modeling

Having shown TASSER's ability to generate models closer to native when threading templates are used as input to provide predicted tertiary restraints and aligned fragments, templates, we next assessed its ability to predict tertiary structure in the absence of any templates, that is, in the *ab initio* or TF

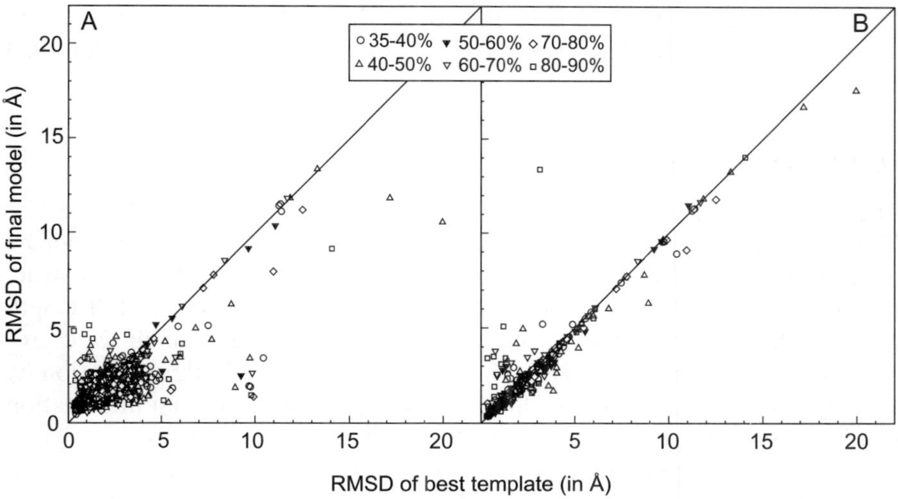

FIGURE 10.4 (A) Scatter plot of the RMSD of the TASSER final model to native versus RMSD of the initial alignment (from PROSPECTOR_3) to native. The same aligned region is used in both the RMSD calculations. (B) Similar data as in (A), but the models are from MODELLER.

limit [68]. The benchmark dataset consisted of 1029 non-homologous sequences from 40 to 200 residues that cover the PDB at 30% sequence identity. For sequences in the 40–100 (101–200) residues range, the best of the top five TASSER models has a global fold that is related to the native structure, as assessed by RMSD, for 25% (16%) of the cases and has a correct core structure for 59% (36%) of the cases [68]. Thus, compared with the case when one can identify suitable templates, the success of TASSER in TF limit is rather modest.

10.4. FURTHER DEVELOPMENTS OF TASSER

The overall performance of TASSER depends strongly on the accuracy of contact and distance restraints derived from threading. The comprehensive benchmarking in the previous section showed that TASSER could fold ~2/3 of all non- or weakly homologous proteins ≤200 residues in length [55]. In addition, TASSER often provides models that are closer to the native tertiary structure than those provided by the initial template alignments. However, for the remaining ~1/3 of proteins, the prediction accuracy of TASSER is significantly worse because PROSPECTOR_3 provides inaccurate template fragments and predicted contact restraints. Thus, improvements in this regime of target difficulty are needed. Toward this goal, we have developed the chunk-TASSER method [69] and an improved side chain contact prediction algorithm using the composite-sequence method [70]. In addition, we have also developed an iterative TASSER procedure [71,72]. We refer the readers to our publications elsewhere [71,72] and also to Chapter 11 for additional details.

10.4.1. Chunk-TASSER

The goal of chunk-TASSER is to improve the accuracy of TASSER for Medium/Hard proteins for which no confident template fragments can be identified from threading. The conventional approach of dealing with structure modeling of these TF proteins is to do *ab initio* folding of full length proteins. As described above, such an approach is limited to small-sized proteins with simple topologies [73]. Here, we attempt to avoid this limitation by folding supersecondary chunks first, and using this as additional information in a later stage. Modeling of protein chunks, that is, three consecutive regular secondary structure segments (helices or strands), is much more efficient than full-length modeling in that the sampling space is much smaller and the possible topologies are much simpler. An overview of methodology is shown in Figure 10.1 where it is shown that chunk-TASSER uses consensus contacts and distance restraints from *ab initio* folded chunks of protein supersecondary structures to provide additional information apart from that extracted from identified threading templates.

10.4.1.1.* Ab initio *Folding of Chunks and Chunk Model Selection. In chunk-TASSER, the protein sequence is divided into sliding windows (chunks) of three consecutive, predicted regular secondary structure segments. Chunk structures are predicted independently by a fragment insertion method as described in Reference [30], but with our own implementation and force field. Each residue consists of its main chain atoms (N, C_α, C, O), the C_β atom and its side chain center of mass. The force field contains: (i) the DFIRE all-atom distance-dependent pair potential for the main chain and C_β atoms [43]; (ii) the distance-dependent pairwise statistical potential DFIRE-SCM for the side chain center of mass [44]; (iii) the TASSER hydrogen bond term based on C_α coordinates [64]; and (iv) an excluded volume term for the main chain and C_β atoms. The fragment library for insertion is generated by extending the SP^3 method [27], where for each position in the sequence, nine-residue long fragments of the top 25 scored templates are selected to form the fragment library for the *ab initio* folding of chunks.

The usual way of selecting models from a set of decoys is to cluster the models and select the most populated clusters. The success of a clustering method depends on the fact that lower free-energy conformations are likely closer to the native structure. Since chunk models are not full-length protein models, this free-energy condition may not be satisfied. Therefore, we developed and tested an alternative way of selecting chunk models that uses the information from the fragment library. For each residue position in the chunk model, a nine-residue fragment with the given residue in the middle is compared using their RMSD to the top 25 corresponding fragments in the fragment library. The average RMSD over the 25 fragments and over all chunk residue positions is computed and combined with the DFIRE energy to give a ranking score. Finally, this score is used to rank the models and those with lower scores are selected.

10.4.1.2. Full-Length Model Assembly by TASSER and Benchmarking. The top 10 models for each chunk are selected and combined with full-length threading templates to derive distance and contact restraints for input into TASSER. By default, chunk-TASSER uses only SP^3 for template identification and alignments. Other than input generation, all procedures are as in TASSER [55].

We benchmarked chunk-TASSER on a nonredundant set (sequence identity less than 35%) of 425 proteins ≤250 residues in length. By removing all structures with a TM-score ≥0.4 [63] to each target from the threading library, we artificially make the benchmark set effectively Hard targets. The average TM-scores of the best of top five models per target are 0.266, 0.336, and 0.362 from the SP^3 threading algorithm, original TASSER [55] and chunk-TASSER, respectively. For the subset of 80 proteins with predicted α–helix content ≥50%, these averages are 0.284, 0.356, and 0.403, respectively. The percentage of proteins with the best of the top five models having a TM-score ≥0.4 (a statistically significant threshold for structural similarity [63]) are 3.76%, 20.94%, and 28.94% by SP^3, TASSER, and chunk-TASSER, respectively. For the subset of 80 predominantly helical proteins, these percentages are 2.50%, 23.75%, and 41.25% [69]. The improvement of chunk-TASSER over TASSER is also evident from Figure 10.5, which shows the comparison of TM-scores of

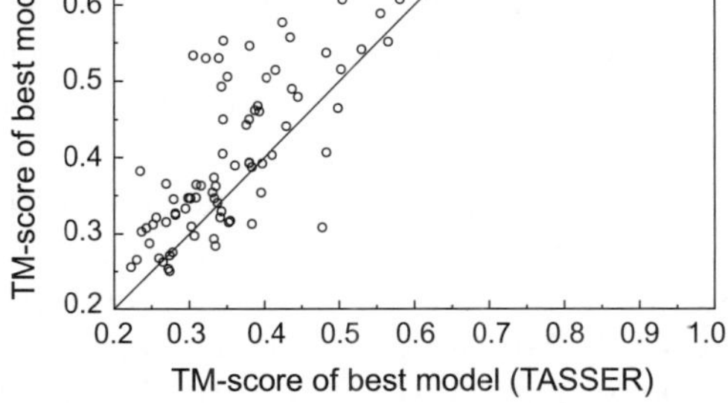

FIGURE 10.5 Comparison of the TM-scores obtained from chunk-TASSER and TASSER on an 80 α-protein benchmark set.

the best model to native from chunk-TASSER and TASSER for the 80 proteins rich in α–helix. Hence, chunk-TASSER shows a significant improvement over TASSER for the structure prediction of Hard targets where no good templates could be identified. The pronounced improvement for targets with high α–helix content is due to the fact that helical proteins have less long-range contacts than β proteins and therefore are easier to fold by an *ab initio* method. We also note that we used chunk-TASSER in the tertiary structure prediction of 21 Medium/Hard proteins in CASP7 [69]. Using chunk-TASSER, the first (best of five) model is ~10% (10%) better than models from TASSER.

10.4.2. TASSER_2.0

The composite-sequence method [70] is based on the idea that consensus contacts from multiple sets of contact predictions will be more accurate than any of the individual input sets. The composite-sequence method was implemented in the next generation of TASSER, referred to as TASSER_2.0 [70]. One set of predicted contacts is generated using the profiles from template sequences that are evolved using a structure-based scoring function that contains secondary structure, burial, and pair interaction potentials. The evolved set of sequences is used to generate sequence profiles in the improved version of threading, PROSPECTOR_3.5 [70] for contact prediction. Another set of contacts is obtained from wild-type template sequences using the original threading PROSPECTOR_3 [58]. We find that the consensus prediction significantly improves the contact accuracy while maintaining sufficient coverage to be effectively used in TASSER_2.0.

10.4.2.1. PROSPECTOR_3.5. As is described in PROSPECTOR_3 [58], PROSPECTOR_3.5 uses a set of four scoring functions, two set of sequence profiles, and multiple iterations. In PROSPECTOR_3, the 3590 set of profiles are derived from all sets of sequences having between 35% and 90% pairwise sequence identity and the e-10 set from sequences whose e-values to the parent (target or template) sequence is ≤10. In the evolved-sequence method, PROSPECTOR_3.5, the identical formalism and parameters are used; but we replaced the 3590 sequence profiles with those generated from the corresponding evolved sequences. In our benchmarking [70], we find that the alignments generated from the evolved-sequence profiles are more accurate than those using the wild-type template sequence profiles, but have less template coverage. The average fraction of correctly predicted side chain contacts for the benchmark set of proteins is 0.46 and the average fraction of predicted contacts per residue is 1.85, while those using the wild-type 3590 profiles are 0.37 and 3.29, respectively. For the consensus contacts between the evolved sequence set and the original PROSPECTOR_3 predicted contacts, the average fraction of accurately predicted contacts increases to 0.60, with the average number of contacts predicted per residue of 1.43.

In TASSER_2.0, the new contact predictions by the composite-sequence method are implemented as an additional energy term defined by

$$E_{\text{add}} = 1 + \left(\frac{r(I,J)}{r_0(I,J)} - 1 \right)^2, \quad r(I,J) > r_0(I,J),$$
$$= 0, \qquad\qquad\qquad\qquad r(I,J) \le r_0(I,J),$$

where $r(I,J)$ is the distance between the side chain centers of mass of the Ith and J^{th} residues and $r_0(I,J)$ is the corresponding cutoff distance for a contact between their side chain centers of mass. This new contact energy term is in addition to, not a replacement of, the original side chain contact energy term.

10.4.2.2. TASSER_2.0 Benchmarking. TASSER_2.0 was benchmarked on a set of 2591 non-homologous, single-domain proteins ≤200 residues that cover the PDB at 35% pairwise sequence identity. We find that the resulting TASSER_2.0 models are closer to their native structures, with an average RMSD of 4.99 Å compared with the 5.31 Å result of TASSER [70]. Defining a successful prediction as a model with an RMSD to native <6.5 Å, the success rate of TASSER_2.0 (TASSER) for Medium targets is 74.3% (64.7%) and 40.8% (35.5%) for Hard targets. For Easy targets, the success rate slightly increases from 86.3% to 88.4% [70].

10.5. TASSER'S PERFORMANCE IN CASP6-CASP8

The status of the field of protein structure prediction has been evaluated on a biannual basis by the CASP experiments, where the sequences of structures that are about to be experimentally determined are made available to the protein structure prediction community; the structures are then blindly predicted and assessed [37]. This offers the advantage that various approaches can be compared for the same set of targets. In the next section, we present the assessment of TASSER's performance in CASP6 to CASP8.

10.5.1. TASSER in CASP 6 and CASP 7

In CASP6, TASSER emerged as one of the most successful structure prediction methods [20,74]. Here, we used PROSPECTOR_3 to generate the threading template alignment, which was followed by structure assembly and refinement with the standard TASSER protocol [75]. The five highest structural density clusters were selected after clustering and used to generate an all-atom model using PULCHRA [76]. Ninety targets/domains were assessed from 64 targets submitted for structure prediction. On average, the best threading templates from PROSPECTOR_3 have an average RMSD to native of 8.4 Å with 79% of the residues aligned. For the same aligned residues, the

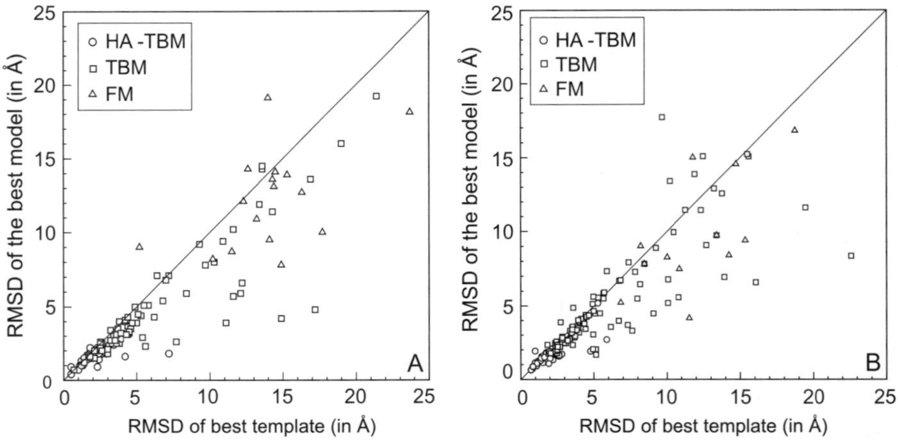

FIGURE 10.6 Assessment of TASSER modeling in various CASPs. (A) Comparison between the best TASSER models and the best threading template for CASP7 targets. (B) Same as in (A), but for CASP8 targets.

average RMSD to the native structure of the best of the top five TASSER models (rank 1 model) is 5.4 (6.4)Å. This shows that TASSER brings the threading templates closer to the native by ~2-3 Å.

In CASP7, we participated in the human prediction category as TASSER and in the server category as METATASSER [60]. For human predictions, we used 3D-jury selected templates followed by multiple instances of TASSER simulations (with varying simulation parameters) for template-based modeling and chunk-TASSER for targets with no reliable templates. For the 124 targets/domain assessed in CASP7, the RMSD to the native structure of our best threading template (from 3D-jury) is 6.2 Å, with an average coverage of 93%. For TASSER, the average RMSD of the best model (first model), over the same aligned residues for all targets is 4.9 (5.5)Å. In Figure 10.6a, using RMSD, we compare the best threading template and final best models. The final models from TASSER for most targets have an RMSD that is closer to native than the RMSD of the best threading template to native. Using the TM-score as the measure of model quality, the cumulative TM-scores for the best threading template, first and best submitted models are 76.47, 82.32, and 84.10, respectively. One particularly interesting target is T0382, a 123-residue protein that is classified in the template-based modeling (TBM) category. However, our 3D-jury procedure did not find any confident templates, and we predicted the structure using chunk-TASSER [69]. The initial template RMSD to native for T0382 is 14.9 Å. The RMSD over the same aligned region improves to 4.2 Å, with a full-length RMSD of 5.6 Å TASSER (Fig. 10.7a). Another interesting case is T0370, a target in the TBM category, where the RMSD to native over the template aligned residues improves from 7.7 Å to 2.6 Å, with a full-length RMSD to native of 2.6 Å as well.

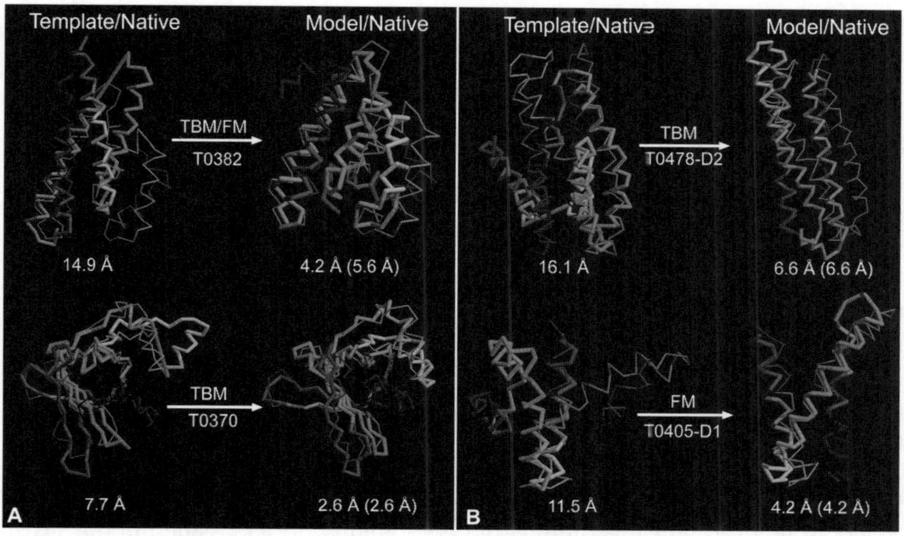

FIGURE 10.7 Successful examples of TASSER modeling in CASP7 and CASP8 are shown in (A) and (B), respectively. For each target, on the left is the superposition of the threading template (thick backbone) and native (thin backbone); on the right is the final model (thick backbone) and native (thin backbone). Blue to red goes from the N- to the C-terminus. The numbers below the superposition are the RMSD over the aligned regions and RMSD over the full-length molecule, respectively. (See color insert.)

METATASSER, in the CASP7 server category, employed 3D-jury to select templates (discussed in Methods section), a limited time TASSER simulation, followed by clustering of structures and submission of the top five models. The average RMSD of the best (first) model to native over the threading aligned region, for all targets, is 5.5 (6.1)Å. There was a technical error in the METATASSER protocol that was fixed after target T0328 was submitted. Using the fixed models for targets T0283-T0328, the average RMSD of the best (first) model to native improves to 5.3 (6.0)Å. The improvement is consistent with the TASSER benchmark results [55,56,69].

10.5.2. TASSER in CASP8

In CASP8, we participated with a newly developed method pro-sp3-TASSER [77] and the previously implemented METATASSER [60] server in the server prediction category. In this CASP, METATASSER implemented chunk-TASSER for Medium/Hard targets and used TASSER-QA for final ranking of the models. Apart from this, we also had a human prediction group, TASSER. The pro-sp3-TASSER method uses five individual threading scores derived from previously developed threading methods, SP^3 [27] and PROSPECTOR_3 [58], for template identification and alignment [77]. These templates are used as input in short TASSER simulations to build full-length models. Of these, the top 20 models are ranked using protein structure quality assessment (QA) method TASSER-QA [78] and are used for final structure refinement using TASSER. For Medium/Hard targets we always use the chunk-TASSER protocol along with pro-sp3-TASSER as well as the side chain contacts predicted by TASSER_2.0 [70]. The ensemble of models generated from TASSER was ranked using TASSER-QA and the top five models were submitted after using PULCHRA to build all-atom models.

Of the 164 targets/domains assessed in CASP8, 162 domains have threading templates as provided by pro-sp3-TASSER. For these, the RMSD to the native structure of the best threading template is 5.2 Å, with an average coverage of 94%. The average RMSD of the best model (first model), over the same aligned residues is 4.3 (4.7) Å. Using RMSD, comparison of the best threading template and best model to the native structure is shown in Figure 10.6b. TASSER models improve over their initial threading templates. Figure 10.7b shows two interesting examples of pro-sp3-TASSER predictions. Target T0478 is an all α-helical protein with two domains, both classified in the TBM category. However, pro-sp3-TASSER could not identify a reliable template and it is treated as a Hard target. The best template for domain T0478-D2 has an RMSD and TM-score to native of 16.1 Å and 0.3, respectively. The best model for T0478-D2 has an RMSD and TM-score to native of 6.6 Å and 0.43, respectively, owing to the good performance of chunk-TASSER on α-helical proteins. Similarly, T0478-D1 showed an improvement over the initial template. Target T0405 is a free-modeling (FM) target with two domains. The best predicted model from pro-sp3-TASSER for T0405-D1 improves over the initial template aligned region from 10.5 Å to 4.2 Å, with a full-length RMSD to native of 4.2 Å as well. Next, we compared the performance of pro-sp3-TASSER and METATASSER servers for all the 164 domains assessed in this CASP. The cumulative TM-score of the best (first) model for pro-sp3-TASSER and METATASSER are 108.2 (102.7) and 106.5 (101.3), respectively. The pro-sp3-TASSER models are closer to native, as assessed by TM-score than METATASSER. In our own and CASP8 official assessment, both servers are among the top five best performing servers.

The TASSER human-expert prediction group implemented TASSER along with the tertiary restraints derived from the models generated by METATASSER, pro-sp3-TASSER and other CASP8 servers. For Easy targets, server models were ranked using TASSER-QA, and the top 20 models were used to derive the distance and contact restraints for TASSER refinement. The final models are the top five centroids models after SPICKER clustering. For Medium/Hard targets, in addition to the above-mentioned protocol, we performed multiple simulations of chunk-TASSER [69] and TASSER 2.0 [70] with the threading templates from METATASSER and pro-sp3-TASSER. The ensemble of models generated using various protocols are then ranked by TASSER-QA, and the top five ranked models are submitted as final models. The whole procedure is automated as in all previous

CASPs. Human intervention was mainly for targets with putative multiple domains. For these targets, domains are modeled separately and subsequently assembled into full-length models. For the 71 targets assessed in CASP8, the average TM-score of the best (first) model is 0.639 (0.607). For the same targets, the average TM-score of best models from pro-sp3-TASSER and METATASSER are 0.620 and 0.598, respectively. The improvement observed for models from TASSER is mainly due to more extensive TASSER simulations and selection of better models from other CASP servers.

10.6. APPLICATIONS OF TASSER

TASSER has been applied in a variety of contexts, both to specific classes of proteins as well as proteomes [55,79,80]. In this section, we will present modeling results for membrane proteins and a subset of the *Escherichia coli* proteome.

G protein-coupled receptors (GPCRs) are an essential class of integral membrane proteins involved in signaling events in various cellular processes such as transmitting stimuli in response to light [81–84]. GPCRs are predicted to comprise about 5% of the human proteome [85,86]. Despite their pharmaceutical importance, there are few experimentally determined GPCRs structures. To address this issue, TASSER was employed to generate protein models for all 907 putative GPCRs in the human proteome [80]. First, we assessed the performance of TASSER on membrane proteins with solved tertiary structures to provide an estimate of the expected success rate. Applying TASSER to a benchmark set of 38-membrane proteins where >30% sequence identity in the aligned region was excluded, 17/38 (45%) of the targets have an RMSD to native structure <6.5 Å and an average improvement over the template alignment of 4.9 Å [80]. Although the overall success rate is lower in comparison to globular proteins, it was sufficiently promising to apply TASSER to model human GPCRs. Based on the benchmarked confidence score [55,80], the predicted models for 820 of the 907 proteins should have correct folds. The majority of GPCR models have the characteristic seven-membrane helix topology. Models of the representative GPCRs were compared with the mutagenesis and affinity labeling experiments, with consistent agreement found. Recently, structures for human β2-andregenic receptor (SWISSPROT: P07550) and A2A adenosine receptor (SWISSPROT: P29274) were experimentally determined. We validated the quality of our protein models (predicted in 2005, well before the experimental structures were determined) for these crystal structures. The best threading template RMSD (backbone only) to the native structures of β2-andregenic receptor (PDB: 2rkh1) [87] and A2A adenosine receptor (PDB: 3eml) [88] are 4.9 Å and 5.1 Å, respectively. For β2-andregenic receptor, the RMSD over the same aligned region improves to 4.1 Å, with a full-length RMSD to native of 4.5 Å after TASSER refinement (Fig. 10.8a). Similar improvement is observed in modeling the A2A receptor, where the

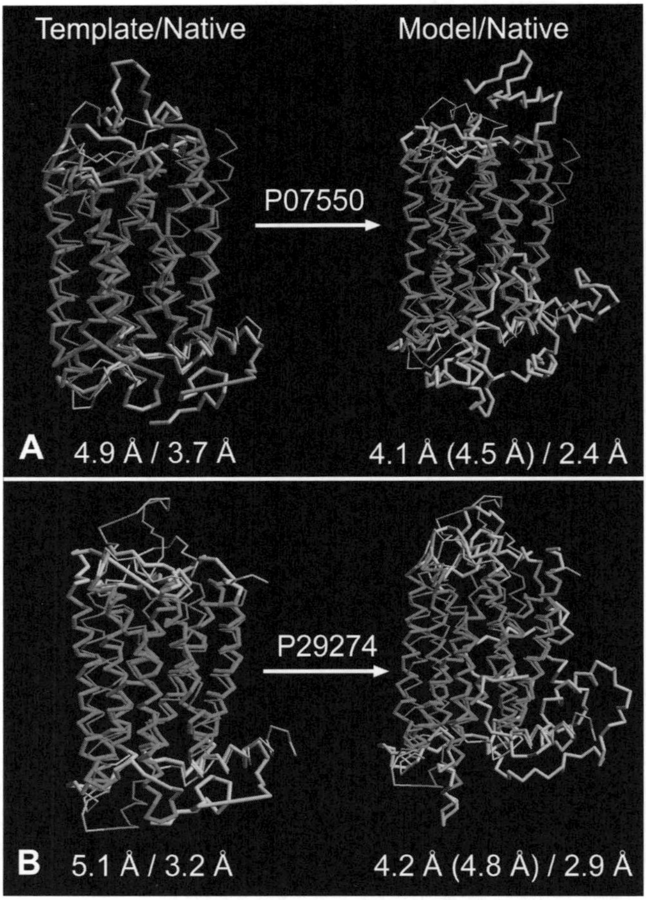

FIGURE 10.8 (A) Comparison between the human β2- andregenic receptor best threading template/model with its native tertiary structure. (B) Comparison between the human A2A adenosine receptor best threading template/model to its native tertiary structure. For each protein, on the left is the superposition of the threading template (thick backbone) and native (thin backbone); on the right is the final model (thick backbone) and native (thin backbone). Transmembrane regions are shown in green. The numbers below the superposition are the RMSD over the aligned region and the RMSD over the transmembrane regions alone, respectively. The RMSD over the full-length protein is shown in parenthesis. (See color insert.)

best model has an RMSD to native over the template aligned residues (full-length) of 4.2 (4.8)Å (Fig. 10.8b). When we consider the transmembrane regions alone, the C_α RMSDs of the model to native for β2-andrenergic receptor and A2A adenosine receptor drop to 2.4 Å and 2.9 Å, respectively, as compared with 3.7 Å and 3.2 Å for the template-aligned residues (shown in Fig. 10.8 on the left hand side of the figure). The higher RMSDs for the full-length models are due to the extracellular loops, which are not modeled correctly.

TASSER was also applied to proteins ≤200 residues in length in the *E. coli* proteome. Of the 1360 proteins modeled, 920 predicted tertiary structures are expected to have an RMSD to native <6.5 Å based on confidence criteria established in a comprehensive benchmark [55].

10.7. CONCLUSIONS

We have developed the structure prediction and modeling program TASSER, which was carefully validated in large-scale benchmarking. We find that TASSER has the ability to refine the threading template and generate final models, which are closer to the native tertiary structure. For single-domain proteins, whose sequence identity to their closest template is on average 18%, ~2/3 of such proteins are foldable [55]. We also demonstrated that in the CM limit, TASSER-Lite generates improved models in comparison with other contemporary CM programs such as MODELLER [54]. More recently, we have improved the TASSER algorithm by incorporation of consensus threading templates (METATASSER), iterative structure modeling ($TASSER^{iter}$) [71], using more accurate fragments (chunk-TASSER) [69], and side chain contacts (TASSER_2.0) [70]. The assessment of TASSER in CASP6-CASP8 shows that it is among the best automated structure prediction algorithms, with results that are consistent with our previous large-scale benchmarking [60,75].

There are, however, a number of areas where TASSER need to be improved. Its ability to generate acceptable models in the TF regime is limited. Given that the PDB is likely complete for single-domain proteins [47–49], there is a clear need for developing better threading algorithms that can identify templates in the very low sequence identity regime. Improvements in the quality of loop predictions are required, especially for the orientation of the loop with respect to the protein core [50,52]. Furthermore, for multi-domain proteins, there is a limited ability to predict the domain-domain interaction and hence, their relative orientation, especially when the relative domain orientation is different from that of the input templates [56] or when there is no identified multi-domain template. This is a significant issue, as it is unlikely that the PDB is complete with respect to the relative domain orientation of multiple domain proteins [89]. A final issue is the need to develop better all-atom refinement protocols, starting from the structures provided by TASSER. To date, the extent of such improvements has been rather limited [90].

In spite of these limitations, predicted structures provided by TASSER have been shown to be quite useful for functional inference, binding-site prediction, small-molecule ligand pose prediction, and ligand ranking [91,92]. Thus, even though considerable work remains before the protein folding problem can be viewed as solved, TASSER's predicted models are of the quality that they can provide significant biological insight for a significant fraction of the proteins in a given proteome.

ACKNOWLEDGEMENTS

This research was supported in part by NIH Grant numbers GM-37408 and GM-48835.

REFERENCES

1. M.D. Adams et al. The genome sequence of Drosophila Melanogaster. *Science*, 287(5461):2185–2195, 2000.
2. D.A. Benson et al. GenBank. *Nucleic Acids Research*, 37(Database issue):D26–D31, 2009.
3. C.M. Fraser et al. The minimal gene complement of Mycoplasma genitalium. *Science*, 270(5235):397–403.1995.
4. J.C. Venter et al. The sequence of the human genome. *Science*, 291(5507):1304–1351, 2001.
5. M. Kanehisa et al. The KEGG resource for deciphering the genome. *Nucleic Acids Research*, 32(Database issue):D277–D380, 2004.
6. S.F. Altschul and E.V. Koonin. Iterated profile searches with PSI-BLAST—A tool for discovery in protein databases. *Trends in Biochemical Science*, 23(10):444–447, 1998.
7. A. Muller, R.M. MacCallum, and M.J. Sternberg. Benchmarking PSI-BLAST in genome annotation. *Journal of Molecular Biology*, 293(5):1257–1271, 1999.
8. W. Tian, A.K. Arakaki, and J. Skolnick. EFICAz: A comprehensive approach for accurate genome-scale enzyme function inference. *Nucleic Acids Research*, 32(21):6226–6239, 2004.
9. R.D. Finn et al. Pfam: Clans, web tools and services. *Nucleic Acids Research*, 34(Database issue):D247–D251, 2006.
10. M. Gerstein. Patterns of protein-fold usage in eight microbial genomes: A comprehensive structural census. *Proteins*, 33(4):518–534, 1998.
11. W. Tian and J. Skolnick. How well is enzyme function conserved as a function of pairwise sequence identity? *Journal of Molecular Biology*, 333(4):863–882, 2003.
12. D. Baker and A. Sali. Protein structure prediction and structural genomics. *Science*, 294(5540):93–96, 2001.
13. S.B. Pandit et al. SUPFAM—a database of potential protein superfamily relationships derived by comparing sequence-based and structure-based families: implications for structural genomics and function annotation in genomes. *Nucleic Acids Research*, 30(1):289–293, 2002.
14. J. Skolnick and J.S. Fetrow. From genes to protein structure and function: Novel applications of computational approaches in the genomic era. *Trends Biotechnology*, 18(1):34–39, 2000.
15. J. Skolnick, J.S. Fetrow, and A. Kolinski. Structural genomics and its importance for gene function analysis. *Nature Biotechnology*, 18(3):283–287, 2000.
16. A. Stark and R.B. Russell. Annotation in three dimensions. PINTS: Patterns in Non-homologous Tertiary Structures. *Nucleic Acids Research*, 31(13):3341–3344, 2003.

17. U. Pieper et al. MODBASE: a database of annotated comparative protein structure models and associated resources. *Nucleic Acids Research*, 34(Database issue):D291–D295, 2006.

18. T. Schwede et al. SWISS-MODEL: An automated protein homology-modeling server. *Nucleic Acids Research*, 31(13):3381–3385, 2003.

19. K.M. Misura and D. Baker. Progress and challenges in high-resolution refinement of protein structure models. *Proteins*, 59(1):15–29, 2005.

20. M. Tress et al. Assessment of predictions submitted for the CASP6 comparative modeling category. *Proteins*, 61(7):27–45, 2005.

21. J.U. Bowie, R. Luthy, and D. Eisenberg. A method to identify protein sequences that fold into a known three-dimensional structure. *Science*, 253(5016):164–170, 1991.

22. M.J. Sippl. Boltzmann's principle, knowledge-based mean fields and protein folding. An approach to the computational determination of protein structures. *Journal of Computer Aided Molecules Design*, 7(4):473–501, 1993.

23. J. Skolnick and D. Kihara. Defrosting the frozen approximation: PROSPECTOR—a new approach to threading. *Proteins*, 42(3):319–331, 2001.

24. H. Zhou and Y. Zhou. Single-body residue-level knowledge-based energy score combined with sequence-profile and secondary structure information for fold recognition. *Proteins*, 55(4):1005–1013, 2004.

25. L.J. McGuffin and D.T. Jones. Improvement of the GenTHREADER method for genomic fold recognition. *Bioinformatics*, 19(7):874–881, 2003.

26. Y. Shan, G. Wang, and H.X. Zhou. Fold recognition and accurate query-template alignment by a combination of PSI-BLAST and threading. *Proteins*, 42(1):23–37, 2001.

27. H. Zhou and Y. Zhou. Fold recognition by combining sequence profiles derived from evolution and from depth-dependent structural alignment of fragments. *Proteins*, 58(2):321–328, 2005.

28. J. Lee, H.A. Scheraga, and S. Rackovsky. New optimization method for conformational energy calculations on polypeptides: Conformational space annealing. *Journal of Computational Chemistry*, 18(9):1222–1232, 1997.

29. J. Pillardy et al. Development of physics-based energy functions that predict medium-resolution structures for proteins of the alpha,beta and alpha/beta structural classes. *Journal of Physical Chemistry B*, 105(30):7299–7310, 2001.

30. K.T. Simons et al. Assembly of protein tertiary structures from fragments with similar local sequences using simulated annealing and Bayesian scoring functions. *Journal of Molecular Biology*, 268(1):209–225, 1997.

31. A. Liwo, M. Khalili, and H.A. Scheraga. Ab initio simulations of protein-folding pathways by molecular dynamics with the united-residue model of polypeptide chains. *Proceedings of the National Academy of Science U S A*, 102(7):2362–2367, 2005.

32. V.S. Pande et al. Atomistic protein folding simulations on the submillisecond time scale using worldwide distributed computing. *Biopolymers*, 68(1):91–109, 2003.

33. A. Sali and T.L. Blundell. Comparative protein modelling by satisfaction of spatial restraints. *Journal of Molecular Biology*, 234(3):779–815, 1993.

34. H. Rangwala and G. Karypis. Profile-based direct kernels for remote homology detection and fold recognition. *Bioinformatics*, 21(23):4239–4247. 2005.

35. J. Skolnick. In quest of an empirical potential for protein structure prediction. *Current Opinion in Structural Biology*, 16(2):166–171, 2006.
36. J. Kopp et al. Assessment of CASP7 predictions for template-based modeling targets. *Proteins*, 69 (8):38–56, 2007.
37. J. Moult. A decade of CASP: Progress, bottlenecks and prognosis in protein structure prediction. *Current Opinion Structural Biology*, 15(3):285–289, 2005.
38. K.T. Simons, C. Strauss, and D. Baker. Prospects for ab initio protein structural genomics. *Journal of Molecular Biology*, 306:1091–1099, 2001.
39. B.N. Dominy and C.L. Books. Identifying native-like protein structures using physics-based potentials. *Journal of Computational Chemistry*, 23:147–160, 2002.
40. S. Odziej et al. Physics-based protein-structure prediction using a hierarchical protocol based on the UNRES force field: Assessment in two blind tests. *Proceedings of the National Academy of Science U S A*, 102:7547–7552, 2005.
41. H. Lu and J. Skolnick. A distance-dependent atomic knowledge-based potential for improved protein structure selection. *Proteins*, 44:223–232, 2001.
42. P.D. Thomas and K.A. Dill Statistical potentials extracted from protein structures: How accurate are they? *Journal of Molecular Biology*, 257(2):457–469, 1996.
43. H. Zhou and Y. Zhou. Distance-scaled, finite ideal-gas reference state improves structure-derived potentials of mean force for structure selection and stability prediction. *Protein Science*, 10:2714–2726, 2002.
44. S. Liu et al. A physical reference state unifies the structure-derived potential of mean force for protein folding and binding. *Proteins*, 56(1):93–101, 2004.
45. F. Melo, R. Sanchez, and A. Sali. Statistical potentials for fold assessment. *Protein Science*, 10(2):430–448, 2002.
46. H.M. Berman et al. The Protein Data Bank. *Nucleic Acids Research*, 28(1):235–242, 2000.
47. D. Kihara and J. Skolnick. The PDB is a Covering Set of Small Protein Structures. *Journal of Molecular Biology*, 334(4):793–802, 2003.
48. Y. Zhang et al. On the origin and highly likely completeness of single-domain protein structures. *Proceeding of the National Academy of Science U S A*, 103:2605–2610, 2006.
49. Y. Zhang and J. Skolnick. The protein structure prediction problem could be solved using the current PDB library. *Proceeding of the National Academy of Science U S A*, 102(4):1029–1034, 2005.
50. M.A. Marti-Renom et al. Comparative protein structure modeling of genes and genomes. *Annual Review of Biophysics and Biomolecular Structure*, 29:291–325, 2000.
51. N. Srinivasan and T.L. Blundell. An evaluation of the performance of an automated procedure for comparative modelling of protein tertiary structure. *Protein Engineering*, 6(5):501–512, 1993.
52. A. Fiser, R.K. Do, and A. Sali. Modeling of loops in protein structures. *Protein Science*, 9(9):1753–1773, 2000.
53. S.Y. Lee, Y. Zhang, and J. Skolnick. TASSER-based refinement of NMR structures. *Proteins*, 63(3):451–456, 2006.
54. S.B. Pandit, Y. Zhang, and J. Skolnick. TASSER-Lite: An automated tool for protein comparative modeling. *Biophysical Journal*, 91:4180–4190, 2006.

55. Y. Zhang and J. Skolnick Automated structure prediction of weakly homologous proteins on a genomic scale. *Proceedings of the National Academy of Science U S A*, 101(20):7594–7599,2004.
56. Y. Zhang and J. Skolnick. Tertiary structure predictions on a comprehensive benchmark of medium to large size proteins. *Biophysical Journal*, 87(4):2647–2655, 2004.
57. V. Grimm, Y. Zhang, and J. Skolnick. Benchmarking of dimeric threading and structure refinement. *Proteins*, 63(3):457–465, 2006.
58. J. Skolnick, D. Kihara, and Y. Zhang. Development and large scale benchmark testing of the PROSPECTOR 3.0 threading algorithm. *Proteins*, 56:502–518, 2004.
59. D.T. Jones. Protein secondary structure prediction based on position-specific scoring matrices. *Journal of Molecular Biology*, 292(2):195–202.1999.
60. H. Zhou et al. Analysis of TASSER-based CASP7 protein structure prediction results. *Proteins*, 69(8):90–97, 2007.
61. K. Ginalski et al. 3D-Jury: A simple approach to improve protein structure predictions. *Bioinformatics*, 19(8):1015–1018, 2003.
62. Y. Zhang and J. Skolnick. TM-align: a protein structure alignment algorithm based on the TM-score. *Nucleic Acids Research*, 33:2302–2309, 2005.
63. Y. Zhang and J. Skolnick. Scoring function for automated assessment of protein structure template quality. *Proteins*, 57(4):702–710, 2004.
64. Y. Zhang, A. Kolinski, and J. Skolnick. TOUCHSTONE II: A new approach to ab initio protein structure prediction. *Biophysical Journal* 85:1045–1064, 2003.
65. Y. Zhang, D. Kihara, and J. Skolnick. Local energy landscape flattening: parallel hyperbolic Monte Carlo sampling of protein folding. *Proteins*, 48(2):192–201, 2002.
66. Y. Zhang and J. Skolnick. SPICKER: A clustering approach to identify near-native protein folds. *Journal of Computational Chemistry*, 25(6):865–871, 2004.
67. A. Sali et al. Evaluation of comparative protein modeling by MODELLER. *Proteins*, 23(3):318–326, 1995.
68. J.M. Borreguero and J. Skolnick. Benchmarking of TASSER in the ab initio limit. *Proteins*, 68(1):48–56, 2007.
69. H. Zhou and J. Skolnick. Ab initio protein structure prediction using chunk-TASSER. *Biophysics Journal*, 93(5):1510–1518, 2007.
70. S. Lee and J. Skolnick. Benchmarking of TASSER_2.0: An improved protein structure prediction algorithm with more accurate predicted contact restraints. *Biophysics Journal*, 95:1956–1964, 2008.
71. S.Y. Lee and J. Skolnick. Development and benchmarking of TASSER(iter) for the iterative improvement of protein structure predictions. *Proteins*, 68(1):39–47, 2007.
72. S. Wu, J. Skolnick, and Y. Zhang. Ab initio modeling of small proteins by iterative TASSER simulations. *BMC Biology*, 5:17, 2007.
73. P. Bradley, K.M. Misura, and D. Baker. Toward high-resolution de novo structure prediction for small proteins. *Science*, 309(5742):1868–1871, 2005.
74. G. Wang, Y. Jin, and R.L. Jr. Dunbrack. Assessment of fold recognition predictions in CASP6. *Proteins*, 61 (7):46–66, 2005.
75. Y. Zhang, A.K. Arakaki, and J. Skolnick. TASSER: An automated method for the prediction of protein tertiary structures in CASP6. *Proteins*, 61 (7):91–98, 2005.

76. P. Rotkiewicz and J. Skolnick. Fast procedure for reconstruction of full-atom protein models from reduced representations. *Journal of Computational Chemistry*, 29:1460–1465, 2008.
77. H. Zhou and J. Skolnick. Protein structure prediction by pro-Sp3-TASSER. *Biophysics Journal*, 96(6):2109–2127, 2009.
78. H. Zhou and J. Skolnick. Protein model quality assessment prediction by combining fragment comparisons and a consensus C(alpha) contact potential. *Proteins*, 71(3):1210–1218, 2008.
79. J.A. Somarelli et al. Structure-based classification of 45 FK506-binding proteins. *Proteins*, 72(1):197–208, 2008.
80. Y. Zhang, M.E. Devries, and J. Skolnick. Structure Modeling of All Identified G Protein-Coupled Receptors in the Human Genome. *PLoS Computational Biology*, 2(2):e13, 2006.
81. Y. Fang, J. Lahiri, and L. Picard. G protein-coupled receptor microarrays for drug discovery. *Drug Discovery Today*, 8(16):755–761, 2003.
82. R.C. Graul and W. Sadee. Evolutionary relationships among G protein-coupled receptors using a clustered database approach., 3(2):E12, 2001.
83. M. Lapinsh et al. Classification of G-protein coupled receptors by alignment-independent extraction of principal chemical properties of primary amino acid sequences. *Protein Science*, 10(4):795–805, 2002.
84. E. Jacoby, A. Schuffenhauer, and P. Floersheim. Chemogenomics knowledge-based strategies in drug discovery. *Drug News Perspect*, 16(2):93–102, 2003.
85. F.S. Collins. Finishing the euchromatic sequence of the human genome. *Nature*, 431(7010):931–945, 2004.
86. S. Takeda et al. Identification of G protein-coupled receptor genes from the human genome sequence. *FEBS Letter*, 520(1–3):97–101, 2002.
87. V. Cherezov et al. High-resolution crystal structure of an engineered human beta2-adrenergic G protein-coupled receptor. *Science*, 318(5854):1258–1265, 2007.
88. V.P. Jaakola et al. The 2.6 angstrom crystal structure of a human A2A adenosine receptor bound to an antagonist. *Science*, 322(5905):1210–1217, 2008.
89. S.J. Littler and S.J. Hubbard. Conservation of orientation and sequence in protein domain–domain interactions. *Journal of Molecular Biology*, 345(5):1265–1279, 2005.
90. A. Jagielska, L. Wroblewska, and J. Skolnick. Protein model refinement using an optimized physics-based all-atom force field. *Proceeding of the National Academy Science U S A*, 105(24):8268–8273, 2008.
91. M. Brylinski and J. Skolnick. Q-Dock: Low-resolution flexible ligand docking with pocket-specific threading restraints. *Journal of Computation Chemistry*, 29(10):1574–1588, 2008.
92. M. Brylinski and J. Skolnick. A threading-based method (FINDSITE) for ligand-binding site prediction and functional annotation. *Proceeding of the National Academy Science U S A*, 105(1):129–134, 2008.

CHAPTER 11

COMPOSITE APPROACHES TO PROTEIN TERTIARY STRUCTURE PREDICTION: A CASE-STUDY BY I-TASSER

AMBRISH ROY, SITAO WU, and YANG ZHANG
Center for Computational Medicine and Bioinformatics
University of Michigan
Ann Arbor, MI

11.1. INTRODUCTION

The post-genomic era is witnessing an upsurge of protein sequences in public databases. By the end of 2009, over 9 million protein sequences had been deposited in the Universal Protein Resource (UniProtKB)/TrEMBL [1]. However, this increase in the amount of sequence data does not necessarily reflect an increase in biological knowledge. One of the most challenging tasks that have emerged in recent years is to functionally characterize these sequences for better understanding of physiological processes and systems [2]. This has motivated computational biologists to develop a variety of fast and accurate methods for quickly characterizing these sequences.

One of the most significant efforts in this regard has been the development of powerful sequence alignment algorithms like Basic Local Alignment Search Tool (BLAST) [3], Position-Specific Iterative-BLAST (PSI-BLAST) [4], and hidden Markov model (HMM) techniques [5,6], which are frequently used for identifying evolutionary homologs and transferring functional annotations. The underlying assumption in these approaches is that evolutionarily related sequences fold similarly [7,8] and the functional similarity between these related proteins can be explored by detecting evolutionary relationship

Introduction to Protein Structure Prediction: Methods and Algorithms,
Edited by Huzefa Rangwala and George Karypis

between them [9,10]. It has been estimated that using these approaches, functional inference can be drawn for nearly 40–60% of the open reading frames (ORFs) in the genome [11]. However, there are numerous cases where functional conservation exists in evolutionarily diverged proteins but annotations cannot be transferred based on evolutionary-based approaches [12,13]. At this juncture, it is apparent that protein sequences are generally insufficient for determining protein functionality and providing support for functional genomics [14].

The three-dimensional (3D) structure of a protein is closely linked to its biological function [15]. As residues located far apart in the primary sequence may be very close in 3D space and only a few spatially conserved residues are generally responsible for a protein's function [16,17], the 3D structure of a protein provides useful insight into the key component(s) of its functionality. This awareness and the limited number of solved protein structures in Protein Data Bank (PDB) [18] have actuated the structural genomics (SG) project to increase the throughput of experimental structure elucidation [19–21] and provide a framework for inferring molecular function [22,23]. While the SG aims to structurally characterize the protein universe by an optimized combination of experimental structure determination and comparative modeling (CM) building, 3D structures of at least 16,000 optimally selected proteins would be required in order that the CM can cover 90% of protein domain families [24] and at the current rate it appears that this goal can be achieved only in the next 10 years [25]. This underscores the need for computational methods for protein structure prediction, so that 3D structural models can be built and can provide insight for functional analysis. Also, the development of better structural refinement and CM methods would dramatically enlarge the scope of structural genomics project.

Historically, protein structure prediction methods have been classified into three categories: CM [26,27], threading [28–33] and *ab initio* modeling [34–38]. In CM, the protein structure is constructed by matching the sequence of the protein of interest (query) to an evolutionarily related protein with a known structure (template) in the PDB [18], where the residue equivalency between query and the template is obtained by aligning sequences or sequence profiles. Threading-based methods match the query protein sequence directly to 3D structures of solved proteins with the goal of recognizing similar protein folds that may have no clear evidence of an evolutionary relationship with the query protein. The last resort for predicting the protein structure, when no good template is detected in the PDB library, is to predict the structure using *ab initio* modeling. Predictions based on this method assume that the native structure of a protein corresponds to its global free-energy minimum [39] and the conformational space is sampled to attain this state as guided by well-designed energy force fields. This is the most difficult category of protein structure prediction and if successful will provide the eventual solution to protein folding problem. However, the success of *ab initio* modeling is currently limited to small proteins with less than 100 amino acids [34–38].

As a general trend in the field of protein structure prediction, the borders between the conventional categories of methods have become blurred. For instance, both CM- and threading-based methods use sequence-profile and profile-profile alignments (PPA) for identifying templates. Similarly, most of the contemporary *ab initio*-based methods often use evolutionary information either for generating sparse spatial restraints or for identifying local structural building blocks. Recent community-wide blind tests have demonstrated significant advantages of the composite approaches in protein structure predictions [40–42], which combines the various techniques from threading, *ab initio* modeling, and atomic-level structure refinements [43,44].

In this chapter, we will focus on the methodology of I-TASSER [35,44,45], which serves as a case study of the composite approach for generating 3D structural models and predicting the function of a given query sequence. The performance of I-TASSER on benchmark tests and in the recent Critical Assessment of Protein Structure Prediction (CASP) experiments [44,46] will be discussed. Finally, in the concluding section, the current status and future perspective are summarized.

11.2. I-TASSER: A COMPOSITE METHOD FOR PROTEIN STRUCTURE PREDICTION

I-TASSER [35,44,45] is a hierarchical protein structure modeling approach based on the multiple threading alignments and an iterative implementation of the Threading ASSEmbly Refinement (TASSER) program [47]. Figure 11.1 represents the schematic diagram of I-TASSER methodology for protein structure and function prediction, which consist of four consecutive steps of threading, structure assembly, structure refinement, and function prediction.

11.2.1. Threading of Query Sequence

Given a query protein, the first step of I-TASSER is to thread the query sequence through a representative PDB structure library (sequence identity cutoff of 70%) with the objective of identifying the global or local threading alignments using either MUSTER [29] (single threading server) or LOMETS [33] (meta-threading server). In this section, we will first describe the methodology of MUSTER threading algorithm and then give an overview and advantage of using LOMETS, a meta-threading server.

11.2.1.1. MUSTER Threading Server. MUSTER is a sequence PPA method assisted by the predicted structural information like secondary structure, structure profiles, solvent accessibility, backbone dihedral torsion angles, and hydrophobic scoring matrix. The scoring function of MUSTER [29] for aligning the *i*th residue of the query and the *j*th residue of the template is defined as

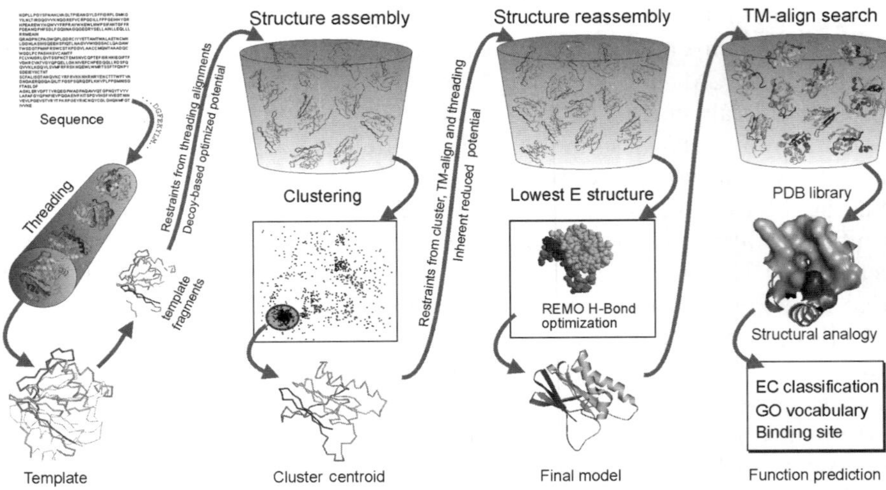

FIGURE 11.1 A schematic diagram of the I-TASSER [35,44,45] protein structure and function prediction protocol. Templates for the query protein are first identified by MUSTER [29] or LOMETS [33], which provide template fragments and spatial restraints. Template fragments are then assembled by modified replica-exchange Monte-Carlo simulations. The conformations generated during the simulation are clustered using SPICKER [48], in order to identify the structure with the lowest free energy. As an iterative refinement strategy, the cluster centroids are then subjected to the second round of simulation for refining the global topology and removing clashes. The final all-atom models are generated by REMO through the optimization of hydrogen-bonding networks [49]. Finally, functional homologs (protein structures with an associated EC number or GO terms) of final models are identified by using a sequence-independent structural alignment tool of TM-align [50] by ranking the hits based on their TM-score [51], RMSD and sequence identity in the structurally aligned region, coverage of the structural alignment, and confidence score (C-score [45]) of the model. (See color insert.)

$$\mathrm{Score}(i,j) = E_{\mathrm{seq\_prof}} + E_{\mathrm{sec}} + E_{\mathrm{struc\_prof}} + E_{\mathrm{sa}} + E_{\mathrm{phi}} + E_{\mathrm{psi}} + E_{\mathrm{hydro}} + E_{\mathrm{shift}}. \quad (1)$$

The first term, $E_{\mathrm{seq\_prof}}$, is the alignment score of the sequence PPA. The second term, E_{sec}, computes the match between the predicted secondary structure of query and secondary structure of templates. The third term, $E_{\mathrm{struc\_prof}}$, calculates the score of aligning the structure-derived profiles of templates to the sequence profile of query. The fourth term, E_{sa}, computes the difference between the predicted solvent accessibility of query and solvent accessibility of templates. The fifth and sixth terms (E_{phi} and E_{psi}) calculate the difference between the predicted torsion angles (phi and psi) of query and those of templates. The experimental torsion angles for templates are calculated using STRIDE [52], while torsion angles of query are predicted by ANGLOR [53]. The seventh term, E_{hydro}, is an element of hydrophobic scoring matrix [54] that encourages the match of hydrophobic residue (V, I, L, F, Y, W, M) in the

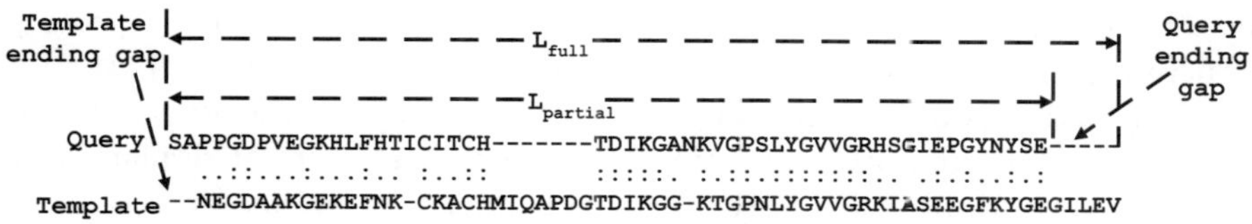

FIGURE 11.2 Illustration of the full (L_{full}) and partial (L_{partial}) alignment lengths used to normalize the threading alignment score (R_{score}). Symbols "-", "." and ":" indicate an unaligned gap, an aligned nonidentical residue pair and an aligned identical residue pair, respectively. The query and template sequences are taken from 1hroA (first 53 residues) and 155c_ (first 61 residues), respectively, as an illustrative example. (From Wu and Zhang [29]).

query and the templates. Finally the last term, E_{shift}, is a constant, which is introduced to avoid alignment of unrelated residues in local regions. While the first term is sequence-based information, the second to seventh terms are related to structural information. If only the first two terms plus E_{shift} in Equation 1 are involved, the corresponding threading program is called PPA [33], which is the precursor of MUSTER.

The sequence and structural information are then combined into a single-body energy term, which can be conveniently used in the Needleman-Wunsch [55] dynamic programming algorithm for identifying the best match between the query and the templates. A position-dependent gap penalty in the dynamic programming is employed, i.e. no gap is allowed inside the secondary structure regions (helices and strands); gap opening (g_o) and gap extension (g_e) penalties apply to other regions; ending gap-penalty is neglected.

Following the dynamic programming alignments, the alignments on different structural templates are ranked based on their alignment score and the length of the alignment. In PPA [33] the templates are ranked based on a raw alignment score (R_{score}) divided by the full alignment length (L_{full}; including query and template ending gaps) as shown in Figure 11.2. In MUSTER, however, $R_{\text{score}}/L_{\text{partial}}$ is used as another possible ranking scheme, where L_{partial} is the partial alignment length excluding query ending gap as shown in Figure 11.2. A combined ranking is then taken as follows: If the sequence identity of the first template selected by $R_{\text{score}}/L_{\text{partial}}$ to the query is higher than that selected by $R_{\text{score}}/L_{\text{full}}$, then the template ranking is done by $R_{\text{score}}/L_{\text{partial}}$. Otherwise, the templates are ranked by $R_{\text{score}}/L_{\text{full}}$.

MUSTER was applied to a benchmark test of 500 non-homologous proteins (Fig. 11.3) and compared with PPA [29] at two different cutoffs: (i) all homologous templates with sequence identity >30% to the query were removed (cutoff 1); (ii) all homologous templates with sequence identity >20% or detectable by PSI-BLAST with an e-value <0.05 were removed (cutoff 2).

Here, the comparison between threading alignments and the native structures was done by evaluating the template-modeling score (TM-score), defined by Zhang and Skolnick [51], to assess the topological similarity of protein structure pairs with a value in the range of [0, 1]. Statistically, a TM-score <0.17 means a randomly selected protein pair with the gapless alignment taken from PDB; TM-score >0.5 corresponds to the protein pairs of similar folds. The statistical meaning of TM-score is independent of protein size [51].

At cutoff 1, the average TM-scores of the best threading alignment generated by PPA and MUSTER are 0.4285 and 0.4503, respectively, which shows that the additional structural information in MUSTER has improved the threading results by approximately 5%. Even at a more stringent cutoff (cutoff 2), MUSTER finds a better threading alignment with an average TM-score of 0.3638, while best threading alignment found by PPA has an average TM-score of 0.3423. Thus, the average TM-score of the first threading alignment by MUSTER at cutoff 2 is about 6% better than that of PPA alignment. The higher TM-score of MUSTER over PPA alignments is due to both the higher alignment coverage and the more accurate alignments as judged by the root-mean-square deviation (RMSD) within the aligned regions.

11.2.1.2. LOMETS: Meta-Threading Server. As shown in Figure 11.3, although MUSTER is better than PPA on average, it cannot outperform PPA on all protein targets. A similar trend has also been observed in CASP/Critical

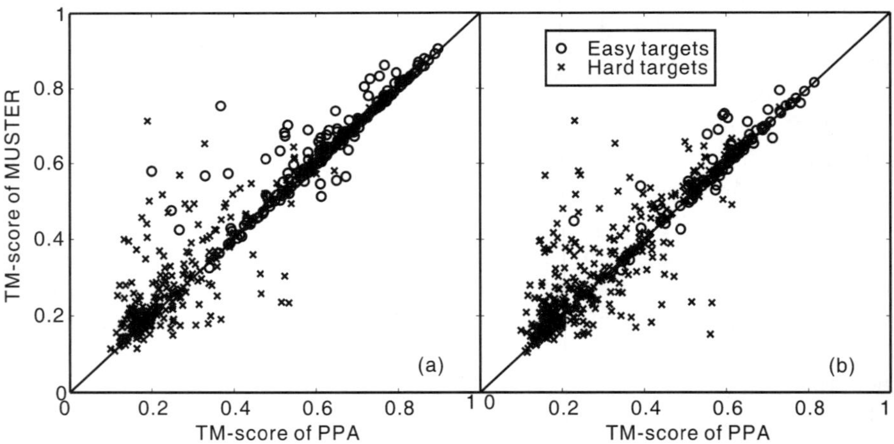

FIGURE 11.3 TM-score comparison between PPA and MUSTER for the first threading alignment of 500 non-homologous proteins. Circles represent the alignments from the "Easy" targets (z-score of alignment by MUSTER >7.5 and z-score of PPA alignment >7.0) and crosses indicate those from the "Hard" targets (z-score of alignment by MUSTER <7.5 and z-score of PPA <7.0). All homologous templates with sequence identity to targets (a) >30% (b) >20%, or detectable by PSI-BLAST with an e-value < 0.05 are excluded in this comparison. After removing homologous templates, the first template alignments by MUSTER for (a) 224 proteins and (b) 137 proteins have a correct fold (TM-score >0.5). (From Wu and Zhang [29]).

Assessment of Fully Automated Structure Prediction (CAFASP) experiments [42,56], where although the average Global Distance Test (GDT) or TM-score of some methods outperform others, there is no single method that can outperform others on all the targets. This inconsistency naturally leads to the prevalence of the metaserver [33,57], which is designed to collect and combine prediction results from a set of individual threading programs.

On the I-TASSER web server (http://zhanglab.ccmb.med.umich.edu/I-TASSER) this idea has been implemented using LOMETS [33], a locally installed meta-threading server. The threading programs in LOMETS represent a diverse set of the state-of-the-art algorithms using different approaches, namely, sequence profile alignments (PPA-I [33], PPA-II [33], SPARKS2 [33], SP^3 [58]), structural profile alignments (FUGUE [59]), pairwise potentials [PROSPECT2] [31], PAINT [33]), and the HMM (HHSearch [5], SAM-T02 [60]).

For each target, LOMETS first threads the query sequence through the PDB library to identify template threading alignments by each threading program and then ranks them purely based on consensus. The idea behind the consensus approach is simple: there are more ways for a threading program to select a wrong template than that to select a right one. Therefore, the chance for multiple threading programs working collectively to make a common wrong selection is lower than the chance to make a common correct selection.

Table 11.1 shows the improvement of LOMETS over individual threading programs. For the purpose of eliminating the dependence on the alignment coverage, the full-length models have been built here by MODELLER [26], using the templates from each threading program. Based on 620 non-homologous testing proteins, the models generated by LOMETS threading

TABLE 11.1 Performance Comparison of Component Threading Programs and LOMETS Metaserver on 620 Non-Homologous Testing Proteins

Threading Servers or Metaservers	TM-score (MODELLER models)		RMSD (Å) (MODELLER models)	
	First Model	Best in Top Five Models	First Model	Best in Top Five Models
PPA-I	0.4117	0.4531	16.66	14.02
SP3	0.4138	0.4551	13.86	12.83
PPA-II	0.4076	0.4512	14.89	13.02
SPARKS2	0.3973	0.4441	13.60	12.23
PROSPECT2	0.3914	0.4384	13.01	12.02
FUGUE	0.3721	0.4173	19.26	15.82
HHSEARCH	0.3827	0.4224	22.38	19.04
PAINT	0.3758	0.4210	15.74	14.21
SAM-T02	0.3575	0.3971	21.75	17.53
LOMETS	0.4434	0.4669	10.99	10.61

alignments achieve an average TM-score of 0.4434, which is at least 8% higher than that by any individual threading program.

11.2.2. Structure Assembly and Refinement

Following the threading procedure, the next step of I-TASSER is to generate the full-length model of the query protein and to refine the structure so that threading-aligned regions move closer to native structure. To achieve this, the complete protein chain in I-TASSER is divided into threading-aligned and unaligned regions, where the continuous fragments are excised from threading alignments, while the threading unaligned regions are build by *ab initio* modeling. The protein chain here is described by a reduced model, that is, a trace of alpha-carbon atoms and side chain center (SC) of mass to reduce the number of explicitly treated freedom and the intra-molecular interactions in the polypeptide chain. We will elaborately describe the whole procedure now.

For a given threading alignment, I-TASSER builds an initial full-length model by connecting the continuous secondary structure fragments (≥five residues) through a random walk of C_α–C_α bond vectors of variable lengths from 3.26 Å to 4.35 Å. To guarantee that the last step of this random walk can quickly arrive at the first C_α of the next template fragment, the distance l between the current C_α and the first C_α of the next template fragment is checked at each step of the random walk, and only walks with $l < 3.54n$ are allowed, where n is the number of remaining C_α–C_α bonds in the walk. If the template gap is too big to be spanned by a specified number of unaligned residues, a big C_α–C_α bond is kept at the end of the random walk and a spring-like force that acts to draw sequential fragments close will be applied in subsequent Monte-Carlo simulations, until a physically reasonable bond length is achieved.

The initial full-length models are then refined by the parallel replica-exchange Monte-Carlo sampling technique [61]. Two kinds of conformational updates (off-lattice and on-lattice) are implemented here: (i) Off-lattice movements of the aligned regions involve rigid fragment translations and rotations that are controlled by the three Euler angles. The fragment length normalizes the movement amplitude so that the acceptance rate is approximately constant for fragments of different sizes. (ii) The lattice-confined residues are subjected to 2–6 bond movements and multi-bond sequence shifts. Overall, the tertiary topology varies by the rearrangement of the continuously aligned substructures, where the local conformation of the off-lattice substructures remains unchanged during the assembly.

The movements in the structure assembly and refinement procedure are guided by an optimized force field that is described in the next section.

11.2.2.1. Force Field. The inherent I-TASSER assembly force field is similar to TASSER [47], which includes a variety of knowledge-based energy terms describing the predicted secondary structure propensities from PSIPRED [62], secondary structure-specific backbone hydrogen bonding, and a variety of statistical short-range and long-range correlation terms that are extracted from multiple threading alignments. Readers are recommended to read Zhang and Skolnick [47,63,64] for further details about these energy terms.

The new potentials terms that have been incorporated in I-TASSER include the predicted accessible surface area (ASA) [35] and sequence-based contact predictions [65]. Both the energy terms have been derived and optimized using machine learning methods.

For the purpose of fast calculations of the ASA effect, the hydrophobic energy in I-TASSER is defined by

$$E_{ASA} = -\sum\left(\frac{x_i^2}{x_0^2}+\frac{y_i^2}{y_0^2}+\frac{z_i^2}{z_0^2}-2.5\right)\times P(i), \quad (2)$$

where (x_i, y_i, z_i) is the coordinate of ith residue at the ellipsoid Cartesian system of the given protein conformation and (x_0, y_0, z_0) is the principal axes length. The constant parameter used for tuning the average depth of the exposed residues is 2.5, while $P(i)$ is the residue exposure index and is defined as $P(i)=\sum_{j=1}^{12} a(j)$, where a_j is the two-state neural network (NN) prediction of exposure ($a_j = 1$) or burial ($a_j = -1$) with the jth ASA fraction cutoff; $P(i)$ has a strong correlation with the real value of ASA. The overall correlation coefficient between the predicted $P(i)$ and the actual exposed area as calculated by STRIDE [52] on a test set of 2234 non-homologous proteins is 0.71. The same correlation for the widely used Hopp-Woods [66] and Kyte-Doolittle hydrophobicity indices [67] are 0.42 and 0.39, respectively. One of the probable reasons for the higher correlation by the NN prediction is because it explores the sequence-profile information, whereas the later methods are sequence-independent.

In the latest version of I-TASSER, sequence-based pairwise residue contact information from SVM-SEQ [65], SVMCON [68], and BETACON [69] are used to constrain the simulation search to a smaller conformational space and improve the minima of the landscape funnel of the overall energy function. Wu and Zhang recently showed that this additional information from SVM-SEQ can significantly increase the contact prediction accuracy in hard targets (when no good template is identified) by about 12–25%, compared with SVM-LOMETS [65], a template-based contact prediction method.

The predicted contacts from SVM-SEQ, SVMCON, and BETACON include contacts for C_α, C_β, and SC at distance cutoffs of 6 Å, 7 Å, and 8 Å. These predicted contacts are implemented as restraints in the I-TASSER simulation in the following way: if two residues i and j are predicted to be in contact by sequence-based methods and they come in contact in the decoys during the course of I-TASSER simulation, then the residue pairs are preferred to keep in contact by giving an energy bonus, which is defined as

$$E_{contact} = -1-(\mathrm{conf}(i,j)-a), \quad (3)$$

where conf(i, j) is the confidence score of the predicted contact pair (i, j) and a ($\in [0,1]$) is an empirically determined score cutoff for each distance cutoff.

11.2.2.2. Iterative Strategy. The trajectories of the low-temperature replicas of the first-round I-TASSER simulations are clustered by SPICKER [48]. The cluster centroids are obtained by averaging all the clustered structures after superposition and are ranked based on the structure density of the clusters. However, the cluster centroids generally have a number of nonphysical steric clashes between C_α atoms and can be overcompressed. Starting from the selected SPICKER cluster centroids, the TASSER Monte-Carlo simulation [61] is performed again (see Fig. 11.1). While the inherent I-TASSER potential remains unchanged in the second run, external constraints are added, which are derived by pooling the initial high-confident restraints from threading alignments, the distance and contact restraints from the combination of the centroid structures, and the PDB structures identified by the structure alignment program TM-align [50] using the cluster centroids as query structures. The conformation with the lowest energy in the second round is selected as the final model.

The main purpose of this iterative strategy is to remove the steric clashes of the cluster centroids. On a benchmark test set of 200 proteins with <300 residues it was found that the average number of steric clashes (residue pairs with C_α distance <3.6 Å) for the cluster centroids of the first cluster dramatically reduces from 79 to 0.8. As strong distance map and contact restraints are implemented in this step, the topology of the models also improves. In these test cases, the average TM-score increased from 0.5734 to 0.5801 (1.2%) and the C_α-RMSD to native decreased from 6.67 Å to 6.52 Å compared with the cluster centroid of the first round.

11.2.3. Reconstruction of Atomic Model

The models generated after I-TASSER Monte-Carlo simulations [61] and SPICKER clustering [48] are reduced models, where each residue is represented by the C_α atom and the SC of mass. To increase the biological usefulness of protein models, all atom models are generated by REMO [49] simulations, which include three general steps: (i) removing steric clashes by moving around each of the C_α atoms that clash with other residues; (ii) backbone reconstruction by scanning a backbone isomer library collected from the solved high-resolution structures in the PDB library; and (iii) hydrogen-bonding network optimization based on predicted secondary structure from PSIPRED [62]. Finally, Scwrl3.0 [70] is used to add the side chain rotamers.

Figure 11.4 shows the performance of REMO [49] on 230 non-homologous test proteins. Figure 11.4a shows the number of the steric clashes (residue pairs with C_α distance <3.6 Å) in the cluster centroids of the test proteins after the first round of I-TASSER simulations (average clash = 119). After the REMO procedure (Fig. 11.4b), only 15 proteins had 2–6 clashes, 44 proteins had one clash, and in the remaining all clashes had been effectively removed.

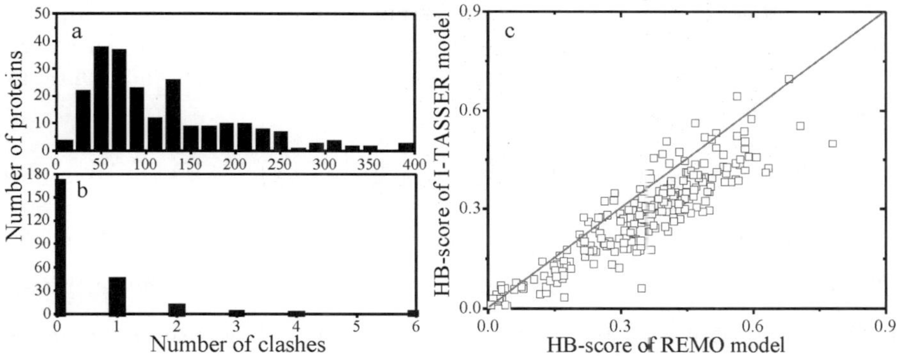

FIGURE 11.4 Histogram of steric clashes in (a) cluster centroids and (b) REMO models of 230 test proteins. (c) Comparison of HB-score of REMO models and I-TASSER models (From Li and Zhang [49]).

Remarkably, although the steric clashes had been completely removed, the clash-removing procedure had no side effect on the global topology of the initial structures. The RMSD change in most cases was <0.9 Å and the average RMSD to the native slightly improved.

Figure 11.4c shows the improvement in the hydrogen bonding (HB) score in REMO models over the I-TASSER models. Here, I-TASSER models refer to the models that had been generated by PULCHRA [71] for adding backbone atoms (N, C, O) and Scwrl3.0 [70] to build side chain atoms. HB-score is defined as the fraction of the common hydrogen bonds between model and the native structure. As shown in the figure, the HB-score of REMO models have dramatically improved in more than 80% (187/230) of the test proteins.

REMO was also used in blind CASP8 experiment for refining the reduced models generated by I-TASSER Monte-Carlo simulations. Based on the 172 released targets/domains, the average TM-score and GDT-score of the I-TASSER (as "Zhang Server") models are higher with a significant margin than that of other groups in the server section (see http://zhanglab.ccmb.med.umich.edu/casp8). In particular, the average HB-score of the I-TASSER, which partially reflects the quality of local structures, is also higher than all other groups, except SAM-08-server [6,72], while in CASP7 the HB-score of the I-TASSER models were much lower than most of other groups [41,42]. These data demonstrate a significant progress in reconstructing and refining atomic models using the REMO protocol. REMO simulations are now a part of I-TASSER methodology for generating atomic level model. The source code and online server of REMO is freely available at http://zhanglab.ccmb.med.umich.edu/REMO.

11.2.4. Function Prediction

One of the main impetuses for predicting the structure is to use it for structure-based functional annotation. To identify the functional homologs of a query protein, the generated models are structurally aligned by TM-align [50] with all known structures in the PDB library that have known functions. The resultant structural alignment is scored based on a Fh-score (Functional homology score), which is defined as [73]:

$$\text{Fh-score} = \text{nC-score} * \left(\text{TM-score} + \tfrac{1}{1+\text{RMSD}_{\text{ali}}} * \text{Cov}\right) + 3 * \text{ID}_{\text{ali}} * \text{Cov}, \qquad (4)$$

where nC-score is the normalized C-score and is defined as $\text{nC}-\text{score} = \dfrac{\text{C}-\text{score}+5}{7}$, which stays in [0, 1] and estimates the quality of I-TASSER protein structure predictions; TM-score [51] measures the global structural similarity between the model and the template proteins; RMSD_{ali} is the RMSD of query model and template structure in the structurally aligned region; Cov represents the coverage of the structural alignment; and ID_{ali} is the sequence identity between query and template based on the alignment by TM-align. For every query protein, predicted functions include both the predicted enzyme commission (EC) number [74] and the Gene Ontology (GO) molecular function [75] terms. While EC number is a commonly used scheme for functional classification of enzymes, GO terms provide a consistent description of function for both enzymatic and nonenzymatic proteins. Accordingly, two independent protein libraries of about 5800 nonredundant enzymatic proteins (pairwise sequence identity <90%) and about 13,500 nonredundant proteins (pairwise sequence identity <90%) with known GO terms have been constructed and are biweekly updated.

Based on a large-scale benchmark test set of 317 non-homologous proteins, it was found that by using the predicted structures (modeled while excluding all the homologous proteins with sequence identity >30% to query protein), the first three digits of EC number and 50% of associated GO terms of query protein could be correctly identified in more than 55% of the test cases from the best identified functional homologs based on Fh-score. Moreover, the true and false positive predictions could be discriminated well and achieved an area of more than 0.80 under the receiver operating characteristic (ROC) curve for both the predictions. For the 196 enzymatic proteins that had another functional homolog (enzymes with same first three EC digits) in the library and having less than 30% sequence identity, Fh-score and PSI-BLAST were able to identify functional homologs with same first three digits of EC number for 107 and 77 proteins, respectively. These data show that the structure-based functional annotations using the I-TASSER models can be about 39% more accurate than the sequence-based approaches (such as PSI-BLAST) [73].

11.3. *AB INITIO* PREDICTION OF I-TASSER ON SMALL PROTEINS

To explore the ability of I-TASSER to fold proteins for which no homologous templates are detected in the PDB, I-TASSER was tested on three sets of non-

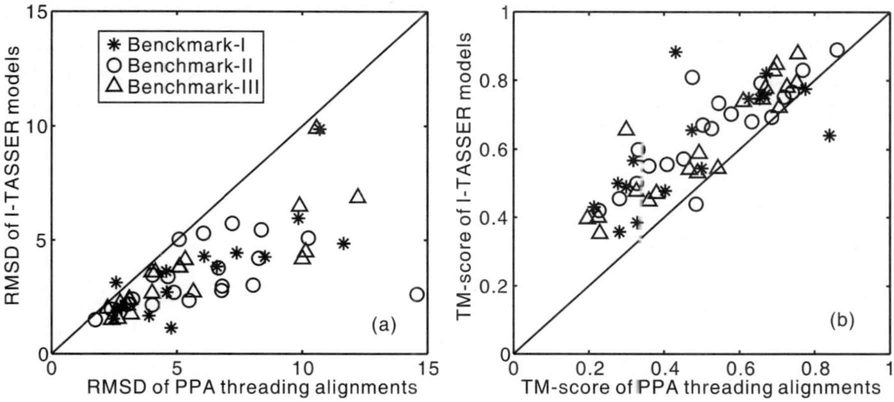

FIGURE 11.5 Comparison of I-TASSER models with the PPA threading alignment results. (a) C_α-RMSD to native of the I-TASSER models versus C_α-RMSD to native of the best threading alignment over the same aligned regions. (b) TM-score of the I-TASSER models versus TM-score of the best threading alignments. (From Wu, Skolnick, and Zhang [35]).

homologous proteins. The test proteins include: (i) Benchmark-I consisting of 16 proteins (<90 residues) that were used by Bradley et al. for testing ROSETTA [76]; (ii) Benchmark-II consisting of 20 proteins (<120 residues) that were used by Zhang et al for testing TOUCHSTONE II [64]; and (iii) Benchmark-III consisting of 20 proteins (<120 residues) selected from PDB [35].

The I-TASSER structure assembly started from PPA threading where all template proteins with a sequence identity >20% to the query or detectable by PSI-BLAST with an e-value <0.05 were excluded Figure 11.5 shows the comparison of the best of the top five I-TASSER models with the initial PPA threading alignments in all three benchmark test sets. As seen in the figure, the global topology of the final models was significantly closer to the native structure than the threading alignments. In Benchmark-I, I-TASSER models have an average C_α-RMSD of 3.8 Å, with six of them having a high-resolution structure with the C_α-RMSD <2.5 Å. On the second set, (Benchmark-II), I-TASSER could fold four of them with a C_α-RMSD <2.5 Å. The average C_α-RMSD of the I-TASSER models in this set of test proteins was 3.9 Å. Average C_α-RMSD of 3.9 Å was obtained for the third benchmark set, with seven cases having a C_α-RMSD <2.5 Å. Overall, the first predicted models had an average C_α-RMSD ranging from 4.3 Å to 4.8 Å and the average TM-score ranged from 0.59 to 0.64 for the three benchmarks. For the best models in the top five predictions, the average C_α-RMSD ranged from 3.8 Å to 3.9 Å and the average TM-score ranged from 0.61 to 0.65.

The first set of proteins was also used for testing ROSETTA [76] and the best of the top five models by ROSETTA had an average RMSD of 3.8 Å; thus, the overall results between the two methods (ROSETTA and I-TASSER) are comparable, but the central processing unit (CPU) time required by

I-TASSER was much shorter (150 CPU days vs. 5 CPU hours). For the second test, the average RMSD by TOUCHSTONE-II [64] is 5.9 Å. These data, together with the significant performance of automated I-TASSER server (the Zhang Server) in the FM section of the CASP experiment [40], demonstrate a new progress in automated *ab initio* model generation.

11.4. BLIND TEST OF I-TASSER IN CASP EXPERIMENTS

CASP [46,77] is a biennial world-wide protein structure prediction experiment, where the organizers release a number of protein sequences for which structure is unknown. The participants are then asked by the organizers to predict the structures of these proteins and submit their predicted models before deadlines. Finally, the experts evaluate the predicted models by comparing them with the structures solved by the X-ray or nuclear magnetic resonance (NMR) experiments.

The seventh CASP experiment was held in 2006, where the performance of I-TASSER was tested in both the human (as "Zhang") and the server section (as "Zhang Server"). The procedure in the server and human predictions are essentially the same and follow the general I-TASSER methodology, except for that the human prediction involved domain border assignment based on visual inspection and used the server predictions from other groups for hard targets; meanwhile, the I-TASSER assembly simulations were done for a longer time in the human prediction.

Ninety-six proteins in CASP7 were split into 124 domains by the assessors. Depending on modeling difficulty (whether or not a good template is present in PDB), these domains can be categorized for simplicity into 105 template-based modeling (TBM) targets and 19 free modeling (FM) targets. Figure 11.6 shows a comparison of the first I-TASSER models and the best threading templates for all these targets in both server and human predictions. Here, the best template refers to the template of the highest TM-score to the native structure among all the templates exploited by I-TASSER. As shown in the figure, although there is a general tendency for better templates resulting in better models, in most cases I-TASSER was able to consistently improve the final model over the templates based on both RMSD and TM-score.

In the TBM category, I-TASSER reassembly resulted in a TM-score increase by ~14%, where about 10% is probably because of the topology reorientation of the secondary structure fragments and the rest may be due to the increase in the size of models when gaps are filled during the reassembly procedure. One of the major reasons for this improvement is because I-TASSER employs consensus spatial constraints from multiple templates that are usually of higher accuracy than that from individual templates. The second driving force for the structure improvement was the optimization of I-TASSER inherent potential from a collection of statistical terms from different resources [35,47,63,64].

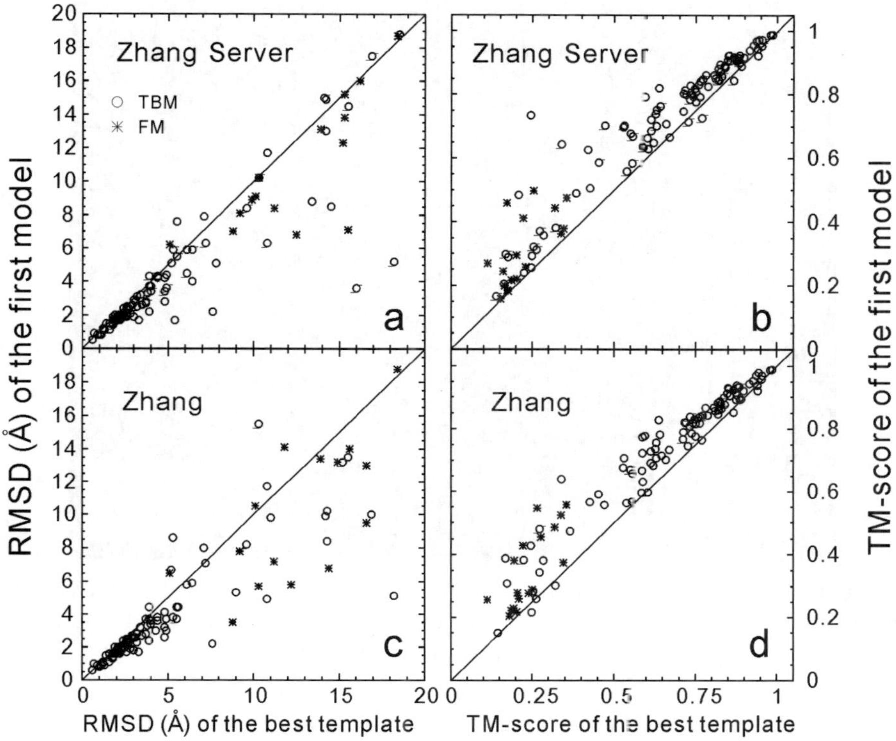

FIGURE 11.6 Comparison of the first predicted models by I-TASSER in human ("Zhang") and server ("Zhang-Server") sections of CASP7 with respect to the best exploited templates. The RMSD is calculated in the same set of aligned residues. The TM-score is calculated in the aligned regions for the templates and in full-length for the models (From Zhang, Y. Proteins 69 (2007): 112).

For the FM targets, I-TASSER was able to fold (RMSD < 6.5 Å or TM-score > 0.5) seven targets (about 1/3) that were up to 155 residues long. Figure 11.7 shows a more detailed analysis of a typical example (target T0382) of the I-TASSER predictions during CASP7. T0382 was a new fold protein (PDB ID: 2I9C) from *Rhodopseudomonas palustris* CGA009 crystallized by the structure genomics project. The topology of T0382 consists of five joggled α-helices. The left panel of Figure 11.7 shows the top five templates hit by the multiple threading programs used by I-TASSER, all having correct local second structure elements but incorrect global topologies with the best RMSD of 9.3 Å from 1xm9A1 (TM-score = 0.28). Contact prediction program generated 148 side chain contacts with 37 correct contacts (accuracy = 25%). The average error of the best predicted C_α distances is 2.2 Å. I-TASSER cuts the fragments from the template alignments and reassembles the topology under the guide of the predicted restraints and the inherent potential, which result in a model of full-length RMSD 3.6 Å and TM-score 0.66 (right panel of Figure

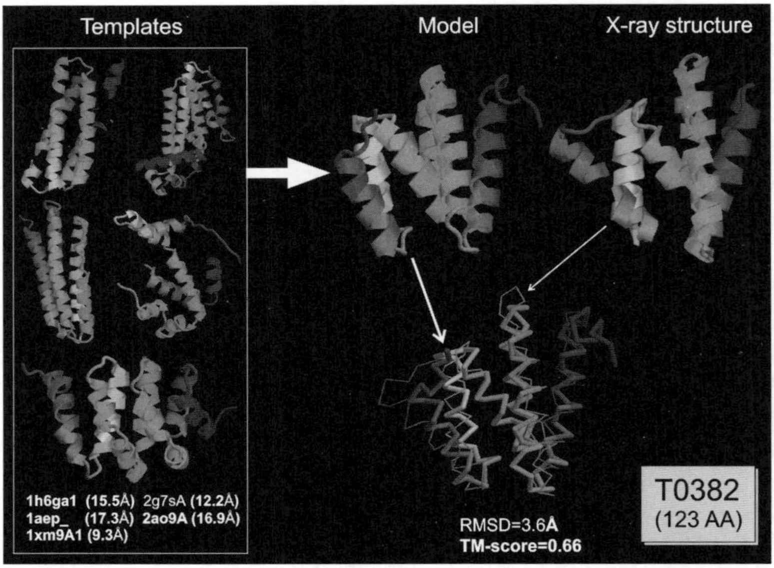

FIGURE 11.7 Structure comparisons of the threading templates, the final I-TASSER model, and the experimental structures for the CASP7 target T0382. Blue to red runs from N- to C-terminals (From Zhang [44]). (See color insert.)

11.7). The correlation of I-TASSER energy and the RMSD of the structure decoys is 0.72, which demonstrates the consistency of the external restraints and the inherent force field.

11.5. CONCLUDING REMARKS

The protein structure prediction problem can be solved in two ways. The first one is to fold all proteins by computationally recovering the nature's protein folding pathway. This task does not appear to be accomplished in foreseeable future, unless a detailed physicochemical description of the intra-protein and protein-solvent interactions are developed, not to mention the delicate interactions of proteins with the associated ligands and chaperones that will dramatically complicate the situation. The second solution is more of an engineering-oriented rather than scientific, in which a selected set of proteins are solved by experiments so that all proteins with unknown structure have at least one neighboring protein with known structure, which can be used as a template in CM; this has been the goal of the SG projects [20]. On the basis of about 40,000 structures in the PDB library (many are redundant) [18], it is estimated that 4 million models/fold assignments can be obtained by a simple combination of the PSI-BLAST search and the CM technique [78]. Development of more sophisticated and automated computer modeling approaches will dramatically enlarge the scope of modelable proteins in the SG projects.

Despite intense efforts and considerable progress in the field [79], the accuracy of protein structure prediction is still largely dictated by the evolutionary distance between the target and the solved proteins in the PDB library. Robust methods that can model proteins that have no or weak structure homologous templates are lacking. Nevertheless, the most efficient approaches to model both homologous and non-homologous proteins are those that combine different algorithms of threading, fragment assembly, *ab initio* modeling, and structural refinements. I-TASSER is one of the successful examples of these composite approaches. The exploitation of multiple threading templates and the optimization of the composite knowledge-based energy terms constitute the two major factors contributing to the success of I-TASSER in refining individual template structures closer to the native.

However, since I-TASSER has a resolution limitation set by its inherent reduced potential, high-resolution models cannot be predicted for most of proteins when a good template is not present. One of the ongoing efforts in this regard is to extend the reduced I-TASSER modeling to the atomic representation with the goal of improving the modeling accuracy at the atomic-level [44]. REMO represents part of our recent efforts to refine atomic models by optimizing the hydrogen-bonding networks. The development of new physics-based force fields in combination with the current I-TASSER knowledge-based potentials as well as the development of the function prediction methodology will be of significant importance in increasing the accuracy and the applicability of these approaches to genome-wide structure and function predictions.

REFERENCES

1. The Universal Protein Resource (UniProt) 2009. *Nucleic Acids Research*, 37:D169–174, 2009.
2. S.R. Wiley. Genomics in the real world. *Current Pharmaceutical Design*, 4(5):417–422, 1998.
3. S.F. Altschul, W. Gish, W. Miller, E.W. Myers, and D.J. Lipman. Basic local alignment search tool. *Journal of Molecular Biology*, 215(3):403–410, 1990.
4. S.F. Altschul, T.L. Madden, A.A. Schaffer, J. Zhang, Z. Zhang, W. Miller, and D.J. Lipman. Gapped BLAST and PSI-BLAST: A new generation of protein database search programs. *Nucleic Acids Research*, 25(17):3389–3402, 1997.
5. J. Soding. Protein homology detection by HMM-HMM comparison. *Bioinformatics*, 21(7):951–960, 2005.
6. K. Karplus, C. Barrett, and R. Hughey. Hidden Markov models for detecting remote protein homologies. *Bioinformatics*, 14(10):846–856, 1998.
7. C. Chothia and A.M. Lesk. The relation between the divergence of sequence and structure in proteins. *EMBO Journal*, 5(4):823–826, 1986.

8. T.C. Wood and W.R. Pearson. Evolution of protein sequences and structures. *Journal of Molecular Biology*, 291(4):977–995, 1999.
9. C.A. Wilson, J. Kreychman, and M. Gerstein. Assessing annotation transfer for genomics: Quantifying the relations between protein sequence, structure and function through traditional and probabilistic scores. *Journal of Molecular Biology*, 297(1):233–249, 2000.
10. D. Pal and D. Eisenberg. Inference of protein function from protein structure. *Structure*, 13(1):121–130, 2005.
11. W.R. Pearson. Effective protein sequence comparison. *Methods Enzymology*, 266:227–258, 1996.
12. W. Tian and J. Skolnick. How well is enzyme function conserved as a function of pairwise sequence identity? *Journal of Molecular Biology*, 333(4):863–882, 2003.
13. B. Rost. Enzyme function less conserved than anticipated. *Journal of Molecular Biology*, 318(2):595–608, 2002.
14. D. Eisenberg, E.M. Marcotte, I. Xenarios, and T.O. Yeates. Protein function in the post-genomic era. *Nature*, 405(6788):823–826, 2000.
15. G. Lopez, A. Rojas, M. Tress, and A. Valencia. Assessment of predictions submitted for the CASP7 function prediction category. *Proteins*, 69(Suppl 8):165–174, 2007.
16. G.J. Kleywegt. Recognition of spatial motifs in protein structures. *Journal of Molecular Biology*, 285(4):1887–1897, 1999.
17. A.C. Wallace, R.A. Laskowski, and J.M. Thornton. Derivation of 3D coordinate templates for searching structural databases: Application to Ser-His-Asp catalytic triads in the serine proteinases and lipases. *Protein Science*, 5(6):1001–1013, 1996.
18. H.M. Berman, J. Westbrook, Z. Feng, G. Gilliland, T.N. Bhat, H. Weissig, I.N. Shindyalov, and P.E. Bourne. The Protein Data Bank. *Nucleic Acids Research*, 28(1):235–242, 2000.
19. M. Gerstein, A. Edwards, C.H. Arrowsmith, and G.T. Montelione. Structural genomics: Current progress. *Science*, 299(5613):1663, 2003.
20. J.M. Chandonia and S.E. Brenner. The impact of structural genomics: Expectations and outcomes. *Science*, 311(5759):347–351, 2006.
21. D. Baker and A. Sali. Protein structure prediction and structural genomics. *Science*, 294(5540):93–96, 2001.
22. J. Skolnick, J.S. Fetrow, and A. Kolinski. Structural genomics and its importance for gene function analysis. *Nature Biotechnology*, 18(3):283–287, 2000.
23. P. Aloy, E. Querol, F.X. Aviles, and M.J. Sternberg. Automated structure-based prediction of functional sites in proteins: applications to assessing the validity of inheriting protein function from homology in genome annotation and to protein docking. *Journal of Molecular Biology*, 311(2):395–408, 2001.
24. D. Vitkup, E. Melamud, J. Moult, and C. Sander. Completeness in structural genomics. *Nature Structural & Molecular Biology*, 8(6):559–566, 2001.
25. Y. Zhang. Protein structure prediction: when is it useful? *Current Opinion in Structural Biology*, 19(2):145–155, 2009.
26. A. Sali and T.L. Blundell. Comparative protein modelling by satisfaction of spatial restraints. *Journal of Molecular Biology*, 234(3):779–815. 1993.

27. A. Fiser, R.K.G. Do, and A. Sali. Modeling of loops in protein structures. *Protein Science*, 9(9):1753–1773, 2000.
28. J.U. Bowie, R. Luthy, and D. Eisenberg. A method to identify protein sequences that fold into a known three-dimensional structure. *Science*, 253(5016):164–170, 1991.
29. S. Wu and Y. Zhang. MUSTER: Improving protein sequence profile-profile alignments by using multiple sources of structure information. *Proteins*, 72(2):547–556, 2008.
30. D.T. Jones, W.R. Taylor, and J.M. Thornton. A New Approach to Protein Fold Recognition. *Nature*, 358(6381):86–89, 1992.
31. Y. Xu and D. Xu, Protein threading using PROSPECT: Design and evaluation. *Proteins*, 40(3):343–354, 2000.
32. J. Skolnick, D. Kihara, and Y. Zhang. Development and large scale benchmark testing of the PROSPECTOR_3 threading algorithm. *Proteins*, 56(3):502–518, 2004.
33. S. Wu and Y. Zhang. LOMETS: A local meta-threading-server for protein structure prediction. *Nucleic Acids Research*, 35(10):3375–3382. 2007.
34. P. Bradley, K.M.S. Misura, and D. Baker. Toward high-resolution de novo structure prediction for small proteins. *Science*, 309(5742):1868–1871, 2005.
35. S. Wu, J. Skolnick, and Y. Zhang. Ab initio modeling of small proteins by iterative TASSER simulations. *BMC Biology*, 5:17, 2007.
36. A. Liwo, J. Lee, D.R. Ripoll, J. Pillardy, and H.A. Scheraga. Protein structure prediction by global optimization of a potential energy function. *Proceedings of the National Academy of Sciences U S A*, 96(10):5482–5485, 1999.
37. K.T. Simons, C. Strauss, and D. Baker. Prospects for ab initio protein structural genomics. *Journal of Molecular Biology*, 306(5):1191–1199, 2001.
38. D. Kihara, H. Lu, A. Kolinski, and J. Skolnick. TOUCHSTONE: An ab initio protein structure prediction method that uses threading-based tertiary restraints. *Proceedings of the National Academy of Sciences U S A*, 98(18):10125–10130, 2001.
39. C.B. Anfinsen. Principles that govern the folding of protein chains. *Science*, 181(96):223–230, 1973.
40. R. Jauch, H.C. Yeo, P.R. Kolatkar, and N.D. Clarke. Assessment of CASP7 structure predictions for template free targets. *Proteins*, 69(S8):57–67, 2007.
41. J. Kopp, L. Bordoli, J.N. Battey, F. Kiefer, and T. Schwede. Assessment of CASP7 predictions for template-based modeling targets. *Proteins*, 69(S8):38–56, 2007.
42. J.N. Battey, J. Kopp, L. Bordoli, R.J. Read, N.D. Clarke, and T. Schwede, Automated server predictions in CASP7. *Proteins*, 69(S8):68–82, 2007.
43. R. Das, B. Qian, S. Raman, R. Vernon, J. Thompson, P. Bradley, S. Khare, M.D. Tyka, D. Bhat, D. Chivian, D.E. Kim, W.H. Sheffler, L. Malmstrom, A.M. Wollacott, C. Wang, I. Andre, and D. Baker. Structure prediction for CASP7 targets using extensive all-atom refinement with Rosetta@home. *Proteins*, 69(S8):118–128, 2007.
44. Y. Zhang. Template-based modeling and free modeling by I-TASSER in CASP7. *Proteins*, 69(8):108–117, 2007.
45. Y. Zhang. I-TASSER server for protein 3D structure prediction. *BMC Bioinformatics*, 9:40, 2008.

46. J. Moult, K. Fidelis, A. Kryshtafovych, B. Rost, T. Hubbard, and A. Tramontano. Critical assessment of methods of protein structure prediction-Round VII. *Proteins*, 69(8):3–9, 2007.

47. Y. Zhang and J. Skolnick. Automated structure prediction of weakly homologous proteins on a genomic scale. *Proceedings of the National Academy of Sciences U S A*, 101(20):7594–7599, 2004.

48. Y. Zhang and J. Skolnick. SPICKER: A clustering approach to identify near-native protein folds. *Journal of Computational Chemistry*, 25(6):865–871, 2004.

49. Y. Li and Y. Zhang. REMO: A new protocol to refine full atomic protein models from C-alpha traces by optimizing hydrogen-bonding networks. *Proteins*, 76(3):665–676, 2009.

50. Y. Zhang and J. Skolnick. TM-align: A protein structure alignment algorithm based on the TM-score. *Nucleic Acids Research*, 33(7):2302–2309, 2005.

51. Y. Zhang and J. Skolnick. Scoring function for automated assessment of protein structure template quality. *Proteins*, 57(4):702–710, 2004.

52. D. Frishman and P. Argos. Knowledge-based protein secondary structure assignment. *Proteins*, 23(4):566–579, 1995.

53. S. Wu and Y. Zhang. ANGLOR: A composite machine-learning algorithm for protein backbone torsion angle prediction. *PLoS ONE*, 3(10):e3400, 2008.

54. P.J. Silva. Assessing the reliability of sequence similarities detected through hydrophobic cluster analysis. *Proteins*, 70(4):1588–1594, 2008.

55. S.B. Needleman and C.D. Wunsch. A general method applicable to the search for similarities in the amino acid sequence of two proteins. *Journal of Molecular Biology*, 48(3):443–453, 1970.

56. D. Fischer, L. Rychlewski, R.L. Jr. Dunbrack, A.R. Ortiz, and A. Elofsson. CAFASP3: The third critical assessment of fully automated structure prediction methods. *Proteins*, 53 (6):503–516, 2003.

57. D. Fischer. 3D-SHOTGUN: A novel, cooperative, fold-recognition meta-predictor. *Proteins*, 51(3):434–441, 2003.

58. H. Zhou and Y. Zhou. Fold recognition by combining sequence profiles derived from evolution and from depth-dependent structural alignment of fragments. *Proteins*, 58(2):321–328, 2005.

59. J. Shi, T.L. Blundell, and K. Mizuguchi. FUGUE: Sequence-structure homology recognition using environment-specific substitution tables and structure-dependent gap penalties. *Journal of Molecular Biology*, 310(1):243–257, 2001.

60. K. Karplus, R. Karchin, J. Draper, J. Casper, Y. Mandel-Gutfreund, M. Diekhans, and R. Hughey. Combining local-structure, fold-recognition, and new fold methods for protein structure prediction. *Proteins*, 53 (6):491–496, 2003.

61. Y. Zhang, D. Kihara, and J. Skolnick. Local energy landscape flattening: Parallel hyperbolic Monte Carlo sampling of protein folding. *Proteins*, 48(2):192–201, 2002.

62. L.J. McGuffin, K. Bryson, and D.T. Jones. The PSIPRED protein structure prediction server. *Bioinformatics*, 16(4):404–405, 2000.

63. Y. Zhang and J. Skolnick. Tertiary structure predictions on a comprehensive benchmark of medium to large size proteins. *Biophysical Journal*, 87:2647–2655, 2004.

64. Y. Zhang, A. Kolinski, and J. Skolnick. TOUCHSTONE II: A new approach to ab initio protein structure prediction. *Biophysical Journal*, 85(2):1145–1164, 2003.
65. S. Wu and Y. Zhang. A comprehensive assessment of sequence-based and template-based methods for protein contact prediction. *Bioinformatics*, 24(7):924–931, 2008.
66. T.P. Hopp and K.R. Woods. Prediction of protein antigenic determinants from amino acid sequences. *Proceedings of the National Academy of Sciences U S A*, 78(6):3824–3828, 1981.
67. J. Kyte and R.F. Doolittle. A simple method for displaying the hydropathic character of a protein. *Journal of Molecular Biology*, 157(1):105–132, 1982.
68. J. Cheng and P. Baldi. Improved residue contact prediction using support vector machines and a large feature set. *BMC Bioinformatics*, 8:113, 2007.
69. J. Cheng and P. Baldi. Three-stage prediction of protein beta-sheets by neural networks, alignments and graph algorithms. *Bioinformatics*, 21 (1):i75–i84, 2005.
70. A.A. Canutescu, A.A. Shelenkov, and R.L. Dunbrack. A graph-theory algorithm for rapid protein side-chain prediction. *Protein Science*, 12(9):2001–2014, 2003.
71. M. Feig, P. Rotkiewicz, A. Kolinski, J. Skolnick, and C.L. Brooks. 3rd. Accurate reconstruction of all-atom protein representations from side-chain-based low-resolution models. *Proteins*, 41(1):86–97, 2000.
72. K. Karplus, S. Katzman, G. Shackleford, M. Koeva, J. Draper, B. Barnes, M. Soriano, and R. Hughey. SAM-T04: What is new in protein-structure prediction for CASP6. *Proteins*, 61 (7):135–142, 2005.
73. A. Roy, A. Kucukural, S. Mukherjee, P.S. Hefty, and Y. Zhang. Large scale benchmarking of protein function prediction using modeled protein structures. *Journal of Molecular Biology*, 2010, submitted.
74. Nomenclature Committee of the International Union of Biochemistry and Molecular Biology (NC-IUBMB). Enzyme supplement 5 (1999). *European Journal of Biochemistry*, 264(2):610–650, 1999.
75. M. Ashburner, C.A. Ball, J.A. Blake, D. Botstein, H. Butler, J.M. Cherry, A.P. Davis, K. Dolinski, S.S. Dwight, J.T. Eppig, M.A. Harris, D.P. Hill, L. Issel-Tarver, A. Kasarskis, S. Lewis, J.C. Matese, J.E. Richardson, M. Ringwald, G.M. Rubin, and G. Sherlock. Gene ontology: Tool for the unification of biology. *The Gene Ontology Consortium. Nature Genetics*, 25(1):25–29, 2000.
76. P. Bradley, K.M. Misura, and D. Baker. Toward high-resolution de novo structure prediction for small proteins. *Science*, 309(5742):1868–1871, 2005.
77. J.N. Battey, J. Kopp, L. Bordoli, R.J. Read, N.D. Clarke, and T. Schwede Automated server predictions in CASP7. *Proteins*, 69 (8):68–82, 2007.
78. U. Pieper, N. Eswar, F.P. Davis, H. Braberg, M.S. Madhusudhan, A. Rossi, M. Marti-Renom, R. Karchin, B.M. Webb, D. Eramian, M.Y. Shen, L. Kelly, F. Melo, and A. Sali. MODBASE: A database of annotated comparative protein structure models and associated resources. *Nucleic Acids Research*, 34(Database issue):D291–D295, 2006.
79. Y. Zhang. Progress and challenges in protein structure prediction. *Current Opinion Structural Biology*, 18(3):342–348, 2008.

CHAPTER 12

HYBRID METHODS FOR PROTEIN STRUCTURE PREDICTION

DMITRI MOURADOV and BOSTJAN KOBE
The University of Queensland
School of Chemistry and Molecular Biosciences
QLD, Australia

NICHOLAS E. DIXON
School of Chemistry
University of Wollongong
NSW, Australia

THOMAS HUBER
The University of Queensland
School of Chemistry and Molecular Biosciences
QLD, Australia

12.1. INTRODUCTION

Structural bioinformatics is a highly cost-efficient solution for accelerated determination of the three-dimensional (3D) structures of proteins. Purely computational prediction methods, such as advanced fold recognition (Chapters 9 and 10), composite approaches (Chapter 12), *ab initio* fragment assembly [1,2], and molecular docking [3] are routinely applied today to extend our knowledge of protein structures, how they interact and what their functional roles are in a biological context. Very often, however, predicted protein structures are not given the same trust as their experimental counterparts. This comes mainly from the need for extensive expertise to produce high-quality models, generally high rates of false predictions, and the difficulty to measure the confidence that can be associated with a structure "solved" algorithmically.

Introduction to Protein Structure Prediction: Methods and Algorithms,
Edited by Huzefa Rangwala and George Karypis

Hybrid approaches are a means to overcome these shortcomings; by incorporating limited experimental measurements, reliable structural models can be computed and unlikely predictions eliminated. Hybrid approaches take advantage of data derived from a range of very different biochemical and biophysical methods, most of which are becoming routinely available in many laboratories. These methods are of increasing interest in view of increasingly easy access to analytical instruments, such as high-resolution mass spectrometers and high-frequency electron paramagnetic resonance (EPR) spectrometers. Similarly, small-angle neutron scattering and small-angle X-ray scattering (SANS/SAXS) data are becoming routinely accessible through advanced neutron and synchrotron light sources. In addition, recent developments in nuclear magnetic resonance (NMR) spectroscopy make large (>100 kDa) protein systems amenable to analysis and, in combination with site-specific isotope labeling, have opened unprecedented possibilities to obtain sparse structural data on selected regions within an entire system.

Moreover, hybrid approaches have shown great promise in complementing high-resolution structural biology. To fully characterize function in dynamically interacting assemblies where both the components and their structures may vary throughout a complex multistep process, structures need to be determined at each step. By using model structures, it is possible to design and analyze new hypothesis-driven experiments and thus significantly speed up high-resolution structure determination.

12.2. SOURCES OF LIMITED STRUCTURAL DATA

A variety of biophysical and biochemical techniques exist that can rapidly give a wealth of information on shape, local structure, residue proximities, and residue environment in macromolecular systems (Table 12.1). These include in-solution scattering measurements where the angular distribution of SAXS and SANS can be fit to yield global information on the structural envelope of a protein in solution. The reliability of SAXS data and computational analysis tools has recently improved dramatically [4,5]. SANS has the added advantage that contrast matching of small-angle and buffers selectively renders parts of the system invisible (to neutrons), and shapes of proteins in a larger assembly can be determined individually [5]. A complementary way to map protein surfaces is by chemical modification (CM) or hydrogen/deuterium (H/D) exchange (DX). After various times of exposure to D_2O or a CM agent that targets side chain functional groups, the protein system is digested with proteases and liquid chromatography-mass spectrometry (LC-MS)/MS fingerprinting is then used to determine where and to what degree CM has occurred. CM and H/D exchange of amide protons is generally several orders of magnitude slower in residues that are buried and/or part of regular secondary structure, thus providing a quantitative measurement of protein structure. Circular dichroism (CD) spectroscopy constitutes a reliable method for mea-

TABLE 12.1 Examples of Methods That Produce Sparse Structural Data and Have Been used in Combination with Molecular Modeling to Compute Structure

Method	Data measured	Structural data generated	Example application
SAXS/SANS	Scattering intensity as a function of momentum transfer	Pair distribution function; shape envelope	[21,22]
CD	Mean residue ellipticity as a function of wavelength	Secondary structure content	[23]
FRET	Yield of fluorescence energy transfer	Distance between donor and acceptor pair	[24]
EPR	Dipole-dipole coupling between electron spins	Spin label environment and distance between pairs of spin labels	[25]
Deuterium exchange-mass spectrometry (DXMS)	Rate constant of H/D exchange	Solvent exposure	[26]
Radical footprinting	Rate constant from dose-response curve	Solvent exposure	[27]
Chemical cross-linking	Mass/charge ratio of joint peptides and fragmentation	Upper limit on pair distance between reacted groups	[28,29]

suring the secondary structure content of a protein. But while secondary structure content information by itself is generally of very limited use for computing structure, in combination with residue by residue secondary structure prediction it can provide important local structure restraints.

Arguably the most powerful constraints in protein structure modeling are measured distances between pairs of residues. Molecular probe techniques, such as fluorescence resonance energy transfer (FRET) [6] and EPR [7], can provide selective distances between specifically labeled parts of a molecule. Paramagnetic or fluorescent labels are generally attached either *via* disulfides or other CM to engineered cysteine mutants, or as more recently demonstrated, can be selectively and efficiently incorporated in the form of noncanonical amino acids into proteins using a cell-free expression system [8]. A clear advantage of these probe techniques is their ability to provide long-range structural information. Among all the lanthanide ions, Gd^{3+} is most highly paramagnetic, and recent work has shown that spin-spin interactions between pairs of Gd^{3+} ions can be used in high-field pulse EPR experiments to accurately determine distances up to 40Å [9]. Similarly, new developments in

NMR spectroscopy employs paramagnetic ions to induce the anisotropic pseudo contact shift (PCS) in NMR active nuclei of a protein molecule [10]. Because of the anisotropy in the PCS, the electron-proton interaction decaying slowly and the gyromagnetic ratio of a free electron being nearly three orders of magnitude larger than that of a proton, the PCS effect provides ample long-range structural information, reporting on interactions over distances up to 40Å.

The chemical analog of these biophysical methods to measure inter-residue distances is chemical cross-linking [11,12], which attempts to covalently connect functional groups with a molecular spacer. A link can only be formed if the functional groups are within the reach of the spacer, thus providing a chemical means to measure (upper bound) distances between groups. This conceptually simple chemical approach has a long history in protein science where it proved to be a useful tool to determine intermolecular interactions [10] and topological information in multi-protein assemblies [13]. More recently, chemical cross-linking has attracted much interest in more detailed structural studies, spurred by improved structural modeling capabilities and rapid advances in high-accuracy multi-dimensional MS, which today allows not only the reliable identification of the chemically modified molecules, but also determination of the exact cross-linker insertion point [14–17]. By combining enzymatic digestion with MS, the cross-linking technique is able to remove size limitations imposed by other techniques as only proteolytic fragments are analyzed. A key advantage of chemical cross-linking over FRET and EPR probe techniques is that multiple distance constraints between several groups within a protein or protein complex can be obtained in a single experiment and no labels need to be incorporated. However, exactly this advantage has also proven to be a major technical challenge, because cross-linked peptides must be identified in an abundance of proteolytically digested native peptides. In the past few years, innovations in cross-linker design and peptide separation methods combined with multi-dimensional MS and new analysis techniques have greatly improved identification [18–20].

12.3. TRANSLATION INTO STRUCTURAL RESTRAINTS

In combination, these approaches yield overlapping and distinct structural information on solvent exposure, inter-amino acid distances, and protein shape. For structure calculations, all this information is gathered in a pseudo-energy function, the global minimum of which corresponds to the structure that best satisfies all restraints:

$$E^{\exp} = \sum_{i}^{N} w_i \left(Q_i^{calc} - Q_i^{\exp}\right)^2 \tag{12.1}$$

where Q_i^{calc} is the ith experimental property calculated from the candidate structure, $Q_i^{\exp}$ is the ith measured property, and N is the total number of experimental data. Different weightings w_i account for the experimental error in each datum.

Depending on the nature of the measurement, Q_i^{calc} itself is a more or less complicated function with respect to the structure. In the case of chemical cross-linking, for example, it is a Boolean function (functional groups in the structure are either within the cross-linker spanning distance, or they are not), rendering the quadratic term into a simple step function. When using different experiments, it appears to be appealing to first transform all measurements into a corresponding structural metric, such as a distance, and then restrain the structures with respect to this new metric. This should be avoided, because the error in the measurement is not necessarily Gaussian with respect to the new transformed metric and correct weighting of each datum becomes difficult.

12.4. USE OF LIMITED EXPERIMENTAL DATA TO ELUCIDATE STRUCTURE

Molecular models are a common way to represent experimental structural measurements.

The classical approach in high-resolution structure determination is to measure sufficient experimental restraints to define a 3D model by the system's 3N Cartesian coordinates, where N is the total number of atoms in the system. What makes hybrid approaches special is that they require only a limited number of explicit experimental constraints to calculate a structure (Fig. 12.1).

Explicit constraints can be used to simply filter through theoretical models, allowing exclusion of models that do not meet the constraints. Taking this a step further, the measurements can be used to refine or even compute models, with direct application of the constraint data during calculations. When

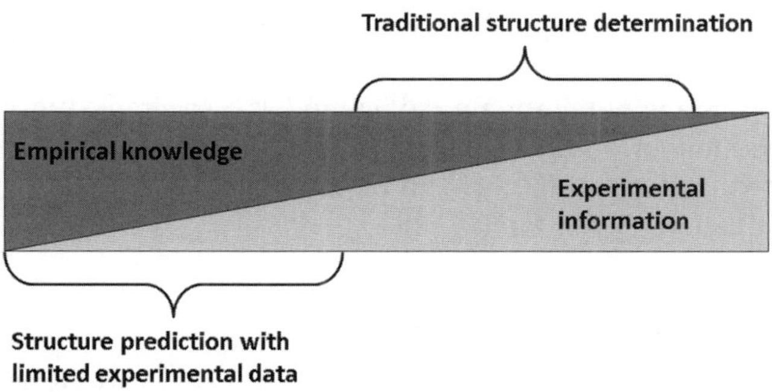

FIGURE 12.1 Limited experimental information is required for structure prediction using hybrid techniques as opposed to traditional techniques for structure determination such as X-ray crystallography and NMR.

combined with molecular modeling/docking [28–31], a limited number of constraints may be sufficient to determine the orientation of components in a protein complex or even of domains in a multi-domain protein.

12.4.1. Re-Scoring Models (Filtering)

At the most basic level, Equation 12.1 can be used to rank a set of theoretical models according to how well they are in agreement with all measurements. Albrecht et al. [32] conducted a theoretical study of how effective distance constraints from chemical cross-linking are at improving the success of fold recognition by threading. The analysis was carried out on 81 single-domain proteins (Hobohm96-25 database) whose pairwise sequence identity does not exceed 25%. Hypothetical cross-linking constraints were generated for all aspartate, glutamate, and lysine residues separated by between 8Å and 12Å based on the known structures of the proteins. These hypothetical cross-links were then used to re-rank theoretical models computed with the 123D threading program [33]. Various ranking functions on validity of distance constraints were applied, including simply counting the number of satisfied constraints and using a more complex scoring function that gives higher scores for satisfied constraints that are conserved among members of the same fold class. The results show that employing sparse constraints from cross-linking studies to re-rank models from fold recognition can improve success rates from 54–65% to 58–73%, depending on the quality of the initial alignment.

A limitation of this study was its use of hypothetical cross-links, equivalent to the outcomes of optimal experiments. Due to large differences in group reactivities, reagent accessibility and competitive suppression of some low-abundance cross-linked peptides in the MS analysis, generally only a small subset of all theoretically possible cross-links can be observed in real experiments. However, studies using data from experimental cross-linking have shown similar improvements to that reported by Albrecht et al. [32]. For example, Young et al. [30] used a similar post-filtering approach on a fibroblast growth factor-2 (FGF-2) protein where cross-links were experimentally identified using MS techniques. Again using the 123D software, the protein was initially incorrectly categorized as belonging to the beta-clip fold family. However, re-ranking of the top threading models by a simple scoring function based on the number of satisfied constraints resulted in the first, second, and fourth ranked structures all correctly identifying the FGF-2 structure as a member of the beta-trefoil family.

12.4.2. Structure Refinement

One inherent limitation of X-ray crystallography is that various proteins, such as membrane channels, may adopt multiple stable conformations that cannot be observed in static crystal structures. In such a case limited experimental constraints can be used to refine an existing crystal structure to show an alternate stable conformational state. FRET spectroscopy has been used to demonstrate the concept by modeling the conformational change involved in channel gating in MscL, a multimeric membrane protein important in releasing pressure during hypo-osmotic stress [34]. A crystal structure had been solved showing a closed conformation of MscL comprised of five identical subunits surrounding a central pore. Site-directed mutagenesis, specifically the insertion of cysteine residues, was used to insert different fluorescent probes into identical sites in all five subunits. Measuring fractions of energy transferred between donor and acceptor probes before and after induction of channel opening correlated with a radius increase of 8Å. As the protein volume remains constant, channel activation must trigger an opening of a large pore, as inferred also by previous studies using paramagnetic resonance spectroscopy and site-directed spin labeling. This radius change was used to model an open-gate conformation of MscL, showing one of the largest conformational changes recorded by any membrane protein. This approach paves the way for probing conformational changes of membrane proteins *in situ*.

12.4.3. Structure Calculations (Docking with Constraints)

A more challenging problem is to directly compute molecular models that satisfy a given set of (sparse) constraints. Xu et al. [35] detail such an approach that uses NMR Nuclear Overhauser Effect (NOE) data to improve threading performance. They employed a divide and conquer strategy, which divides the structures into substructures (cores) each comprised of only one secondary structure element, then optimally aligns substructures with subsequences. Two conditions must always be met to incorporate the distance constraints: (i) a link must be present for a constraint to be aligned to two cores and (ii) linked cores must not be aligned to sequence positions that violate constraints. Results show that even a small number of NOEs were sufficient to improve threading success in difficult to predict proteins. Even though NOEs provide a medium density network of distance restraints, the same approach can also be applied to the more sparse constraints derived from chemical cross-linking.

12.4.4. Multi-Domain Proteins and Multimeric Assemblies

The success of structural genomics has brought about the systematic determination of many structures of individual proteins and protein domains. This has led frequently to the situation where although individual domains in multi-domain proteins are structurally characterized, their relative orientations are not. Similarly, structures of constituents of many multimeric protein assemblies have been solved by X-ray crystallography or NMR spectroscopy, but due to technical difficulties the whole assemblies could not be structurally determined using these techniques. While using purely computational techniques to predict relative orientations of domains and proteins in complexes often results in an inaccurate conformation [36], hybrid approaches have been

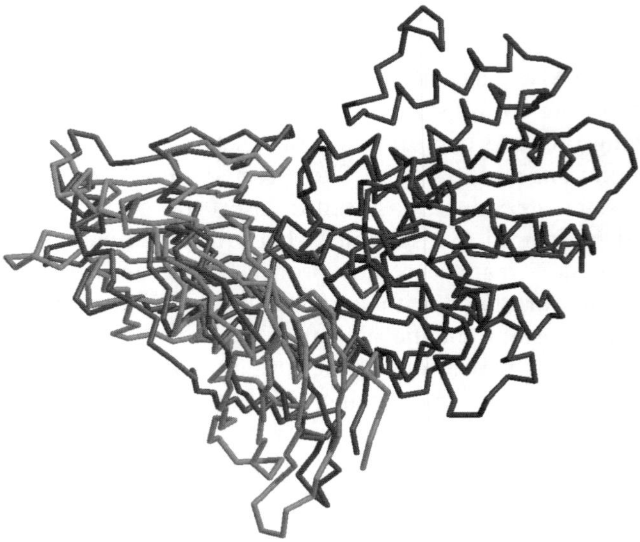

FIGURE 12.2 The crystal structure of the latexin (green)—CPA1 (blue) complex, overlaid with the orientation of the latexin molecule revealed by the top-scoring docked structure (red). (See color insert.)

shown to complement high-resolution structure determination in these cases. In combination with molecular docking, they can provide sufficient additional inter-domain or inter-protein constraints to allow the positioning of components in an overall structure at high resolution.

As an example, we demonstrated this concept by combining molecular modeling with a very small number of experimental distance restraints from chemical cross-linking to establish the structure of the protein complex between a mammalian carboxypeptidase A (CPA1) and its protein inhibitor latexin [28]. Three distance constraints were identified using a combination of chemical cross-linking and MS technology. Rigid body docking with a simple scoring function was employed to calculate the structure of latexin:CPA1 to within a C_α root-mean-square deviation (RMSD) of 3.74Å relative to the crystal structure (Fig. 12.2). The elucidated structure defines the interface between the two molecules accurately enough to guide mutagenesis experiments to probe the contribution of interacting residues and to provide reagents for use to probe the cellular functions of the proteins.

Structure determination of multi-domain proteins using the hybrid approach was assessed using acyl-CoA thioesterase (Acot7), which contains two hotdog fold domains, both of which are required to catalyse the hydrolysis of fatty acyl-CoA into CoA and free fatty acids [29]. While the two hotdog domains were able to be crystallized and their structures solved separately, no diffraction quality crystals were obtained for full-length Acot7. Using two separate techniques, molecular docking and homology modeling, together with identification of seven inter-domain cross-links, the orientation of the two hotdog

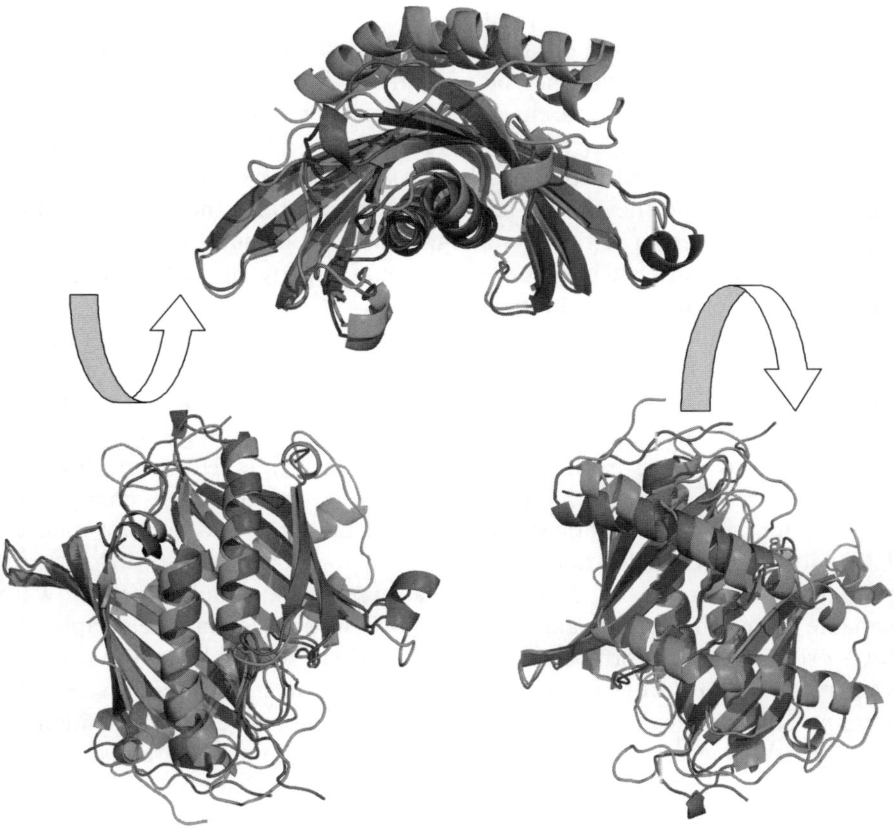

FIGURE 12.3 Multiple views of the superimposition of the predicted Acot7 model using chemical cross-linking and molecular modeling (green) with the crystal structure of ACOT12 (red). The structure suggests that the two hotdog fold domains adopt a similar structure to a dimer of single hotdog domains from other species. (See color insert.)

domains in the full-length Acot7 was predicted. A consistent model of a double-hotdog structure was observed using both docking and modeling. Superimposition of the model onto the a recently solved crystal structure of the human homolog ACOT12 shows remarkable structure similarity with C_α RMSD of 1.6Å over 242 amino acids (Fig. 12.3).

12.5. CONCLUSIONS AND FUTURE OUTLOOK

A clear advantage of hybrid approaches is that they offer means to combine very different sources of structural information and to integrate them iteratively into a single structural model (but not necessarily a single structure). Even when the amount of experimental information is still insufficient to determine a structure, it will already provide structural models that enable new hypotheses to be tested experimentally, for example, better paramagnetic labeling sites will be identified, and putative interaction surfaces will be identified at which labels for distance measurements can be site-specifically incorporated. This is quite different from structure determination with X-ray crystallography, where a high-resolution structure is almost guaranteed if one is able to grow well-diffracting protein crystals, but is left in despair if this is not the case. While the hybrid technique cannot attain the high-resolution structures observed in X-ray crystallography, it is currently seen as a rescue strategy for protein systems where traditional structure characterization techniques do not provide a solution.

Despite the clear appeal of hybrid approaches and the enthusiasm of its proponents, they have not been widely used. For example, no protein structure has been determined completely *de novo* using this method. This is partly due to various experimental obstacles and the need for combining a diverse range of expertise to derive explicit experimental constraints. However, development of novel low-cost techniques has the potential to quickly change this. Specifically, the incorporation of subsequences of amino acids as fluorophores, EPR labels, and cross-linkable groups has the potential to drastically simplify the process of deriving experimental constraints. When coupled with software capable of automated data analysis, the hybrid techniques can yield rapid and low-cost structural information. As the hybrid approach allows the use of a wide range of sources of structural information, one must consider the strengths, weaknesses, and peculiarities of each method on a case-specific basis to identify which method will yield the most promising experimental constraints.

The hybrid method's strongest point may lie in its potential to determine structures of protein complexes through a bottom-up integration of atomic-detail crystallographic or NMR structures with explicit experimental constraints to obtain missing pieces of information, for example, the relative orientation of the individual components as we demonstrated with the complex of CPA1 and latexin. This approach is not limited to high-affinity complexes. By introducing strong bonds formed on contact it is in principle possible to trap short-lived protein-protein interactions in the cross-linking process. Detailed structural information about such transient complexes would be invaluable for our functional understanding of biological process, since it is currently not directly accessible by any other methods. Similarly this technique is amenable to accurately define global orientation of structural domains in large molecular assemblies, where the structures of individual domains have already been determined, for example, by X-ray crystallography, NMR, or molecular modeling.

ACKNOWLEDGEMENT

This work was supported by an Australian Research Council (ARC) grant to NED and TH. BK is an ARC Federation Fellow and a National Health and

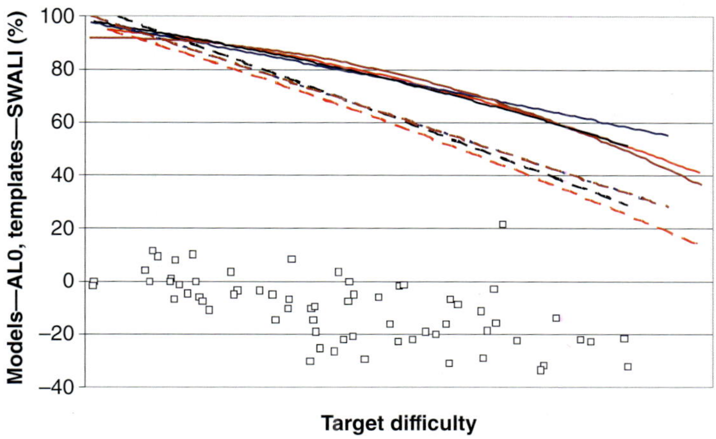

FIGURE 2.4 Maximum template-imposed alignability (SWALI, solid lines) and alignment accuracy of the best template-based models (AL0, dashed lines) from CASP5–8 as a function of target difficulty. Maximum alignability is defined as the fraction of equivalent residues in superposition of the target and best template structure; target difficulty combines coverage of the target structure by the best template and target-template sequence identity. CASP8—black lines; CASP7—blue; CASP6—brown; CASP5—red. Squares represent the difference between alignment quality and maximum alignability for CASP8 targets. Points over the 0% level represent targets where alignment accuracy was better than maximum alignability.

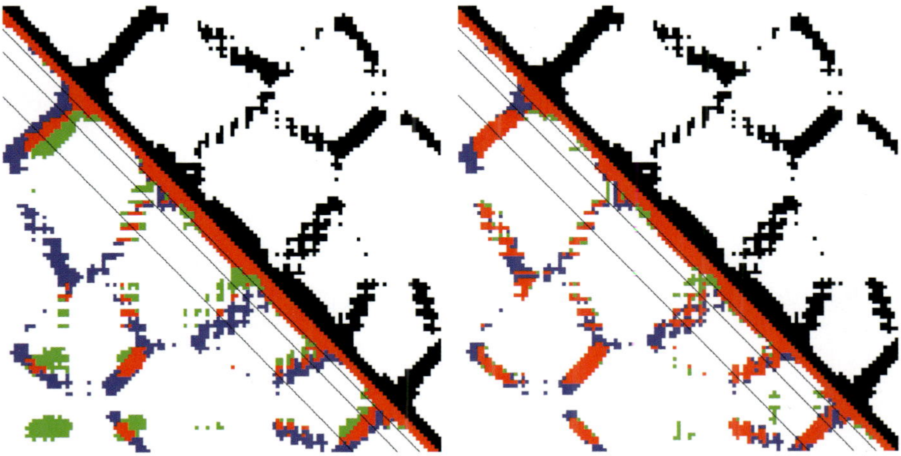

Figure 7.9 Protein 1B9LA 12 Å contact maps for *ab initio* (left) and template-based (right) predictions. The best template sequence identity is 22.7%. The top-right of each map is the true map and the bottom-left is predicted. In the predicted half white and red are true negative and positive respectively, blue and green are false negative and positive, respectively. The three black lines correspond to $|i - j| \geq 6, 12, 24$.

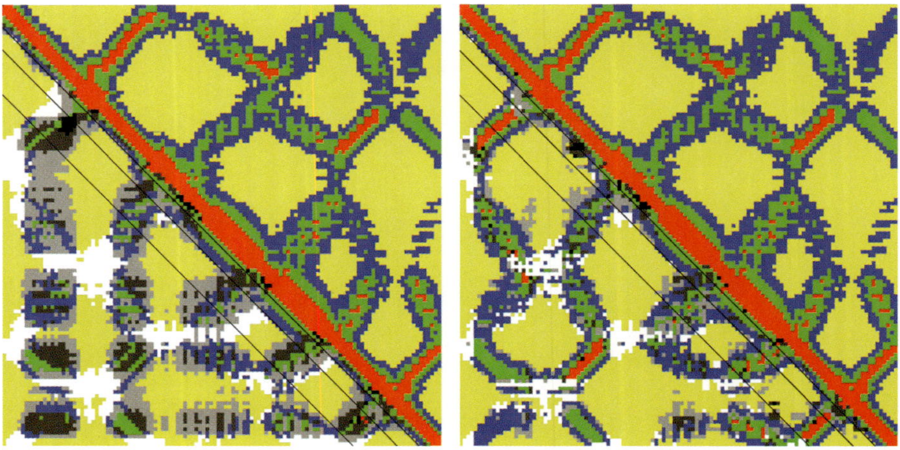

FIGURE 7.10 Protein 1B9LA multi-class contact maps for *ab initio* (left) and template-based (right) predictions. The best template sequence identity is 22.7%. The top-right of each map is the true map and the bottom left is predicted. In the predicted half red, blue, green and yellow correspond to class 0, 1, 2, and 3, respectively. The gray scale in the predicted half corresponds to falsely predicted classes. The three black lines correspond to $|i - j| \geq 6, 12, 24$.

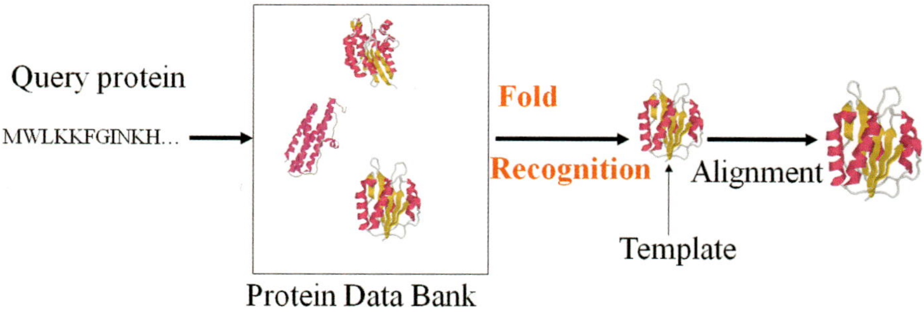

FIGURE 9.1 The basic procedure of the template-based protein structure prediction. Given a query protein without a solved structure, template-based methods try to identify a template protein structure in the Protein Data Bank (PDB) [34], which is assumed to have the similar structure with the query protein. A query-template alignment is then generated by alignment methods. Finally, based on the query-template alignment, the structure of the template protein is transferred to generate the structure model of the query protein. The simplest approach is to copy the coordinates of the residues in the template structure to their aligned counterparts of the query protein. In this procedure, the important step of identifying a template structure for the query protein is called *fold recognition*.

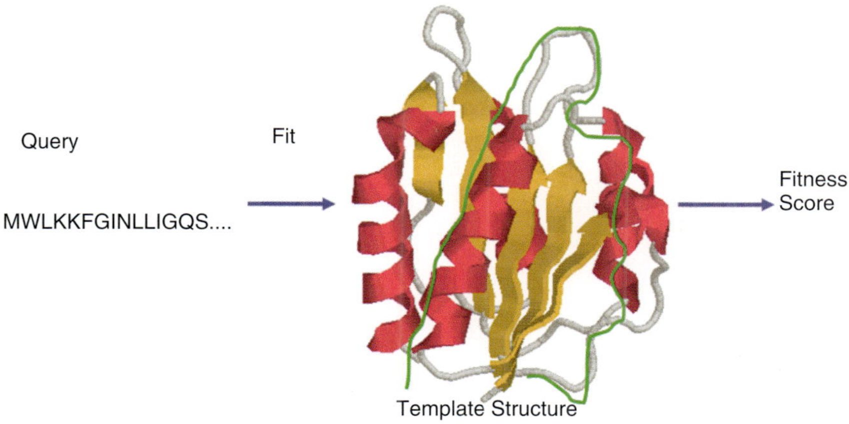

FIGURE 9.4 Sequence-structure alignment (threading) of a query protein against a known structure database.

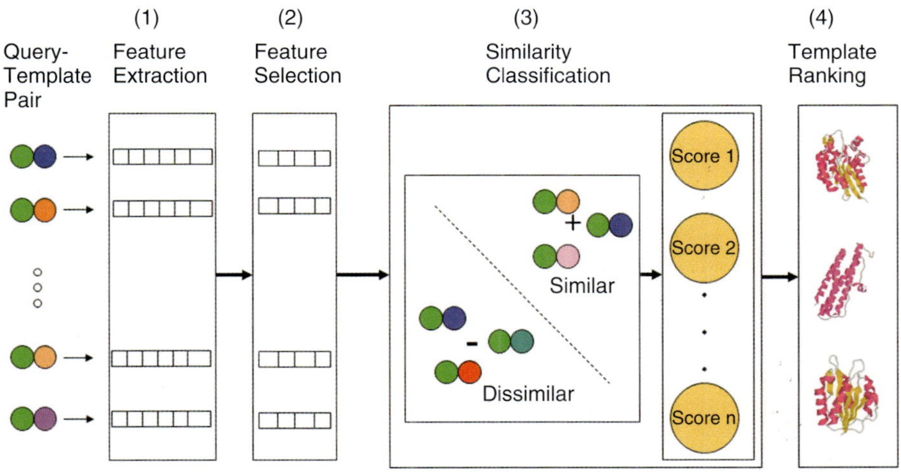

FIGURE 9.7 The machine learning framework for protein fold recognition. Step 1. Pair a query protein with template proteins and generate pairwise similarity features for each query-template pair. Step 2. Select a set of informative features. Step 3. Classify structure similarity into two categories (similar or dissimilar) using the features with machine learning methods. Step 4. Rank template proteins using classification scores with respect to the same query.

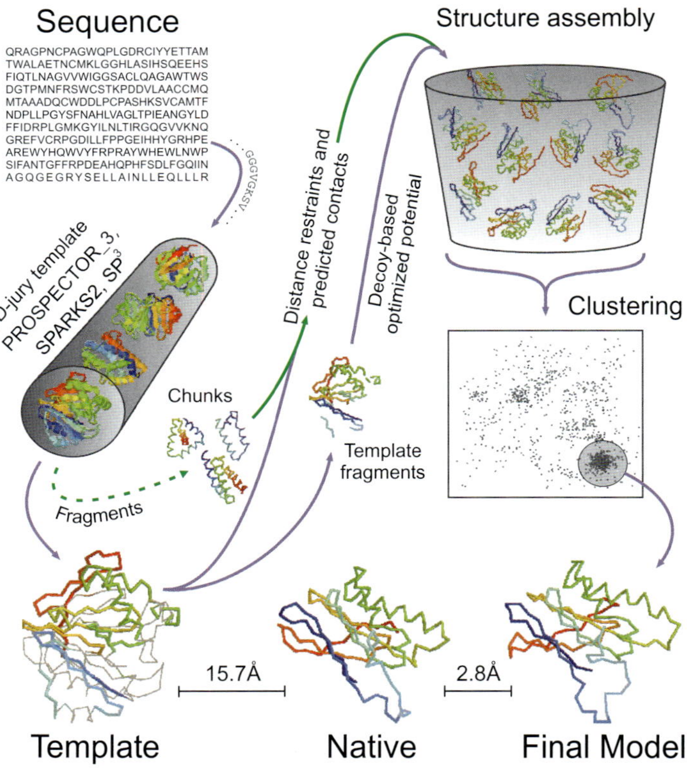

FIGURE 10.1 Overview of the TASSER/chunk-TASSER methodology.

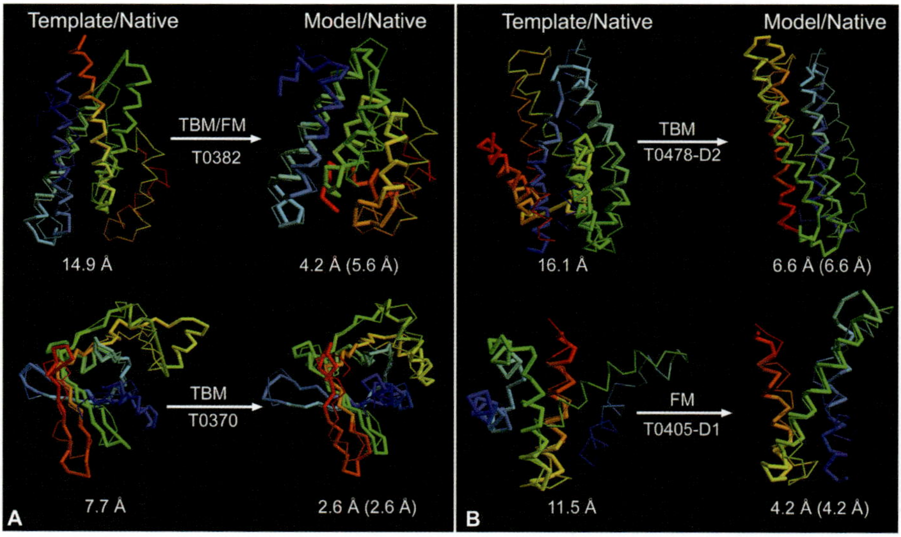

FIGURE 10.7 Successful examples of TASSER modeling in CASP7 and CASP8 are shown in (A) and (B), respectively. For each target, on the left is the superposition of the threading template (thick backbone) and native (thin backbone); on the right is the final model (thick backbone) and native (thin backbone). Blue to red goes from the N- to the C-terminus. The numbers below the superposition are the RMSD over the aligned regions and RMSD over the full-length molecule, respectively.

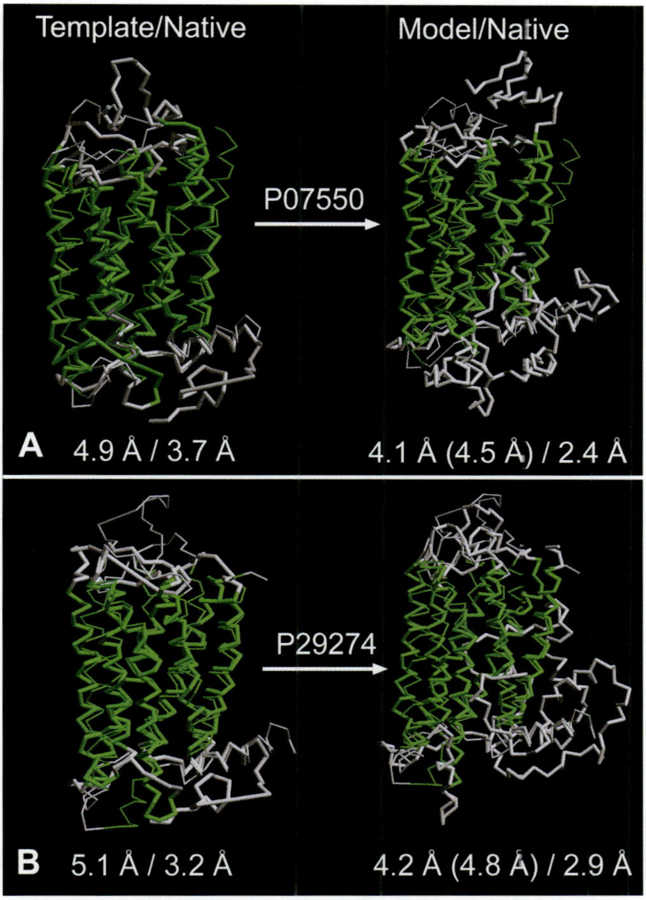

FIGURE 10.8 (A) Comparison between the human β2- aadregenic receptor best threading template/model with its native tertiary structure. (B) Comparison between the human A2A adenosine receptor best threading template/model to its native tertiary structure. For each protein, on the left is the superposition of the threading template (thick backbone) and native (thin backbone); on the right is the final model (thick backbone) and native (thin backbone). Transmembrane regions are shown in green. The numbers below the superposition are the RMSD over the aligned region and the RMSD over the transmembrane regions alone, respectively. The RMSD over the full-length protein is shown in parenthesis.

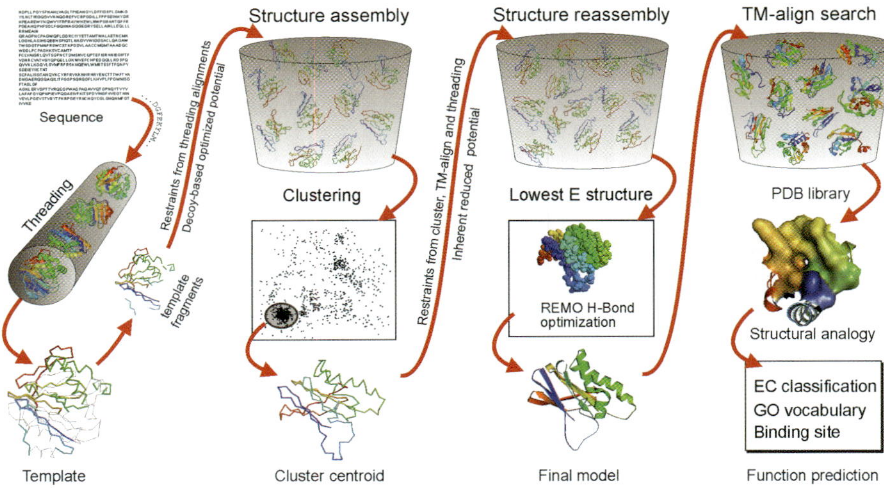

FIGURE 11.1 A schematic diagram of the I-TASSER [35,44,45] protein structure and function prediction protocol. Templates for the query protein are first identified by MUSTER [29] or LOMETS [33], which provide template fragments and spatial restraints. Template fragments are then assembled by modified replica-exchange Monte-Carlo simulations. The conformations generated during the simulation are clustered using SPICKER [48], in order to identify the structure with the lowest free energy. As an iterative refinement strategy, the cluster centroids are then subjected to the second round of simulation for refining the global topology and removing clashes. The final all-atom models are generated by REMO through the optimization of hydrogen-bonding networks [49]. Finally, functional homologs (protein structures with an associated EC number or GO terms) of final models are identified by using a sequence-independent structural alignment tool of TM-align [50] by ranking the hits based on their TM-score [51], RMSD and sequence identity in the structurally aligned region, coverage of the structural alignment, and confidence score (C-score [45]) of the model.

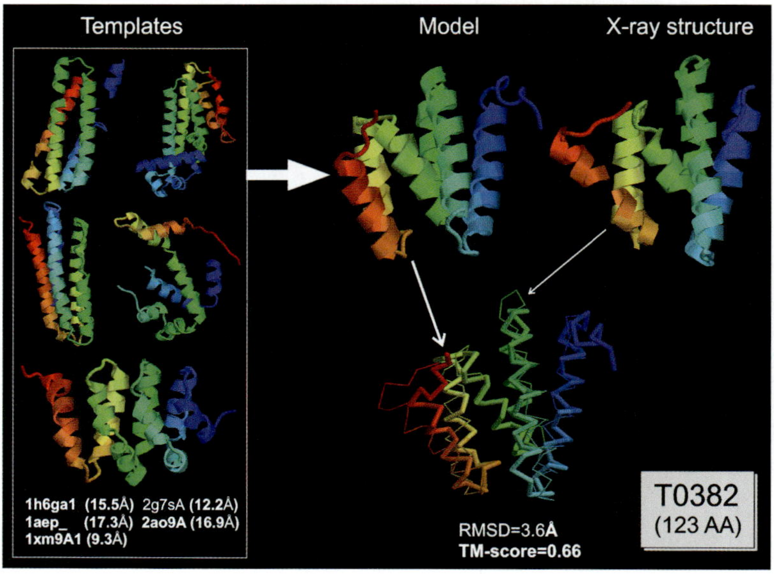

FIGURE 11.7 Structure comparisons of the threading templates, the final I-TASSER model, and the experimental structures for the CASP7 target T0382. Blue to red runs from N- to C-terminals (From Zhang [44]).

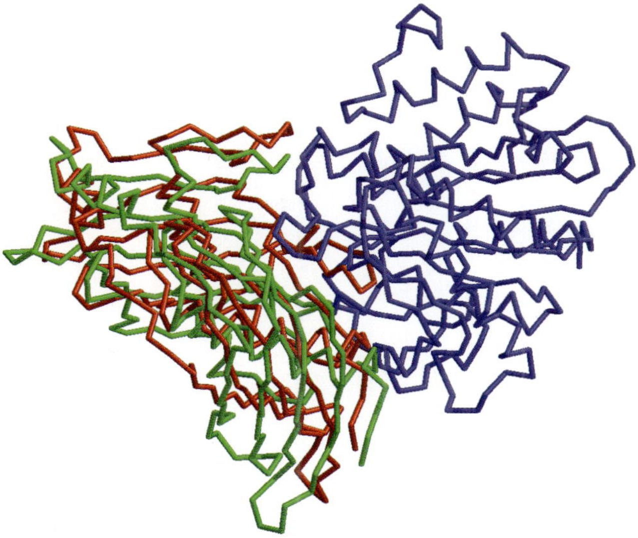

FIGURE 12.2 The crystal structure of the latexin (green)—CPA1 (blue) complex, overlaid with the orientation of the latexin molecule revealed by the top-scoring docked structure (red).

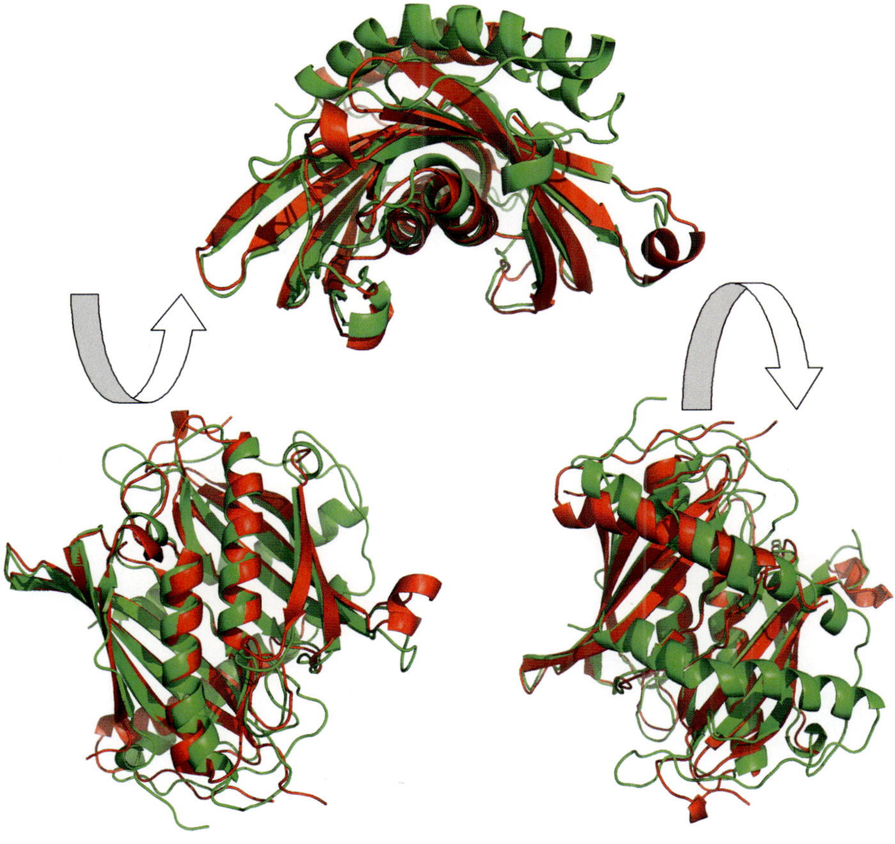

FIGURE 12.3 Multiple views of the superimposition of the predicted Acot7 model using chemical cross-linking and molecular modeling (green) with the crystal structure of ACOT12 (red). The structure suggests that the two hotdog fold domains adopt a similar structure to a dimer of single hotdog domains from other species.

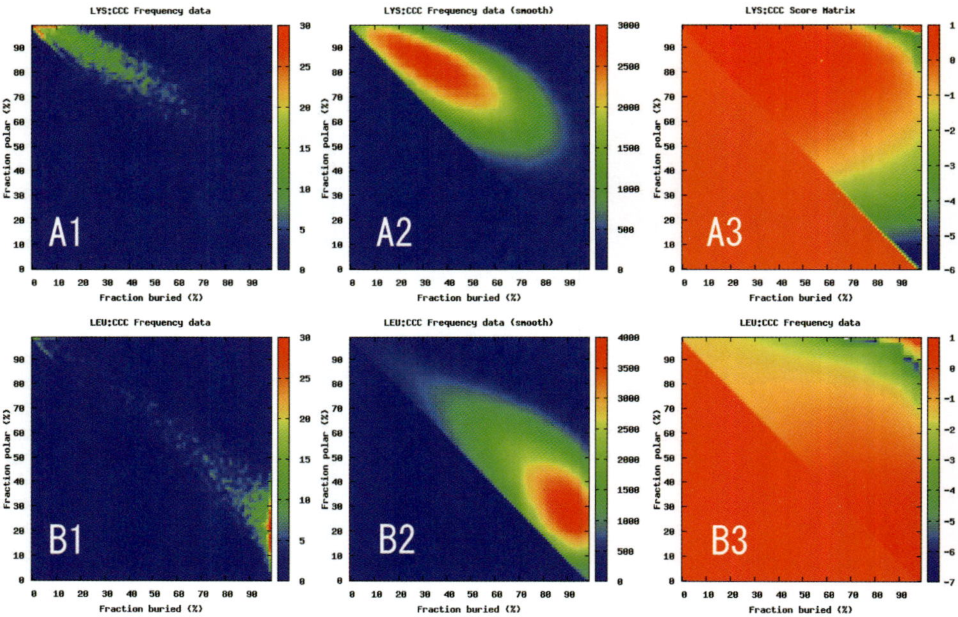

FIGURE 14.1 Matrices data used in the scoring function of CIRCLE. Sets (A1, B1) show the frequencies of residues LYS and LEU observed in the PDB dataset, respectively, as described by Equation 14.6. Sets (A2, B2) show converted data from the A1, B1 matrices using Gaussian weightings as described by Equation 14.7. Sets (A3, B3) show the scoring matrices of LYS and LEU according to the environment of the side chains as described by Equation 14.9.

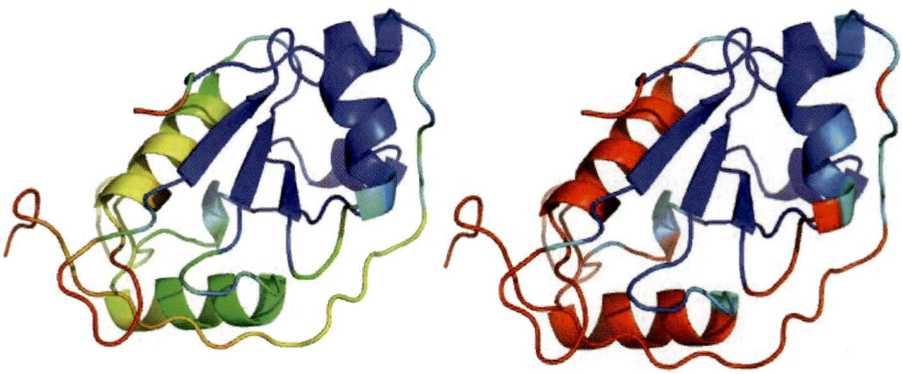

FIGURE 15.6 The ModFOLDclust predicted per-residue quality (left) for a model of CASP8 target T0389 is compared with the observed quality obtained from the alignment to the native structure (right). Each image was rendered using Pymol (http://www.pymol.org). The colors represent the residue accuracy according to the temperature scheme (blue indicates residues closest to the native structure; red, those furthest from the native structure). The blue regions that are predicted to be correct (left) are shown to be correct according to the observed accuracy (right); however, the incorrect regions are less accurately predicted.

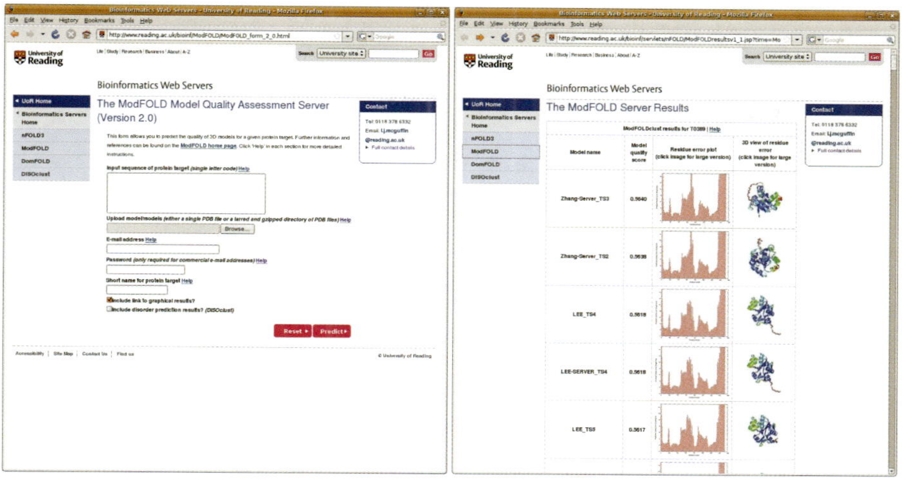

FIGURE 15.7 Screen shots of the ModFOLD server version 2.0 (http://www.reading.ac.uk/bioinf/ModFOLD/). The web interface (left) allows users to upload either single or multiple models for quality assessment. The graphical results page (right) can be accessed by following a link in the results email. Per-residue accuracy plots are provided for each model along with color-coded graphical depictions. Users may also download PDB files for their models with the predicted residue errors listed in the B-factor column.

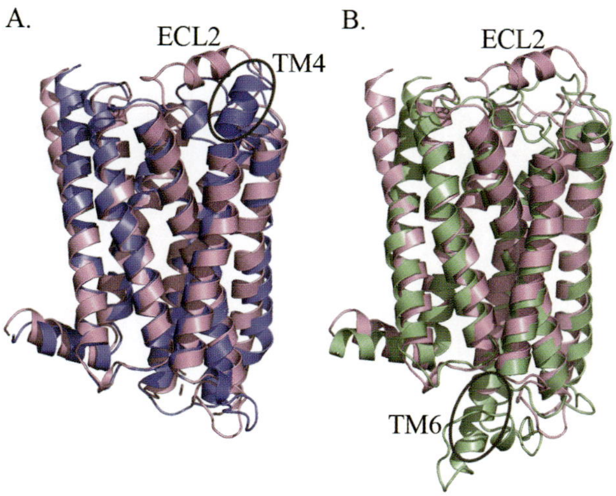

FIGURE 17.2 Performance of Rosetta and TASSER in predicting the structure of the human β2-adrenergic receptor. In both panels, the crystal structure of the β2-adrenergic receptor [45] is shown as pink ribbons; the cytoplasmic region is downward and the short helix in ECL2 is marked. The Rosetta model (Barth and Baker, unpublished results) (purple) and the best TASSER model [44] (green) are superimposed on the native structure in panels A and B, respectively. The prolonged segments in predicted TM4 and TM6 are marked, along with ECL2 of the native structure.

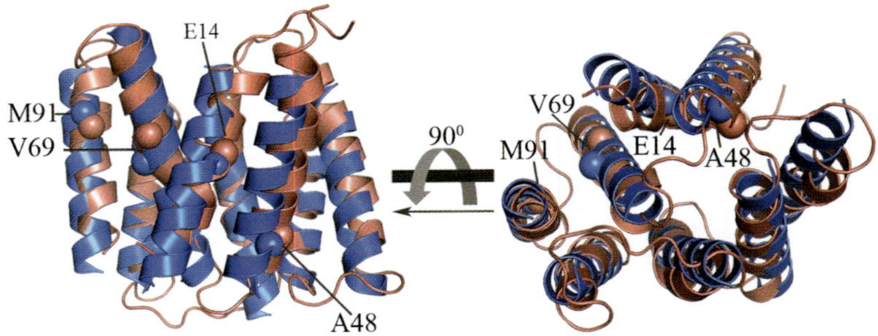

FIGURE 17.5 Model versus crystal structure of EmrE. A model of the EmrE homodimer (blue) was derived using a cryo-EM map and the computational approach of Fleishman et al. [95]. The crystal structure of EmrE (pink) was solved later [107]. The model and structure are aligned and viewed from the side (left) and top (right). The 3D location of specific Cα atoms (marked as spheres on one monomer of the model and structure) demonstrate the similarity between the model and native structure.

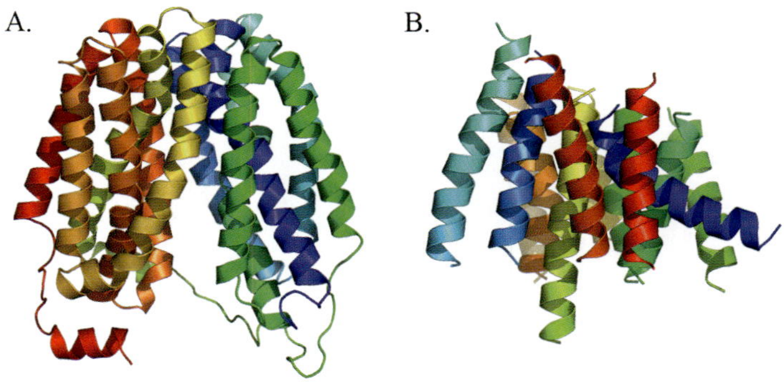

FIGURE 17.6 Comparison of the lacY model, produced via experimental constraints, and the solved crystal structure. In both panels the cytoplasmic side points downward. The lacY crystal structure (A) [112] and computational model (B) [98] are colored by rainbow. Although the overall fold and helix organization are quite distinct, there are regions of similarity, especially between the helices that contribute to the cytoplasmic-facing pore.

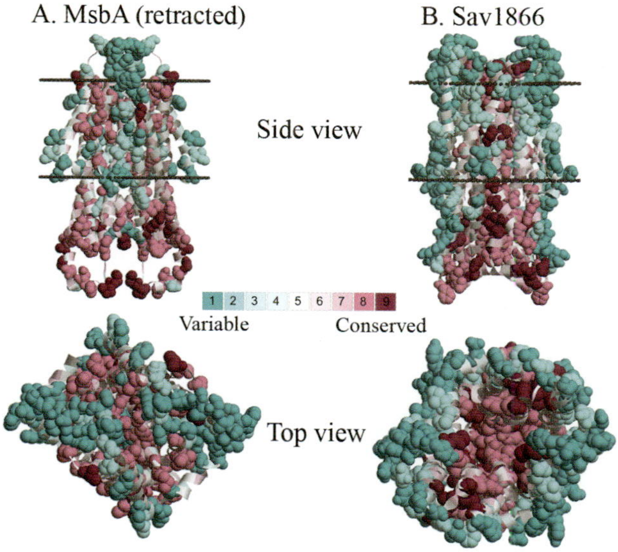

FIGURE 17.7 Conservation analysis of erroneous and correct structures of ABC transporters. The retracted structure of MsbA [131] (panel A) and the structure of sav1866 [133] (panel B) are colored according to conservation, using the ConSurf color scale [134]. Highly conserved residues, receiving grades of 8 or 9, along with the outermost variable (grades of 1 or 2) are shown as spheres. The two upper panels show a side view of the two proteins with their cytoplasmic sides facing down. Approximated membrane boundaries are shown in gray. The nucleotide-binding cytoplasmic domains of both MsbA and sav1866 were omitted for clarity. The two lower panels show a top (and closer) view of the same proteins.

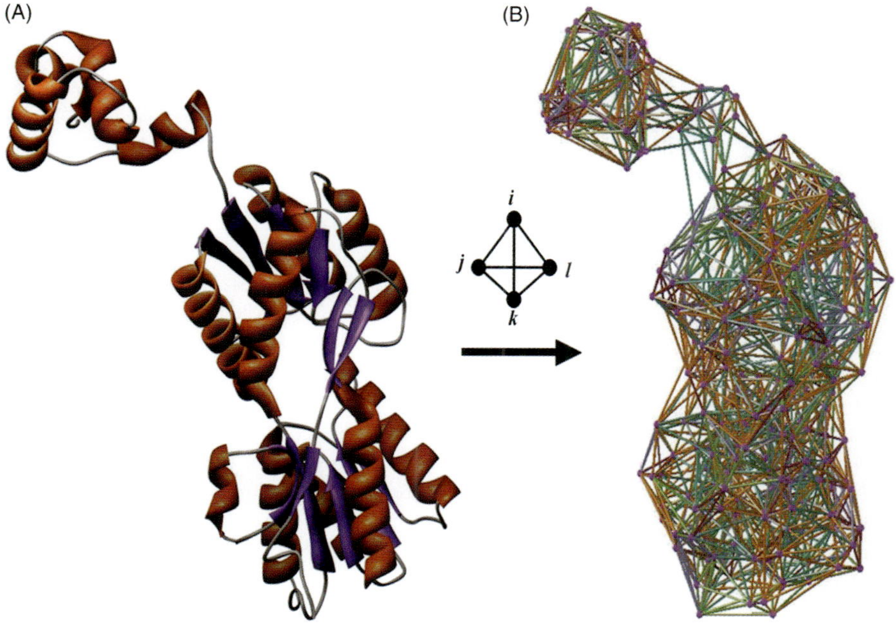

FIGURE 18.2 (A) Ribbon diagram of a single chain of the lac repressor homotetramer. (B) Delaunay tessellation of the same monomer of lac repressor, subject to a 12 Angstrom edge-length filter. PDB accession file: 1efa, chain B.

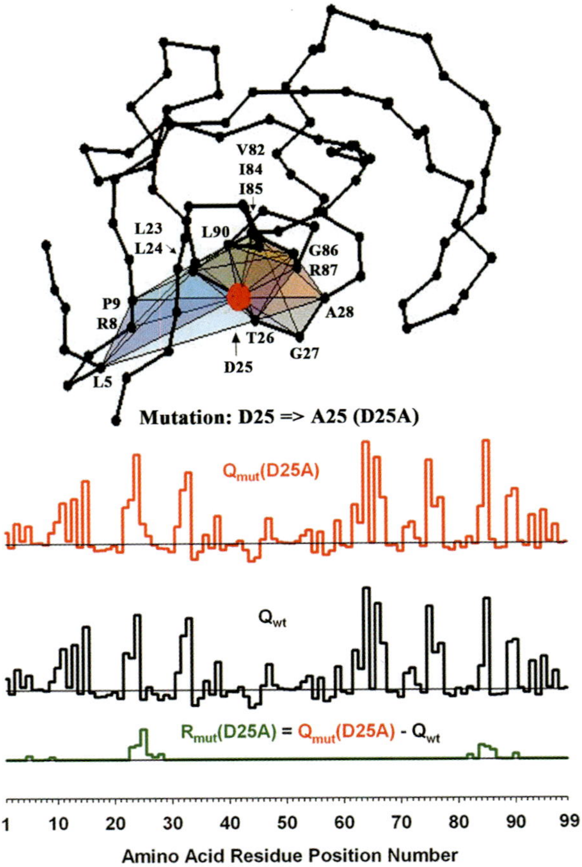

$EC_5 = 0.43$, $EC_8 = 0.01$, $EC_9 = 0.31$, $EC_{23} = 1.63$, $EC_{24} = 1.87$, $EC_{25} = 3.83$, $EC_{26} = 0.67$, $EC_{27} = 0.34$, $EC_{28} = 0.67$, $EC_{82} = 0.49$, $EC_{84} = 1.76$, $EC_{85} = 1.52$, $EC_{86} = 1.14$, $EC_{87} = 4.60 \times 10^{-5}$, $EC_{90} = 0.63$, $EC_i = 0$ for all other positions i

FIGURE 18.3 Representation of the D25A mutation in HIV-1 protease (PDB ID: 3phv). Top: A Cα trace of HIV-1 protease and a subset of its Delaunay tessellation highlighting only the tetrahedral simplices that all share the point representing residue D25 as a vertex (Cα enlarged and colored red). All nearest neighbor residues to D25, whose Cα coordinates participate as the additional vertices in these simplices, are labeled on the trace. Bottom: The 3D-1D potential profiles (Q) of the mutant and wild type protein are shown in red and black, respenctively. Their difference is the mutant residual profile (R), shown in green, whose components are EC scores. $EC_{25} = 3.83$ is the residual score of the D25A HIV-1 protease mutant and provides an empirical scalar measure of the change in sequence-structure compatibility relative to the native protein.

FIGURE 19.2 One hundred forty-four conformations computed in Reference [74] are superimposed in transparent over the first one (in opaque) of ubiquitin. The ensemble is obtained from the Protein Data Bank (PDB) under id 2nr2. Reproduced with permission of Michele Vendruscolo.

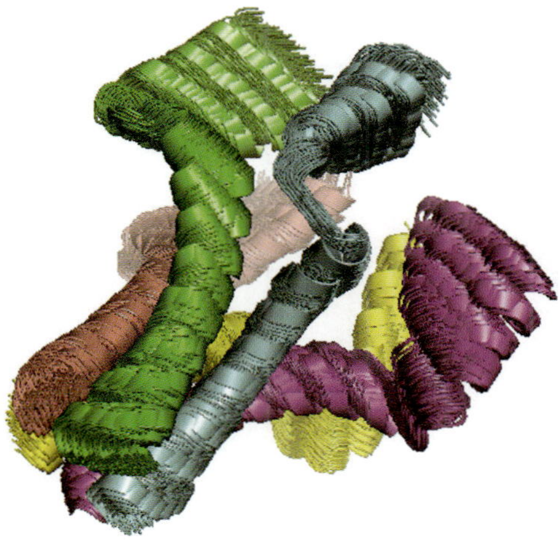

FIGURE 19.3 One hundred eighty-four phospholamban conformations (under id 2hyn in the PDB) computed in Reference [77] are shown superimposed over one another. The five monomers of the complex are shown in different colors. Courtesy of Chris Bailey-Kellogg.

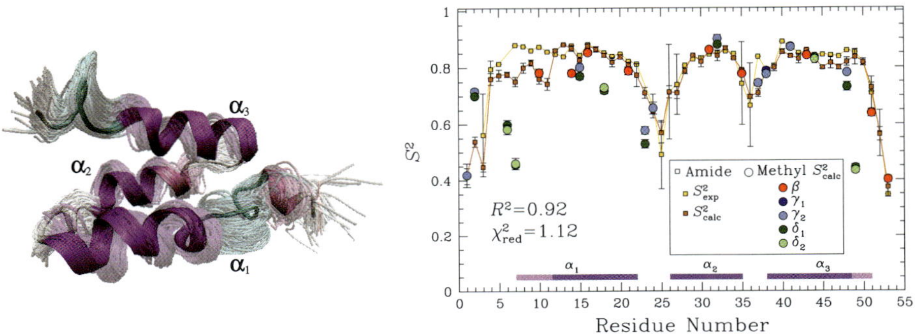

FIGURE 19.4 Left: The lowest-energy conformations computed with the method described in Reference [81] are drawn in transparent over the opaque X-ray structure of ALB8-GA. Right: Amide S^2_{calc} data (orange squares) calculated over the ensemble are compared with the available NMR S^2_{exp} data (yellow squares). Methyl S^2_{calc} data are shown in colored circles (no NMR data are available for comparison). Horizontal bars on the x-axis show the position of the three α-helices (also annotated over the ensemble). The parts of the bars in lighter colors indicate amino acids found in unfolded configurations. Reprinted from Reference [81] with permission.

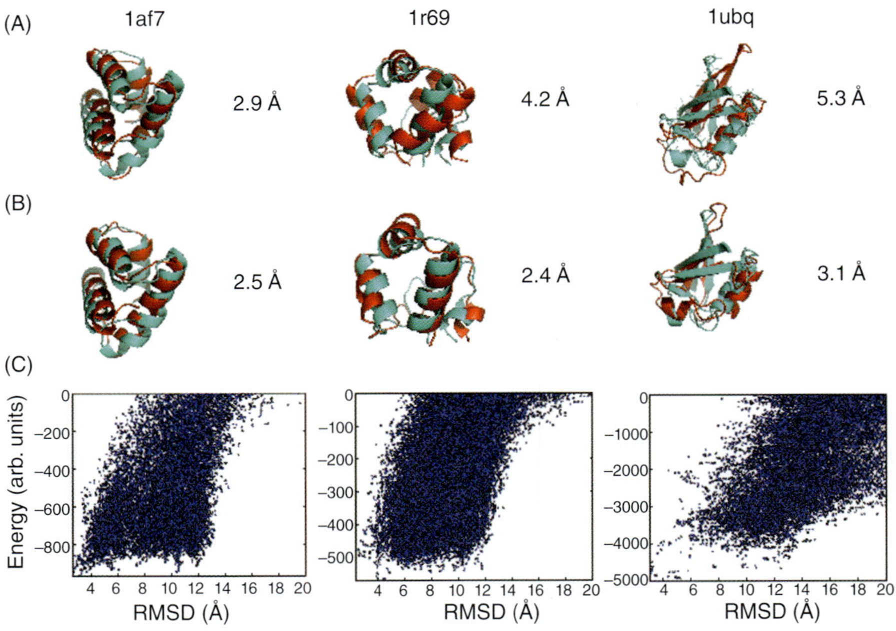

FIGURE 19.5 Alignments of the predictions generated in Reference [45] (in red) have (A) the lowest energy and (B) the lowest least root-mean-squared deviation (lRMSD) to the native structure (blue) for three proteins (PDB codes at the top, lRMSD values indicated). Images are created using the PyMol visualization software. (C) Scatter plots of lRMSD versus energy. Courtesy of Tobin Sosnick.

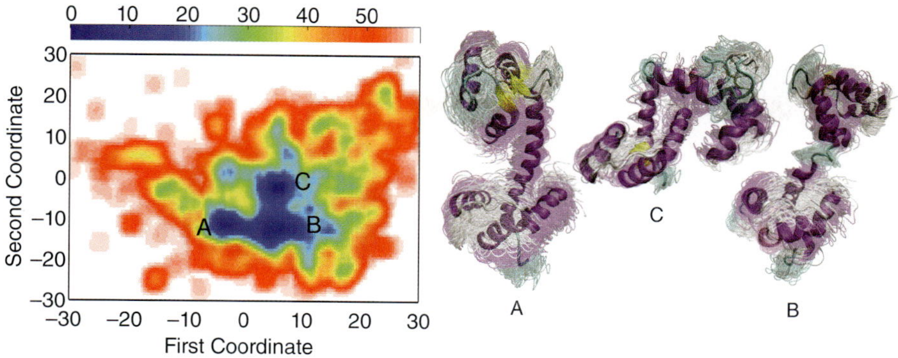

FIGURE 19.6 Left: The 2D energy pseudo-free energy landscape obtained for the calmodulin sequence with the method described in Reference [42] is shown in a red to blue color scheme that denotes high-to-low energy values. The deepest minima are labeled A, B, and C. Right: Computed conformational ensembles that correspond to the minima are shown and labeled accordingly. Conformations are superimposed in transparent over lowest-energy ones drawn in opaque. Reproduced from Reference [42] with permission.

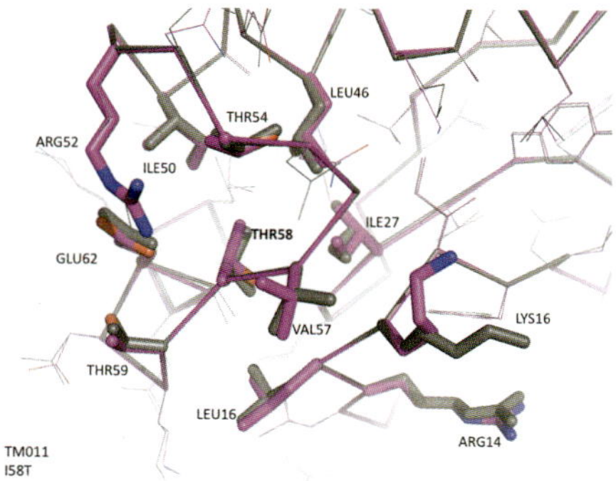

FIGURE 20.1 Comparison of computational model and crystal structure of single mutation I58T. The crystal structure of the mutant is shown in gray color, and the model in magenta.

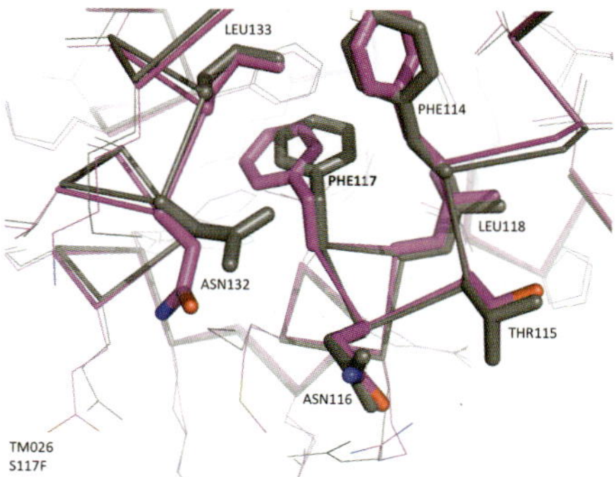

FIGURE 20.2 Comparison of computational model and crystal structure of single mutation S117F.

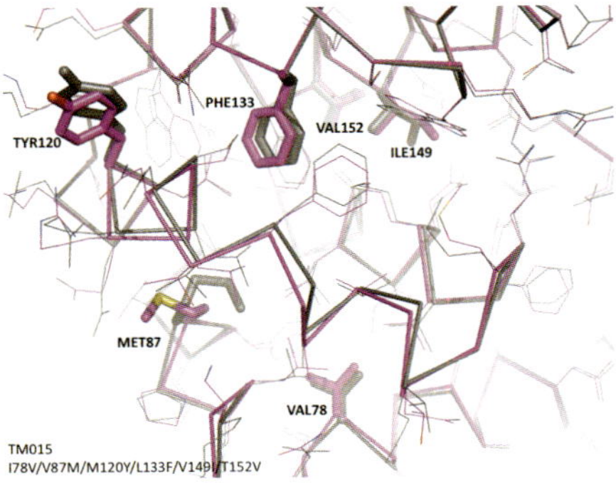

FIGURE 20.3 Comparison of computational model and crystal structure of multiple mutation I78V V87M M120Y L133F V149I T152V.

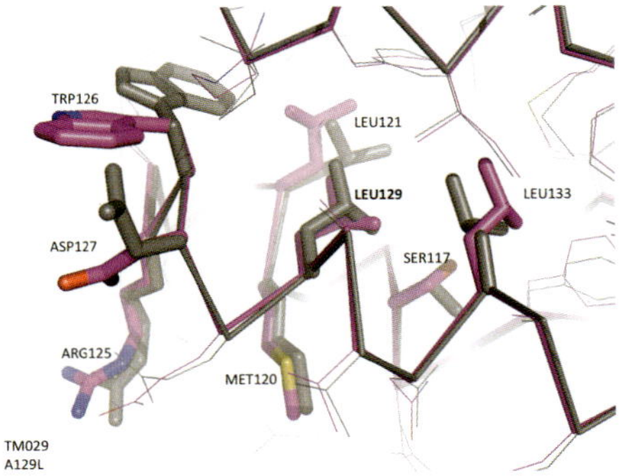

FIGURE 20.4 Comparison of computational model and crystal structure of single mutation A129L.

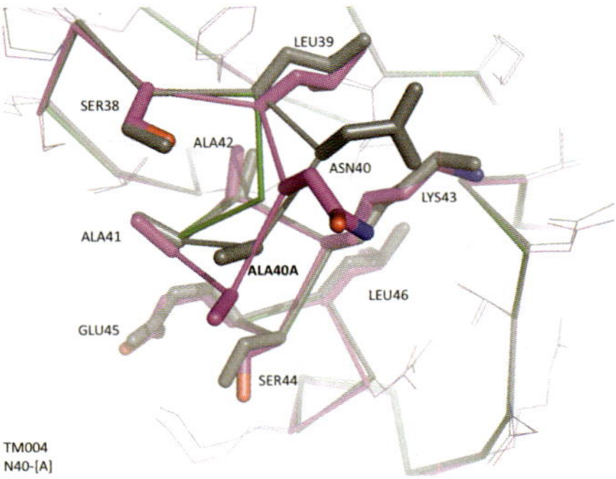

FIGURE 20.5 Comparison of computational model and crystal structure of single insertion mutation N40-[A]. The backbone of the wild-type protein is shown in green.

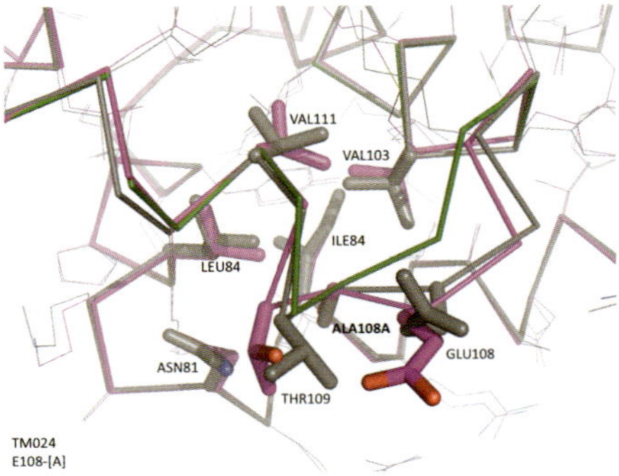

FIGURE 20.6 Comparison of computational model and crystal structure of mutation E108-[A].

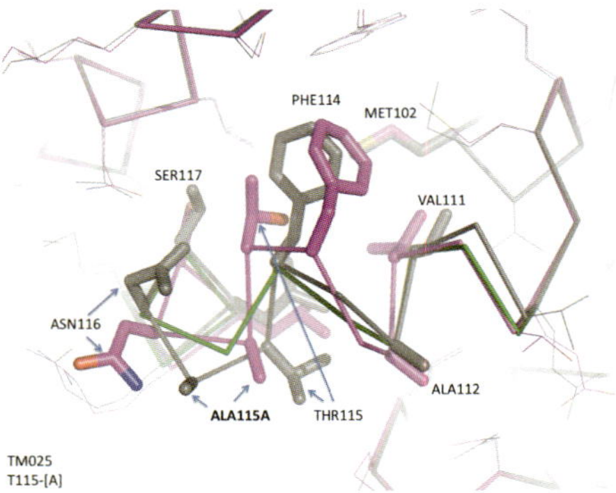

FIGURE 20.7 Comparison of computational model and crystal structure of mutation E115-[A].

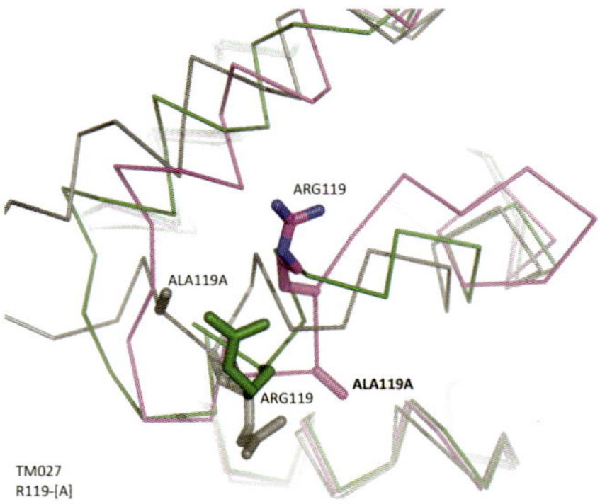

FIGURE 20.8 Comparison of computational model and crystal structure of single insertion mutation R119-[A].

Medical Research Council Honorary Research Fellow, and NED is an ARC Professorial Fellow.

REFERENCES

1. K.T. Simons, C. Kooperberg, E. Huang, and D. Baker. Assembly of protein tertiary structures from fragments with similar local sequences using simulated annealing and Bayesian scoring functions. *Journal of Molecular Biology*, 268:209–225, 1997.
2. Z. Yang, K.A. Adrian, and S. Jeffrey. TASSER: An automated method for the prediction of protein tertiary structures in CASP6. *Proteins: Structure, Function, and Bioinformatics*, 61:91–98, 2005.
3. G.R. Smith and M.J.E. Sternberg. Prediction of protein-protein interactions by docking methods. *Current Opinion in Structural Biology*, 12:28–35, 2002.
4. C.D. Putnam, M. Hammel, G.L. Hura, and J.A. Tainer. X-ray solution scattering (SAXS) combined with crystallography and computation: Defining accurate macromolecular structures, conformations and assemblies in solution. *Quarterly Reviews of Biophysics*, 40:191–285, 2007.
5. C. Neylon. Small angle neutron and X-ray scattering in structural biology: Recent examples from the literature. *European Biophysics Journal*, 37:531–541, 2008.
6. E.R. Goedken, S.L. Kazmirski, G.D. Bowman, M. O'Donnell, and J. Kuriyan. Mapping the interaction of DNA with the *Escherichia coli* DNA polymerase clamp loader complex. *Nature Structural & Molecular Biology*, 12:183–190, 2005.
7. S. Popp, L. Packschies, N. Radzwill, K.P. Vogel, H.-J. Steinhoff, and J. Reinstein. Structural dynamics of the DnaK-peptide complex. *Journal of Molecular Biology*, 347:1039–1052, 2005.
8. K. Ozawa, M.J. Headlam, D. Mouradov, S.J. Watt, J.L. Beck, K.J. Rodgers, R.T. Dean, T. Huber, G. Otting, and N.E. Dixon. Translational incorporation of L-3,4-dihydroxyphenylalanine into proteins. *FEBS Journal*, 272:3162–3171, 2005.
9. A.M. Raitsimring, C. Gunanathan, A. Potapov, I. Efremenko, J.M.L. Martin, D. Milstein, and D. Goldfarb. Gd3+ complexes as potential spin labels for high field pulsed EPR distance measurements. *Journal of the American Chemical Society*, 129:14138–14139, 2007.
10. M.J. Hunter and M.L. Ludwig. Reaction of imidoesters with proteins and related small molecules. *Journal of the American Chemical Society*, 84:3491–3504, 1962.
11. J.B. Swaney, J. Segrest, P., and J.J. Albers. Use of cross-linking reagents to study lipoprotein structure. *Methods in Enzymology*, 128:613–626, 1986.
12. P. Friedhoff. Mapping protein–protein interactions by bioinformatics and cross-linking. *Analytical and Bioanalytical Chemistry*, 381:78–80, 2005.
13. S.C. Liu and J. Palek. Metabolic dependence of protein arrangement in human erythrocyte membranes. II. Crosslinking of major proteins in ghosts from fresh and ATP-depleted red cells. *Blood*, 54:1117–1130, 1979.
14. K.M. Pearson, L.K. Pannell, and H.M. Fales. Intramolecular cross-linking experiments on cytochrome c and ribonuclease A using an isotope multiplet method. *Rapid Communications in Mass Spectrometry*, 16:149–159, 2002.

15. G.H. Dihazi and A. Sinz. Mapping low-resolution three-dimensional protein structures using chemical cross-linking and Fourier transform ion-cyclotron resonance mass spectrometry. *Rapid Communications in Mass Spectrometry*, 17:2005–2014, 2003.

16. G.H. Kruppa, J. Schoeniger, and M.M. Young. A top down approach to protein structural studies using chemical cross-linking and Fourier transform mass spectrometry. *Rapid Communications in Mass Spectrometry*, 17:155–162, 2003.

17. X.H. Chen, Y.H. Chen, and V.E. Anderson. Protein cross-links: Universal isolation and characterization by isotopic derivatization and electrospray ionization mass spectrometry. *Analytical Biochemistry*, 273:192–203, 1999.

18. G.J. King, A. Jones, B. Kobe, T. Huber, D. Mouradov, D.L. Hume, and I.L. Ross. Identification of disulfide-containing chemical cross links in proteins using MALDI-TOF/TOF-mass spectrometry. *Analytical Chemistry*, 80:5036–5043, 2008.

19. D.R. Muller, P. Schindler, H. Towbin, U. Wirth, H. Voshol, S. Hoving, and M.O. Steinmetz. Isotope tagged cross linking reagents. A new tool in mass spectrometric protein interaction analysis. *Analytical Chemistry*, 73:1927–1934, 2001.

20. J.W. Back, V. Notenboom, L.J. de Koning, A.O. Muijsers, T.K. Sixma, C.G. de Koster, and L.Z. de Jong. Identification of cross-linked peptides for protein interaction studies using mass spectrometry and ^{18}O labeling. *Analytical Chemistry*, 74:4417–4422, 2002.

21. W. Zheng and S. Doniach. Protein structure prediction constrained by solution X-ray scattering data and structural homology identification. *Journal of Molecular Biology*, 316:173–187, 2002.

22. W. Zheng and S. Doniach. Fold recognition aided by constraints from small angle X-ray scattering data. *Protein Engineering, Design and Selection*, 18:209–219, 2005.

23. J. Lees and R. Janes. Combining sequence-based prediction methods and circular dichroism and infrared spectroscopic data to improve protein secondary structure determinations. *BMC Bioinformatics*, 9:24, 2008.

24. G.F. Schröder and H. Grubmüller. FRETsg: Biomolecular structure model building from multiple FRET experiments. *Computer Physics Communications*, 158:150–158, 2004.

25. N. Alexander, A. Al-Mestarihi, M. Bortolus, H. McHaourab, and J. Meiler. De novo high-resolution protein structure determination from sparse spin-labeling EPR data. *Structure*, 16:181–196, 2008.

26. Y. Hamuro, L.L. Burns, J.M. Canaves, R.C. Hoffman, S.S. Taylor, and V.L. Woods. Domain organization of -AKAP2 revealed by enhanced deuterium exchange-mass spectrometry (DXMS). *Journal of Molecular Biology*, 321:703–716, 2002.

27. A.J.K. Kamal and M.R. Chance. Modeling of protein binary complexes using structural mass spectrometry data. *Protein Science*, 17:79–94, 2008.

28. D. Mouradov, A. Craven, J.K. Forwood, J.U. Flanagan, R. Garcia-Castellanos, F. X. Gomis-Ruth, D.A. Hume, J.L. Martin, B. Kobe, and T. Huber. Modelling the structure of latexin-carboxypeptidase A complex based on chemical cross-linking and molecular docking. *Protein Engineering, Design and Selection*, 19:9–16, 2006.

29. J.K. Forwood, A.S. Thakur, G. Guncar, M. Marfori, D. Mouradov, W. Meng, J. Robinson, T. Huber, S. Kellie, J.L. Martin, D.A. Hume, and B. Kobe. Structural basis for recruitment of tandem hotdog domains in acyl-CoA thioesterase 7 and

its role in inflammation. *Proceedings of the National Academy of Sciences U S A*, 104:10382–10387, 2007.

30. M.M. Young, N. Tang, J.C. Hempel, C.M. Oshiro, E.W. Taylor, I.D. Kuntz, B.W. Gibson, and G. Dollinger. High throughput protein fold identification by using experimental constraints derived from intramolecular cross-links and mass spectrometry. *Proceedings of the National Academy of Sciences U S A*, 97:5802–5806, 2000.
31. D.M. Schulz, S. Kalkhof, A. Schmidt, C. Ihling, C. Stingl, K. Mechtler, O. Zschoernig, and A. Sinz. Annexin A2/P11 interaction: New insights into annexin A2 tetramer structure by chemical crosslinking, high-resolution mass spectrometry, and computational modeling. *Proteins: Structure Function, and Bioinformatics*, 69:254–269, 2007.
32. M. Albrecht, D. Hanisch, R. Zimmer, and T. Lengauer. Improving fold recognition of protein threading by experimental distance constraints. *In Silico Biology*, 2:325–337, 2002.
33. A.N. Nickolai, N. Ruth, and Z.M. Ralf. Fast protein fold recognition via sequence to structure alignment and contact capacity potentials. In *Pacific Symposium on Biocomputing '96*, pp. 53–72. Singapore: World Scientific Publishing Co, 1996.
34. B. Corry, P. Rigby, Z.-W. Liu, and B. Martinac. Conformational changes involved in MscL channel gating measured using FRET spectroscopy. *Biophysical Journal*, 89:L49–L51, 2005.
35. Y. Xu, D. Xu, O.H. Crawford, J.R. Einstein, and E. Serpersu. Protein structure determination using protein threading and sparse NMR data (extended abstract). *Proceedings of the Fourth Annual International Conference on Computational Molecular Biology*. Tokyo, Japan: ACM Press, 2000.
36. R.X. Wang, Y.P. Lu, and S.M. Wang. Comparative evaluation of 11 scoring functions for molecular docking. *Journal of Medicinal Chemistry*, 46:2287–2303, 2003.

CHAPTER 13

MODELING LOOPS IN PROTEIN STRUCTURES

NARCIS FERNANDEZ-FUENTES
Leeds Institute of Molecular Medicine
University of Leeds
Leeds, UK

ANDRAS FISER
Department of Systems and Computational Biology
Department of Biochemistry
Albert Einstein College of Medicine
Bronx, NY

13.1. INTRODUCTION

The first three-dimensional (3D) structures of proteins [1,2] confirmed the existence of two dominant conformations in protein structures, α-helices and β-strands, just as predicted earlier by Linus Pauling, Robert Corey, and Herman Brason [3]. These regions were characterized by their translational symmetry, which is supported by conserved bond patterns between backbone hydrogen bond donors and acceptors. However, certain parts of protein structures that connect α-helices and β-strands do not follow repetitive patterns; these are the loop regions.

Originally, loops were considered to have an "irregular conformation" and sometimes misnamed as "random coils." Due to their flexibility and nonperiodic nature, loops long escaped structural classification. It was not until 1968 when loops were first classified, by introducing three categories of four residue β-turns that follow a conserved structural pattern [4]. As more structures were solved, new structurally conserved patterns were found, and the classifications were subsequently refined and extended [5–7]. Three residue chain reversals,

Introduction to Protein Structure Prediction: Methods and Algorithms,
Edited by Huzefa Rangwala and George Karypis

termed γ-turns, first identified in 1972 by Matthews [8], were also classified [9,10]. Analysis of ββ loops connecting two adjacent antiparallel strands revealed specific preferences for certain β-turn families and allowed the definition of novel, common substructures [11–13].

As the number of solved protein structures grew and more and more loop conformations became available, it was possible to find structural patterns in longer loops, which are geometrically less restricted. Several studies identified commonly occurring structural families, often accompanied by sequence patterns, for αα, αβ, βα, and ββ arches [14–19]. More general classifications of short to medium loop lengths were also proposed [20–24]. A recent release of ArchDB database (http://sbi.imim.es/archdb/) [25,26] contains more than 100,000 classified loops with lengths ranging from 1 to 36 residues long.

Loops are common in protein structures and are often essential for both function and structure. In 1986, the composition of 60 new protein structures was analyzed, and it was found that 47% of the residues were located in loop regions [27]. Updating these figures in the current Protein Data Bank (PDB, 54,076 structures as of November 2008), the percentage of residues that are found in loops, α-helices, and β-strands are 41%, 31%, and 29%, respectively. But loops are not only important because they represent a substantial part of protein structures; they also often play an essential role in protein function, as illustrated in the next section of the chapter. Yet loops are often not visible in experimental solved protein structures either because the solvent exposed part of the structure is not well defined or because the experimental preparation of protein structures required the cleavage of certain loops. In addition, in the case of structure modeling studies, the core of the protein is often well conserved within the same protein family and therefore can be modeled with high accuracy using comparative or homology modeling techniques. However, these related structures often differ in their function or specificity, and these differences frequently materialize as structural differences in the loop regions. Therefore, identifying, predicting the most likely conformation of loops at atomic details, that is, modeling of loops in incomplete experimental structures, or adding or improving the prediction of loop conformations in models produced by computational modeling is critical to properly understand proteins both at functional and structural levels.

13.2. STRUCTURAL AND FUNCTIONAL ROLES OF LOOPS IN PROTEINS

13.2.1. Loops and Protein Function

Loops often play a pivotal role in protein function. There are numerous examples in the scientific literature relating loops and protein function. Figure 13.1 shows some examples of functional important loops such as the complementarity-determining regions (CDRs) [28], calcium-binding loops

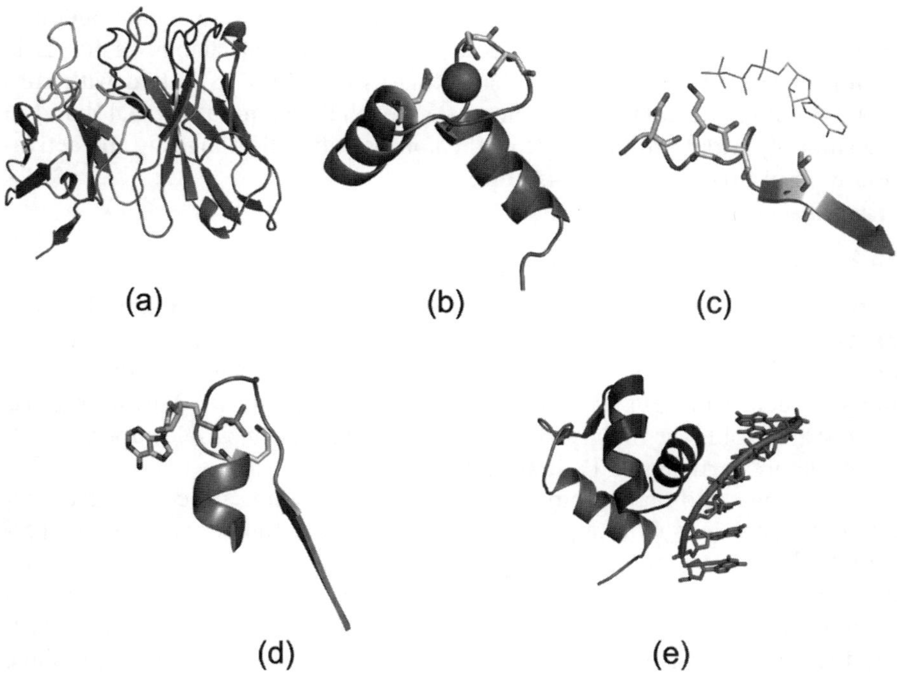

FIGURE 13.1 Functionally important loop regions. Some examples of functional loops are shown: (a) CDRs of immunoglobulins; (b) EF hand; (c) catalytic loop of Ser/Thr kinases; (d) p-loop; (e) helix-turn-helix DNA interaction motif.

(EF hands) [29], catalytic loop in Ser/Thr kinases [30], the p-loop [31], and the helix-turn-helix DNA interaction motif [32]. Other examples include loops involved in signaling processes [33,34], protein–protein interactions [35,36], cofactors (NAD(P)-binding loops) [37], glycine-rich loop [38], and catalytic loops such as the ones in serine proteases [39].

Loops are often the place where functional differences of otherwise homologous proteins are manifested. Functional differences can arise as a consequence of amino acid substitutions, insertions, and deletions that typically happen in loops. As a result, loops often determine the functional specificity of a given protein fold by introducing structural variability to the active and binding sites [40].

On the other hand, enzyme function often involves conformational changes in protein structures. Flexibility of loops plays a vital role in correctly positioning catalytically important residues. Motions range from a simple bending and stretching of bonds to subunit rotations and translations. There are several examples of loops that are involved in protein motion to modulate function. For instance, so-called triggering loops were identified, whose conformational change is required for the catalytic process of β1,4-galactosyltransferase [41]. Zgiby et al. described a case where the flexibility of a loop was important for the correct functioning of the class II fructose 1,6-bisphosphate aldolase (FBPA) [42]. It is well known that the conformational change in the LID region of nucleotide monophosphate kinases (NMPKs) is linked with substrate binding and catalysis and in triosephosphate isomerases [43,44] and in the similarly widely studied case of the conformational changes in the activation loop of protein kinases [45,46].

13.2.2. Loops and Protein Structure

Experimental and theoretical evidence suggests that local structural determinants are mostly encoded in short segments of the protein sequence; therefore, analyses of local sequence–structure relationships could significantly enhance the accuracy of protein structure prediction methods [47,48]. Various reports suggest that folds are mainly made up from a number of simple supersecondary structures articulated by loops [49–53].

Loops may also play an important role in the folding process of proteins. Single substitutions in a loop residue can destabilize the entire protein structure [54], whereas the stabilizing role played by surface loops has been described in eightfold α/β proteins [55]. Other studies have shown that in some proteins, the formation of a single loop is the rate-determining step in the folding process, while in others, a loop can misfold to serve as the hinge loop region for domain swapping [56]. The relationship between loop elongation and stability was demonstrated in a fibronectin type III domain [57]. Fersht reviewed the importance of loop–loop long-range interactions in the folding process of proteins [58]. An active role of loops was found in the folding/unfolding process of cytochrome c [59] and in phosphoglycerate kinases [60].

Loops are also important in the thermostability of proteins. The stability of proteins has been studied from two points of view: (1) to study and compare protein structures in the folded state and (2) to study the chain entropy and its reduction to favor the folded state. In both cases, the knowledge of loops is of great interest. In the first case, the stability of thermophilic proteins has been attributed, among other reasons, to the shortening of loops and an increased occurrence of proline residues in loops [61]. In connection with the second issue, Zhou reviewed the roles that loops play in entropy-based strategies, since loop length greatly affects the total value of chain entropy [62].

13.3. PREDICTION OF LOOP CONFORMATIONS

Despite recent technical improvements and the sharp reduction of price of structure determination at the atomic level using X-ray diffraction and nuclear magnetic resonance (NMR) spectroscopy [63], there is an ever-growing gap between the number of known protein sequences and structures. In the absence of an experimentally determined structure, *ab initio* and threading or comparative modeling methods can sometimes provide a useful 3D model. Currently, approximately 60% of all protein sequences can have at least one domain modeled on a related, known protein structure [64]; thus, their structures can be described computationally. In general, these methods tend to correctly predict the protein core when the structure of a close homolog of the target protein is available, but not the loop regions. Moreover, at least two-thirds of the comparative modeling cases are based on <40% sequence identity between the target and the templates, and thus generally require loop modeling [65]. Even at levels of overall sequence identity >35%, loops among the homologs vary while the core regions are still relatively conserved and aligned accurately [66].

Similar to the prediction of whole protein structures, there are two basic approaches in loop structure prediction: first, database search methods, also known as knowledge-based methods, and second, *ab initio* or conformational search methods. Finally, there are also procedures that combine the two basic approaches, that is, that use both database searches and *ab initio* predictions. Each of these approaches has its advantages and disadvantages, as described later in the chapter.

13.3.1. Knowledge-Based Approaches for Loop Modeling

The knowledge-based methods consist of finding a main-chain segment extracted from known protein structures that fits two stem regions of the loop. The stems are defined as the main-chain atoms that precede and follow the loop, but are not part of it, that is, are part of the core of the spanned secondary structures. The search is performed through a database of many known protein structures, not only homologs of the modeled protein. Usually, many different alternative segments that fit the stem residues are obtained and sorted according to geometric criteria or sequence similarity between the template and target loop sequences. The selected segments are then superposed and annealed on the stem regions.

One of the key issues in knowledge-based methods in loop structure prediction is the availability of suitable fragments or the so-called *database completeness*, and this issue is more obvious as the length of the target loop increases. In 1994, Fidelis et al. explored the relevance of database search methods in the loop structure prediction field [67]. They assessed the database completeness by studying the frequency of distribution of repeat conformations. All conformations related to each other by a root-mean-square deviation (RMSD) of 1 Å or less using Cα atoms were considered to have the same conformation. The sampling of different loop lengths was explored, and it was concluded that the database search method was useful only for short fragments (up to four residues long). Later, Lessel and Schomburg explored the completeness of fragments between length of 3 and 12 in the PDB using a clustering approach [68]. Fragments were grouped in clusters according to a similarity measure: two fragments were grouped together if the distance between the first and last

Cα atoms was smaller than 1.6 Å and the RMSD considering only Cα atoms was smaller than 0.8 Å. Lessel and Schomburg's conclusions supported the ones of Fidelis et al., as they found that only short fragments made of three and four residues were well sampled.

In recent years, the number of solved proteins has grown almost exponentially. This is due in part to advances in X-ray and NMR techniques and also to structural genomic efforts that aim to solve at least one structure for each protein family [63]. The question of database completeness was revisited in 2003 by Du et al. [69]. Their conclusion was more promising: they reported that even for long loops (15 residues), there is a >90% probability that a nonhomologous structure exists within 2 Å RMSD considering Cα atoms.

Fernandez-Fuentes and Fiser explored the question of what fraction of loops extracted from all known protein sequences, the so-called *sequence space*, are represented by loops extracted from protein structures, or *structure space* [70]. Fragments from structure space were structurally clustered after an all-to-all comparison, and sequence identity cutoffs assuring structural similarities were identified for each loop length. Next, all possible loop fragments from clusters of sequence space were matched with the sequences from structure space, and the coverage was assessed. The results showed that loops up to 10 residues had a related (i.e., at least 50% sequence identity) fragment in PDB, and despite the sixfold increase in sequence data bank size and the doubling of PDB since 2002, there was not a single unique loop conformation entered in the PDB or sequence segment observed that shares <50% sequence identity to a PDB fragment, which indicates that newly sequenced proteins keep recycling the same set of already known short structural segments. All sequence segments up to 10–12 residues have at least one corresponding structural segment that shares at least 50% identity, thus ensuring structural similarity, except a very few notable exceptions [70–73]. These results suggest that knowledge-based prediction methods are no longer limited by the completeness of fragments in data banks but rather by the effective scoring and search algorithms to locate them.

Knowledge-based loop structure prediction methods were initiated by Greer [74] and Jones and Thirup [75] in the 1980s. As reviewed above, the applicability of knowledge-based methods was seriously impaired by the lack of suitable fragments in protein structure data banks, although good performances were achieved for cases where canonical loop conformation existed, such as in the case of the prediction of CDR loops in immunoglobulins [76]. During the late 1990s, several loop structure classification databases were established [20,22,24,77,78]. Such loop classifications were used to model protein loops by using sequence patterns derived from the structural clusters. Wojcik et al. constructed a library of 13,563 loops from three to eight residues long [22]. Loops of the same length were clustered into families, and sequence signatures were derived. The method showed competitive results for loops up to five to six residues long. Oliva et al. [23] and Martin and Thornton [79] employed loop structure classifications to predict the structure of CDRs in immunoglobulins. In 2001, Burke and Deane [80] presented a method to exploit the environment-dependent sequence profiles [81] derived from the structural classes in the Sloop database [21], showing reasonable predictions for loops up to six residues long. More recently, Fernandez-Fuentes et al. employed the ArchDB database [25,26] to derive Hidden Markov Models (HMM) sequence profiles that were then used to predict the conformational class (loop conformation only) and subclass (loop conformation and geometry of regular flanking secondary elements) of loops using only sequence information [82]. In the case of class prediction (i.e., when accurate structural information about the stem residues is known and only the conformation of the loops is predicted), the average RMSD for loops of length 8 was around 2.0 Å, which can be considered a good prediction [83]. Finally, another recently published loop prediction approach first predicts conformation for a query loop sequence and then structurally aligns the predicted structural fragments to a set of non-redundant loop structural templates. These sequence-template loop alignments are then quantitatively evaluated with an artificial neural network model trained on a set of predictions with known outcomes [84].

In addition to profiles derived from structural clusters, large libraries of protein fragments have also been used to model loops [85,86]. A recently developed approach, ArchPRED [87], also uses a library of protein fragments. ArchPRED exploits a hierarchical and multidimensional library that has been set up to classify about 300,000 loop fragments and loop flanking secondary structures. Besides the length of the loops and types of bracing secondary structures, the database is organized along four internal coordinates: distance and three angles characterizing the geometry of stem regions [24]. Candidate fragments are selected from this library by matching the length and types of bracing secondary structures of the query and satisfying the geometrical restraints of the stem residues; these are subsequently inserted into the query protein framework where their fit is assessed by the RMSD of stem regions and by the number of rigid body clashes with the environment. In the final step, the remaining candidate loops are ranked by a *Z*-score that combines information on sequence similarity and fit of predicted and observed phi/psi main chain dihedral angle propensities. Confidence *Z*-score cutoffs were determined for each loop length, which identify those predicted fragments that outperform a competitive *ab initio* method. A Web server implements the method [88], regularly updates the fragment library, and performs predictions. Predicted segments are returned, or, optionally, these can be completed with side chain reconstruction and subsequently annealed in the environment of the query protein by conjugate gradient minimization.

13.3.2. *Ab Initio* Methods for Loop Modeling

Ab initio methods are based on a conformation search or enumeration of conformations, or decoys, in a given environment, guided by a scoring or pseudo-energy function to identify or optimize the conformation of a loop.

Consequently, *ab initio* methods overcome the main limitation of the database search methods, that is, *database completeness*. In addition, *ab initio* predictions are applicable both to simultaneous modeling of several loops and to those loops interacting with ligands, neither of which is straightforward for the database search approaches, where fragments are collected from unrelated structures with different environments. There are, however, limitations that come from the sampling and scoring processes. The conformational sampling, depending on the loop length, can be very time-consuming. In addition, conformational sampling methods are not sensitive enough to bias the conformational sampling toward native-like conformations; therefore, a large proportion of the time is spent sampling non-native-like conformations. The second limitation comes from the scoring or pseudo-energy functions that are used to approximate the free energy of loops and usually are not accurate enough to properly rank the many alternative conformations, especially in the case of longer loops. Thus, the sorting of candidate loops is not trivial, and usually, native conformation does not correspond to the lowest energy, which is a clear indication of the inaccuracy of these pseudo-energy functions and of our limited understanding and ability to represent the free energy in the applied scoring functions [89,90].

Ab initio predictions of loops were pioneered by Moult and James [91], Fine et al. [92], and Bruccoleri and Karplus [93]. Initially restricted to short loops, due to the computational limitations, these works were revisited and extended, including novel search algorithms and scoring functions, and their applicability to longer loops has been explored. Currently, there are many such methods exploiting different protein representations, such as unified atoms, all non-hydrogen atoms, non-hydrogen and polar hydrogen atoms, and all atoms; as well as implicit and explicit solvent models, sampling methods, energy function terms, and optimization or enumeration algorithms. The research efforts in this area have focused on addressing the two main issues: improving the sampling of loop conformations, or improving the pseudo-energy or scoring functions that rank these sampled conformations.

A number of conformational search algorithms aiming to improve and speed up the conformational sampling have been proposed in the past. These include sampling of main-chain dihedral angles biased by their distributions in known protein structures [91,94,95]; local optimization sampling such as minimum perturbation random tweak [90,92,96], local adjustments of phi/psi angles [97], local minimization of randomly generated conformations [98], and global energy minimization by mapping a trajectory of local minima [99,100]. Other methods rely on the optimization of an energy scoring function with a number of variations aiming to comprehensively sample the energetic landscape. These include molecular dynamics (MD) simulations [101–103], high-temperature MD [104], temperature replica exchange MD [105,106], low-barrier MD [107], Monte Carlo (MC) simulations combined with simulated annealing [108–112], MC and scaled collective variable techniques [113], MC and MD [114], biased probability MC [115–117], scaling relaxation and multiple copy sampling [118–121], and self-consistent mean field optimization [122]. Other approaches explored the use of systematic conformational searches [93,123–125], genetic algorithms [126,127], random sampling relying on dimers from known protein structures [128], mutually orthogonal Latin square sampling [129], a "divide and conquer" approach [130], and dynamic programming combined with 1D statistical mechanics [131,132]. Finally, some methods explored the use of robotic algorithms such as inverse kinematics [133,134] and graph theory [135] to sample the possible conformations of a given loop sequence.

There has also been active research to improve and develop novel pseudo-energy functions to score loop conformations. Some of these efforts have focused on exploring the inclusion of novel terms. The inclusion of an entropy-like term to the scoring function, the "colony energy," derived from geometrical comparisons and clustering of sampled loop conformations, proved to improve the selection of loop decoys [136,137]. In the case of long loops, the inclusion of a hydrophobic term has a positive effect [138], whereas the role of solvation terms and its impact on scoring has been explored in several studies [66,139–142]. All atom statistical potential methods such as DFIRE [143,144] and ROSETTA [145] have also been applied successfully in loop selection, and in the specific case of ROSETTA with encouraging results in predicting very long loops. As an anecdotal case, ROSETTA was able to correctly predict the structure of a 39-residue insertion of the protein target T0186 submitted to the fifth Critical Assessment of Techniques for Protein Structure Prediction (CASP) exercise.

Several of the *ab initio* methods described above are accessible online (Table 13.1); one of them is MODLOOP [83,146], which is explained here in more detail. MODLOOP is part of the comparative modeling suite MODELLER [147]. Loop optimization in MODLOOP relies on conjugate gradients and MD with simulated annealing. The pseudo-energy function is a sum of many terms, including some terms from the CHARMM-22 molecular mechanics force field [148], spatial restraints based on distributions of distances [149,150], and dihedral angle frequencies derived from known protein structures. MODLOOP has been optimized and evaluated on a large number of loops of known structures in both native and only approximately correct environments. Further improvements in the original method have been made by using CHARMM molecular force with Generalized Born (GB) solvation term to rank final conformations [66]. A Web server implements the method [146], where predicted loops are returned to the user by e-mail.

13.3.2. Combined Methods for Loop Modeling

Combined methods use both database search and *ab initio* methods. The underlying idea is to use database search methods to find candidate loops for a given target loop, which are subsequently evaluated and optimized in the target protein. The rational behind is that—especially in the case of longer

TABLE 13.1 Collection of Internet Resources for Loop Prediction

Description	Name	URL	Reference
Ab initio loop modeling server	MODLOOP	http://www.salilab.org/modloop/modloop.html	[83]
	Rapper	http://mordred.bioc.cam.ac.uk/~rapper/	[124]
	Loopy	http://wiki.c2b2.columbia.edu/honiglab_public/index.php/Software	[136]
	Robetta	http://robetta.bakerlab.org/	[145]
Knowledge-based loop modeling server	ArchPRED	http://www.fiserlab.org/servers_table.htm	[87]
	Wloop	http://psb00.snv.jussieu.fr/wloop/	[22]
	ArchFIT	http://sbi.imim.es/archdb/	[26]
	SuperLooper	http://bioinformatics.charite.de/superlooper/sllt.php	
	PrISM	http://cmb.genomics.sinica.edu.tw/~px172/Loop2/	[84]
Loop structure classification	ArchDB	http://sbi.imim.es/archdb/	[25]
	ArchKI	http://sbi.imim.es/archdb/	[158]
	Loop database	http://mdl.ipc.pku.edu.cn/moldes/oldmem/liwz/home/loop.htm	[159]
	Loop database	http://bmm.cancerresearchuk.org/loop/	[24]

loops—if a suitable fragment is available, then one can restrict the risky *ab initio* optimization within the conformational vicinity of the fragment, and therefore, better exploring and scoring a smaller sampling space deliver a more accurate model. The downside of this attractive idea is that a suitable fragment may not often be available.

An example of a combined algorithm is that of Martin et al. [151], in which antibody hypervariable loops were predicted using a database search followed by the reconstruction of sections of the predicted loops *ab initio* and addition of side chains using the CONGEN conformational searching algorithm [93]. This idea has also been applied by van Vlijmen and Karplus [152] by selecting loops from a fragment data bank followed by optimizing and ranking of the possible fragments using the CHARMM energy function [148]. The method has been tested for loops with different lengths (4–16 residues) showing competitive performance for up to nine residue loops. Deane and Blundell [153] presented the CODA algorithm, which combines two different algorithms: FREAD, a knowledge-based, and PETRA [94], an *ab initio* loop structure prediction method. CODA was benchmarked against loops ranging from three to eight residues long, and CODA predictions showed a clear improvement when compared against FREAD and PETRA predictions separately. ArchPRED [88] also allows user to select database-driven loop predictions for further refinement in the context of the target protein environment. Once the candidate fragments have been selected, a short conjugated gradient minimization optimizes the annealing of the loop to the stem residues while preserving the overall conformation of the loop structure.

13.4. APPLICATION OF LOOP PREDICTIONS

Loop modeling is used in a variety of applications in Biology and Bioinformatics, including interpreting and solving fuzzy or incomplete X-ray crystallographic maps, in homology modeling, in protein–ligand and protein–protein interaction studies, and in rational drug design.

Several examples reported in the scientific literature demonstrate the usefulness of comparative models in drug discovery process (reviewed in Reference [154]) and in particular the need of accurate prediction of loops that are part of the active site. The lack of selectivity of some dipeptidyl pepdidases 4 (DP4) inhibitors can be explained by the conformation adopted by loops P and R [155]. Through the modeling of the second extracellular loop (e2), the interaction with substituted benzamides by the dopamine receptor D2R could be explained [113]. A comparative model of the 1-deoxy-D-xylulose-5-phosphate reductoisomerase, a potential antimalaria target, including the modeling of a 10-residue-long loop that contains the residues in the active site, has been validated for the early design and development of inhibitors against this target [156]. In protein kinases, the conformation and orientation of the activation loop is critical to distinguish between active and inactive forms of the protein [157].

A restricted loop modeling may be used in X-ray crystallography during data fitting process when low electron density regions make the interpretation of X-ray crystallographic maps difficult. Usually, low electron density regions correspond to mobile regions, that is, loops. Loop modeling also plays an important role in the last stages of the homology modeling process. As mentioned above, even at a high level of sequence identity between the target sequence and template structure, there can still be regions that are not present in the template, hence require loop modeling. On the other hand, the accuracy of loop modeling is a very important factor when it comes to the study of protein–ligand(s) interactions. Loops are often part of active and binding sites; therefore, the structure-based studies of the interaction between a protein and its ligands often require loop modeling. Finally, loop modeling is used to improve the quality of protein–protein docking by modeling alternative conformation of protein–protein interacting loops.

REFERENCES

1. J.C. Kendrew, G. Bodo, H.M. Dintzis, R.G. Parrish, H. Wyckoff, and D.C. Phillips. A three-dimensional model of the myoglobin molecule obtained by x-ray analysis. *Nature*, 181:662–666, 1958.

2. C.C. Blake, D.F. Koenig, G.A. Mair, A.C. North, D.C. Phillips, and V.R. Sarma. Structure of hen egg-white lysozyme. A three-dimensional Fourier synthesis at 2 Angstrom resolution. *Nature*, 206:757–961, 1965.
3. D. Eisenberg. The discovery of the alpha-helix and beta-sheet, the principal structural features of proteins. *Proc Natl Acad Sci U S A*, 100:11207–11210, 2003.
4. C.M. Venkatachalam. Stereochemical criteria for polypeptides and proteins. V. Conformation of a system of three linked peptide units. *Biopolymers*, 6:1425–1436, 1968.
5. J.S. Richardson. The anatomy and taxonomy of protein structure. *Adv Protein Chem*, 34:167–339, 1981.
6. C.M. Wilmot and J. Thornton. Analysis and prediction of the different types of beta-turns in proteins. *J Mol Biol*, 203:221–232, 1988.
7. E.G. Hutchinson and J.M. Thornton. A revised set of potentials for beta-turn formation in proteins. *Protein Sci*, 3:2207–2216, 1994.
8. B.W. Matthews. The gamma turn. Evidence for a new folded conformation in proteins. *Macromolecules*, 5:818–819, 1972.
9. E. Milner-White, B.M. Ross, R. Ismail, K. Belhadj-Mostefa, and R. Poet. One type of gamma-turn, rather than the other gives rise to chain-reversal in proteins. *J Mol Biol*, 204:777–782, 1988.
10. G.D. Rose, L.M. Gierasch, and J.A. Smith. Turns in peptides and proteins. *Adv Protein Chem*, 37:1–109, 1985.
11. B.L. Sibanda and J.M. Thornton. Conformation of beta hairpins in protein structures: Classification and diversity in homologous structures. *Methods Enzymol*, 202:59–82, 1991.
12. B.L. Sibanda, T.L. Blundell, and J.M. Thornton. Conformation of beta-hairpins in protein structures. A systematic classification with applications to modelling by homology, electron density fitting and protein engineering. *J Mol Biol*, 206:759–777, 1989.
13. E.J. Milner-White and R. Poet. Four classes of beta-hairpins in proteins. *Biochem J*, 240:289–292, 1986.
14. A.V. Efimov. Patterns of loops regions in proteins. *Curr Opin Struct Biol*, 3:379–384, 1993.
15. A.V. Efimov. Structure of coiled beta-beta-hairpins and beta-beta-corners. *FEBS Lett*, 284:288–292, 1991.
16. N. Colloch and F.E. Cohen. Beta-breakers: An aperiodic secondary structure. *J Mol Biol*, 221:603–613, 1991.
17. P.A. Rice, A. Goldman, and T.A. Steitz. A helix-turn-strand structural motif common in alpha-beta proteins. *Proteins*, 8:343–340, 1990.
18. J.M. Thornton, B.L. Sibanda, M.S. Edwards, and D.J. Barlow. Analysis, design and modification of loop regions in proteins. *Bioessays*, 8:63–69, 1988.
19. M. Edwards, M.J.E. Sternberg, and J. Thornton. Structural and sequence patterns in the loops of beta alpha beta units. *Protein Eng*, 1:173–181, 1987.
20. L.E. Donate, S.D. Rufino, L.H. Canard, and T.L. Blundell. Conformational analysis and clustering of short and medium size loops connecting regular secondary structures: A database for modeling and prediction. *Protein Sci*, 5:2600–2616, 1996.

21. D. Burke, C. Deane, and T. Blundell. Browsing the Sloop database of structurally classified loops connecting elements of protein secondary structure. *Bioinformatics*, 16:513–516, 2000.

22. J. Wojcik, J.P. Mornon, and J. Chomilier. New efficient statistical sequence-dependent structure prediction of short to medium-sized protein loops based on an exhaustive loop classification. *J Mol Biol*, 255:235–253, 1999.

23. B. Oliva, P.A. Bates, E. Querol, F.X. Aviles, and M.J. Sternberg. Automated classification of antibody complementarity determining region 3 of the heavy chain (H3) loops into canonical forms and its application to protein structure prediction. *J Mol Biol*, 279:1193–1210, 1998.

24. B. Oliva, P.A. Bates, E. Querol, F.X. Aviles, and M.J. Sternberg. An automated classification of the structure of protein loops. *J Mol Biol*, 266:814–830, 1997.

25. J. Espadaler, N. Fernandez-Fuentes, A. Hermoso, E. Querol, F.X. Aviles, M.J. Sternberg, and B. Oliva. ArchDB: Automated protein loop classification as a tool for structural genomics. *Nucleic Acids Res*, 32:D185–D188, 2004.

26. A. Hermoso, J. Espadaler, E. Querol, F. Aviles, J.M.E. Sternberg, B. Oliva, and N. Fernandez-Fuentes. Including functional annotations and extending the collection of structural classification of protein loops. *Bioinform Biol Insights*, 1:1–14, 2007.

27. J.F. Leszczynski and G.D. Rose. Loops in globular proteins: A novel category of secondary structure. *Science*, 234:849–855, 1986.

28. S.T. Kim, H. Shirai, N. Nakajima, J. Higo, and H. Nakamura. Enhanced conformational diversity search of CDR-H3 in antibodies: Role of the first CDR-H3 residue. *Proteins*, 37:683–696, 1999.

29. H. Kawasaki and R.H. Kretsinger. Calcium-binding proteins 1: EF-hands. *Protein Profile*, 2:297–490, 1995.

30. L.N. Johnson, E.D. Lowe, M.E. Noble, and D.J. Owen. The Eleventh Datta Lecture. The structural basis for substrate recognition and control by protein kinases. *FEBS Lett*, 430:1–11, 1998.

31. M. Saraste, P.R. Sibbald, and A. Wittinghofer. The P-loop—A common motif in ATP- and GTP-binding proteins. *Trends Biochem Sci*, 15:430–434, 1990.

32. J.A. Tainer, M.M. Thayer, and R.P. Cunningham. DNA repair proteins. *Curr Opin Struct Biol*, 5:20–26, 1995.

33. L.S. Bernstein, S. Ramineni, C. Hague, W. Cladman, P. Chidiac, A.I. Levey, and J.R. Hepler. RGS2 binds directly and selectively to the M1 muscarinic acetylcholine receptor third intracellular loop to modulate Gq/11alpha signaling. *J Biol Chem*, 279:21248–21256, 2004.

34. E. Zomot and B.I. Kanner. The interaction of the gamma-aminobutyric acid transporter GAT-1 with the neurotransmitter is selectively impaired by sulfhydryl modification of a conformationally sensitive cysteine residue engineered into extracellular loop IV. *J Biol Chem*, 278:42950–42958, 2003.

35. K. Fritz-Wolf, T. Schnyder, T. Wallimann, and W. Kabsch. Structure of mitochondrial creatine kinase. *Nature*, 381:341–345, 1996.

36. W. Feng, Y. Shi, M. Li, and M. Zhang. Tandem PDZ repeats in glutamate receptor-interacting proteins have a novel mode of PDZ domain-mediated target binding. *Nat Struct Biol*, 10:972–978, 2003.

37. R.K. Wierenga, P. Terpstra, and W.G. Hol. Prediction of the occurrence of the ADP-binding beta alpha beta-fold in proteins, using an amino acid sequence fingerprint. *J Mol Biol*, 187:101–107, 1986.
38. P.W. Schenk and B.E. Snaar-Jagalska. Signal perception and transduction: The role of protein kinases. *Biochim Biophys Acta*, 1449:1–24, 1999.
39. A. Wlodawer, M. Miller, M. Jaskolski, B.K. Sathyanarayana, E. Baldwin, I.T. Weber, L.M. Selk, L. Clawson, J. Schneider, and S.B. Kent. Conserved folding in retroviral proteases: Crystal structure of a synthetic HIV-1 protease. *Science*, 245:616–621, 1989.
40. J.S. Fetrow. Omega loops: Nonregular secondary structure significant in protein function and stability. *FASEB J*, 9:708–717, 1995.
41. K. Gunasekaran, B. Ma, and R. Nussinov. Triggering loops and enzyme function: Identification of loops that trigger and modulate movements. *J Mol Biol*, 332:143–159, 2003.
42. S. Zgiby, A.R. Plater, M.A. Bates, G.J. Thomson, and A. Berry. A functional role for a flexible loop containing Glu182 in the class II fructose-1,6-biphosphate aldolase from *Escherichia coli*. *J Mol Biol*, 315:131–140, 2002.
43. D. Joseph, G.A. Petsko, and M. Karplus. Anatomy of a conformational change: hinged "lid" motion of the triosephosphate isomerase loop. *Science*, 249:1425–1428, 1990.
44. M.A. Sinev, E.V. Sineva, V. Ittah, and E. Haas. Domain closure in adenylate kinase. *Biochemistry*, 35:6425–6437, 1996.
45. J.A. Adams. Activation loop phosphorylation and catalysis in protein kinases: Is there functional evidence for the autoinhibitor model? *Biochemistry*, 42:601–607, 2003.
46. L.N. Johnson, M.E. Noble, and D.J. Owen. Active and inactive protein kinases: Structural basis for regulation. *Cell*, 85:149–158, 1996.
47. A.S. Yang and L.Y. Wang. Local structure prediction with local structure-based sequence profiles. *Bioinformatics*, 19:1267–1274, 2003.
48. A. Fiser, Z. Dosztanyi, and I. Simon. The role of long-range interactions in defining the secondary structure of proteins is overestimated. *Comput Appl Biosci*, 13:297–301, 1997.
49. G.M. Salem, E.G. Hutchinson, C.A. Orengo, and J.M. Thornton. Correlation of observed fold frequency with the occurrence of local structural motifs. *J Mol Biol*, 287:969–981, 1999.
50. T. Wood and W. Pearson. Evolution of protein sequences and structures. *J Mol Biol*, 291:977–995, 1999.
51. I.N. Shindyalov and P.E. Bourne. An alternative view of protein fold space. *Proteins*, 38:247–260, 2000.
52. A.N. Lupas, C.P. Ponting, and R.B. Russell. On the evolution of protein folds: Are similar motifs in different protein folds the result of convergence, insertion, or relics of an ancient peptide world? *J Struct Biol*, 134:191–203, 2001.
53. A.V. Tendulkar, A.A. Joshi, M.A. Sohoni, and P.P. Wangikar. Clustering of protein structural fragments reveals modular building block approach of nature. *J Mol Biol*, 338:611–629, 2004.

54. F.J. Hoedemaeker, R.R. van Eijsden, C.L. Diaz, B.S. de Pater, and J.W. Kijne. Destabilization of pea lectin by substitution of a single amino acid in a surface loop. *Plant Mol Biol*, 22:1039–1046, 1993.
55. R. Ulfer and K. Kirschner. The importance of surface loops for stabilizing an eightfold beta alpha protein. *Protein Sci*, 1:31–45, 1992.
56. A. Linhananta, H. Zhou, and Y. Zhou. The dual role of a loop with loop contact distance in folding and domain swapping. *Protein Sci*, 11:1695–1701, 2002.
57. V. Batori, A. Koide, and S. Koide. Exploring the potential of the monobody scaffold: Effects of loop elongation on the stability of a fibronectin type III domain. *Protein Eng*, 15:1015–1020, 2002.
58. A.R. Fersht. Transition-state structure as unifying basis in protein-folding mechanisms: Contact order, chain topology, stability, and the extended nucleus mechanism. *Proc Natl Acad Sci U S A*, 15:1525–1529, 2000.
59. M.M.G. Krishna, Y. Lin, J.N. Rumbley, and S. Walter Englander. Cooperative omega loops in cytochrome c: Role in folding and function. *J Mol Biol*, 331:29–36, 2003.
60. B. Collinet, P. Garcia, P. Minard, and M. Desmadril. Role of loops in the folding and stability of yeast phosphoglycerate kinase. *Eur J Biochem*, 268:5107–5118, 2001.
61. S. Kumar and R. Nussinov. How do thermophilic proteins deal with heat? *Cell Mol Life Sci*, 58:1216–1233, 2001.
62. H.X. Zhou. Loops, linkage, rings, catenanes, cages, and crowders: Entropy-based strategies for stabilizing proteins. *Acc Chem Res*, 37:123–130, 2004.
63. M.R. Chance, A. Fiser, A. Sali, U. Pieper, N. Eswar, G. Xu, J.E. Fajardo, T. Radhakannan, and N. Marinkovic. High-throughput computational and experimental techniques in structural genomics. *Genome Res*, 14:2145–2154, 2004.
64. U. Pieper, N. Eswar, H. Braberg, M.S. Madhusudhan, F.P. Davis, A.C. Stuart, N. Mirkovic, A. Rossi, M.A. Marti-Renom, A. Fiser, B. Webb, D. Greenblatt, C.C. Huang, T.E. Ferrin, and A. Sali. MODBASE, a database of annotated comparative protein structure models, and associated resources. *Nucleic Acids Res*, 32:D217–D222, 2004.
65. R. Sanchez and A. Sali. Advances in comparative protein-structure modelling. *Curr Opin Struct Biol*, 7:206–214, 1997.
66. A. Fiser, M. Feig, C.L. 3rd Brooks, and A. Sali. Evolution and physics in comparative protein structure modeling. *Acc Chem Res*, 35:413–421, 2002.
67. K. Fidelis, P.S. Stern, D. Bacon, and J. Moult. Comparison of systematic search and database methods for constructing segments of protein structure. *Protein Eng*, 7:953–960, 1994.
68. U. Lessel and D. Schomburg. Creation and characterization of a new, non-redundant fragment data bank. *Protein Eng*, 10:659–664, 1997.
69. P. Du, M. Andrec, and R.M. Levy. Have we seen all structures corresponding to short protein fragments in the Protein Data Bank? An update. *Protein Eng*, 16:407–414, 2003.
70. N. Fernandez-Fuentes and A. Fiser. Saturating representation of loop conformational fragments in structure databanks. *BMC Struct Biol*, 6:15, 2006.

71. M. Mezei. Chameleon sequences in the PDB. *Protein Eng*, 11:411–414, 1998.
72. B.I. Cohen, S.R. Presnell, and F.E. Cohen. Origins of structural diversity within sequentially identical hexapeptides. *Protein Sci*, 2:2134–2145, 1993.
73. W. Kabsch and C. Sander. On the use of sequence homologies to predict protein structure: Identical pentapeptides can have completely different conformations. *Proc Natl Acad Sci U S A*, 81:1075–1078, 1984.
74. J. Greer. Comparative model-building of the mammalian serine proteases. *J Mol Biol*, 153:1027–1042, 1981.
75. T.A. Jones and S. Thirup. Using known substructures in protein model building and crystallography. *EMBO J*, 5:819–822, 1986.
76. C. Chothia and A.M. Lesk. Canonical structures for the hypervariable regions of immunoglobulins. *J Mol Biol*, 196:901–917, 1987.
77. J.M. Kwasigroch, J. Chomilier, and J.P. Mornon. A global taxonomy of loops in globular proteins. *J Mol Biol*, 259:855–872, 1996.
78. R.T. Wintjens, M.J. Rooman, and S.J. Wodak. Automatic classification and analysis of alpha alpha-turn motifs in proteins. *J Mol Biol*, 255:235–253, 1996.
79. A.C. Martin and J.M. Thornton. Structural families in loops of homologous proteins: Automatic classification, modelling and application to antibodies. *J Mol Biol*, 263:800–815, 1996.
80. D.F. Burke and C.M. Deane. Improved protein loop prediction from sequence alone. *Protein Eng*, 14:473–478, 2001.
81. C.M. Topham, A. McLeod, F. Eisenmenger, J.P. Overington, M.S. Johnson, and T.L. Blundell. Fragment ranking in modelling of protein structure. Conformationally constrained environmental amino acid substitution tables. *J Mol Biol*, 229:194–220, 1993.
82. N. Fernandez-Fuentes, E. Querol, F.X. Aviles, M.J. Sternberg, and B. Oliva. Prediction of the conformation and geometry of loops in globular proteins: Testing ArchDB, a structural classification of loops. *Proteins*, 60:746–757, 2005.
83. A. Fiser, R.K. Do, and A. Sali. Modeling of loops in protein structures. *Protein Sci*, 9:1753–1773, 2000.
84. H.P. Peng and A.S. Yang. Modeling protein loops with knowledge-based prediction of sequence-structure alignment. *Bioinformatics*, 23:2836–2842, 2007.
85. E. Michalsky, A. Goede, and R. Preissner. Loops in proteins (LIP)—A comprehensive loop database for homology modelling. *Protein Eng*, 16:979–985, 2003.
86. P. Heuser, G. Wohlfahrt, and D. Schomburg. Efficient methods for filtering and ranking fragments for the prediction of structurally variable regions in proteins. *Proteins*, 54:583–595, 2004.
87. N. Fernandez-Fuentes, B. Oliva, and A. Fiser. A supersecondary structure library and search algorithm for modeling loops in protein structures. *Nucleic Acids Res*, 34:2085–2097, 2006.
88. N. Fernandez-Fuentes, J. Zhai, and A. Fiser. ArchPRED: A template based loop structure prediction server. *Nucleic Acids Res*, 34:W173–W176, 2006.
89. J.L. Pellequer and S.W. Chen. Does conformational free energy distingish loop conformations in proteins? *Biophys J*, 73:2359–2375, 1997.
90. K.C. Smith and B. Honig. Evaluation of the conformation free energies of loops in proteins. *Proteins*, 18:119–132, 1994.

91. J. Moult and M.N. James. An algorithm for determining the conformation of polypeptide segments in proteins by systematic search. *Proteins*, 1:146–163, 1986.
92. R.M. Fine, H. Wang, P.S. Shekin, D.L. Yarmush, and C. Levinthal. Predicting antibody hypervariable loop conformations. II: Minimization and molecular dynamics studies of MPCPC603 from many randomly generated loop conformations. *Proteins*, 1:342–362, 1986.
93. R.E. Bruccoleri and M. Karplus. Prediction of the folding of short polypeptide segments by uniform conformational sampling. *Biopolymers*, 26:137–168, 1987.
94. C.M. Deane and T.L. Blundell. A novel exhaustive search algorithm for predicting the conformation of polypeptide segments in proteins. *Proteins*, 40:135–144, 2000.
95. H. Jiang and C. Blouin. Ab initio construction of all-atom loop conformations. *J Mol Model*, 12:221–228, 2006.
96. P.S. Shenkin, D.L. Yarmush, R.M. Fine, H.J. Wang, and C. Levinthal. Predicting antibody hypervariable loop conformation. I. Ensembles of random conformations for ringlike structures. *Biopolymers*, 26:2053–2085, 1987.
97. A. Elofsson, S.M. Le Grand, and D. Eisenberg. Local moves: An efficient algorithm for simulation of protein folding. *Proteins*, 23:73–82, 1995.
98. M.H. Lambert and H.A. Scheraga. Pattern recognition in the prediction of protein structure. II. Chain conformation from a probability-directed search procedure. *J Comput Chem*, 10:798–816, 1989.
99. M. Dudek, K. Ramnarayan, and J. Ponder. Protein structure prediction using a combination of sequence homology and global energy minimization: II. Energy functions. *J Comput Chem*, 19:548–573, 1998.
100. M. Dudek and H. Scheraga. Protein structure prediction using a combination of sequence homology and global energy minimization I. Global energy minimization of surface loops. *J Comput Chem*, 11:121–151, 1990.
101. J.J. Tanner, L.J. Nell, and J.A. McCammon. Anti-insulin antibody structure and conformation. II. Molecular dynamics with explicit solvent. *Biopolymers*, 32:23–32, 1992.
102. N. Nakajima, J. Higo, A. Kidera, and H. Nakamura. Free energy landscapes of peptides by enhanced conformational sampling. *J Mol Biol*, 296:197–216, 2000.
103. U. Rao and M.M. Teeter. Improvement of turn structure prediction by molecular dynamics: A case study of alpha 1-purothionin. *Protein Eng*, 6:837–847, 1993.
104. R.E. Bruccoleri and M. Karplus. Conformational sampling using high-temperature molecular dynamics. *Biopolymers*, 29:1847–1862, 1990.
105. A.K. Felts, E. Gallicchio, D. Chekmarev, K.A. Paris, R.A. Friesner, and R.M. Levy. Prediction of protein loop conformations using the AGBNP implicit solvent model and torsion angle sampling. *J Chem Theory Comput*, 4:855–868, 2008.
106. M.A. Olson, M. Feig, and C.L. 3rd Brooks. Prediction of protein loop conformations using multiscale modeling methods with physical energy scoring functions. *J Comput Chem*, 29:820–831, 2008.
107. V. Hornak and C. Simmerling. Generation of accurate protein loop conformations through low-barrier molecular dynamics. *Proteins*, 51:577–590, 2003.
108. G. Vasmatzis, R. Brower, and C. Delisi. Predicting immunoglobulin-like hypervariable loops. *Biopolymers*, 34:1669–1680, 1994.

109. V. Collura, J. Higo, and J. Garnier. Modeling of protein loops by simulated annealing. *Protein Sci*, 2:1502–1510, 1993.

110. L. Carlacci and S. Englander. Loop problem in proteins: Developments on Monte Carlo simulated annealing approach. *J Comput Chem*, 17:1002–1012, 1996.

111. L. Carlacci and S.W. Englander. The loop problem in proteins: A Monte Carlo simulated annealing approach. *Biopolymers*, 33:1271–1286, 1993.

112. J. Higo, V. Collura, and J. Garnier. Development of an extended simulated annealing method: Application to the modeling of complementary determining regions of immunoglobulins. *Biopolymers*, 32:33–43, 1992.

113. S. Kortagere, A. Roy, and E.L. Mehler. Ab initio computational modeling of long loops in G-protein coupled receptors. *J Comput Aided Mol Des*, 20:427–436, 2006.

114. C.S. Rapp and R.A. Friesner. Prediction of loop geometries using a generalized born model of solvation effects. *Proteins*, 35:173–183, 1999.

115. N. Thanki, J.P. Zeelen, M. Mathieu, R. Jaenicke, R.A. Abagyan, R.K. Wierenga, and W. Schliebs. Protein engineering with monomeric triosephosphate isomerase (monoTIM): The modelling and structure verification of a seven-residue loop. *Protein Eng*, 10:159–167, 1997.

116. J.S. Evans, A.M. Mathiowetz, S.I. Chan, and W.A. 3rd Goddard. De novo prediction of polypeptide conformations using dihedral probability grid Monte Carlo methodology. *Protein Sci*, 4:1203–1216, 1995.

117. R. Abagyan and M. Totrov. Biased probability Monte Carlo conformational searches and electrostatic calculations for peptides and proteins. *J Mol Biol*, 235:983–1002, 1994.

118. D. Rosenbach and R. Rosenfeld. Simultaneous modeling of multiple loops in proteins. *Protein Sci*, 4:496–505, 1995.

119. Q. Zheng, R. Rosenfeld, C. DeLisi, and D.J. Kyle. Multiple copy sampling in protein loop modeling: Computational efficiency and sensitivity to dihedral angle perturbations. *Protein Sci*, 3:493–506, 1994.

120. Q. Zheng and D.J. Kyle. Accuracy and reliability of the scaling-relaxation method for loop closure: An evaluation based on extensive and multiple copy conformational samplings. *Proteins*, 24:209–217, 1996.

121. R. Rosenfeld, Q. Zheng, S. Vajda, and C. DeLisi. Computing the structure of bound peptides. Application to antigen recognition by class I major histocompatibility complex receptors. *J Mol Biol*, 234:515–521, 1993.

122. P. Koehl and M. Delarue. A self consistent mean field approach to simultaneous gap closure and side-chain positioning in homology modelling. *Nat Struct Biol*, 2:163–170, 1995.

123. R.C. Brower, G. Vasmatzis, M. Silverman, and C. Delisi. Exhaustive conformational search and simulated annealing for models of lattice peptides. *Biopolymers*, 33:329–334, 1993.

124. M.A. DePristo, P.I. de Bakker, S.C. Lovell, and T.L. Blundell. Ab initio construction of polypeptide fragments: Efficient generation of accurate, representative ensembles. *Proteins*, 51:41–55, 2003.

125. V.Z. Spassov, P.K. Flook, and L. Yan. LOOPER: A molecular mechanics-based algorithm for protein loop prediction. *Protein Eng Des Sel*, 21:91–100, 2008.

126. D. McGarrah and R. Judson. Analysis of the genetic algorithm method of molecular conformation determination. *J Comp Chem*, 14:1385–1395, 1993.

127. C.S. Ring and F.E. Cohen. Conformational sampling of loop structures using genetic algorithm. *Isr J Chem*, 34:245–252, 1994.

128. S. Sudarsanam, R.F. DuBose, C.J. March, and S. Srinivasan. Modeling protein loops using a phi i + 1, psi i dimer database. *Protein Sci*, 4:1412–1420, 1995.

129. V. Kanagasabai, J. Arunachalam, P.A. Prasad, and N. Gautham. Exploring the conformational space of protein loops using a mean field technique with MOLS sampling. *Proteins*, 67:908–921, 2007.

130. S.C. Tosatto, E. Bindewald, J. Hesser, and R. Manner. A divide and conquer approach to fast loop modeling. *Protein Eng*, 15:279–286, 2002.

131. A.V. Finkelstein and B.A. Reva. Search for the stable state of a short chain in a molecular field. *Protein Eng*, 5:617–624, 1992.

132. S. Vajda and C. Delisi. Determining minimum energy conformations of polypeptides by dynamic programming. *Biopolymers*, 29:1755–1772, 1990.

133. R. Kolodny, L. Guibas, M. Levitt, and P. Koehl. Inverse kinematics in biology: The protein loop closure problem. *Int J Rob Res*, 24:151–163, 2005.

134. A.A. Canutescu and R.L. Jr. Dunbrack. Cyclic coordinate descent: A robotics algorithm for protein loop closure. *Protein Sci*, 12:963–972, 2003.

135. R. Samudrala and J. Moult. An all-atom distance-dependent conditional probability discriminatory function for protein structure prediction. *J Mol Biol*, 275:895–916, 1998.

136. Z. Xiang, C.S. Soto, and B. Honig. Evaluating conformational free energies: The colony energy and its application to the problem of loop prediction. *Proc Natl Acad Sci U S A*, 99:7432–7437, 2002.

137. F. Fogolari and S.C. Tosatto. Application of MM/PBSA colony free energy to loop decoy discrimination: Toward correlation between energy and root mean square deviation. *Protein Sci*, 14:889–901, 2005.

138. K. Zhu, D.L. Pincus, S. Zhao, and R.A. Friesner. Long loop prediction using the protein local optimization program. *Proteins*, 65:438–452, 2006.

139. P.I. de Bakker, M.A. DePristo, D.F. Burke, and T.L. Blundell. Ab initio construction of polypeptide fragments: Accuracy of loop decoy discrimination by an all-atom statistical potential and the AMBER force field with the Generalized Born solvation model. *Proteins*, 51:21–40, 2003.

140. B. Das and H. Meirovitch. Optimization of solvation models for predicting the structure of surface loops in proteins. *Proteins*, 43:303–314, 2001.

141. L.R. Forrest and T.B. Woolf. Discrimination of native loop conformations in membrane proteins: Decoy library design and evaluation of effective energy scoring functions. *Proteins*, 52:492–509, 2003.

142. B. Das and H. Meirovitch. Solvation parameters for predicting the structure of surface loops in proteins: Transferability and entropic effects. *Proteins*, 51:470–483, 2003.

143. C. Zhang, S. Liu, and Y. Zhou. Accurate and efficient loop selections by the DFIRE-based all-atom statistical potential. *Protein Sci*, 13:391–399, 2004.

144. C.S. Soto, M. Fasnacht, J. Zhu, L. Forrest, and B. Honig. Loop modeling: Sampling, filtering, and scoring. *Proteins*, 70:834–843, 2008.

145. C.A. Rohl, C.E. Strauss, D. Chivian, and D. Baker. Modeling structurally variable regions in homologous proteins with rosetta. *Proteins*, 55:656–677, 2004.

146. A. Fiser and A. Sali. ModLoop: Automated modeling of loops in protein structures. *Bioinformatics*, 19:2500–2501, 2003.
147. A. Sali and T.L. Blundell. Comparative protein modeling by satisfaction of spatial restraints. *J Mol Biol*, 234:779–815, 1993.
148. B. Brooks, R.E. Bruccoleri, B.D. Olafson, D.J. States, S. Swaminathan, and M. Karplus. A program for macromolecular energy minimisation and dynamics calculations. *J Comp Chem*, 4:187–217, 1983.
149. F. Melo and E. Feytmans. Novel knowledge-based mean force potential at atomic level. *J Mol Biol*, 267:207–222, 1997.
150. M.J. Sippl. Knowledge-based potentials for proteins. *Curr Opin Struct Biol*, 5:229–235, 1995.
151. A.C. Martin, J.C. Cheetham, and A.L. Rees. Modeling antibody hypervariable loops: A combined algorithm. *Proc Natl Acad Sci U S A*, 86:9268–9272, 1989.
152. H.W. van Vlijmen and M. Karplus. PDB-based protein loop prediction: Parameters for selection and methods for optimization. *J Mol Biol*, 267:975–1001, 1997.
153. C.M. Deane and T.L. Blundell. CODA: A combined algorithm for predicting the structurally variable regions of protein models. *Protein Sci*, 10:599–612, 2001.
154. M. Jacobson, A. Sali, and A.M. Doherty. Comparative protein structure modeling and its applications to drug discovery. *Annu Rep Med Chem*, 39:259–276, 2004.
155. C. Rummey and G. Metz. Homology models of dipeptidyl peptidases 8 and 9 with a focus on loop predictions near the active site. *Proteins*, 66:160–171, 2007.
156. N. Singh, G. Cheve, M.A. Avery, and C.R. McCurdy. Comparative protein modeling of 1-deoxy-D-xylulose-5-phosphate reductoisomerase enzyme from *Plasmodium falciparum*: A potential target for antimalarial drug discovery. *J Chem Inf Model*, 46:1360–1370, 2006.
157. A.P. Kornev, N.M. Haste, S.S. Taylor, and L.F. Eyck. Surface comparison of active and inactive protein kinases identifies a conserved activation mechanism. *Proc Natl Acad Sci U S A*, 103:17783–17788, 2006.
158. N. Fernandez-Fuentes, A. Hermoso, J. Espadaler, E. Querol, F.X. Aviles, and B. Oliva. Classification of common functional loops of kinase super-families. *Proteins*, 56:539–555, 2004.
159. W. Li, Z. Liu, and L. Lai. Protein loops on structurally similar scaffolds: Database and conformational analysis. *Biopolymers*, 49:481–495, 1999.

CHAPTER 14

MODEL QUALITY ASSESSMENT USING A STATISTICAL PROGRAM THAT ADOPTS A SIDE CHAIN ENVIRONMENT VIEWPOINT

GENKI TERASHI, MAYUKO TAKEDA-SHITAKA, KAZUHIKO KANOU, and HIDEAKI UMEYAMA
School of Pharmacy
Kitasato University
Tokyo, Japan

14.1. INTRODUCTION

The accurate prediction of protein structures represents one of the major challenges in the bioinformatics field. An accurate prediction depends on three factors: (i) the technique employed to search for sequences representing a family of protein structures; (ii) accurate modeling methods for constructing high-quality protein models and; (iii) model quality assessment approaches that distinguish between the near native models (high-quality models) and the inferior "decoy" models.

In our laboratory, the interactive modeling system CHIMERA [1] and the Fully Automated Modeling System (FAMS [2]) have been developed and, as a result, the accurate predictions of three-dimensional (3D) protein models have been achieved. CHIMERA and FAMS perform well, especially in the Critical Assessment of Protein Structure Prediction (CASP [3]) rounds 3, 4, 5, and 6 [4–7]. However, room for improving the program for model quality assessment was noted. Although high-quality models were obtained, consistent selection of these models from a set of predicted structures was challenging.

Introduction to Protein Structure Prediction: Methods and Algorithms,
Edited by Huzefa Rangwala and George Karypis

In this work, we have developed the Model Quality Assessment Programs (MQAPs) CIRCLE [8] and META selector and the modeling server fams-ace2 (which includes the CIRCLE scoring function and consensus method). A feature of the fams-ace2 server is that the processes are fully automated and do not require human intervention. In this chapter, we introduce both MQAPs methods, CIRCLE and META selector, the server fams-ace2, and discuss the successes and failures of these approaches during CASP7 [9] and CASP8.

14.2. BACKGROUND

14.2.1. MQAPs

Many of the scoring functions for evaluating protein structures are founded on knowledge-based potentials [10], clustering methods [11], structural energies using molecular mechanics force fields[12], and the profile of the sequence or structure (e.g., Verify3D [13,14], ProQ [15]). These scoring functions are used to assess the model quality and ultimately select the best model among a set of predicted protein 3D models.

During the Fourth Critical Assessment of Fully Automated Structure Prediction (CAFASP4) [16], the MQAPs contest category [17] was employed to assess the accuracy based on an evaluation using the predicted models produced by participating CAFASP4 servers. Several scoring functions of MQAPs were able to identify correct and incorrect protein models and consistently selected the best models among the candidate models. According to the report from Daniel Fischer [18], the CAFASP4 organizer, the best performing MQAPs were Verify3D, Victor/FRST [19], and MODCHECK [20]. The Verify3D methodology uses 18 discrete environmental classes for each amino acid residue and uses the buried area and the fraction of polar area for assessing model quality. Victor/FRST combines four knowledge-based potentials (pairwise, solvation, hydrogen bonds, and torsion angle potentials) and performed consistently well, especially for comparative modeling targets. MODCHECK is based on threading potentials (pairwise and solvation potentials) and calculates the quality score by summing the pairwise and solvation Z-scores that are obtained by extensive sequence shuffling trials. These three methods consider only the quality of the target model.

In contrast, other research groups use consensus methods (e.g., Pcons [21], 3D-Jury [22], and Threading ASSembly Refinement quality assessment [TASSER-QA] [23]). Consensus methods require various candidate models to calculate the consensus value. Pcons combines the confidence score reported by each server and the similarity between models. The 3D-Jury technique is a fully automated protein structure meta prediction system that is efficient in the event of a high correlation between the accuracy of the model and the confidence score that is calculated by the similarity between models. The

TASSER-QA combines the fragment comparison score and consensus residue-residue contact potentials. Pcons and TASSER-QA (according to their article because the TASSER-QA group did not participate in the QA category) were the top performing quality assessment predictors for the correlation coefficients averages and the total Global Distance Test Total Score (GDT_TS) values in the CASP7 targets.

14.2.2. CASP

CASP is used in order to establish the capabilities and limitations of current methods of modeling protein structures from sequences. The methods are assessed on the basis of the analysis of a large number of blind predictions of protein structures. From 1994 to 2008, there have been eight CASP experiments [24] and our group has participated five times. In CASP7 [9] and CASP8 experiments, there are human predictor groups and server predictor groups that correspond to the manual and automatic server predictions, respectively. The human predictor groups were required to submit predicted models of the target structures by a deadline, usually within a 3-week prediction window. The server predictor groups were required to respond within 48 hours. The server predictions were made available 3 days after the targets were released.

14.2.3. GDT_TS

In CASP experiments, the GDT_TS [25] was used as the measure to represent the structural similarities between the models and the native structures. A high GDT_TS indicates that the models and the native structures have high similarities. The GDT_TS is calculated as:

$$GDT\_TS = \frac{GDT\_P_1 + GDT\_P_2 + GDT\_P_4 + GDT\_P_8}{4} \tag{14.1}$$

where GDT_P_n denotes the percentage of residues under a distance cutoff <= n Å, when the superimposed structures depend on the position of the Cα atoms between a model and the experimentally determined structure. In CASP7, the GDT High Accuracy (GDT_HA) value was introduced. The GDT_HA value is calculated by the following equation:

$$GDT\_HA = \frac{GDT\_P_{0.5} + GDT\_P_1 + GDT\_P_2 + GDT\_P_4}{4} \tag{14.2}$$

where GDT_P_n denotes the percentage of residues under a distance cutoff <= n Å. The GDT_HA value measures the structural similarity between the models and the native structure (as in GDT_TS). Due to the stricter distance cutoffs when compared with GDT_TS, the aim of GDT_HA value is to evaluate small improvements in easy targets.

14.3. THE CIRCLE ALGORITHM: AN MQAP

In this section, the method of the quality assessment program CIRCLE is presented. This program considers the environment of the side chains and the secondary structures of the models. The results of CIRCLE in CASP7 and CASP8 experiments are presented.

14.3.1. Definition of the Side Chain Environments

In order to calculate the stability of residues according to the amino acid type and position in the protein structure, the side chain environment for each residue is determined from three parameters: (i) the fraction of the molecular surface area of a side chain covered by polar atoms (Equation 14.3); (ii) the fraction of a side chain area buried by any other atoms (Equation 14.4); and (iii) the secondary structure.

$$polar = \frac{n_p}{N_{all}} \tag{14.3}$$

$$buried = \frac{n_b}{N_{all}} \tag{14.4}$$

In Equations 14.3 and 14.4, N_{all} is the total number of points that are homogeneously distributed on the solvent accessible surface of the amino acid side chain. n_p is the number of points that contact with polar atoms or the number of points located on the surface of the protein structure within N_{all}. n_b is the number of points that contact with any atom of the protein structure within N_{all}. A high *buried* value corresponds to an environment where the side chain is positioned within the core of the protein structure.

The secondary structure was described using a sliding window around each residue in order to classify the secondary structure in more detail. For easy (which can be predicted by comparative modeling [CM] methods) and hard (non-CM) targets, the window sizes were set to three and one, respectively. For example, the secondary structure classified as "CCC" indicates that the center residue of the three residues exists in a coil structural region. The secondary structure represents the local conformation of the main chain.

14.3.2. Scoring Function Based on the Side Chain Environment

CIRCLE considers two terms for the model quality: (i) model quality calculated from the side chain environment of each residue; and (ii) similarity between the secondary structure propensities predicted for an amino acid sequence by PSIPRED [26] and the secondary structure of the 3D model.

Design considerations are performed to compensate for the insufficient number of residues in the Protein Data Bank (PDB), according to Equations

14.5 to 14.9. The score describing the quality of the model is calculated by the following function.

$$w(polar, buried, p, b) = \exp\left(\frac{-p^2}{2\times\sigma_{polar}^2}\right)\times\exp\left(\frac{-b^2}{2\times\sigma_{buried}^2}\right) \tag{14.5}$$

$$N(AA|ss, polar, buried) \tag{14.6}$$

$$\sum_{m,n} w(polar, buried, p, b)\, N(AA|ss, polar+m, buried+n) \tag{14.7}$$

$$P(AA|ss, polar, buried) = \frac{\sum_{m,n} w(polar, buried, m, n)\, N(AA|ss, polar+m, buried+n)}{\sum_{aa}\sum_{m,n} w(polar, buried, m, n)\, N(aa|ss, polar+m, buried+n)} \tag{14.8}$$

$$SideChain(AA|env) = \log\left(\frac{P(AA|env)}{P(AA)}\right) = \log\left(\frac{P(AA|ss, polar, buried)}{P(AA)}\right) \tag{14.9}$$

N(*AA*|*ss*,*polar*,*buried*) in Equation 14.6 is the number of residues, *AA*, observed in an environment, *env*, from the PDB dataset. As mentioned above, the environment corresponds to the fraction of polar area (*polar*), the fraction of buried area (*buried*) and the secondary structure (*ss*). P(*AA*|*env*) in Equation 14.9 is the probability of amino acid residue *AA* being present in the environment *env*. *aa* is a variable for the 20 amino acids. P(*AA*) in Equation 14.9 is the probability of finding residue *AA* in all the amino acids. The P(*AA*|*env*) contains the Gaussian weighted function corresponding to Eequation 14.5 and considers the variability of the frequencies in the side chain environments (e.g., when *AA* is "LYS" and *ss* is "CCC", A2 in Fig. 14.1). The standard deviations ($\sigma_{polar}, \sigma_{buried}$), according to the environment of the side chain (fraction of polar area, buried area) in Equation 14.5, are calculated from the virtual mutation dataset. A total of 504,716 datasets was constructed by using homology modeling methods. The letters *b* and *p* in the two Gaussian weighted functions represent the buried ratio and the polar ratio axes, respectively, and describe the distance from the center location of each Gaussian weighted function. Examples of the scoring matrices of hydrophobic (leucine) and hydrophilic (lysine) residues are shown in Figure 14.1.

14.3.3. Scoring Function Based on the Secondary Structure

We predicted the secondary structure for a target amino acid sequence and compared the prediction for each residue with the observed secondary structure of the model. The measure of the similarity in the secondary structure is based on the following scoring function.

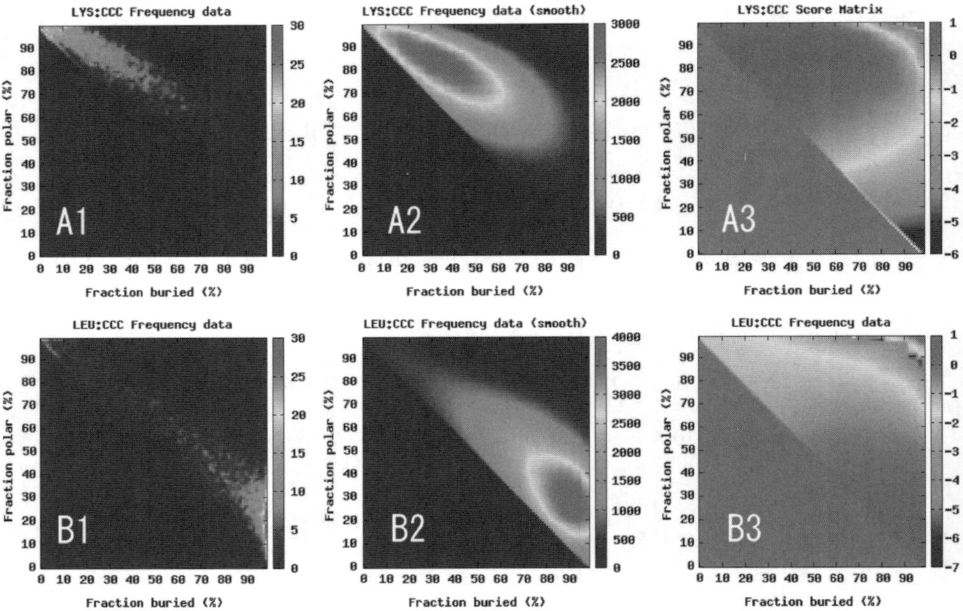

FIGURE 14.1 Matrices data used in the scoring function of CIRCLE. Sets (A1, B1) show the frequencies of residues LYS and LEU observed in the PDB dataset, respectively, as described by Equation 14.6. Sets (A2, B2) show converted data from the A1, B1 matrices using Gaussian weightings as described by Equation 14.7. Sets (A3, B3) show the scoring matrices of LYS and LEU according to the environment of the side chains as described by Equation 14.9. (See color insert.)

$$SSscore(i,j,conf) = \log\left(\frac{P(i,j|conf)}{P(i|conf)P(j|conf)}\right) \tag{14.10}$$

i represents the secondary structure of the target sequence predicted by PSIPRED. j is the secondary structure observed in the model assigned by STRIDE [27]. *conf* is the value of the confidence (0, 1, 2, … , 9) calculated by PSIPRED. $P(i|conf)$ is the probability of the secondary structure i, which was predicted by PSIPRED, when a value of confidence is *conf*. $P(j|conf)$ is the probability of the secondary structure j observed in the model when the value of confidence is *conf*. $P(i, j|conf)$ is the probability of the secondary structures i and j corresponding to *conf*.

14.3.4. Prediction of the Target Difficulty

This step evaluates the feasibility of the model construction with the template protein coordinates from the previous CASP experiments. CIRCLE employs two scoring functions in which the use of either is dependent on the target difficulty. In order to predict the target difficulty, the support vector machines (SVMs) program [28] was used. Classification based on SVMs has been used in several applications of bioinformatics and computational biology [29]. The training datasets consisted of CASP6 targets classified as CM targets, as positive, or fold recognition (FR) and new fold (NF), as negative. Score and homology (%) values of the best alignments resulting from the SPARKS2 program [30] were used as vectors for SVM classification. SPARKS2 performs alignments using a knowledge-based energy score with sequence-profile and secondary structure information. For comparative modeling in CASP6, SPARKS2 was the most accurate server among all the servers (by official ranking). Consequently, we assumed that if SPARKS2 cannot find reasonable alignments, the target(s) must be significantly difficult. The sensitivity of classification, defined as (TP/(TP+FN)), was 93.0% (40/(40+3)) in the CASP6 targets. TP, FP, TN, and FN denote True Positive, False Positive, True Negative, and False Negative, respectively. The specificity (TN/(TN+FP)) was 93.6% (44/(44+3)). The sensitivity and specificity calculated by SVMs using the Position-Specific Iterative-Basic Local Alignment Search Tool (PSI-BLAST) output (e-value and homology value) were 79.1% and 78.7%, respectively. The predicted classification for target difficulty was then used to select one of the two scoring functions in the next section.

14.3.5. Total Score

According to the predicted target difficulty, a total score is calculated as:

$$TotalScore = \begin{cases} \sum_{n}^{length} (0.35 \times SSscore(i_n, j_n, conf_n) + SideChain(AA_n, env_n)) & \text{CMtargets} \\ \sum_{n}^{length} (0.75 \times SSscore(i_n, j_n, conf_n) + SideChain(AA_n, env_n)) & \text{nonCMtargets} \end{cases} \tag{14.11}$$

The coefficients for the measure of the similarity of the secondary structures (*SSscore*) were optimized from the CASP6 targets in order to choose the best model among all server models. The similarity score of the secondary structures (SSscore, Equation 14.10 is emphasized in difficult targets. For this optimization, we used the GDT_TS value as a guide to the structural similarity between the model and the native structure.

14.3.6. Results in CASP7

The results of CIRCLE-QA, which participated in the QA of CASP7 and corresponds to the CIRCLE scoring function, are shown in Table 14.1 and Figure 14.2. The data were obtained from the CASP7 web site [31]. In the QA

TABLE 14.1 Average of the Correlation Coefficients and GDT-Scores in CASP7 for 98 Targets (Top Three Groups)

Group ID	Number of Targets	Pearson[a]	Spearman[b]	GDT_TS[c]
QA634[d]	94	**0.80**	0.61	53.8
QA556[e]	93	**0.80**	**0.76**	55.5
QA713[f]	94	0.73	0.68	57.2

The data were obtained and calculated from the CASP7 web site. The quality of the model corresponds to the GDT_TS. Bold values represent the best groups among all QA groups.
[a]Average of the Pearson linear correlation coefficient per target. The Pearson linear correlation is calculated by:

$$\frac{\sum_{i=1}^{n}(Q_i-\overline{Q})(GDT_i-\overline{GDT})}{\sqrt{\sum_{i=1}^{n}(Q_i-\overline{Q})^2}\sqrt{\sum_{i=1}^{n}(GDT_i-\overline{GDT})^2}}$$

where n is the total number of models that were evaluated by the QA groups. Q_i is the predicted quality value of the ith model. $\overline{Q}$ is the average value of the predicted quality values. GDT_i corresponds to the GDT_TS of the ith model. $\overline{GDT}$ is an average GDT_TS value of the evaluated models.
[b]The average of the Spearman rank correlation coefficient per target. The Spearman rank correlation coefficient is calculated by:

$$1-\frac{6\sum_{i=1}^{n}(\Delta rank_i)}{n(n^2-1)}$$

where $\Delta rank_i$ is the difference between each rank of the GDT_TS and the predicted model quality value of the ith model.
[c]Average GDT_TS of the server models that are ranked as the best quality model by the methods of the QA groups.
[d]Pcons.
[e]LEE.
[f]CIRCLE-QA groups.

category, predictor groups provided quality estimates with scores ranging between 0.0 and 1.0 for each protein structure model produced by the server groups participating in CASP7. A good MQAP can assign the quality scores that correlate well with the real quality score (such as GDT_TS) of the models and can select the best model that has the highest GDT_TS. The results of CIRCLE-QA in the QA category provided the performance of the CIRCLE scoring function. This performance measure did not depend on the quality of the alignment or the modeling technique.

As shown in Figure 14.2 and Table 14.1, CIRCLE-QA was ranked third and second among all QA groups according to the value of the Pearson linear correlation coefficient and the Spearman rank correlation coefficient, respectively. For the quality of the models that were selected as the best models by the QA methods, the CIRCLE-QA had the second highest GDT_TS.

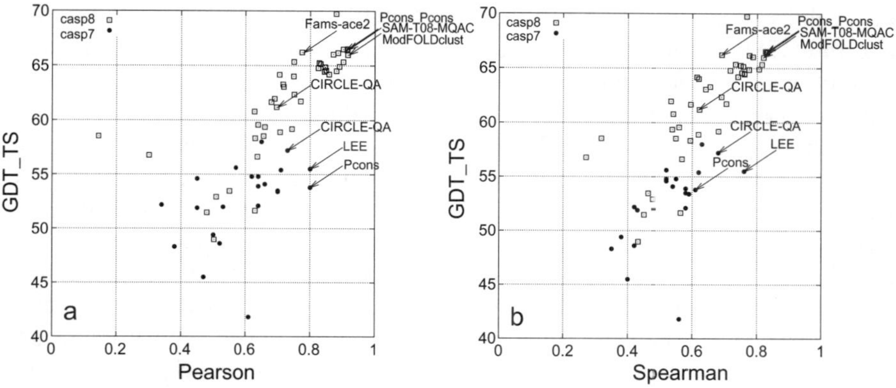

FIGURE 14.2 Results of the QA categories at CASP8 and CASP7.
The data were obtained and calculated from the CASP8 and CASP7 web sites. The squares and circles represent the results of CASP8 and CASP7, respectively. The quality of the models corresponds to the GDT_TS values. Some groups are discounted because the GDT_TS were below 40. Both y-axes represent the average of the GDT_TS of the server models that are ranked as the best quality model by each QA group. (a) The x-axis is an average of the Pearson linear correlation coefficient per target. (b) x-axis is an average of the Spearman rank correlation coefficient per target.

Moreover, Figure 14.2 also shows that Pcons (QA634), LEE (QA556), and CIRCLE-QA (QA713) are the best groups for the Pearson and Spearman coefficients. The strategies adopted by these three groups are different from each other. As described earlier, the methodology used by the Pcons group is the consensus method. Pcons predicts the quality of the model by assigning a score reflecting the average similarity to the entire ensemble of the server models. The LEE group produced very good models for most easy targets and then compared all target models with their own predictions. Therefore, the LEE group assigns a value corresponding to the distance of the target model from their own model, which was submitted as the best model by the LEE group [32]. Both Pcons and LEE use the relative similarity of the target model to the ensemble or one of the other predicted models as a quality score. As such, this approach cannot be used for assessing the quality of a model on its own. In contrast, our CIRCLE-QA is based on the structural features obtained from the coordinates of the target model. Furthermore, in the GDT_TS, CIRCLE-QA shows good performance, selecting near-native models from the sets of models. This is a valuable characteristic of CIRCLE-QA. Consequently, from the viewpoint of methodology, even though CIRCLE-QA did not perform as well as the Pcons and LEE groups, CIRCLE-QA is the best method for solely assessing the quality of a model among the QA groups that participated in CASP7.

In 2007, Anna Tramontano (an assessor of the QA category in CASP7) presented the evaluation [33] of all QA groups. She calculated the distribution of r values (Pearson's correlation coefficient) between the predicted quality and the GDT_TS value for each target, and assigned a Z-score to each of the predictions. She then calculated the sum of the Z-scores for each predictor. Negative Z-scores were set to zero. According to the report by Anna Tramontano, Pcons (634) and LEE (556) groups outperformed the others for 90 overall targets (68 single-domain targets and 22 multi-domains targets whose domains were assigned to the same category). In this evaluation, CIRCLE-QA was ranked third. Additionally, she provided detailed results by dividing the targets into template-based modeling (TBM) and template-free modeling (FM). In the TBM category, LEE, Pcons, and CIRCLE-QA were ranked as first, second, and third, respectively. In the FM category, Pcons and CIRCLE-QA stand out among the others and were ranked first and second, respectively. The reports indicated that CIRCLE-QA did not perform better than either of the Pcons and LEE group; however, it will be interesting to follow the strategy employed by CIRCLE-QA in the future.

14.3.7. Results in CASP8

Recently the CIRCLE-QA has been tested at the eight CASP experiment (CASP8). The methodology and evaluation methods of the QA category of CASP8 remained the same as those employed in CASP7. The data were obtained from the CASP8 web site [34]. Among all the QA groups, CIRCLE-QA was middle-ranking according to the Pearson's and the Spearman's correlation coefficients despite the results of CIRCLE-QA (QA396) being

TABLE 14.2 The Average of the Correlation Coefficients and the GDT-Score in CASP8 for 121 Targets (the Top Three Groups, CIRCLE-QA and Fams-Ace2)

Group ID	Number of Targets	Pearson[a]	Spearman[b]	GDT_TS[c]
QA239[d]	121	**0.92**	0.82	66.0
QA031[e]	121	**0.92**	**0.83**	66.5
QA056[f]	120	0.91	**0.83**	66.4
QA434[g]	120	0.78	0.69	66.2
QA396[h]	121	0.70	0.62	61.2

The data were obtained and calculated from the CASP8 web site. The quality of the model corresponds to the GDT_TS. Bold values represent the best groups among all the QA groups.

[a]The average of the Pearson linear correlation coefficient per target.

[b]The average of the Spearman rank correlation coefficient per target.

[c]The average GDT_TS value of the server models that are ranked as the best quality model by the methods of the QA groups.

[d]Pcons_Pcons group.

[e]ModFOLDclust group.

[f]SAM-T08-MQAC group.

[g]Fams-ace2 group.

[h]CIRCLE-QA group.

similar to the results obtained in CASP7 (see Fig. 14.2, Tables 14.1 and 14.2). Top performing groups for Pearson's and Spearman's correlation coefficients were Pcons_Pcons (QA239), ModFOLDclust (QA031), and SAM-T08-MQAC (QA056). The Pcons_Pcons (QA239) used a consensus method. ModFOLDclust (QA031) used the global clustering score based on the 3D-Jury method. The global clustering score is calculated by a pairwise comparison of the models using a template-modeling score (TM-score) [35]. SAM-T08-MQAC (QA056) used consensus terms (median TM-score, median root-mean-square deviation [RMSD], median GDT_TS, and median MaxSub) and Undertaker's cost functions, which use either evolutionary or physics-like terms [35,36].

It is clear that the consensus and clustering methods are essential techniques for ensuring an accurate model quality assessment and these methods have become a basic feature during CASP7 and CASP8.

14.4. THE FAMS-ACE2 (CIRCLE + CONSENSUS METHOD) ALGORITHM

The good performance of the consensus methods, such as the top three QA groups (Pcons_Pcons, ModFOLDclust, and SAM-T08-MQAC) in CASP8, was dependent on the variety of the set of models derived from the servers. In circumstances where there are no high-quality models or there exists a minority of good models, the consensus methods do not perform well. Moreover, the general consensus methods such as 3D-jury use the median of the global similarities of the models to be assessed. In particular, for hard targets the similarities that are calculated by comparing the full chain of the models are not clearly observed. However, in particular local space, the similarities can be observed. After CASP7, in order to improve these problems, we applied two different scoring functions for the fully automated model prediction server, fams-ace2: (i) the local consensus score from comparing the local folding of the server models with each other; and (ii) the model quality score based on the classification of the side chain environment for each residue (CIRCLE scoring function).

We participated in a fully automated model prediction server, fams-ace2, in the human predictions of the tertiary structure and QA categories of CASP8. Fams-ace2 involves structure remodeling of the server models, a consensus method and model evaluation by CIRCLE. The fams-ace2 methodology is composed of four steps as illustrated in Figure 14.3: (i) calculate "Local Consensus Total Score" (LOC_TS) for all first models of each server; (ii) select the top 10% models according to the LOC_TS; (iii) rebuild and refine passed models; and (iv) select final models using CIRCLE.

Additionally, following the structure prediction by the fams-ace2 method, we measured the similarity between the first fams-ace2 prediction and the models to be assessed for the QA category. Unlike other MQAPs, the fams-ace2 was optimized in order to generate and select the best model that has

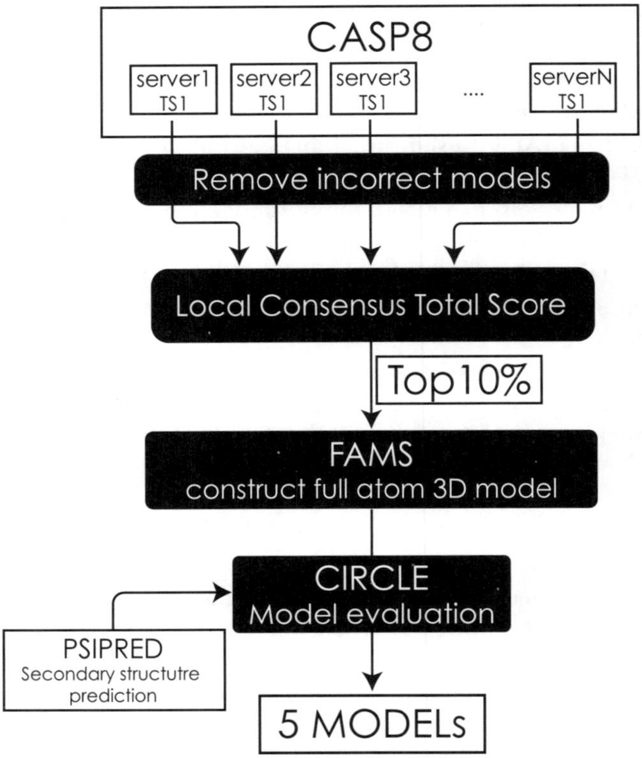

FIGURE 14.3 Methodology of the fams-ace2.
A flowchart illustrating the key steps of the fams-ace2 method and describing the first tertiary structure (TS1) of the servers to participate in CASP8, the selection of the top 10% models according to the local consensus total score, the models refined by FAMS, and selection of the final five models by the CIRCLE score.

the highest GDT_TS score from the server models. Therefore we did not consider the Pearson's and Spearman's correlation coefficients. Some details are discussed in the following sections.

14.4.1. LOC_TS

Initially, 3D models submitted by the automatic servers within 48 hours of the target sequence publication were obtained from the CASP8 web site. Subsequently, in order to avoid bad influences for the calculation of the consensus value (LOC_TS), the models that had serious physical clashes between the main chains were eliminated. The LOC_TS is the sum of the local structural conservations of local structures among all amino acid residues according to the set of the server models.

$$LOC\_TS_k = \frac{\sum_{n}^{N}\sum_{r}^{R_n} SIM_{3.0}(LOC_{k,r}, LOC_{n,r})}{N} \tag{14.12}$$

where LOC_TS_k is the Local Consensus Total Score of the kth model. N is the total number of the first models from each server. R_n is the total number of residues of the nth model. $LOC_{k,r}$ is the local structure that exists around the rth residue of the kth model within a 13Å range. $SIM_{3.0}(LOC_{k,r}, LOC_{n,r})$ is equal to the number of C-alpha atom pairs that are within 3.0Å following superposition. According to the LOC_TS, the top 10% of the set of server models are selected for the next step.

In general, the consensus score (clustering score or median score) is described by Equation 14.13.

$$consensus\_score(M_a) = \frac{\sum_{i,a \neq i}^{N} sim(M_a, M_i)}{1+N} \tag{14.13}$$

where M_a is the predicted structure model of server a. N is the number of servers. $sim(M_a,M_i)$ is the similarity between model M_a and M_i. In the 3D-Jury protocols, $sim(M_a,M_i)$ is the number of C-alpha atom pairs that are within 3.5Å following superimposition of models M_a and M_i [37] If the $sim(M_a,M_i)$ is below 40, the $sim(M_a,M_i)$ is set to zero.

14.4.2. Rebuilding and Refinement of the Server Models

The selected models that have high LOC_TS were rebuilt to full atom 3D models with our FAMS. Detailed information on the FAMS process has previously been published by our laboratory [2]. Short contacts that were observed in the server models were removed by optimization of the main chain coordinates using simulated annealing of the main chain with the conservation of the side chain conformations for each residue. Side chain atom coordinates were optimized by iterative cycles of side chain generation and main chain optimization. If regions were missing or had discontinuities in the main chain model structure, the missing regions were constructed by FAMS using a loop search process to obtain an energetically stable structure. Thus, this FAMS remodeling is an essential step in the energy evaluation of models. In the next step, our scoring method uses the environment of the side chain that is described by the fraction of the buried area and the fraction of the area that is covered by polar atoms. Consequently, even if the coordinates of the main chain are close to the native structure, a model that has many short contacts in the side chains will be rejected by the selection process.

14.4.3. Selection of the Five Best Models using CIRCLE

After the remodeling step, the five best models were selected according to the score calculated by the evaluation program, CIRCLE. There was no human intervention in any of the steps and the fams-ace2 process did not consider either server name or server performance. Therefore, a server with exceptional performance is not considered as an "excellent server" before commencement of our model selection process.

14.4.4. Results and Discussion

In the following sections, we present the results of fams-ace2 and discuss its successes and failures. The GDT_TS, described by Equation 14.1, was used as the quality score of the models. In CASP7 and CASP8, the targets were classified by the following four categories according to prediction difficulty [38,39]: (i) TBM-High Accuracy (TBM-HA): target domains in which suitable template structures (local-global alignment [LGA] score [25] >80) were available and the best prediction gave a GDT_TS score of ≥80. (ii) TBM: target domains in which at least one structurally similar template was available and a template had been used in at least one prediction. (iii) TBM/FM: target domains in which a significant fraction of the secondary structure elements could not be modeled with the correct topology based on a single template (by visual inspection). (iv) FM: target domains in which no structurally similar templates were identified or no template-based predictions were submitted.

14.4.5. Overall Results of Fams-Ace2 in the Model Prediction of the TBM Targets

The overall results that were assigned by sequence dependent measures (GDT_TS and GDT_HA), a sequence independent measure (AL0), and the cumulative Z-scores of these measures were available on the CASP8 web site [40]. Table 14.3 shows the results of the top 10 groups among all human and server predictor groups. The ranking is according to the cumulative Z-score of the GDT_TS. Except for the Zhang Server group [41], almost all of the top 10 groups were human predictor groups that used the predictions of the modeling servers and human intervention. The results showed that while fams-ace2 does not depend on human intervention, it can provide high-quality models that compare favorably with other expert human groups.

Additionally, Table 14.3 shows that fams-ace2 performed slightly better than the Zhang Server, which was the best performing individual modeling server. The plots of the GDT_TS show characteristic distributions in the single- and multi-domains when comparing the fams-ace2 and Zhang Server (Fig. 14.4). Although the performance of fams-ace2 was not better than the Zhang Server in the total number of domains, the fams-ace2 showed good performance in the Medium and Hard multi-domain targets. These are

TABLE 14.3 The Top 10 Best Groups among the 159 Server and Human Groups in the 64 TBM Domains from the Human/Server Targets of CASP8

Rank	Group ID	Group Name	Number of Predictions	Cumulative Z-score (GDT_TS)[a]	Average GDT_TS[b]	Cumulative Z-score (AL0P)[c]	Average AL0P[d]	Cumulative Z-score (GDT_HA)[e]	Average GDT_HA
1.	283	IBT_LT	64	67.383	64.834	66.333	62.949	71.870	46.855
2.	489	DBAKER	64	64.115	64.134	62.873	61.472	67.443	45.832
3.	071	Zhang	64	56.457	63.614	53.553	60.590	57.774	45.487
4.	434	fams-ace2	64	52.278	62.681	53.837	60.514	51.731	44.417
5.	426 s	Zhang Server	64	51.667	62.581	47.851	59.260	51.401	44.577
6.	057	TASSER	64	51.466	62.624	50.412	59.170	52.063	44.892
7.	046	SAM-T08-human	62	50.489	61.816	52.484	59.197	51.884	44.100
8.	196	ZicoFullSTP	64	50.374	61.396	47.756	57.501	50.898	43.754
9.	299	Zico	64	48.469	61.321	44.913	57.321	49.381	43.754
10.	453	MULTICOM	64	47.747	60.890	44.482	56.289	50.147	43.633

The data were obtained from the CASP8 web site (http://predictioncenter.org/casp8/groups_analysis.cgi). Groups with fewer than 20 predictions were removed. Models that had negative Z-scores and physically impossible structures were assigned a Z-score of zero.

[a]The cumulative Z-score of the GDT_TS.

[b]The average of the GDT_TS.

[c]The cumulative Z-score of AL0P. AL0P is defined as the percentage of correctly aligned residues in the 5Å LGA sequence-independent superposition of the model and experimental structure of the target.

[d]Average of AL0P.

[e]The cumulative Z-score of the GDT_HA value.

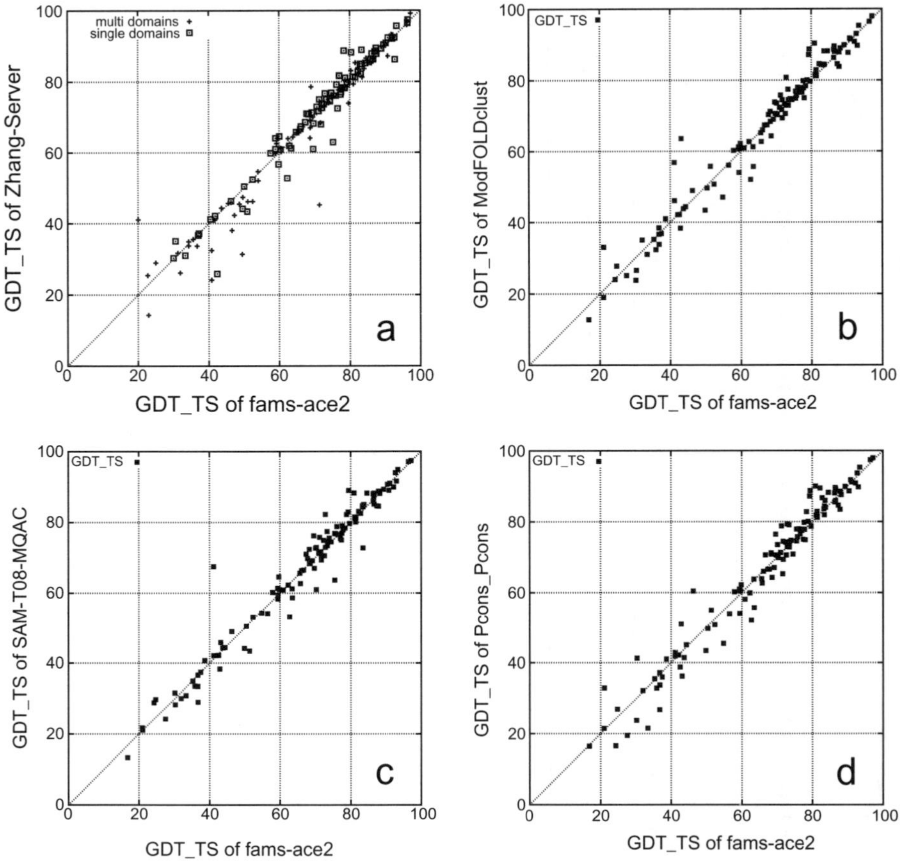

FIGURE 14.4 (a) Comparison of the GDT_TS scores between fams-ace2 and the best server (Zhang Server). The square and cross plots represent the multi- and single-domains, respectively. (b-d) Comparison of the GDT_TS values between the fams-ace2 and the QA groups: (b) ModFOLDclust, (c) SAM-T08-MQAC, and (d) Pcons_Pcons.

summarized in Table 14.4. This advantage of fams-ace2 on Medium and Hard targets gives rise to the good performance observed for GDT_TS and the cumulative Z-score of the GDT_TS (Table 14.3).

14.4.6. Results of Fams-Ace2 in the QA Category of CASP8

The fams-ace2 also participated in the QA category of CASP8. The methodology employed is very simple. Here, the 3D model of the target is initially predicted and subsequently the similarity that is based on the GDT_TS value is calculated between the first model of fams-ace2 and the server models to be assessed (e.g., LEE group). Subsequently the estimated quality score of the

TABLE 14.4 Comparison of the GDT_TS between Fams-Ace2 and the Zhang Server

	Easy[a]		Medium and Hard[b]		Easy		Medium and Hard	
	Win_5[c]	Lose_5[d]	Win_5	Lose_5	Win_0[e]	Lose_0[f]	Win_0	Lose_0
Single domains	3	4	4	0	16	48	9	7
Multi-domain	1	1	8	1	14	31	17	9

$GDT\_TS_{fams\text{-}ace2}$ is the GDT_TS of the model that was selected as the best from the server models by fams-ace2. $GDT\_TS_{Zhang\ Server}$ is the GDT_TS of the model that was selected as the best by the Zhang Server.
[a]$(GDT\_TS_{fams\text{-}ace2} + GDT\_TS_{Zhang\text{-}Server})*0.5 >= 60$.
[b]$(GDT\_TS_{fams\text{-}ace2} + GDT\_TS_{Zhang\text{-}Server})*0.5 < 60$.
[c]$GDT\_TS_{fams\text{-}ace2} - GDT\_TS_{Zhang\text{-}Server} > 5.0$
[d]$(GDT\_TS_{Zhang\ Server} - GDT\_TS_{fams\text{-}ace2} > 5.0$
[e]$GDT\_TS_{fams\text{-}ace2} - GDT\_TS_{Zhang\ Server} > 0.0$
[f]$GDT\_TS_{Zhang\ Server} - GDT\_TS_{fams\text{-}ace2} > 0.0$.

first model derived from fams-ace2 was set to 1.0 and the scores of the other models were normalized according to the ratio of their GDT_TS to that of the first model of fams-ace2.

In the QA category of CASP8, the fams-ace2 was not optimized for obtaining high correlations with GDT_TS. Consequently the fams-ace2 performance was poor for the average Pearson and Spearman correlations (Table 14.2, Fig. 14.2). In contrast, fams-ace2 showed relatively better average GDT_TS values (i.e., equivalent to the top three groups) when compared with QA groups that had similar Pearson and Spearman correlations. Table 14.5 and Figure 14.4b–d show comparisons of the quality of the best selected models at CASP8 between fams-ace2 and the top three QA groups (ModFOLDclust, SAM-T08-MQAC, and Pcons_Pcons). Fams-ace2 showed better GDT_TS values for the Medium and Hard targets when compared with the three top QA groups.

14.4.7. Comparison between the Native Structure and the Server Models

Following the CASP8, the local structural conservations between the native structures and all server models were compared. According to Equation 14.12, the local structural conservation of the kth model in the rth position was calculated as:

$$LOC\_S_{k,r} = \frac{\sum_{n}^{N} SIM(LOC_{k,r}, LOC_{n,r})}{N} \tag{14.14}$$

TABLE 14.5 Comparison of the GDT_TS of the Top Ranked Model by Fams-Ace2 and the Top Three QA Groups

	Easy[a]		Medium and Hard[b]		Easy		Medium and Hard	
	Win_5[c]	Lose_5[d]	Win_5	Lose_5	Win_0[e]	Lose_0[f]	Win_0	Lose_0
ModFOLDclust	0	4	6	4	19	53	20	13
SAM-T08-MQAC	3	4	4	2	26	45	17	12
Pcons_Pcons	1	8	11	4	28	49	21	15

$GDT\_TS_{fams\text{-}ace2}$ is the GDT_TS of the model that was selected as the best from the server models by fams-ace2.
GDT_TS_{QA} is the GDT_TS of the model that was selected as the best by the QA groups.
[a]$(GDT\_TS_{fams\text{-}ace2} + GDT\_TS_{QA})*0.5 >= 60$.
[b]$(GDT\_TS_{fams\text{-}ace2} + GDT\_TS_{QA})*0.5 < 60$.
[c]$GDT\_TS_{fams\text{-}ace2} - GDT\_TS_{QA} > 5.0$
[d]$GDT\_TS_{QA} - GDT\_TS_{fams\text{-}ace2} > 5.0$
[e]$GDT\_TS_{fams\text{-}ace2} - GDT\_TS_{QA} > 0.0$
[f]$GDT\_TS_{QA} - GDT\_TS_{fams\text{-}ace2} > 0.0$.

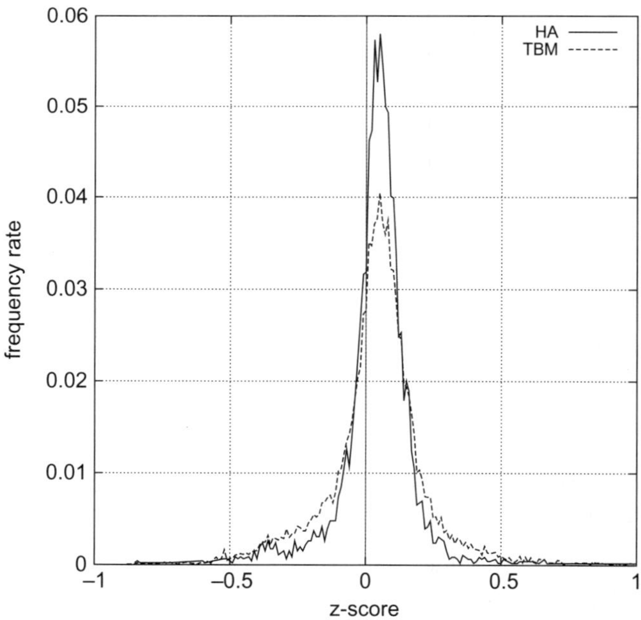

FIGURE 14.5 Comparison of the local structural conservation of the native structure with the server models. Solid and broken lines are the Z-scores of the HA targets and the TBM targets, respectively. The vertical line at the Z-score = 0 represents the average of the *LOC_S*.

where $LOC\_S_{k,r}$ is the Local Consensus Score of the kth model in the rth position. Figure 14.5 shows the distribution of the Z-score of $LOC\_S_{native,r}$. The Z-score is defined by

$$\text{Z}-\text{score}_{native,r} = \frac{LOC\_S_{native,r} - \overline{LOC\_S_r}}{\sigma_r} \tag{14.15}$$

where $LOC\_S_{native,r}$ is the Local Consensus Score of the native structure in the rth position. $\overline{LOC\_S_r}$ is the mean value of LOC_S in the rth position. σ_r is the standard deviation of LOC_S in the rth position. Figure 14.5 shows that 69 and 75% of the local structures in the native fold had higher LOC_S values than the average in the HA and TBM targets, respectively.

14.4.8. What Went Right and What Went Wrong?

Despite the poor performance with respect to the correlation scores, the fams-ace2 method selected models that had a relatively high similarity with the native structure (GDT_TS). The success of fams-ace2 can be explained by two factors. First, the fams-ace can manage the combination of the consensus method and the CIRCLE scoring function. Figure 14.5 shows a comparison

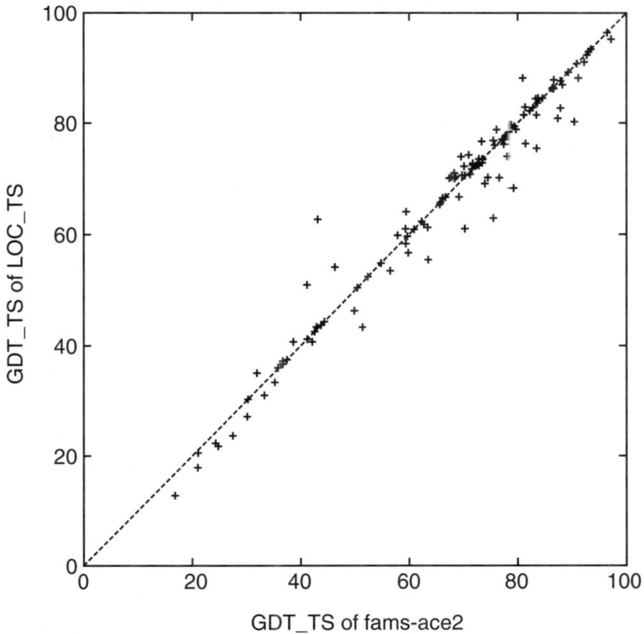

FIGURE 14.6 Comparison of the GDT_TS of the selected models by fams-ace2 with LOC_TS.
Each point corresponds to a target at CASP8.

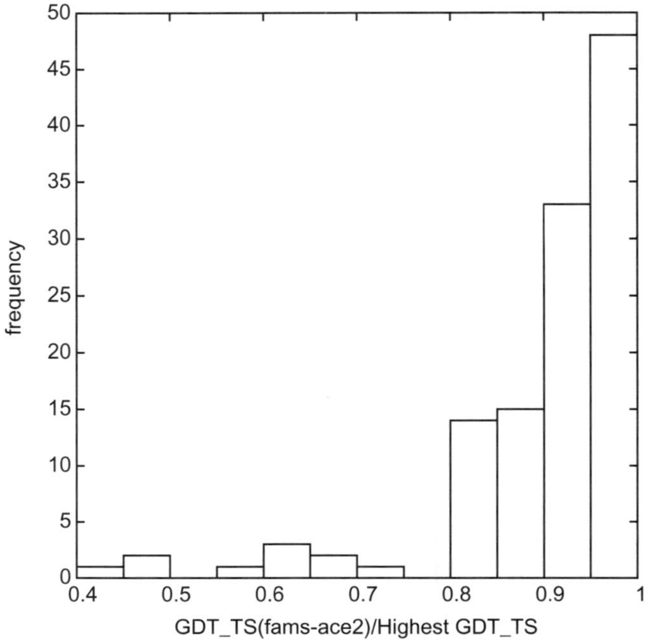

FIGURE 14.7 GDT_TS distribution of the top ranked models by fams-ace2 compared with the highest GDT_TS among all server models of each target.

of the quality (GDT_TS) of the top ranked model by fams-ace2 and LOC_TS. For 68.6% (83/121) of the CASP8 targets, fams-ace2 can select better models than the LOC_TS selected models. Additionally, for 11 targets, fams-ace2 improved the GDT_TS by more than five points. For only four targets, the fams-ace2 gave a GDT_TS score that was worse by five points or more. These data indicate that the CIRCLE method, based on the side chain environment, improved the consensus method that uses only the C-alpha coordinates. In the results of the 121 targets on CASP8, the average GDT_TS of the fams-ace2 and LOC_TS based method were 66.2 and 65.35, respectively. Second, since the consensus method was used for selecting a set of good models, fams-ace2 did not select outstanding models or appropriate models in negative outcomes. For 66.9% (81/121) of the CASP8 targets, fams-ace2 could select models that have a GDT_TS within 90% of the highest GDT_TS among all server models (Fig. 14.7).

However, during CASP8, several problems in the fams-ace2 method were found. In comparison with other QA groups (Fig. 14.4), our fams-ace2 did not perform well with easy targets. In the case when the set of models to be assessed includes near-native models that have very high GDT_TS, the consensus methods outperform the side chain-based scoring function (CIRCLE) or other knowledge-based, physical-based energy functions.

14.5. CONCLUSIONS

In this chapter, we have shown the methodology and performance of the quality assessment program CIRCLE, which considers the environments of the side chains and the secondary structures of models used in CASP7 and CASP8. In the QA category at CASP7, the CIRCLE program showed a relatively high correlation between the assigned score and the real model quality, and the performance of CIRCLE was found to be essentially equivalent to the top performing QA groups. This observation on performance is despite the fact that CIRCLE uses a single model to be assessed, whereas the other top performing QA programs used a consensus approach. Although CIRCLE showed good performance in CASP7, it did not perform well in CASP8. As such, using only the side chain environment-based score is not a satisfactory approach for model quality assessment. In other words, if the set of models includes the high-quality model, the consensus methods performs extremely well.

We also presented the fams-ace2 method that combines side chain refinement, the CIRCLE scoring function, and a consensus method. This combination performs better than methods solely reliant on consensus (LOC_TS-based method; Fig. 14.6) and CIRCLE (Fig. 14.2). This combination used in fams-ace2 is a powerful tool for selecting high-quality models. In CASP8, the fams-ace2 had a tendency to perform better than other QA groups for difficult targets. Consequently, the set of models to be assessed (i.e., high or low quality) will be crucial for improving our QA method and improvements will most likely arise through using homologous protein structures. The performance of the selection of only good models has been presented. We believe that our CIRCLE and fams-ace2 have room to improve the correlation metrics and this challenge will be solved by optimization of the method that estimates the quality of the non-top ranked models rather than using the similarity with the top ranked model.

The latest version of CIRCLE is available at http://www.pharm.kitasato-u.ac.jp/biomoleculardesign/files/circle_server.html.

REFERENCES

1. M. Takeda-Shitaka, G. Terashi, D. Takaya, K. Kanou, M. Iwadate, and H. Umeyama. Protein structure prediction in CASP6 using CHIMERA and FAMS. *Proteins*, 61(7):122–127, 2005.
2. K. Ogata and H. Umeyama. An automatic homology modeling method consisting of database searches and simulated annealing. *Journal of Molecular Graphics and Modeling*, 18(3):258–272, 305–256, 2000.
3. J. Moult, J.T. Pedersen, R. Judson, and K. Fidelis. A large-scale experiment to assess protein structure prediction methods. *Proteins*, 23:ii–v, 1995.

4. J. Moult, T. Hubbard, K. Fidelis, and J.T. Pedersen. Critical assessment of methods of protein structure prediction (CASP): Round III. *Proteins*, 37(3):2–6, 1999.
5. J. Moult, K. Fidelis, A. Zemla, and T. Hubbard. Critical assessment of methods of protein structure prediction (CASP): Round IV. *Proteins*, 45(5):2–7, 2001.
6. J. Moult, K. Fidelis, A. Zemla, and T. Hubbard. Critical assessment of methods of protein structure prediction (CASP): Round V. *Proteins*, 53(6):334–339, 2003.
7. J. Moult, K. Fidelis, B. Rost, T. Hubbard, and A. Tramontano. Critical assessment of methods of protein structure prediction (CASP): Round 6. *Proteins*, 61(7):3–7, 2005.
8. G. Terashi, M. Takeda-Shitaka, K. Kanou, M. Iwadate, D. Takaya, A. Hosoi, K. Ohta, and H. Umeyama. Fams-ace: A combined method to select the best model after remodeling all server models. *Proteins*, 69(8):98–107, 2007.
9. J. Moult, K. Fidelis, A. Kryshtafovych, B. Rost, T. Hubbard, and A. Tramontano. Critical assessment of methods of protein structure prediction: Round VII. *Proteins*, 69(8):3–9, 2007.
10. M.J. Sippl. Knowledge-based potentials for proteins. *Current Opinion in Structural Biology*, 5(2):229–235, 1995.
11. Y. Zhang and J. Skolnick. SPICKER: A clustering approach to identify near-native protein folds. *Journal of Computational Chemistry*, 25(6):865–871, 2004.
12. M.R. Lee, J. Tsai, D. Baker, and P.A. Kollman. Molecular dynamics in the endgame of protein structure prediction. *Journal of Molecular Biology*, 313(2):417–430, 2001.
13. R. Luthy, J.U. Bowie, and D. Eisenberg. Assessment of protein models with three-dimensional profiles. *Nature*, 356(6364):83–85, 1992.
14. D. Eisenberg, R. Luthy, and J.U. Bowie. VERIFY3D: Assessment of protein models with three-dimensional profiles. *Methods in Enzymology*, 277:396–404, 1997.
15. B. Wallner and A. Elofsson. Can correct protein models be identified? *Protein Science*, 12(5):1073–1086, 2003.
16. http://www.cs.bgu.ac.il/~dfischer/CAFASP4/
17. http://www.cs.bgu.ac.il/~dfischer/CAFASP4/mqap.html
18. D. Fischer. Servers for protein structure prediction. *Current Opinion in Structural Biology*, 16:178–182, 2006.
19. S.C. Tosatto. The Victor/FRST function for model quality estimation. *Journal of Computational Biology*, 12:1316–1327, 2005.
20. C.S. Pettitt, L.J. McGuffin, and D.T. Jones. Improving sequence-based fold recognition by using 3D model quality assessment. *Bioinformatics*, 21:3509–3515, 2005.
21. J. Lundstrom, L. Rychlewski, J. Bujnicki, and A. Elofsson. Pcons: A neural-network-based consensus predictor that improves fold recognition. *Protein Science*, 10(11):2354–2362, 2001.
22. K. Ginalski, A. Elofsson, D. Fischer, and L. Rychlewski. 3D-Jury: A simple approach to improve protein structure predictions. *Bioinformatics*, 19(8):1015–1018, 2003.
23. H. Zhou and J. Skolnick. Protein model quality assessment prediction by combining fragment comparisons and a consensus C(alpha) contact potential. *Proteins*, 2007, [Epub ahead of print].

24. http://predictioncenter.org/index.cgi
25. A. Zemla. LGA: A method for finding 3d similarities in protein structures. *Nucleic Acids Research*, 31:3370–3374, 2003.
26. D.T. Jones. Protein secondary structure prediction based on position-specific scoring matrices. *Journal of Molecular Biology*, 292(2):195–202, 1999.
27. D. Frishman and P. Argos. Knowledge-based protein secondary structure assignment. *Proteins: Structure, Function, and Genetics*, 23:566–579, 1995.
28. D. Anguita, A. Boni, S. Ridella, F. Rivieccio, and D. Sterpi. Theoretical and practical model selection methods for support vector classifiers. In L. Wang (Ed.), *Support Vector Machines: Theory and Applications*, vol. 177, pp. 159–179. Berlin: Springer, 2005.
29. K.J. Park, M.M. Gromiha, P. Horton, and M. Suwa. Discrimination of outer membrane proteins using support vector machines. *Bioinformatics*, 21:4223–4229, 2005.
30. H. Zhou and Y. Zhou. Single-body residue-level knowledge-based energy score combined with sequence-profile and secondary structure information for fold recognition. *Proteins*, 55(4):1005–1013, 2004.
31. http://predictioncenter.org/casp7/
32. K. Joo, J. Lee, S. Lee, J-H. Seo, S.J. Lee, and J. Lee. High accuracy template based modeling by global optimization. *Proteins*, 69(8):83–89, 2007.
33. D. Cozzetto, A. Kryshtafovych, M. Ceriani, and A. Tramontano. Assessment of predictions in the model quality assessment category. *Proteins*, 69(8):175–183, 2007.
34. http://predictioncenter.org/casp8/
35. http://www.predictioncenter.org/casp8/doc/CASP8_book.pdf
36. M. Paluszewski and K. Karplus. Model quality assessment using distance constraints from alignments. *Proteins*, 75(3):540–549, 2009.
37. N. Siew, A. Elofsson, L. Rychlewski, and D. Fischer. MaxSub: An automated measure for the assessment of protein structure prediction quality. *Bioinformatics*, 16(9):776–785, 2000.
38. http://www.predictioncenter.org/casp8/doc/Target_classification_1.html
39. J. Kopp, L. Bordoli, J.N.D. Battey, F. Kiefer, and T. Schwede. Assessment of CASP7 predictions for template-based modeling targets. *Proteins*, 69(8):38–56, 2007.
40. http://www.predictioncenter.org/casp8/groups_analysis.cgi
41. Y. Zhang. Template-based modeling and free modeling by I-TASSER in CASP7. *Proteins*, 69(8):108–117, 2007.

CHAPTER 15

MODEL QUALITY PREDICTION

LIAM J. MCGUFFIN
School of Biological Sciences
The University of Reading
Reading, UK

Since the early 1990s, a plethora of template-based and free-modeling methods have been developed that aim to predict the tertiary structures of proteins from their sequences. Often these methods will produce a number of alternative three-dimensional (3D) models of a protein with various conformations, but how do you select the most accurate conformation? The role of Model Quality Assessment Programs (MQAPs) is to provide scores that aim to predict the overall quality, or per-residue quality, of each 3D model. These scores help us to accurately estimate how much a model may deviate from the native structure and they greatly aid the process of selecting between multiple alternative models. In this chapter, we will discuss the history of MQAPs, from the early use of energy potentials to evaluate threading and homology models, through to the recent introduction of a specialized Quality Assessment (QA) category in the biennial Critical Assessment of Protein Structure Prediction (CASP) experiment. Furthermore, we will give a short review of the current state-of-the-art methods, including consensus and clustering-based methods, many of which are now available via online servers. Finally, we will contemplate the future directions for developers of methods that aim to predict 3D model quality.

15.1. INTRODUCTION

The prediction of the 3D structure of a protein from its amino acid sequence remains a difficult task for two main reasons. The first problem is one of

Introduction to Protein Structure Prediction: Methods and Algorithms,
Edited by Huzefa Rangwala and George Karypis

developing an efficient strategy in order to search through the large number of potential conformations that could be formed by a given sequence. The second problem is how to properly evaluate all of the alternative conformations in order to identify the fold that is likely to be closest to the native structure.

A number of different methods now exist that are able to generate a number of likely alternative conformations for a protein, given only information about the sequence. The most successful methods have been those that attempt to identify known structures in the Protein Data Bank [1], which can then be used as templates in order to model the structure. Using these so-called template-based methods, we are able to reduce the search space to relatively few alternative conformations. However, in the absence of homologous structures, which can be used as templates to restrict the search space, template-free modeling methods are often required to generate several thousand alternative 3D models in an attempt to sample all likely folds.

Regardless of whether template-based or template-free modeling strategies (or a mixture of both) are used for tertiary structure prediction, the result is often the generation of many alternative 3D models for each protein target. The selection of the highest quality 3D model from among numerous alternatives remains a fundamental challenge and is important for a successful prediction strategy. Over the last decade several different programs that aim to distinguish between near-native-like models of structures and so-called "decoy" structures have been developed.

The title of this chapter is "Model Quality Prediction," however, the methods that we will discuss are often popularly referred to as Model Quality Assessment Programs, or MQAPs for short. The use of the term "Assessment" in this acronym requires some clarification. MQAPs are computational methods that produce scores that aim to assess the quality of 3D models of proteins *prior* to the availability of their experimental structures, in other words they are *predictive* methods. On the other hand, model quality scoring methods that are used for benchmarking of tertiary structure predictions, such as the Global Distance Test Total Score (GDT_TS) [2] used in the CASP experiments [3], produce scores that relate to the *observed* accuracy of models by comparing them to the known experimental structures when they become available.

15.2. A BRIEF HISTORY OF MODEL QUALITY ASSESSMENT

Since the first theoretical models of proteins structures were made, structural biologists have been developing computational methods that attempt to assess their quality, prior to the availability of experimental data. Early methods ranged from basic stereochemical checks to more complex methods, which attempted to evaluate models based on their predicted free energy.

15.2.1. Basic Stereochemical Testing Scores

A popular traditional technique for evaluating structural models has been to carry out stereochemical tests using methods such as PROCHECK [4], WHAT-CHECK [5], and more recently, MolProbity [6]. These methods carry out a number of basic checks in order to determine how much a model deviates from validated crystal structures. A series of physics-and knowledge-based algorithms produces scores that are designed to check how various features in the model differ from those features observed in solved structures. Using these scores modelers are able to identify any unusual geometric features such as Ramachandran outliers, incorrect H-bonds, steric clashes, or any bond angle distortions that are either not possible or highly unlikely to occur in "real" protein structures.

These methods may be useful for providing a simple "reality check" for protein models, however they are often insufficient to use for discriminating between stereochemically correct comparative models. Furthermore, models built from distant templates may have a very accurate backbone topology, yet may fail basic checks due to the inaccurate placement of side chains or slight bond distortions. It may be unhelpful to discard these models outright in favor of more stereochemically correct models that have incorrect backbone traces. In addition, these methods often output a number of alternative scores that cannot readily be combined in order to form a single score that directly relates to overall model quality. So, in this sense they cannot be considered as MQAP methods per se, although many single-model MQAP methods do include some of these basic checks.

15.2.2. Physical and Statistical Energy Functions

Many of the pioneering model quality assessment programs were based on the premise that an accurate measurement of the free energy of a protein 3D model will enable the determination of its distance from the native structure. Figure 15.1 shows a simplified diagram of the energy funnel concept of protein folding. At the top of the funnel an unfolded protein is thought to have a high free energy as it is unfolded. This energy begins to decrease as more residues are folded, until the protein adopts its native folded structure where it is assumed to be its lowest energy state. However, the energy funnel is not completely smooth and proteins will often adopt intermediate states, falling into local energy minima in the energy landscape.

If this concept is valid, and we are able to develop a score that accurately reflects the energy of a protein, then we should be able to determine which models are likely to be closest to the native structure.

Since the early 1990s, a variety of "energy-based" programs have been developed that are specifically tuned for the discrimination of native-like models from decoy structures. These programs have been based either on statistical potentials derived from the analysis of known structures or on

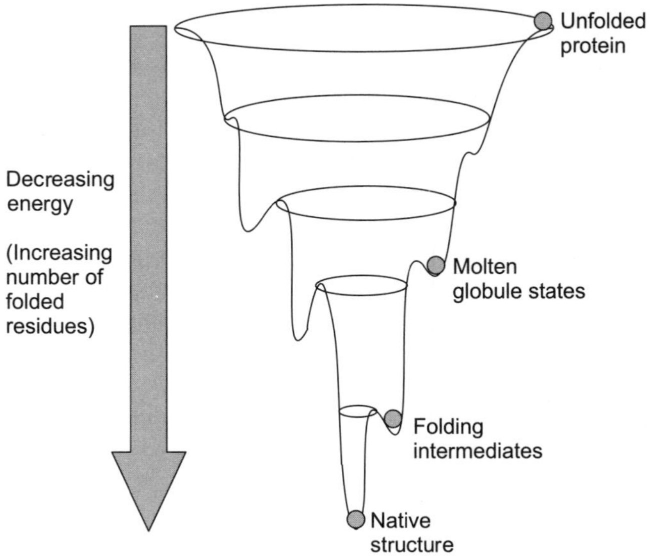

FIGURE 15.1 Energy funnel concept of protein folding. At high energy the protein is less stable and unfolded. As the protein moves down the energy landscape it becomes more stable and approaches its native conformation.

empirically derived physical effective energy functions [7]. For some time, methods employing Sippl's statistically derived energy potentials [8], such as ProSA [9], have been in popular use for rating model quality and ultimately allowed for the development of successful threading algorithms in the early 1990s [10].

Several different programs have been produced that provide alternative statistically derived energy functions for model quality assessment, for example, ANOLEA [11,12], DFIRE [13] and FRST [14], while the CHARMM [15] and Amber [16] programs provide a number of physics-based energy functions based on molecular mechanics force fields. Recently, the use of threading-based energy scores for quality assessment has been revisited with the development of the MODCHECK [17] algorithm, which uses scores similar to those used in the GenTHREADER method [18,19].

MQAPs, such as the ProSA method, introduced the concept of algorithms producing single scores representing the overall quality of the predicted structures. In some cases these pioneering energy-based MQAPs have been extremely effective at assessing model quality, however it must be stated that the development of suitable energy functions remains a major challenge. While the energetics of protein folding are to some extent understood [20,21], current simulations of folding presently require intractable amounts of computational power.

15.2.3. Other Model Quality Scoring Methods

Often the discovery of the most plausible 3D model of a protein structure may not be possible using energy-based methods. In many cases, both physics-based and statistically derived energy functions will fail to identify the best models from a number of alternatives [22]. The VERIFY3D [23] method was one of the first MQAPs, which used an alternative knowledge-based approach to statistically derived energy potentials. The VERIFY3D methods work by examining the compatibility of the structural environments of each residue in a predicted model. This was done by comparing the residue with amino acid preferences for solvent accessibility, contact with polar atoms and secondary structure types that were observed to occur in known structures.

More recently, machine learning-based MQAPs, such as ProQ, have proved to be more effective than traditional methods, such as ProSA and VERIFY3D, at enhancing model selection [24]. The ProQ method consisted of an artificial neural network that was trained in order to distinguish between correct and incorrect models using various scores—such as the frequency of atom-atom contacts and solvent accessibility—as inputs. Wallner and Elofsson demonstrated that using a neural network that was trained to recognize observed model quality—as defined by the MaxSub [25] or LG score [26]—increased the accuracy of predicted model quality over using each score individually.

A couple of independent studies have led to the discovery that very simple scores based purely on the compatibility of the secondary structure of the model compared with the predicted secondary structure of the sequences can be very effective in model quality assessment [22,27]. Eramian et al. (2006) found that a simple percentage agreement in the DSSP [28] secondary structure assignment of the models and the PSIPRED [29] predicted secondary structure of the sequence, proved to be more effective than numerous physical- and statistical-based energy potentials. Comparable results were shown in a study by McGuffin (2007), where the similar secondary structure element alignment method (ModSSEA) was benchmarked against various MQAPs using the CASP7 dataset of server models.

Although from an academic viewpoint the use of simple bioinformatics-based MQAPs methods may be less satisfying than obtaining a realistic physical energy potential, which would help us to understand the fundamental nature of protein folding, from a pragmatic viewpoint they are often shown to be more effective than current energy functions for selecting higher quality 3D models.

15.3. STATE-OF-THE-ART METHODS

Up until this point in the chapter we have considered MQAPs in terms of the analysis of individual 3D models; in other words we have been focusing on situations where we may take a single model and then use a program to predict how close it may be to the native structure. However, in many circumstances multiple alternative models for a given protein target may exist. In these situations the most accurate methods for model quality prediction often involve the comparison of multiple models, using some variation of clustering that is based on structural distances.

Whether or not clustering-based methods should be considered as "true" MQAPs is a contentious issue. Arguably, clustering-based methods will be of little use if few models exist from one protein fold recognition server. In addition, in the case of template-free modeling, clustering may be prohibitively CPU-intensive, if many thousands of models are available for each target. Clearly, there is a need to distinguish between clustering-based methods and single-model methods when discussing the state-of-the-art MQAPs.

15.3.1. Single-Model Methods

Single-model MQAPs consider each model individually in order to produce global scores for predicted model quality. In the past few years, numerous single-model MQAPs have been developed, the most successful of which use machine learning algorithms, such as artificial neural networks [30] and support vector machines [31], that are trained to recognize the observed quality based on a number of input features from a given model. Often these input scores are rescaled output scores from other individual MQAPs and therefore the output can be seen as the weighted consensus of individual methods. The ModFOLD method is one such attempt at combining scores from several individual methods [27]. The idea behind the ModFOLD method follows on from the ProQ method, in that it uses an artificial neural network to combine multiple scores, however in this case the ProQ output scores themselves are used as additional inputs along with the MODCHECK score and several secondary structure-based scores to form a meta-prediction.

Alternative methods based on support vector machines have also been developed, which combine a number of input MQAP scores to perform meta-predictions [22]. In addition, the meta-MQAP [32] and QMEAN [33] methods combine multiple MQAP scores using multivariate linear regression. Regardless of the methods used to combine scores, each of these studies has demonstrated that a greater accuracy can be gained if a consensus model quality prediction is made.

These consensus single-model methods, or meta-MQAPs, may be the best option if you have few models from an individual fold recognition server and require a prediction of global model quality [27]. However, if you are able to generate multiple models using several alternative fold recognition servers, then clustering-based methods will often produce better predictions than single-model methods.

15.3.2. Clustering-Based Methods

Clustering-based model quality prediction methods require multiple alternative 3D models as inputs, which are then compared against one another in order to determine individual model quality scores. Perhaps the first popularly used clustering-based method for model quality prediction was the 3D-Jury metaserver [34,35]. The 3D-Jury metaserver obtains individual 3D models from various independent fold recognition servers; the server then carries out all-against-all structural comparisons in order to identify the best model. The concept of clustering 3D models is represented in 2D in Figure 15.2. Each point in the figure represents a model and the space between the points indicates the structural distance from one model another. The model, which has, on average, the least structural separation from all other models in the cluster (i.e., the model close to the centroid of the cluster), is often found to be close to the native structure. However, clustering methods do not always identify the best model for a given target and in many cases the closest model to the native structure will not be the top scoring model.

The clustering of models by their structural distance is also used by the ModFOLDclust method [36,37] in order to produce a global score, as shown in the following equation:

$$S = \frac{1}{N-1} \sum_{m \in M} T_m$$

where S is the quality score for model, N is number of models ($N - 1$ is the number of pairwise structural alignments), M is the set of structural

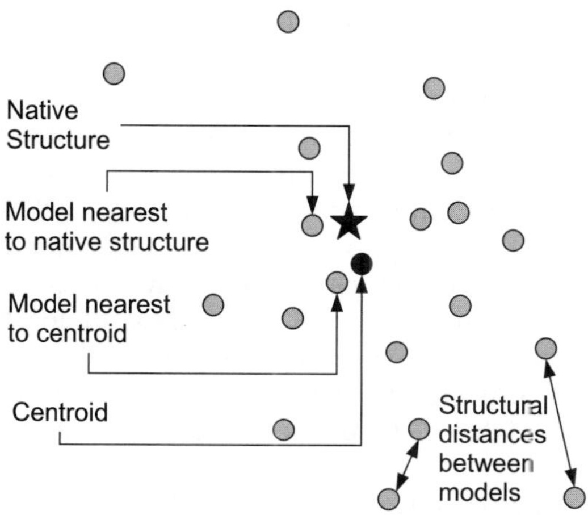

FIGURE 15.2 Concept of clustering 3D models based on their structural distances represented in 2D. The nearest model to the centroid of the cluster is often close to the native structure.

alignments and T_m is the template-modeling score (TM-score) [38] for the pairwise comparison of models. A TM-score cut-off is implemented so that alignments with scores < 0.2 are not included in the calculation. Therefore the size of set M is always equal to the number of alignments with TM-scores ≥0.2.

The ModFOLDclust method was perhaps the most successful MQAP tested in the recent CASP8 experiment in terms of global model quality scores [37]. However, similar 3D-Jury-based analysis is also incorporated into several other successful clustering-based MQAPs such as the Pcons metaserver [39]. The major difference between clustering methods lies in the type of structural alignment scores that are used to compare models. For example, Pcons uses LG-scores to compare models, 3D-Jury uses MaxSub scores and ModFOLDclust uses TM-scores. More subtle differences between methods include the parameters that are used to choose which comparisons of models are included in the final calculations of quality scores. In addition, some clustering-based methods make use of additional scores from single-model-based methods in an attempt to improve predictions, for example, ModFOLD 2.0 [37].

All-against-all structural comparisons of 3D models are computationally intensive, particularly when many thousands of models are available for each target, such as in the case of template-free modeling. In order to speed up the process more efficient clustering algorithms have been developed such as the SPICKER method [38], which uses a variation on the k-means algorithm for making structural comparisons of multiple models.

15.4. PER-RESIDUE MODEL QUALITY PREDICTION

Until now we have focused on methods that aim to produce scores that relate to the global quality of the 3D model. However, a number of MQAPs also include quality scores for each individual residue in a model. In fact, several of the traditional MQAP methods such as Verify3D and ProSA also include scores pertaining to the per-residue accuracy.

The per-residue scores in these early single-model-based methods were often found to be inaccurate and poorly correlated to the actual distance of residues in the model of a native structure [40]. As a result various attempts have recently been made at improving the per-residue accuracy using single-model methods. Currently, the most successful single-model per-residue methods are those that make use of a consensus of local scores such as ProQlocal [40] and the machine learning-based methods developed by Honig et al. [41]. However, as is the case with the global MQAP scores, the current state-of-the-art per-residue methods are based on the clustering of multiple 3D models.

Perhaps the first per-residue clustering-based MQAP method was Pcons-local [40]. In the Pcons-local method, the residue accuracy is calculated by comparing the position of each residue in a model with the positions of equivalent residues in every other model. Thus for each pairwise per-residue comparison the S-score is calculated, which relates to how close the residues are in a structural superposition. The predicted accuracy score of a residue is then calculated as mean S-score, which relates to the conservation of the residue position in 3D space. A variation on this technique has been incorporated into the ModFOLDclust method [36,37].

In the ModFOLDclust method, the S-score between structurally aligned residues is calculated as follows:

$$S_i = \frac{1}{1 + \left(\dfrac{d_i}{d_0}\right)^2}$$

where S_i is S-score for residue i, d_i is distance between residue i and the aligned residue, according to the TM-score superposition, and d_0 is a distance threshold (e.g., 3.9 Å).

The calculation of the predicted per-residue accuracy is then calculated using the following equation:

$$S_r = \frac{1}{N-1} \sum_{a \in A} S_{ia}$$

where S_r is the predicted residue accuracy for the model, N is the number of models for the target, A is the set of alignments, and S_{ia} is the S_i score for a residue in a structural alignment (a). The size of set A is equal to $N - 1$.

The predicted per-residue accuracy score is then simply converted to the predicted error—or the distance of the residue from the native structure in Angstroms (d_r)—by using the following equation:

$$d_r = d_0 \sqrt{\left(\left(\frac{1}{S_r}\right) - 1\right)}$$

An upper limit of 15 Å is set for d_r.

Histograms of the predicted per-residue error in a model are available to download from the ModFOLD server [36]. An example plot is shown in Figure 15.3.

15.5. THE ASSESSMENT OF MODEL QUALITY ASSESSMENT

Perhaps the first independent blind assessment of model quality prediction methods was carried out during the Fourth Critical Assessment of Fully Automated Structure Prediction (CAFASP4) experiment, which was conducted in parallel with CASP6 [42]. The CAFASP4 assessors coined the abbreviation MQAP, which has become popularly used to describe methods that predict model quality.

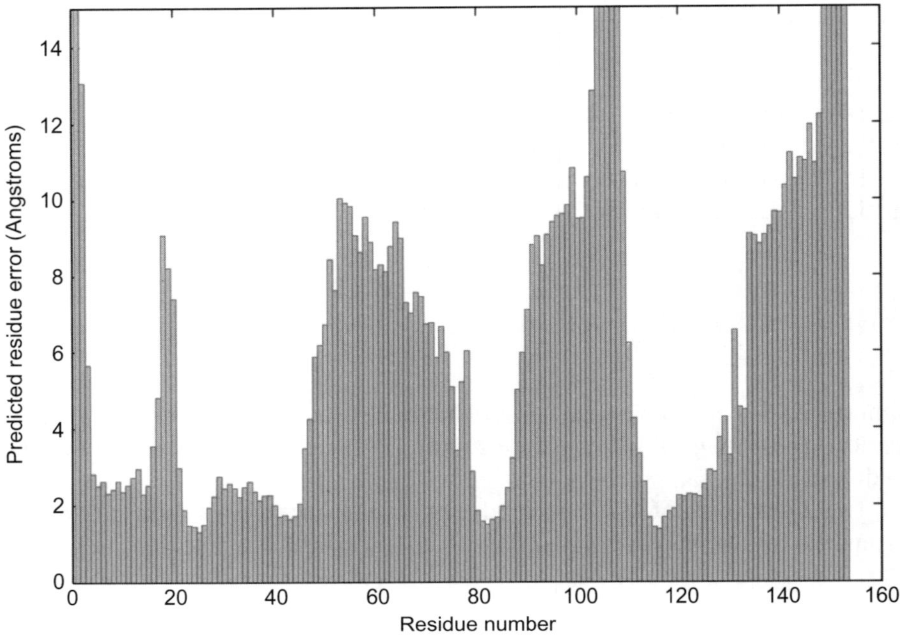

FIGURE 15.3 A per-residue error plot from the ModFOLD server (http://www.reading.ac.uk/bioinf/ModFOLD/). Plots for individual models are available to download as PostScript files.

Until CAFASP4, model quality prediction methods were benchmarked on various artificially generated "decoy" sets, using a number of alternative criteria, and most of these results had not been independently verified. The assessment of MQAP performance at CAFASP4 proposed a set of rules for the assessment of methods and benchmarked all methods on a common subset of more realistic 3D models—the actual tertiary structure predictions for the CASP6 targets from fully automated servers.

The independent blind assessment of MQAP methods was continued in CASP7, when the assessment of fully automated methods was incorporated into the main experiment [43]. The inclusion of a QA category at CASP has been extremely important for driving the development of novel MQAPs. Furthermore, in order to discover which aspects of tertiary structure prediction methods are the most successful it is often helpful to break down predictive methods into their separate stages. The introduction of the QA category also allows for partial differentiation between tertiary structure prediction methods that aim to generate novel 3D models and those that purely focus on selecting the highest quality models produced by many alternative modeling methods.

15.5.1. QA Categories at CASP

For the purposes of CAFASP4, an MQAP was defined as a program that took, as its input, a single model and then produced a single score representing the global quality of that model. Developers were encouraged to submit MQAPs as executables, which were subsequently used by the assessors to predict the quality of each model individually.

However, when the QA category was officially introduced in the CASP7 experiment, predictors were instead required to download tarballs (gzipped tar files) of tertiary structure prediction server predictions (following the server submission deadline), carry out their predictions of model quality in-house and then manually submit their scores for each model. This allowed for the participation of a wider variety of methods, such as clustering-based methods, and allowed predictors to have full control over their methods. In CASP8, a facility for assessing fully automated quality assessment methods was introduced whereby server tarballs were submitted by the assessors to individual MQAP web servers and predictions were returned via email within 3 days of submission.

The CASP7 QA category introduced a number of data formats and standards for evaluating and categorizing MQAP methods. The QA category was divided into two subcategories: QMODE1 for the assessment of the global prediction of model quality and QMODE2 for the assessment of per-residue quality predictions. Predictors are therefore required to submit their predictions in one of the two data formats. Methods that only produce global quality scores are required to use the QMODE1 data format, which includes a list of models on separate lines with their associated predicted model quality scores ranging between 0 and 1. Methods that also produce per-residue scores are required to use to QMODE2 format, whereby each global score is followed by a series of scores that relate to the predicted model quality expressed as distances in Angstroms. Developers of new methods are encouraged to refer to the CASP web site for the latest formatting rules.

15.5.2. Assessment of Global Model Quality Predictions (QMODE1)

The assessment of global model quality prediction is often carried out by comparing the output scores of each predictive method against the observed model quality scores, obtained by structurally aligning each model to the native structure. Typically observed model quality scores such as the GDT_TS [2] are used for the official CASP assessment. However, a number of alternative metrics are available in order to measure observed global model quality. For example, the TM-score program [44] is intuitive to use and provides output for three popular alternative measurements including the MaxSub score [25] and GDT_TS, as well as the TM-score.

Figure 15.4 shows scatter plots of the ModFOLDclust predicted model quality versus the observed quality, according to the GDT_TS scores, for

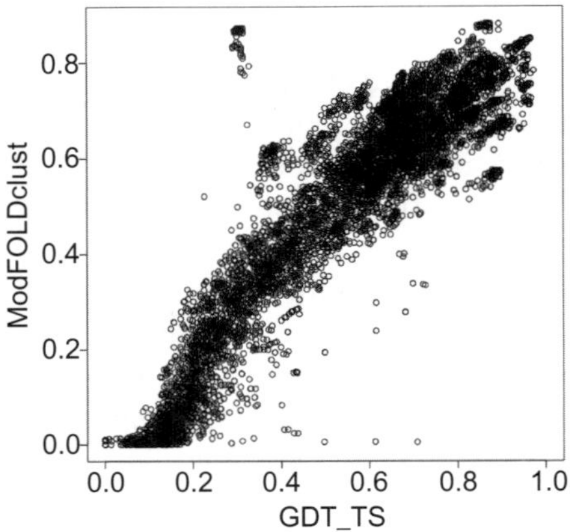

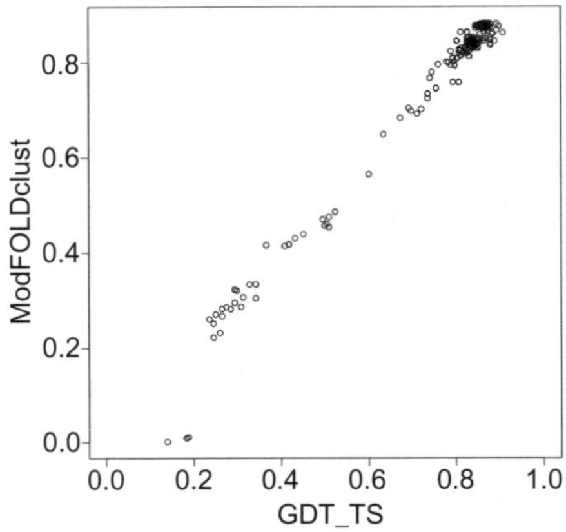

FIGURE 15.4 The global predicted model quality according to the ModFOLDclust score (y-axis) is plotted against the observed model quality according to GDT_TS (x-axis) based on structural alignment to native structures using the TM-score method. The plot on the left is for all CAPS8 server TS1 models on all targets (Pearson's $R = 0.918$, Spearman's $Rho = 0.910$, Kendall's $Tau = 0.763$). The plot on the right shows particularly accurate model quality predictions for CASP8 target T0499 ($R = 0.991$, $Rho = 0.831$, $Tau = 0.670$).

tertiary structure prediction server models submitted for each CASP8 target. The plot on the left of the figure shows the global correlation between predicted and observed quality, that is, models for all targets are pooled together. The plot on the right just shows the results for target T0499.

Typically, the global prediction of model quality is assessed using correlation coefficients. However, correlations are useful only if there is a direct linear relationship between scores, few outliers and scores are normally distributed. For nonlinear relationships between scores, correlations may be measured using Spearman's Rho or Kendall's Tau, which may be more appropriate than the Pearson's R as a measure of the ability of the method to correctly rank models. However, no correlation measures are completely robust to outliers and individual scatter plots should probably be examined before any conclusions should be drawn about the relative performances of methods.

As an alternative, MQAP methods can be directly compared with individual tertiary structure prediction servers using cumulative model quality scores over all protein targets, such as that used for the assessment of the main TS category in CASP. Most biologists may only be interested in obtaining one good quality model for their protein rather than being concerned about the complete ranking of models. Thus, the selection of the highest quality model for each target may also be a fairly pragmatic way to measure the success of methods.

15.5.3. Assessment of Per-Residue Model Quality Predictions (QMODE2)

The assessment of per-residue model quality is also carried out by making a structural superposition of the model with the native structure. The distance in Angstroms between equivalent residues in the model and the native structure is taken as the observed per-residue error. Figure 15.5 shows a comparison of the ModFOLDclust predicted per-residue error with the observed error in Angstroms for a model of CASP8 target T0389. Residues in the model that are predicted to be close to the native structure according to ModFOLDclust are shown to coincide with those that are observed to be close to native structure.

Scatter plots may also be produced of predicted versus observed per-residue error and the resulting correlations may be observed. Typically correlation coefficients are used to benchmark methods, but again these alone may be insufficient to draw useful conclusions. For example, the troughs match up very well in Figure 15.5, and Pearson's R between predicted and observed per-residue accuracy is 0.93. In Figure 15.6 if we examine the blue regions that are predicted to be accurate (left) we see that these are closely matched if we

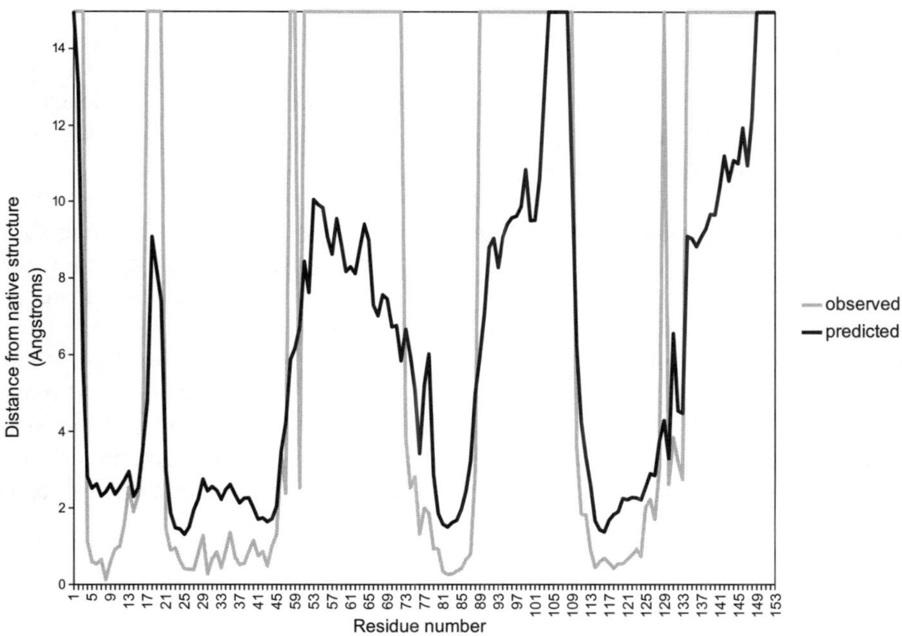

FIGURE 15.5 The ModFOLDclust predicted per-residue model quality is compared with the observed per-residue quality for a model of CASP8 target T0389. The observed model quality is measured as the distance in Angstroms of each residue from the native structure determined using the TM-score structural alignment method (the maximum distance has been set to 15 Å).

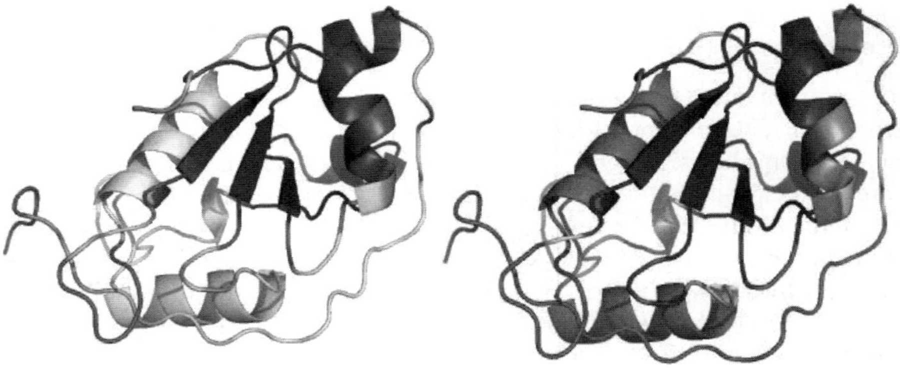

FIGURE 15.6 The ModFOLDclust predicted per-residue quality (left) for a model of CASP8 target T0389 is compared with the observed quality obtained from the alignment to the native structure (right). Each image was rendered using Pymol (http://www.pymol.org). The colors represent the residue accuracy according to the temperature scheme (blue indicates residues closest to the native structure; red, those furthest from the native structure). The blue regions that are predicted to be correct (left) are shown to be correct according to the observed accuracy (right); however, the incorrect regions are less accurately predicted. (See color insert.)

color the model by the observed error (right). On the other hand, if we focus our attention to the regions in the model predicted to be inaccurate, they range from red through yellow to green (left), however if we color by the observed accuracy then these residues are shown to be mostly colored red (right) and therefore further from the native structure than predicted.

Clearly there is a high correlation between predicted and observed distances but the predicted *range* of distances appears to be smaller than the observed range. Of course the actual values for distances may not be relevant if you are just concerned with measures of the relative accuracy. Therefore, it may be useful to determine the true positive and false positive rates for per-residue accuracy prediction at a certain distance cutoffs. Receiver Operating Characteristic (ROC) plots will allow such scores to be determined and may provide more useful evaluations of predictions than using correlations alone. Examples of where such ROC analysis has been useful for benchmarking methods include the study by Wallner and Elofsson [40] and the official CASP7 assessment of quality prediction methods [43].

15.6. ONLINE RESOURCES

A number of web servers that implement some of the MQAPs that have been discussed in this chapter have been developed. Table 15.1 lists several alternative web server implementations of published MQAPs, which are both freely available for academic use and active at the time of writing. These web servers

TABLE 15.1 A Selection of Publicly Available Prediction Servers for Model Quality Assessment

Server name	URL	Clustering-Based Methods Available?	Multiple Models As Well As Single Models?	Per-Residue Accuracy Predictions?	Global Accuracy Predictions?	Graphical Results Available?[a]
3D-Jury[b]	http://meta.bioinfo.pl/	Yes	No	Yes	Yes	No
Anolea	http://protein.bio.puc.cl/cardex/servers/anolea/index.html	No	No	Yes	Yes	No
DFIRE	http://sparks.informatics.iupui.edu/hzhou/dfire.html	No	No	No	Yes	No
MetaMQAP	https://genesilico.pl/toolkit/mqap	No	No	Yes	Yes	Yes
ModFOLD	http://www.reading.ac.uk/bioinf/ModFOLD/	Yes	Yes	Yes	Yes	Yes
ProQ	http://www.sbc.su.se/~bjornw/ProQ/ProQ.html	No	No	No	Yes	No
ProSA	https://prosa.services.came.sbg.ac.at/prosa.php	No	No	Yes	Yes	Yes
VERIFY3D	http://nihserver.mbi.ucla.edu/Verify_3D/	No	No	Yes	No	Yes

[a]Graphical results can be in the form of plots or 3D models colored by residue error.
[b]This method requires initial submission of the sequence to the 3D-Jury metaserver prior to model quality assessment.

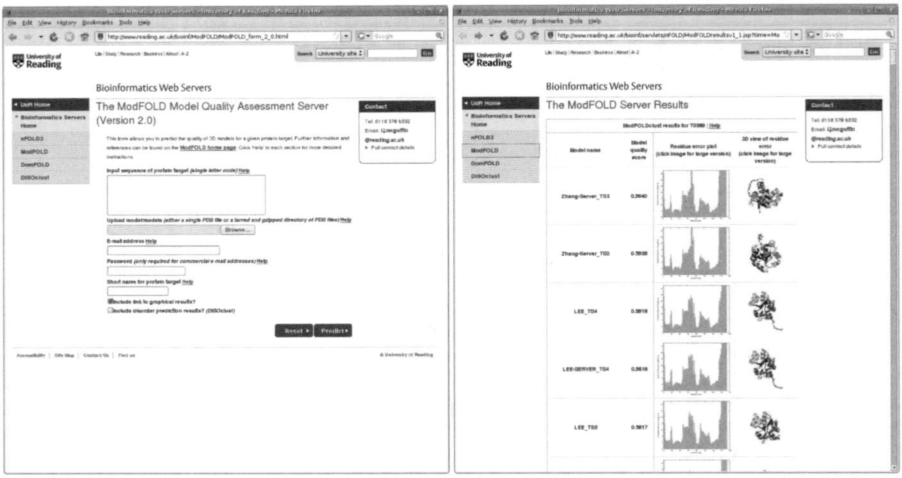

FIGURE 15.7 Screen shots of the ModFOLD server version 2.0 (http://www.reading.ac.uk/bioinf/ModFOLD/). The web interface (left) allows users to upload either single or multiple models for quality assessment. The graphical results page (right) can be accessed by following a link in the results email. Per-residue accuracy plots are provided for each model along with color-coded graphical depictions. Users may also download PDB files for their models with the predicted residue errors listed in the B-factor column. (See color insert.)

vary in their use; however, they all require the submission of the 3D model in Protein Data Bank (PDB) format and/or the target sequence. Some servers display results of the model quality predictions as single scores on a web page, others may send back predictions by email or produce results in a graphical format, for example, as per-residue error plots.

The latest version of the ModFOLD server aims to be user friendly in both the job submission and the interpretation of results. Users receive the results for their prediction via an email file attachment in machine readable QMODE2 format; however, graphical results are also accessed by following a link in the email. Screen shots of the ModFOLD web server submission form and an example of the graphical results page are shown in Figure 15.7.

15.7. CONCLUSIONS AND FUTURE DIRECTIONS FOR DEVELOPERS

The results of the recent CASP experiments (CASP7 and CASP8) have shown that clustering-based quality assessment methods are presently superior to single-model methods, when multiple models are available for a given target [43]. Furthermore, the simplest 3D-Jury-like [34] clustering-based methods, such as ModFOLDclust [36,37], currently seem to be the optimal methods for predicting both global and per-residue model quality. Methods that attempt to carry out more complex clustering or those that incorporate single-model-based scores do not seem to gain any significant advantage and may even produce worse results. Despite the success of clustering-based methods, it is clear from the recent blind assessments that there remains room for improvement—clustering methods often produce inconsistent ranges of scores from target-to-target and often do not correctly identify the best model. There is also the need for the development of faster clustering algorithms that are able to maintain high accuracy with reduced CPU requirements.

The single-model MQAP methods may be of more use than clustering-based methods where few models are available from a single server [27]. It is therefore important to encourage developers to continue to pursue the development of single-model-based methods. Perhaps single-model methods could be treated separately in future CASP assessments in order to properly credit any novel developments. However, assessors would have to either find a way to identify true single-model methods or develop an alternative assessment strategy.

Developers of individual MQAP servers should also be encouraged to pursue novel methods for combining clustering approaches with single-model methods, as most recent attempts at combining scores have made little improvement to overall accuracy. However, a number of methods have been developed, which may blur the distinction between single-model methods and clustering-based methods. These methods are technically single-model-based methods as they only require single model as their input; however, they produce scores by comparing the input model against fold recognition models generated for the sequence [35,43].

Finally, developers of tertiary structure prediction servers should be encouraged to make use of model quality assessment methods in order to add value to their predictions both through the re-ranking of 3D models and through the incorporation of per-residue accuracy scoring into their PDB files.

REFERENCES

1. H.M. Berman, J. Westbrook, Z. Feng, G. Gilliland, T.N. Bhat, H. Weissig, I.N. Shindyalov, and P.E. Bourne. The protein data bank. *Nucleic Acids Research*, 28:235–242, 2000.
2. A. Zemla, C. Venclovas, J. Moult, and K. Fidelis. Processing and analysis of CASP3 protein structure predictions. *Proteins*, 3:22–29. 1999.
3. J. Moult, K. Fidelis, A. Kryshtafovych, B. Rost, T. Hubbard, and A. Tramontano. Critical assessment of methods of protein structure prediction: Round VII. *Proteins*, 69(8):3–9, 2007.
4. R.A. Laskowski, M.W. McArthur, D.J. Moss, and J.M. Thornton. PROCHECK: A program to check the stereochemical quality of protein structures. *Journal of Applied Crystallography*, 26:283–291, 1993.
5. R.W. Hooft, G. Vriend, C. Sander, and E.E. Abola. Errors in protein structures. *Nature*, 381:272, 1996.

6. I.W. Davis, L.W. Murray, J.S. Richardson, and D.C. Richardson. MOLPROBITY: Structure validation and all-atom contact analysis for nucleic acids and their complexes, *Nucleic Acids Research*, 32:W615–W619, 2004.
7. T. Lazaridis and M. Karplus. Effective energy functions for protein structure prediction. *Current Opinion in Structural Biology*, 10:139–145, 2000.
8. M.J. Sippl. Calculation of conformational ensembles from potentials of mean force. An approach to the knowledge-based prediction of local structures in globular proteins. *Journal of Molecular Biology*, 213:859–883, 1990.
9. M.J. Sippl. Recognition of errors in three-dimensional structures of proteins. *Proteins*, 17:355–362, 1993.
10. D.T. Jones, W.R. Taylor, and J.M. Thornton. A new approach to protein fold recognition. *Nature*, 358:86–89, 1992.
11. F. Melo, D. Devos, E. Depiereux, and E. Feytmans. ANOLEA: A www server to assess protein structures. *Proceedings of the International Conference on Intelligent Systems for Molecular Biology*, 5:187–190, 1997.
12. F. Melo and E. Feytmans. Novel knowledge-based mean force potential at atomic level. *Journal of Molecular Biology*, 267:207–222, 1997.
13. H. Zhou and Y. Zhou. Distance-scaled, finite ideal-gas reference state improves structure-derived potentials of mean force for structure selection and stability prediction, *Protein Science*, 11:2714–2726, 2002.
14. S.C. Tosatto. The victor/FRST function for model quality estimation. *Journal of Computational Biology*, 12:1316–1327, 2005.
15 B.R. Brooks, R.E. Bruccoleri, D.J. Olafson, D.J. States, S. Swaminathan, and M. Karplus. CHARMM: A program for macromolecular energy, minimization, and dynamics calculations. *Journal of Computational Chemistry*, 4:187–217, 1983.
16. S.J. Weiner, P.A. Kollman, D.A. Case, C. Singh, C. Ghio, G. Alagona, S. Profeta, and P. Weiner. A new force field for molecular mechanical simulation of nucleic acids and proteins. *Journal of the American Chemical Society*, 106:765–784, 1984.
17. C.S. Pettitt, L.J. McGuffin, and D.T. Jones. Improving sequence-based fold recognition by using 3D model quality assessment. *Bioinformatics*, 21:3509–3515, 2005.
18. D.T. Jones. GenTHREADER: An efficient and reliable protein fold recognition method for genomic sequences. *Journal of Molecular Biology*, 287:797–815, 1999.
19. L.J. McGuffin and D.T. Jones. Improvement of the GenTHREADER method for genomic fold recognition. *Bioinformatics*, 19:874–881, 2003.
20. L. Mirny and E. Shakhnovich. Protein folding theory: from lattice to all-atom models. *Annual Review of Biophysics and Biomolecular Structure*, 30:361–396, 2001.
21. J.N. Onuchic, H. Nymeyer, A.E. Garcia, J. Chahine, and N.D. Socci. The energy landscape theory of protein folding: insights into folding mechanisms and scenarios. *Advances in Protein Chemistry*, 53:87–152, 2000.
22. D. Eramian, M.Y. Shen, D. Devos, F. Melo, A. Sali, and M.A. Marti-Renom. A composite score for predicting errors in protein structure models. *Protein Science*, 15:1653–1666, 2006.
23. D. Eisenberg, R. Luthy, and J.U. Bowie. VERIFY3D: Assessment of protein models with three-dimensional profiles. *Methods in Enzymology*, 277:396–404, 1997.

24. B. Wallner and A. Elofsson. Can correct protein models be identified? *Protein Science*, 12:1073–1086, 2003.
25. N. Siew, A. Elofsson, L. Rychlewski, and D. Fischer. MaxSub: An automated measure for the assessment of protein structure prediction quality. *Bioinformatics*, 16:776–785, 2000.
26. S. Cristobal, A. Zemla, D. Fischer, L. Rychlewski, and A. Elofsson. A study of quality measures for protein threading models. *BMC Bioinformatics*, 2:5, 2001.
27. L.J. McGuffin. Benchmarking consensus model quality assessment for protein fold recognition. *BMC Bioinformatics*, 8:345, 2007.
28. W. Kabsch and C. Sander. Dictionary of protein secondary structure: Pattern recognition of hydrogen-bonded and geometrical features. *Biopolymers*, 22:2577–2637, 1983.
29. D.T. Jones. Protein secondary structure prediction based on position-specific scoring matrices. *Journal of Molecular Biology*, 292:195–202, 1999.
30. A. Krogh. What are artificial neural networks? *Nature Biotechnology*, 26:195–197, 2008.
31. W.S. Noble. What is a support vector machine? *Nature Biotechnology*, 24:1565–1567, 2006.
32. M. Pawlowski, M.J. Gajda, R. Matlak, and J.M. Bujnicki. MetaMQAP: A meta-server for the quality assessment of protein models. *BMC Bioinformatics*, 9:403, 2008.
33. P. Benkert, S.C. Tosatto, and D. Schomburg. QMEAN: A comprehensive scoring function for model quality assessment. *Proteins*, 71:261–277, 2008.
34. K. Ginalski, A. Elofsson, D. Fischer, and L. Rychlewski. 3D-Jury: A simple approach to improve protein structure predictions. *Bioinformatics*, 19:1015–1018, 2003.
35. L. Kajan and L. Rychlewski. Evaluation of 3D-Jury on CASP7 models. *BMC Bioinformatics*, 8:304, 2007.
36. L.J. McGuffin. The ModFOLD server for the quality assessment of protein structural models. *Bioinformatics*, 24:586–587, 2008.
37. L.J. McGuffin. Prediction of global and local model quality in CASP8 using the ModFOLD server. *Proteins*, 2009, in press.
38. Y. Zhang and J. Skolnick. Scoring function for automated assessment of protein structure template quality. *Proteins*, 57:702–710, 2004.
39. B. Wallner and A. Elofsson. Prediction of global and local model quality in CASP7 using Pcons and ProQ. *Proteins*, 69(8):184–193, 2007.
40. B. Wallner and A. Elofsson. Identification of correct regions in protein models using structural, alignment, and consensus information. *Protein Science*, 15:900–913, 2006.
41. M. Fasnacht, J. Zhu, and B. Honig. Local quality assessment in homology models using statistical potentials and support vector machines. *Protein Science*, 16:1557–1568, 2007.
42. J. Moult, K. Fidelis, B. Rost, T. Hubbard, and A. Tramontano. Critical assessment of methods of protein structure prediction (CASP): Round 6. *Proteins*, 61(7):3–7, 2005.

43. D. Cozzetto, A. Kryshtafovych, M. Ceriani, and A. Tramontano. Assessment of predictions in the model quality assessment category. *Proteins*, 69(8):175–183, 2007.

44. Y. Zhang and J. Skolnick. SPICKER: A clustering approach to identify near-native protein folds. *Journal of Computational Chemistry*, 25:865–871, 2004.

CHAPTER 16

LIGAND-BINDING RESIDUE PREDICTION

CHRIS KAUFFMAN and GEORGE KARYPIS
Department of Computer Science
University of Minnesota
Minneapolis, MN

16.1. INTRODUCTION

In this chapter, we explore means for predicting protein residues that interact with small molecules. We will motivate the problem by describing potential uses for such information and proceed to discuss methods advanced for prediction. We describe our sequence-based approach and contrast it with another current method that relies on predicted protein structure to help identify ligand-binding residues. In the last part of the chapter, we employ sequence-based predictions in a homology modeling task that shows that the predictions are presently accurate enough to improve downstream performance.

16.1.1. Background on Ligand Binding

Recent advances in high-throughput sequencing technologies have continued to increase the gap between the number of proteins whose function is well-characterized and the proteins for which there is no experimental functional data. As a result, life sciences researchers are becoming increasingly more dependent on computational methods to infer the function of proteins. To address this challenge, a number of novel and sophisticated methods have been developed within the field of computational biology that are designed to predict the different aspects of a protein's function.

Our focus is on methods that predict, from a protein's primary sequence, the *ligand-binding residues* that bind to small molecules. Small molecules

Introduction to Protein Structure Prediction: Methods and Algorithms,
Edited by Huzefa Rangwala and George Karypis

interact with proteins in regions that are accessible and that provide energetically favorable contacts. Geometrically, these binding sites are generally deep, concave-shaped regions on the protein surface, referred to alternately as clefts or pockets.

Identifying ligand-binding residues reliably aids the overall understanding of the role and function of a protein by using them to subsequently predict the types of ligands to which they bind and, in the case of enzymes, the types of reactions that are catalyzed. Moreover, knowledge of the residues involved in protein-ligand interactions has broad applications in drug discovery and chemical genetics, as it may be used to better virtually screen large chemical compound libraries [1] and to aid the process of lead optimization [2,3]. In addition, the ligand-binding residues of a protein can be used to influence target-template sequence alignment in comparative protein modeling that has been shown to improve the quality of the three-dimensional (3D) models produced for the target's binding site [4]. These quality improvements in the binding site's 3D model are critical to docking-based approaches for virtual screening [5].

16.1.2. Overview of Methods

Predicting ligand-binding site residues from sequence information is similar to several site interaction prediction problems involving DNA [6–8], RNA [9,10], and other proteins [11–13]. Existing approaches for identifying ligand-binding residues can be broadly classified into two groups that alternately use machine learning and sequence homology to solve the problem.

16.1.2.1. Machine Learning Approaches. A number of groups have employed supervised machine learning techniques for binding residue prediction. This involves using some proteins to develop a model of what constitutes a binding residue and then testing the model on an independent set of proteins. A variety of features and techniques have been explored but the consensus seems to be that sequence profiles and conservation are the important features and support vector machines (SVMs) provide the best discrimination.

Fischer et al. presented a method for functional residue prediction based on sequence features [14]. They studied prediction for residues contacting ligands and also for the more restrictive catalytic site residues as defined in the Catalytic Site Atlas [15]. A Bayesian-type learner was trained to produce the probability of a residue being a binder with the primary feature of interest being residue conservation. The authors introduced a new conservation measure, FRcons, which proved the most effective in their benchmark but achieved a precision of less than 30% at sensitivity equal to 50%.

Petrova and Wu performed a fairly comprehensive evaluation of machine learning algorithms and features useful for direct prediction of catalytic residues in a small set of proteins [16]. They found that SVMs were the most powerful method for this task. The features they found to be most important were residue conservation, amino acid identity, entropy, and characteristics of the nearest cleft to a residue. The first three of these are sequence features that may be utilized even when no structure is available for the target. Features of clefts necessitate the target structure to be either known or predicted.

Youn et al. also studied the use of various features with SVMs for catalytic residue prediction [17]. Their evaluation encompassed a large set of 987 protein domains from the Structural Classification of Proteins (SCOP) [18,19] which they analyzed at the family, superfamily, and fold levels. They achieved a receiver operating characteristic (*ROC*) of 0.866 for feature-only predictions at the family level. However, catalytic residues are a more restricted set than general ligand binders: only 1.1% of the residues are in the positive class in their study while 8.6% of residues in our data were in the positive class. The precision and recall reported at the family level by Youn et al. [17] is quite low: 16.6% precision at 15.1% recall. Feature ranking was done and they found that the Position-Specific Scoring Matrices (PSSMs) and the information per position (IPP) reported by the Position-Specific Iterative-Basic Local Alignment Search Tool (PSI-BLAST) were most useful for prediction. Structural conservation was found to be the next best feature.

16.1.2.2. Homology-Based Approaches. The transfer of sequence properties, such as ligand-binding status, to the target based on its alignment to templates is a common method for prediction. These techniques are often referred to as as *homology transfer* (HT) as properties of the target sequence are predicted by transferring them from homologous templates. HT is a close relative of nearest neighbor methods frequently employed for machine learning tasks. The primary difference is that nearest neighbor methods typically deal with individual objects with feature representations, while in HT predictions are made on a per-residue basis but similarity search is done using whole sequences with the sequence alignments determining individual residue relations.

The *firestar* algorithm of López et al. [20] utilizes HT and conservation scores to make binding residue predictions from sequence. A profile is calculated for a target using PSI-BLAST and significant alignments are searched in their FireDB, which is composed of ligand-binding proteins. The resulting multiple sequence alignment is used to estimate conservation of target residues that are then predicted to be binders if they align to template residues that are binders. In *firestar*, profiles are used to estimate the reliability of alignments between target and templates to determine when transfer should occur, but not to directly characterize ligand-binding residues.

Brylinski and Skolnick recently introduced FINDSITE as a method for making predictions about protein-ligand interactions [21,22]. The method belongs to the HT category but uses structural measures of similarity rather than sequence alignment. FINDSITE identifies templates by threading the target sequence through candidate template structures and retaining high-scoring templates. The accumulated templates are then structurally aligned to the target structure. If the target structure is not available, it is predicted using one of several methods. The binding status of template residues is then transferred to target residues based on this structural correspondence.

The drawback of FINDSITE is that the target structure is required for the alignment of templates. In cases where the target structure is available, FINDSITE can exploit it well to make binding site predictions. However, when it is not available, predicting the structure of the target-protein can be a computationally expensive proposition with no guarantees on quality.

16.2. EVALUATION OF LIGAND-BINDING PREDICTION METHODS

16.2.1. Methods

In this section we describe prototype algorithms that represent the basic ideas behind most sequence-based binding residue predictors. We begin by discussing relevant features to both types of algorithms. Sequence alignment plays a central role in the homology-based method and is described subsequently. With these tools laid out, two prototype prediction methods are described: homology-based transfer and machine learning with SVMs. LIBRUS combines these two approaches and is described in the last section. We also briefly discuss FINDSITE, which uses predicted structures to make its predictions rather than direct predictions from sequence.

16.2.1.1. Relevant Sequence Features. The primary source of information about proteins of unknown structure is their amino acid sequence. Evolutionary information may be inferred from the sequence using PSI-BLAST, which computes a substitution profile for each residue in the protein sequence [23]. This profile has two parts: a PSSM and a Position-Specific Frequency Matrix (PSFM). The PSSM is a real-valued matrix of dimension $n \times 20$ where n is the length of the protein. A row represents the log-odds probability of each of the 20 amino acids occurring at that sequence position. The row of a PSSM may be used directly as a feature vector for a residue as is done in the machine learning case or may be utilized along with the PSFM in alignment scoring schemes as will be done for the homology-based transfer method.

Secondary structure in proteins are locally recurring structures that are commonly divided into three major classes: α-helices (H), β-sheets (S), and unstructured coils (C). Each residue of the protein may be assigned one of these classes based on its tertiary structure, a feature referred to as secondary structure elements (SSE) and encoded as an $n \times 3$ matrix. A popular and long-standing means of assigning SSE is the DSSP program of Kabsch and Sander [24]. Many methods have been studied to predict secondary structure from protein sequence and some studies have shown that these methods can positively impact downstream prediction tasks [25,26]. A relatively recent approach using SVMs is YASSPP [27], which produces, for each residue of a protein, a likelihood of being helix, sheet, and coil. This predicted secondary structure (SSP) is used as a surrogate for SSE when the true secondary structure is unavailable.

16.2.1.2. Alignment Techniques. Given two protein sequences, a core problem is to construct a sequence alignment. The scoring mechanism used to construct this alignment can have drastic effects on the constructed alignment similarity score assigned to two sequences. The profile-based alignment scoring scheme that we used is derived from the work on PICASSO [28], which was shown to be very sensitive in subsequent studies [29,30]. Our own work aligns sequences by computing an optimal alignment using an affine gap model with aligned residues i and j in sequences X and Y, respectively, scored using a combination of profile-to-profile scoring and secondary structure matching. The score is given by

$$
\begin{aligned}
S(X_i, Y_j) = & \sum_{k=1}^{20} PSSM_X(i,k) \times PSFM_Y(j,k) \\
& + \sum_{k=1}^{20} PSSM_Y(j,k) \times PSFM_X(j,k) \\
& + w_{SSE} \sum_{k=1}^{3} SSE_X(j,k) \times SSE_Y(j,k),
\end{aligned}
\tag{16.1}
$$

where $PSSM$, $PSFM$, and SSE are the profile matrices and SSE described in Section 16.2.1.1. We will frequently deal with the case of aligning a target of unknown structure with a template of known structure. In this situation, SSP is used in place of true secondary structure for the target. The parameter w_{SSE} is the relative weighting of the secondary structure score that is set to $w_{SSE} = 3$ based on our experience [4].

16.2.1.3. Homology-Based Transfer. Alignment of protein sequences is a powerful tool that allows characteristics of one to be inferred from the other. This is the crux of homology-based methods. Given a target protein, a database of template sequences with known binding information is searched for high scoring alignments to the target. Once good templates are identified, a score is assigned to each residue in the target based on the number of template residues that aligned against it and are known to be ligand binders. This score is referred to as the *homology-based transfer score* or HTS.

There are a number of dimensions along which alignment and prediction may be adjusted including the scoring mechanism and weighting of template contributions to the final prediction. The alignment scoring of Section 16.2.1.2 may be applied in many alignment frameworks (see Chapter 11 of Reference [31]). Our experience has been that local alignments provide the best results due to reporting only the best matching target-template subsequences that can increase the reliability of prediction. The top 20 alignments should be used with weighting for each residue based on the alignment score in a window of seven residues. See our previous work for additional details [32].

16.2.1.4. SVM Prediction.

In this method, the prediction problem is treated as a supervised learning problem whose goal is to build a model that can predict whether a residue is ligand-binding or not, a binary classification problem. In supervised learning, each object of interest is encoded by a feature vector and a model is learned that can predict the class based on those features.

Recent research on building models for predicting various structural and functional properties of protein residues in References [27] and [8] has suggested training SVMs [33] on sequence features of each residue to classify the residue as a ligand binder or nonbinder. Effective features include PSSM and SSP in a window around each residue. Sliding windows are an easy way to expand feature vectors. The results shown later use a window of nine residues centered on the residue of interest and concatenated the PSSMs and SSPs of adjacent residues for a total of 207 features per residue ($9 \times [20 + 3]$). Window features that extended beyond the first or last residue of the sequence were assigned zero values. This feature representation is closest to that of Reference [17] where PSSMs in a sliding window of size 21 were employed in one of their methods for the related problem of predicting a protein's catalytic residues.

One important aspect of combining different types of features is providing proper weights on them as their numerical ranges may vary greatly. In the results reported later, we combined features by weighting them to have equal norm. Examples of the norms and weighting are given in Table 16.1. Properly weighting the combination of features significantly enhanced the performance of the final model.

16.2.1.5. LIBRUS: Combining SVM and Homology-Based Transfer.

Direct prediction by SVMs and prediction by homology-based transfer utilize training information in different ways to make their predictions. SVMs utilize intrinsic features of the residue represented as PSSMs and SSPs with little context for the residue within the whole protein nor any relation of the con-

TABLE 16.1 Average Norms of Residue Features

Statistic	PSSM	SSP	HTS
Average	13.53	2.00	0.07
Standard deviation	3.88	0.53	0.11
Weight	1.00	6.75	207.00

Columns are position specific scoring matrices (PSSM), predicted secondary structure vector (SSP), and HT scores (HTS). The bottom row is the weight used on these features in SVMs for sequence-based predictions.

taining protein to other proteins in the training set. Conversely, homology-based transfer solely relies on the global context of the residue: where it is located in alignments of the containing protein against other proteins and how many ligand-binding residues align against it. The different characteristics of the information utilized by the two approaches suggest that their combination can lead to a better overall predictor.

A simple linear combination of SVM and HTS may be used. With proper weights set on the two scores, this approach works rather well as will be seen in the results.

Alternatively, an SVM may be trained on the PSSMs and SSEs of the direct prediction method and the HTS of the HT method. The resulting hybrid predictor utilizes both types of features. We have built such a predictor called LIBRUS [32], which uses a total of $9 \times (20 + 3 + 1) = 216$ features weighted according to Table 16.1.

16.2.1.6. FINDSITE. The methods mentioned in the previous sections solely utilize sequence information for targets of unknown structure to directly predict ligand-binding residues. Alternatively, the target structure can be predicted and then utilized to identify binding residues. This is the approach taken in FINDSITE, which is a recent approach to binding site identification [21]. The results of this method on one dataset are provided later to contrast the direct predictions made by sequence-based methods.

FINDSITE identifies a number of predicted binding sites with associated binding residues for each target. The prediction values for these correspond to the fraction of template structure residues that were identified as ligand binding and aligned against the target residue. Up to the first five predicted binding sites are reported in the results section. Some residues appear as part of multiple binding sites in the FINDSITE predictions and have different scores associated with them in the different sites. In those cases, the score from the first binding site a residue occurred in was used as this was typically the largest and most well-defined predicted binding site.

16.2.2. Experimental Setup

16.2.2.1. Sequence Data. The sequence-based methods were evaluated on a dataset referred to as DS1, which consists of 885 protein chains (268,699 residues) that were derived from the RCSB Protein Data Bank (PDB) in October of 2008 [34]. The set of proteins in DS1 were selected so that they satisfy the following constraints: (i) has better than 2.5 Å resolution, (ii) is longer than 100 residues, (iii) has an unbroken backbone, and (iv) has at least five residues in contact with a ligand. Finally, the dataset was culled so that no two sequences have above a 30% sequence identity according to the National Center for Biotechnology Information's (NCBI) blastclust program.

Ligands in our datasets were small molecules in contact with proteins identified by scanning the PDB using the "has ligand" search option. DNA, RNA, and other large proteins were excluded as candidate ligands as were ligands with fewer than eight heavy (non-hydrogen) atoms. We required proteins to have ligand-binding residues with a heavy atom within 5 Å of a ligand. By this definition, 8.6% of DS1 residues are ligand-binding residues (positive class). In-house software was developed to identify ligands and ligand-binding residues.

Protein sequences were derived directly from the structures using in-house software. When nonstandard amino acids appeared in the sequence, the three-letter to one-letter conversion table from ASTRAL [35] version 1.55 was used to generate the sequence (http://astral.berkeley.edu/seq.cgi?get=release-notes;ver=1.55). When multiple chains occurred in a PDB file, the chains were treated separately from one another. Profiles for each sequence were generated using PSI-BLAST version 2.2.13 [23] and the NCBI NR database (version 2.2.12 with 2.87 million sequence, downloaded August 2005). PSI-BLAST produces a PSSM and a PSFM for a query protein, both of which are employed for our sequenced-based prediction and alignment methods. Three iterations were used in PSI-BLAST with the default e-value threshold for inclusion in the profile and default expectation value (options -j 3 -h 2e-3 -e 10).

True SSE for each protein of DS1 was obtained using the DSSP program [24] while SSP was obtained using YASSPP [27]. YASSPP predicted the correct secondary structure for 83% of the residues in DS1.

In the homology-based transfer method, template proteins are assumed to have known structure and therefore SSE is available for them while the targets must use SSP as they have unknown structure. Care must be taken so that the encoding of SSE is compatible with SSP. A straightforward means of defining the SSE is, for each residue, assign 1 to the dimension corresponding to its true state and 0 to the other dimensions, for example, for a true helix, the encoding would be [1, 0, 0], a true sheet [0, 1, 0], and true coil [0, 0, 1]. Our experience has been that a better means of encoding true SSE to compare it to YASSPP's SSP is the following. The average YASSPP vector of all true helices was computed. For a true helix, the SSE is assigned this average vector. Similar averaging steps for sheets and coils were computed and used for true secondary structure. This ensures that SSE and SSP are scaled similarly.

A second dataset, referred to as DS2, was derived from the set of proteins used to evaluate FINDSITE in Reference [21]. DS2 consists of 564 proteins (136,316 residues) after eliminating those sequences with 35% identity or better to any sequence in DS1 according to BLAST. This dataset was used to illustrate the relative performances of LIBRUS and FINDSITE with LIBRUS using DS1 as training data. Sequence features for the members of DS2 were derived as they are in Section 16.2.2.1.

16.2.2.2. Evaluation Metrics. Three-fold cross-validation is used on DS1 to assess how well the methods generalize. In each step, two sets of the data were used to learn a model and predictions were made on the remaining set of targets. This generated a single prediction of binding/nonbinding for each residue that was subsequently used in evaluation.

To generate HTS, all targets in set one used sets two and three as the template database and similarly for sets two and three. This amounts to having two-thirds of the data as templates for training with the remaining third as the test set. This allows us to directly compare the performance achieved by direct SVM predictions, homology-based transfer, and LIBRUS as all methods use identical training and testing data. The same cross-validation approach was also used to compute the predictions for linear combination of homology-based transfer and SVM scores (Section 16.2.1.5).

We evaluated the performance of the different methods using the ROC curve [36]. This is obtained by varying the threshold at which residues are considered ligand-binding or not according to the value provided by the predictor. In the case of the SVM predictions, a continuous prediction value is produced, which is the distance from a hyperplane optimized to separate the positive and negative classes. This is the threshold that is varied to produce the ROC curve. For HTS, the threshold to be assigned a ligand-binding residue is varied to produce the ROC curve. The area under the ROC curve, abbreviated *ROC* (note italics), summarizes the predictor behavior: a random predictor has $ROC = 0.5$ while a perfect predictor has $ROC = 1.0$ so that a larger *ROC* indicates better predictive power.

For any binary predictor, the number of true positives (TP), false positives (FP), true negatives (TN), and false negatives (FN) determines the standard classification statistics that we use later for comparison. These are

$$\text{Precision} = \frac{TP}{TP + FP}, \text{ and} \tag{16.2}$$

$$\text{Recall} = \text{Sensitivity} = \frac{TP}{TP + FN}. \tag{16.3}$$

Fischer et al. noted in their study of functional residue predictions that analyzing only an ROC curve can be misleading in terms of the performance of the predictor [14]. As an alternative, they present precision versus recall plots (called precision-sensitivity plots in their work, referred to as PR curves here) as a means to compare performance. We provide this measure as well, both graphically and summarized by the area under the PR curve, abbreviated *PR* (note italics).

Performance differences between FINDSITE and LIBRUS on DS2 are illustrated using the Welch's *t*-test. This test assumes that the populations are normally distributed with potentially unequal variance and calculates a *p*-value that the mean of one is higher than the other. In our case, this corresponds to one method outperforming another. Welch's *t*-test was used in favor of Student's *t*-test as the latter assumes equal variance of the populations that may not be the case for the methods under consideration. The populations we analyzed are the *ROC* and *PR* scores of each protein according to the predictions of LIBRUS and FINDSITE. The test allows us to determine whether, on average, one of the two methods outperforms the other on per-protein identification of ligand-binding residues.

16.2.3. Results

16.2.3.1. Performance of Direct Sequence-Based Predictors. The performance of the prototype methods described in Section 16.2.1 on dataset DS1 is shown in Table 16.2 and Figure 16.1. The methods are grouped into three classes:

TABLE 16.2 Cross-Validation Results on the DS1 Dataset

Method	Overall		Per Protein			
	ROC	PR	μ_{ROC}	σ_{ROC}	μ_{PR}	σ_{PR}
SVM with PSSM	0.7545	0.2637	0.7487	0.1492	0.2930	0.1722
SVM with PSSM, SSE	0.7737	0.2942	0.7648	0.1532	0.3177	0.1886
Homology Transfer	0.7845	0.4516	0.7581	0.1811	0.4024	0.2971
Linear SVM and HTS	0.8259	0.4792	0.8030	0.1666	0.4342	0.2838
LIBRUS	0.8334	0.4807	0.8066	0.1686	0.4374	0.2809

Three-way cross-validation was used on the set of 885 proteins of the DS1 dataset. The overall area under curve is given for ROC and precision/recall (PR) curves in the first two columns. The per protein averages, μ, and standard deviation, σ, for these two statistics are also given.

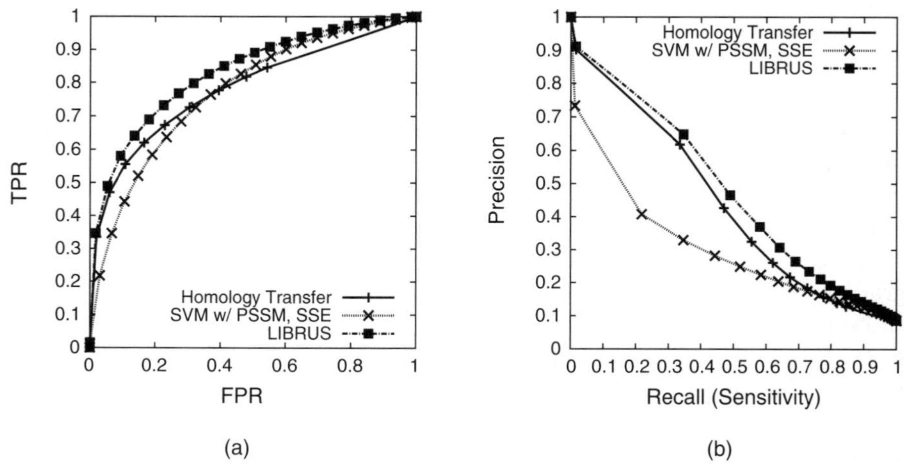

FIGURE 16.1 ROC and PR curves of some sequence-based predictors. Curves are given for the overall performance on the DS1 dataset. (a) ROC curves and (b) Precision versus Recall.

SVM prediction, HT, and combined. Comparing the best performance achieved by each of the classes, we see that the combined methods achieve the best overall results. Among the two methods that fall in that category, LIBRUS, which uses SVM to combine this information, achieves the best overall results. Specifically, it achieves an overall $ROC = 0.8334$, which is better than the ROCs of 0.7737 and 0.7849 that were obtained by the SVM and homology-based transfer methods, respectively. Its performance in terms of the overall PR is also better, achieving a $PR = 0.4807$ compared with the PRs of 0.2942 and 0.4516 achieved by the other two classes of methods. These relative performance also gains hold when the experiments are evaluated in terms of the average per-protein ROC and PR. The performance of the simple linear combination of SVM and HTS scores also performs quite well, further reenforcing the fact that coupling the two sources of information leads to a better overall predictor.

Comparing the other two classes of methods, we see that homology-based transfer outperforms the direct SVM-based approach that utilizes PSSM- and SSE-based features. The performance difference between these two schemes is more pronounced when the methods are evaluated in terms of their PR (both overall and per protein).

Finally, the results of Table 16.2 show that when predicted secondary structure information is used to augment the PSSM-based features, the performance of the SVM-based method improves. This fact is in agreement with a number of studies that have shown that the inclusion of this type of information helps the performance of supervised learning methods [25,26].

16.2.3.2. Performance of LIBRUS and FINDSITE. Performance measures for FINDSITE and LIBRUS predictions on the proteins in dataset DS2 are summarized in Table 16.3 while Figure 16.2 plots the ROC and PR curves obtained. Note that Tables 16.3–16.4 and Figure 16.2 also contain results for the scheme that combines the LIBRUS and FINDSITE predictions, which are discussed later in Section 16.2.3.3. Table 16.4 shows the results of a paired Welch's t-test comparing the methods. Comparisons on both ROC and PR are done in parts (a) and (b) of Table 16.4, respectively.

Examining the predictions of the various versions of FINDSITE and LIBRUS, in Table 16.3 we see that their overall prediction performance is quite close. The FINDSITE results using one site achieve the best PR (0.4955), whereas the FINDSITE results using three sites achieve the best ROC (0.8216). However, compared with the former method, LIBRUS achieves a better ROC (0.8169 vs. 0.8088), whereas compared with the latter method, LIBRUS achieves a better PR (0.4565 vs. 0.3760). The difference between FINDSITE and LIBRUS is somewhat more consistent when the per-protein results are considered, in which case the FINDSITE results using two sites lead to average ROC and PR (0.8043 and 0.4360) that are better than those produced by LIBRUS (0.7982 and 0.4165).

TABLE 16.3 Results on the DS2 Dataset

Method	Overall		Per Protein			
	ROC	PR	μ_{ROC}	σ_{ROC}	μ_{PR}	σ_{PR}
FINDSITE 1 Site	0.8088	0.4955	0.7981	0.2040	0.4841	0.2978
FINDSITE 2 Sites	0.8187	0.4258	0.8043	0.1935	0.4360	0.2697
FINDSITE 3 Sites	0.8216	0.3760	0.8034	0.1852	0.3957	0.2436
FINDSITE 4 Sites	0.8182	0.3370	0.7970	0.1808	0.3620	0.2228
FINDSITE 5 Sites	0.8155	0.3074	0.7918	0.1716	0.3340	0.2055
LIBRUS	0.8169	0.4565	0.7982	0.1600	0.4165	0.2550
Combined	0.8617	0.5618	0.8410	0.1741	0.5324	0.2991

The performance of FINDSITE considering the first five binding sites and the best SVM method, LIBRUS, are shown. The dataset comprised 564 proteins from the FINDSITE benchmark that were sequence independent from the DS1 dataset that was used to train LIBRUS. The last row shows the results obtained by linearly combining the predictions produced by LIBRUS and FINDSITE 1 Site. For column descriptions, see Table 16.2.

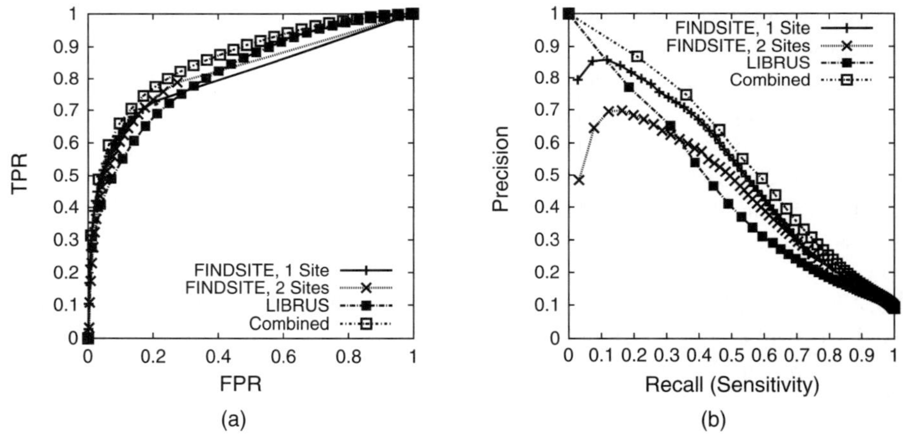

FIGURE 16.2 Comparison of FINDSITE and LIBRUS.
Overall comparison of FINDSITE to the sequence-only SVM learner developed in this work on the 564 independent proteins from the FINDSITE benchmark. (a) ROC curves of FINDSITE based on the top binding sites, the SVM approach, and the combined predictor. (b) Precision versus Recall of the methods.

Figure 16.2 shows the ROC and PR plots graphically. According to Figure 16.2a, the strength of LIBRUS is at higher false positive rates where it exceeds the TPR of FINDSITE. At low FPR, FINDSITE dominates LIBRUS with the crossing point at FPR = 0.35 and FPR = 0.40 for one and two sites, respectively. In Figure 16.2b, LIBRUS is seen to have better precision at very low recall, but falls below FINDSITE at 11% recall for one site and at 34% recall for two sites. At 50% recall, LIBRUS achieves 40% precision while FINDSITE achieves 55% and 49% precision for one and two sites, respectively.

TABLE 16.4 Statistical Comparison of Methods on the DS2 Dataset

(a) Per Protein *ROC* *p*-values

	FS 1	FS 2	FS 3	FS 4	FS 5	LIB.	Comb.
FS 1	0.500	0.701	0.675	0.464	0.289	0.503	1.000
FS 2	0.299	0.500	0.466	0.257	0.126	0.281	1.000
FS 3	0.325	0.534	0.500	0.281	0.140	0.308	1.000
FS 4	0.536	0.743	0.719	0.500	0.310	0.545	1.000
FS 5	0.711	0.874	0.861	0.690	0.500	0.740	1.000
LIB.	0.496	0.719	0.692	0.455	0.260	0.500	1.000
Comb.	0.000	0.000	0.000	0.000	0.000	0.000	0.500

(b) Per Protein *PR* *p*-values

	FS 1	FS 2	FS 3	FS 4	FS 5	LIB.	Comb.
FS 1	0.500	0.002	0.000	0.000	0.000	0.000	0.997
FS 2	0.998	0.500	0.004	0.000	0.000	0.106	1.000
FS 3	1.000	0.996	0.500	0.008	0.000	0.919	1.000
FS 4	1.000	1.000	0.992	0.500	0.014	1.000	1.000
FS 5	1.000	1.000	1.000	0.986	0.500	1.000	1.000
LIB.	1.000	0.893	0.081	0.000	0.000	0.500	1.000
Comb.	0.003	0.000	0.000	0.000	0.000	0.000	0.500

Performance of the methods is compared via *p*-values on Welch's *t*-test. For the entry at row *i*, column *j* of the table, the alternate hypothesis that Method *i* has a higher mean than method *j* is tested as an alternative to the methods having equal means. A low *p*-value indicates that method *i* has better performance than method *j*. Part (a) of the table shows performance comparisons in terms of per protein *ROC* while part (b) shows per protein *PR* comparisons. FINDSITE for various number of sites are reported in the FS row/columns, LIBRUS in LIB, and the combined FINDSITE/LIBRUS predictor in Comb.

One aspect that we have not touched on empirically so far is the time required to make predictions. According to communications with the FINDSITE authors, running their program for a protein takes from 30 minutes to several hours. This is not surprising as FINDSITE needs to initially predict the structure of the protein and also identify good templates from their database. The amount of time required by LIBRUS to predict the ligand-binding residues of a protein is much lower. Based on the average performance over many proteins, LIBRUS predictions can be made in under 10 min, which encompasses profile generation, secondary structure prediction, alignment to the database, and final SVM prediction. A larger template database will lengthen this process somewhat, but we expect it to remain faster.

16.2.3.3. Complementary Nature of Sequence and Structure Predictions. While analyzing the nature of the predictions produced by FINDSITE and LIBRUS, we noticed that, although there is agreement on many of the

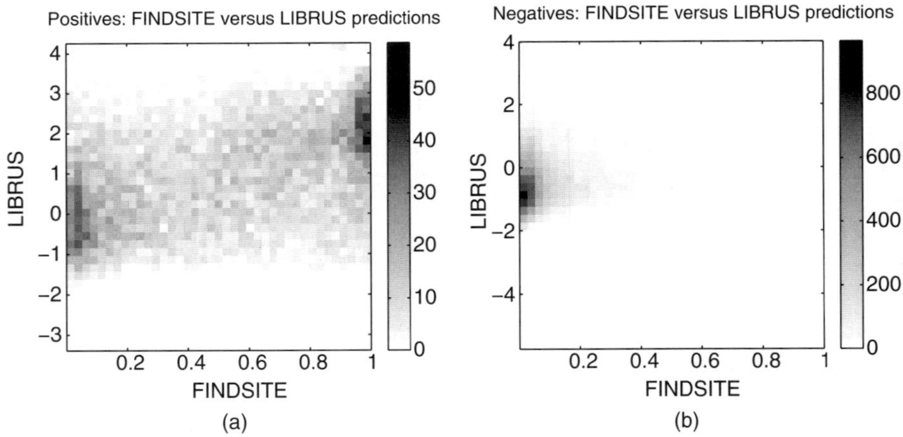

FIGURE 16.3 Heatmap illustrating FINDSITE and LIBRUS prediction values. Heatmap illustrating FINDSITE and LIBRUS values on the positive class (a) and the negative class (b). The positive LIBRUS predictions on some mis-predicted FINDSITE residues indicate that LIBRUS may provide additional information in some cases. The correlations between FINDSITE and LIBRUS are 0.52 on the positive class, 0.27 on the negative class, and 0.48 overall. Note that residues that had FINDSITE predictions of zero were eliminated as they dominate the nonzero predictions.

residues they identified as being ligand binding, there are enough differences to merit further inquiry. Figure 16.3 illustrates these differences by plotting the prediction scores produced by LIBRUS and FINDSITE (using one site) for the positive instances (ligand-binding residues) and the negative instances (nonbinding residues). In Figure 16.3a (positive class) we see that there are two clusters, one on the right and one on the left of the plot. The cluster on the right contains residues that FINDSITE predicts correctly, whereas the cluster on the left contains residues that FINDSITE mis-predicts. The predictions produced by LIBRUS are, to a large extent, in agreement for the right cluster (even though LIBRUS mis-predicts some of these residues) but are split for the left cluster. LIBRUS predicts correctly (i.e., positive SVM score) a noticeable fraction of the residues that are falsely predicted as negative by FINDSITE. Overall, the Pearson correlation coefficient between FINDSITE predictions and LIBRUS predictions is 0.48.

Figure 16.4a illustrates how the above trend carries over to the whole protein. It plots the per-protein *ROC*s of LIBRUS and FINDSITE (one site) on DS2 against one another. The greatest density lies in the upper right corner where both methods achieve high *ROC*s. Points below the main diagonal indicate where LIBRUS outperforms FINDSITE while points above indicate the opposite. The large number of off-diagonal points shows that if information from both predictors can be exploited, overall predictions may be improved.

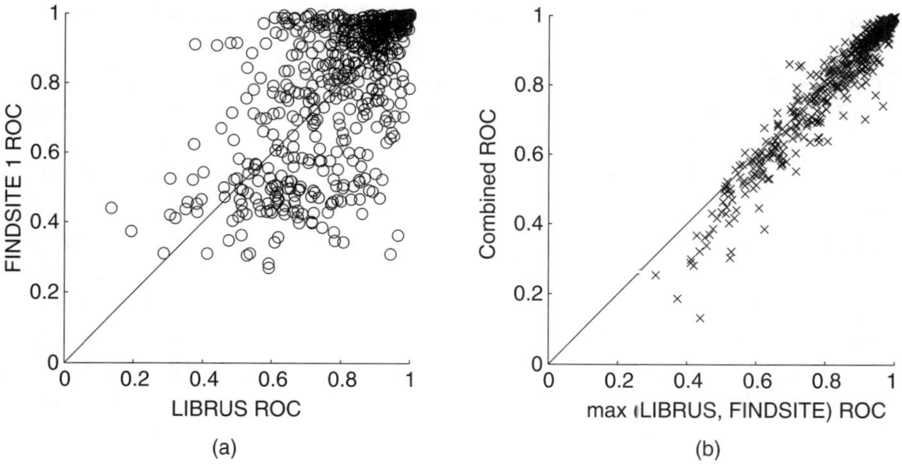

FIGURE 16.4 Complementary nature of FINDSITE and LIBRUS predictions. (a) LIBRUS versus FINDSITE. The abundance of off-diagonal entries indicates LIBRUS and FINDSITE outperform one another on certain proteins and must be exploiting different signals for those proteins. (b) The *ROC* of the combined method is plotted against the maximum of LIBRUS and FINDSITE and achieves nearly the same performance.

Motivated by the above differences, we linearly combined the prediction scores of LIBRUS and FINDSITE. The results of this combined predictor are reported at the bottom of Table 16.3, and in Figure 16.2. The combined predictor achieves higher overall *ROC* and *PR* than either approach on its own. Also notable is the superior per-protein prediction rate of both *ROC* and *PR* for the combined method, which is statistically significant (Table 16.4, row/column Comb). This improvement is apparent in Figure 16.4b in which the combined method achieves performance close to the maximum of both LIBRUS and FINDSITE.

16.2.3.4. Sequence and Structure Carry Nearly the Same Amount of Predictive Information. Table 16.4a shows that there is no statistical difference between LIBRUS and FINDSITE in terms of per-protein *ROC* performance. This is seen in the LIB row and column of the table in which no small p-values occur. This lack of significance is interesting as it shows sequence and predicted structure carry approximately equal amounts of information that may be used to identify ligand-binding residues. In terms of *PR* (Table 16.4b), examining a single FINDSITE site outperforms LIBRUS at a statistically significant level ($p = 0.002$) while examining two FINDSITE sites is not significantly better than LIBRUS ($p = 0.106$). LIBRUS is nearly better than FINDSITE with three sites at a significant level ($p = 0.081$), and better than four and five sites ($p = 0.000$ for both).

16.3. APPLICATION: HOMOLOGY MODELING OF BINDING SITES

The preceding sections have shown that binding residues can be identified from sequence alone with reasonable accuracy. The sequence-based LIBRUS achieves close to the same accuracy as structure-based FINDSITE. The next logical step is to put those sequence-based predictions to use in some application.

In this section we explore such an application. Binding residue predictions are exploited to aid the development of a homology model of a protein. In drug discovery applications, the primary interest is in the binding site of the protein. By allowing predicted binding labels to influence the target-template alignment, the quality of the resulting predicted binding site structure is improved. This effect is most prevalent when the homology modeling problem is difficult, that is, there is little relation between target and template.

16.3.1. Background on Homology Modeling

Accurate modeling of protein-ligand interactions is an important step to understanding many biological processes. For example, many drug discovery frameworks include steps where a small molecule is docked with a protein to measure binding affinity [5].

A frequent approximation is to keep the protein rigid, necessitating a high-quality model of the binding site. Such models can be onerous to obtain experimentally.

Computational techniques for protein structure prediction provide an attractive alternative for this modeling task [37]. Protein structure prediction accuracy is greatly improved when the task reduces to homology modeling [38]. These are cases in which the unknown structure, the target, has a strong sequence relationship to another protein of known structure, referred to as the template. Such a template can be located through structure database searches. Once obtained, the target sequence is mapped onto the template structure and then refined. A detailed discussion of homology modeling appears in Chapters 8 and 9 of this book.

A number of authors have studied the use of homology modeling in predicting the structure of clefts and pockets, the most common interaction site for ligand binding [39–41]. Their consensus observation is that modeling a target with a high sequence similarity template is ideal for model quality while a low sequence similarity template can produce a good model provided alignment is done correctly. This sensitivity calls for special treatment of the interaction site during sequence alignment assuming ligand-binding residues can be discerned a priori.

The factors involved in modeling protein interaction sites have received attention from a number of authors. These studies tend to focus on showing relationships between target-template sequence identity and the model quality of surface clefts/pockets.

DeWeese-Scott and Moult made a detailed study of CASP targets (http://predictioncenter.org) that bind to ligands [39]. Their primary interest was in atom contacts between the model protein and its ligand. They measured deviations from true contact distances in the crystal structures of the protein-ligand complexes. Although the number of complexes they examined was small, they found that errors in the alignment of the functional region between target and template created problems in models, especially for low sequence identity pairs.

Chakravarty et al. did a broad study of various structural properties in a large number of homology models including surface pockets [40]. They noted that in the case of pockets, side chain conformations had a high degree of variance between predicted and true structures. Due to this noise, we will measure binding site similarity using the α-carbons of backbone residues. They also found that using structure-induced sequence alignments improved the number of identical pockets between model and true structures over sequenced-only alignments. This point underscores the need for a good alignment that is sensitive to the functional region. It also suggests using structure alignments as the baseline to measure the limits of homology modeling.

Finally, Piedra et al. executed an excellent large-scale study of protein clefts in homology models [41]. To assess the difficulty of targets, the true structure was used as the template in their homology models and performance using other templates was normalized against these baseline models. Although a good way to measure the individual target difficulty, this approach does not represent the best performance achievable for a given target-template pair. This led us to take a different approach for normalization. We follow their convention of assessing binding site quality using only the binding site residues rather than all residues in the predicted structure. As their predecessors noted, Piedra et al. point to the need for very good alignments between target and template when sequence identity is low.

The suggestion from these studies, that quality sequence alignments are essential, led us to employ sensitive alignment methods discussed in Section 16.3.4.

16.3.2. Homology Modeling with Binding Residue Predictions

Assuming that the ligand-binding residues of all template proteins are known, we illustrate a method to modify alignments of target and template. The modification influences ligand-binding residues to align to one another and discourages the alignment of binders to nonbinders. Once the target-template alignment is constructed, standard homology modeling techniques are employed to produce the target structure prediction. An analysis of the ligand-binding site shows that these modified alignments improve the accuracy of this part of the model over standard alignment techniques.

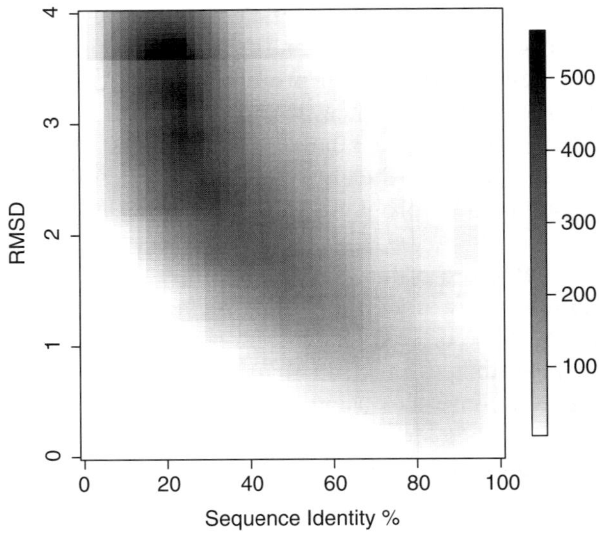

FIGURE 16.5 Distribution of Homology Pairs.
The heatmap varies in intensity based on the number of homology modeling pairs that have the sequence/structure relationship at the center pixel. A sliding window of 20% sequence identity and 0.8 Å is used to create the counts. Darker colors correspond to more pairs.

16.3.3. Experimental Setup

In homology modeling experiments, target-template pairs are required. We used the set of 885 proteins in DS1 as the targets (structures to be predicted). We used the MAMMOTH structure alignment program to search the PDB for other proteins that had a significant structure alignment [42]. Of these, we kept templates that had a bound ligand that would allow ligand-binding residues to be used to influence the target-template alignment. We then proceeded to generate homology models for each target template pair using techniques described below. The final result included 2045 homology pairs and 862 individual target proteins. The distribution of these pairs in the sequence and structure relation space is given in Figure 16.5.

16.3.4. Alignment Modification by Binding Prediction

The basic framework for sequence alignment is identical to that of Section 16.2.1.2. As special attention needs to be given to the ligand-binding residues, an additional term is incorporated into Equation 16.1 to reflect this goal.

Each residue is labeled either as ligand-binding or not. In the case of the targets, these labels the sequence-predicted labels obtained from LIBRUS. Templates always used true labels. Binding residue predictions that come from LIBRUS are a continuous valued numbers with positive values indicating stronger confidence that the residue is a binding residue. To convert this into a discrete label, thresholding can be used. In the following results, a threshold of 0.7 was used so that residues above this value were labeled as predicted binders and those below were labeled nonbinders.

A very simple approach to influence target-template alignments with predicted ligand-binding labels is to add a constant m_{bb} whenever a predicted and binding residue in the target aligned with a true ligand-binding residue in the template. Setting $m_{bb} = 0$ gives *standard alignments* that do not incorporate the predictions while setting $m_{bb} > 0$ gives a *modified alignment*. Setting $m_{bb} > 0$ encourages the alignment of binding residues and for the results reported below, $m_{bb} = 15$.

16.3.5. Homology Model Generation

Once a sequence alignment has been determined between target and template, homology modelling may be used to predict the target structure using a variety of standard tools described elsewhere. The results shown here employed version 9.2 of the MODELLER package, which is freely available [43]. As input, MODELLER takes a target-template sequence alignment and the structure of the template. An optimization process ensues in which the predicted coordinates of the target are adjusted to violate, as little as possible, spatial constraints derived from the template. MODELLER offers a high degree of flexibility and automation through a programmable interface. Modeling can be done using only a target sequence and a database of known structures. However, the comments by the software authors and numerous studies indicate that a crucial step in the problem is aligning target and template sequences. This is where predicted binding residues can be useful to influence the proper alignment of target and template.

16.3.6. Evaluation Metrics for Homology Modeling

The root-mean-square deviation (RMSD) is a standard metric used to compare two protein structures. A low RMSD between target and template indicates similarity between two structures. Typically, only the α-carbon coordinates are used for the RMSD computation. Our interest is in the binding site and thus only a good measure of success is to consider the RMSD between the ligand-binding residues in the true and predicted structures that follow the convention of Piedra et al. [41]. For brevity, this will be called the *lig*RMSD for ligand-binding residues RMSD.

Student's *t*-test is used on the *lig*RMSD of the standard alignment predictions paired with the corresponding *lig*RMSD of modified alignments to show when their performance differs significantly. The null hypothesis is that the two have equal mean while the alternative hypothesis is that the modified alignments produce models with a lower mean *lig*RMSD (a one-tailed test). We report *p*-values for the comparisons noting that a *p*-value smaller than

0.05 is typically considered statistically significant. We also report the mean improvement (gain) from using modified alignments. If the mean of all *lig*RMSD for the standard alignments is $\bar{R}_{stand}$ and that of a modified alignment is $\bar{R}_{\text{mod}}$, the percent gain is

$$\%\text{Gain} = \frac{\bar{R}_{stand} - \bar{R}_{mod}}{\bar{R}_{stand}}. \tag{16.4}$$

A positive gain indicates improvement through the use of the ligand-binding residue predictions while a negative gain indicates using predictions degrades the homology models.

Finally, a permutation test can be used to assure us that the observed gains are not tied too tightly to the particular data being used. For the sequence/structure subgroups of interest, the permutation test examines a random subsets one-third the size of the subgroup and performs a paired Student's t-test on the standard and modified *lig*RMSDs. The mean p-value over 100 random subsets is reported as μ_p and may be used as an indication of how well the parameters are expected to perform on future data. The standard deviation of the permutation test p-values is also given as σ_p.

16.3.7. Model Quality Improvements

We are interested in knowing when it is worth the extra effort to predict ligand-binding residues from sequence. For the homology modeling task, we would not expect the alignment of very similar target and template to benefit much from the additional knowledge of ligand-binding residues—as long as the alignment method is sensitive good correspondence should be obtainable solely from sequence similarity. However, when the target and template are sufficiently different, ligand-binding residues have more potential to influence the proper alignment of binding residues.

Table 16.5 shows the results of homology modeling experiments restricted to different regions of target-template relationship. A t-test is conducted to determine if the average *lig*RMSD of models produced using LIBRUS-predicted binding labels is lower than for models produced using standard alignments. A small p-value indicates significant improvement in *lig*RMSD. The percentage improvement (gain as defined in Equation 16.4) is given for each subgroup along with the size of the subgroup. A negative gain indicates models using predicted labels were worse than those using standard alignments. The final two columns describe the mean and standard deviation of p-values for permutation tests on the subgroups.

Our intuition on the effectiveness of predicted binding labels is confirmed in Table 16.5. The regions with low sequence identity and high structure difference between target and template see the most improvement. For pairs with less than 30% sequence identity and more than 2 Å RMSD, we can expect

TABLE 16.5 Results of Homology Model Experiment

SeqID	RMSD	N	p-val	%Gain	μ_p	σ_p
$0 \leq 30$	$0 \leq 2$	27	0.8210	–3.61	0.5467	0.2679
$0 \leq 30$	$2 \leq 4$	1078	0.0000	2.61	0.0003	0.0009
$30 \leq 60$	$0 \leq 2$	347	0.9516	–1.44	0.8145	0.2070
$30 \leq 60$	$2 \leq 4$	438	0.0321	–0.83	0.2417	0.2011
$60 \leq 100$	$0 \leq 2$	166	0.9437	–8.14	0.7496	0.2157
$60 \leq 100$	$2 \leq 4$	35	0.7908	–0.33	0.6655	0.2564

Results of the homology modeling experiment are divided into regions according to sequence identity and RMSD relations between the target and template. The p-value indicates whether gains from using predicted binding labels are statistically significant: smaller p-values correspond to greater significance. Gain is defined in Equation 16.4. N is the number of homology pairs satisfying the sequence/RMSD relationship and are used to compute the statistics. The final two columns are the mean (μ_p) and standard deviation (σ_p) of p-values in a permutation test that measures robustness of the results. A smaller μ_p indicates the results are robust.

to get around 2.61% improvement in RMSD. These results appear highly robust in the permutation test ($\mu_p = 0.0000$). For pairs with a close structure relationship (0 to 2 Å RMSD), it does not appear predicted labels are useful as the gains are all negative in these cases (note, however, the small sample size for low sequence identity in the first line).

Figure 16.6 graphically represents the homology modeling results. In Figure 16.6a, the intensity of each pixel of the figure corresponds to the p-value of a t-test on a subgroup of the dataset. The position along the Sequence Identity and RMSD axes indicates which pairs are used in the comparison. Subgroups are comprised of pairs in a window of 20% sequence identity and 0.8 Å RMSD around the center. For example, at sequence identity of 20% and RMSD of 3.0 Å, target-template pairs related by 10–30% sequence identity and 2.6–3.4 Å RMSD are used to compute the p-value. The same approach is used in Figure 16.6b, which shows the subgroup percentage gain.

The pattern in Figure 16.6 follows that of Table 16.5—the region of low sequence and structure similarity (upper left corner) produces the significant results and positive gains. There are some large positive gains in a few other regions of the similarity space, particularly 50–60% sequence identity for high RMSD, but they are not statistically significant.

Practical lessons can be drawn from this experiment. When faced with generating a homology model of a ligand-binding site, one should consider the available templates carefully as this is the most critical step. Once selected, the template(s) should be aligned to the target sequence using the most sensitive alignment approach available. If it is found that the sequences are very similar, modeling can proceed as normal. If they are dissimilar, it is likely worth the effort to predict the ligand-binding residues of the target using a method such as LIBRUS and then recompute the alignment. Alternatively,

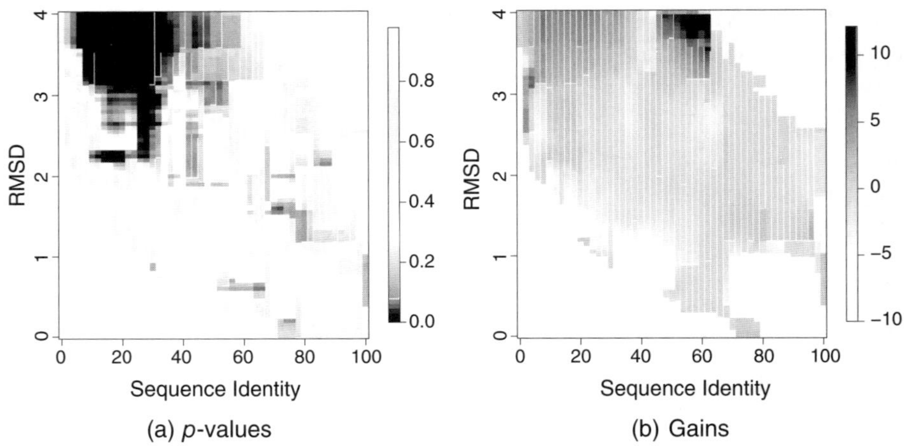

FIGURE 16.6 Homology Model Improvements.
(a) Statistical significance of homology model improvements. Pixels denote whether predicted ligand-binding residues improve homology models of the binding site. Pixel intensity corresponds to the p-value of a t-test measuring whether the mean *lig*RMSD of models that used predicted labels is lower than that of standard alignments. Dark pixels represent low p-values and statistical significance. Significant improvements are achieved when the target and template have low sequence identity and large RMSD (upper left corner). (b) Percentage of improvement (gain). The intensity of each pixel represents a lower *lig*RMSD using predicted labels in modified alignments versus using standard alignments. The gains are small but statistically significant in the region of low sequence identity and high RMSD between target and template. Greater gains occur in a few other regions but are not statistically significant.

the modeler may wish to first generate the usual homology model, use a structure-based method such as FINDSITE to predict the binding site, and then possibly realign target and template to produce a better model. As mentioned in Section 16.2.3.4 it is not clear whether this latter approach will improve the binding-site predictions significantly. This is a matter that will require further study.

16.4. CONCLUSION AND FUTURE OUTLOOK

This chapter has discussed the identification of protein residues involved in ligand binding. Identification may be done based solely on the protein sequence or by utilizing structure information when it is available. There are several downstream applications of this capability and we have illustrated that sequence-based predictions are presently accurate enough to impact homology modeling of the binding site in a positive fashion.

Although we have seen that the accuracy of binding site homology models increases by leveraging predicted binding residues, examining how these models actually affect docking experiments is unexplored territory. A simple benchmark would measure the docking scores of ligands using the true structure of the protein as the baseline and test whether homology models that use binding residues behave more or less closely to the baseline than models that do not use such predictions. An alternative approach is to modify the scoring function or energy measure in docking experiments to favor locations with predicted residues. This may improve accuracy or intelligently bias the search space of docking locations. Success on any of these experiments would have a positive impact on docking-based virtual screening.

Another potential application of binding residues is to compare protein structures based on binding site and potential ligands. This is most applicable when structures are available and are thus appropriate for structure-based methods. Discovering proteins with a similar binding site to a particular target can help elucidate side effects of introducing a small molecule. FINDSITE has already developed some methodologies to determine a ligand profile for a target protein and was utilized to examine function prediction of the protein based on the ligand profile. With the need for automated function assignment for proteins on the rise, it is likely that this trend will continue and develop additional sophistication.

Finally, recent work has used generic machine learning models that incorporate protein similarity to determine the structure-activity relationship of small molecules [44]. In this setting, the set of positive ligands for a target protein can be expanded by identifying other similar targets and adopting their positive ligands. Several methods of target similarity are developed from the standpoint of having no target structures. Sequence-based binding residue predictions may be leveraged in such cases to aid in determining the similarity of two protein targets. In cases where a structure is available, protein similarity for this application should likely be based upon binding sites that require identification of binding residues by either sequence or structure means.

With a good body of foundational work and a variety of downstream applications, the ligand-binding residue identification problem is likely to remain a topic of interest for bioinformatics and cheminformatics researchers for some time to come. We hope that this chapter has provided a sufficient overview to guide readers to future advances in the area.

REFERENCES

1. J.R. Bock and D.A. Gough. Virtual screen for ligands of orphan g protein-coupled receptors. *Journal of Chemical Information and Modeling*, 45(5):1402–1414, 2005.
2. K.H. Bleicher, H-J. Bohm, K. Muller, and A.I. Alanine. Hit and lead generation: Beyond high-throughput screening. *Nature Reviews Drug Discovery*, 2:369–378, 2003. 10.1038/nrd1086.
3. D. Weber, C. Berger, T. Heinrich, P. Eickelmann, J. Antel, and H. Kessler. Systematic optimization of a lead-structure identities for a selective short peptide

agonist for the human orphan receptor brs-3. *Journal of Peptide Science*, 8(8):461–475, 2002.

4. C. Kauffman, H. Rangwala, and G. Karypis. Improving homology models for protein-ligand binding sites. *Computational Systems Bioinformatics Conference*. San Francisco, CA, August 26–29, 2008. Available at http://www.cs.umn.edu/karypis, last accessed October 12, 2009.
5. N. Moitessier, P. Englebienne, D. Lee, J. Lawandi, and C.R. Corbeil. Towards the development of universal, fast and highly accurate docking//scoring methods: A long way to go. *British Journal of Pharmacology*, 153(S1):S7–S26, 2007.
6. Y. Ofran, V. Mysore, and B. Rost. Prediction of dna-binding residues from sequence. *Bioinformatics*, 23(13):i347–i353, 2007.
7. S. Ahmad and A. Sarai. Pssm-based prediction of dna binding sites in proteins. *BMC Bioinformatics*, 6:33, 2005.
8. H. Rangwala, C. Kauffman, and G. Karypis. A generalized framework for protein sequence annotation. *Proceedings of the NIPS Workshop on Machine Learning in Computational Biology,* Vancouver, B.C., Canada. December 7, 2007.
9. M. Terribilini, J-H. Lee, C. Yan, R.L. Jernigan, V. Honavar, and D. Dobbs. Prediction of RNA binding sites in proteins from amino acid sequence. *RNA*, 12(8):1450–1462, 2006.
10. M. Kumar, M.M. Gromiha, and G.P.S. Raghava. Prediction of rna binding sites in a protein using svm and pssm profile. *Proteins*, 71(1):189–194, 2008.
11. Y. Ofran and B. Rost. Predicted protein-protein interaction sites from local sequence information. *FEBS Letters*, 544(1–3):236–239, 2003.
12. M-H. Li, L. Lin, X-L. Wang, and T. Liu. Protein protein interaction site prediction based on conditional random fields. *Bioinformatics*, 23(5):597–604, 2007.
13. A. Koike and T. Takagi. Prediction of protein-protein interaction sites using support vector machines. *Protein Engineering, Design and Selection*, 17(2):165–173, 2004.
14. J.D. Fischer, C.E. Mayer, and J. Söding. Prediction of protein functional residues from sequence by probability density estimation. *Bioinformatics*, 24(5):613–620, 2008.
15. C.T. Porter, G.J. Bartlett, and J.M Thornton. The catalytic site atlas: A resource of catalytic sites and residues identified in enzymes using structural data. *Nucleic Acids Research*, 32(Database issue):D129–D133, 2004.
16. N.V. Petrova and C.H. Wu. Prediction of catalytic residues using support vector machine with selected protein sequence and structural properties. *BMC Bioinformatics*, 7:312, 2006.
17. E. Youn, B. Peters, P. Radivojac, and S.D. Mooney. Evaluation of features for catalytic residue prediction in novel folds. *Protein Science*, 16(2):216–226, 2007.
18. A.G. Murzin, S. E. Brenner, T. Hubbard, and C. Chothia. Scop: A structural classification of proteins database for the investigation of sequences and structures. *Journal of Molecular Biology*, 247(4):536–540, 1995.
19. A. Andreeva, D. Howorth, J-M. Chandonia, S.E. Brenner, T.J.P. Hubbard, C. Chothia, and A.G. Murzin. Data growth and its impact on the scop database: New developments. *Nucleic Acids Research*, 36(Database issue):D419–D425, 2008.

20. G. López, A. Valencia, and M.L. Tress. firestar–prediction of functionally important residues using structural templates and alignment reliability. *Nucleic Acids Research*, 35(Web Server issue):W573–W577, 2007.
21. M. Brylinski and J. Skolnick. A threading-based method (findsite) for ligand-binding site prediction and functional annotation. *Proceedings of the National Academy of Sciences U S A*, 105(1):129–134, 2008.
22. J. Skolnick and M. Brylinski. Findsite: A combined evolution/structure-based approach to protein function prediction. *Briefings in Bioinformatics*, 15:378–391, 2009.
23. S.F. Altschul, T.L. Madden, A.A. Schaffer, J. Zhang, Z. Zhang, W. Miller, and D.J. Lipman. Gapped blast and psi-blast: A new generation of protein database search programs. *Nucleic Acids Research*, 25(17):3389–3402, 1997.
24. W. Kabsch and C. Sander. Dictionary of protein secondary structure: Pattern recognition of hydrogen-bonded and geometrical features. *Biopolymers*, 22:2577–637, 1983.
25. K. Chen and L. Kurgan. Pfres: Protein fold classification by using evolutionary information and predicted secondary structure. *Bioinformatics*, 23(21):2843–2850, 2007.
26. K. Ginalski, J. Pas, L.S. Wyrwicz, M. von Grotthuss, J.M. Bujnicki, and L. Rychlewski. Orfeus: Detection of distant homology using sequence profiles and predicted secondary structure. *Nucleic Acids Research*, 31(13):3804–3807, 2003.
27. G. Karypis. Yasspp: Better kernels and coding schemes lead to improvements in svm-based secondary structure prediction. *Proteins: Structure, Function, and Bioinformatics*, 64(3):575–586, 2006.
28. D. Mittelman, R. Sadreyev, and N. Grishin. Probabilistic scoring measures for profile-profile comparison yield more accurate short seed alignments. *Bioinformatics*, 19(12):1531–1539, 2003.
29. A. Heger and L. Holm. Picasso: Generating a covering set of protein family profiles. *Bioinformatics*, 17(3):272–279, 2001.
30. H. Rangwala and G. Karypis. Frmsdpred: Predicting local rmsd between structural fragments using sequence information. *Computational Systems Bioinformatics Conference*, 6:311–322, 2007.
31. D. Gusfield. *Algorithms on Strings, Trees, and Sequences—Computer Science and Computational Biology*. New York: Cambridge University Press, 1997.
32. C. Kauffman and G. Karypis. Librus: Combined machine learning and homology information for sequence-based ligand-binding residue prediction. *Bioinformatics*, 25(23):3099–3107, 2009.
33. V.N. Vapnik. *The Nature of Statistical Learning Theory*. New York: Springer Verlag, 1995.
34. H.M. Berman, J. Westbrook, Z. Feng, G. Gilliland, T.N. Bhat, H. Weissig, I.N. Shindyalov, and P.E. Bourne. The protein data bank. *Nucleic Acids Reseach*, 28(1):235–242, 2000.
35. J-M. Chandonia, N.S. Walker, L.L. Conte, P. Koehl, M. Levitt, and S.E. Brenner. Astral compendium enhancements. *Nucleic Acids Research*, 30(1):260–263, 2002.
36. T. Fawcett. An introduction to roc analysis. *Pattern Recognition Letters*, 27(8):861–874, 2006.

37. P. Ferrara and E. Jacoby. Evaluation of the utility of homology models in high throughput docking. *Journal of Molecular Modeling*, 13:897–905, 2007. 10.1007/s00894-007-0207-6.
38. D. Baker and A. Sali. Protein structure prediction and structural genomics. *Science*, 294(5540):93–96, 2001.
39. C. DeWeese-Scott and J. Moult. Molecular modeling of protein function regions. *Proteins*, 55(4):942–961, 2004.
40. S. Chakravarty, L. Wang, and R. Sanchez. Accuracy of structure-derived properties in simple comparative models of protein structures. *Nucleic Acids Research*, 33(1):244–259, 2005.
41. D. Piedra, S. Lois, and X. de la Cruz. Preservation of protein clefts in comparative models. *BMC Structural Biology*, 8(1):2, 2008.
42. A.R. Ortiz, C.E.M. Strauss, and O. Olmea. Mammoth (matching molecular models obtained from theory): An automated method for model comparison. *Protein Science*, 11(11):2606–2621, 2002.
43. A. Sali and T.L. Blundell. Comparative protein modelling by satisfaction of spatial restraints. *Journal of Molecular Biology*, 234(3):779–815, 1993.
44. X. Ning, H. Rangwala, and G. Karypis. Multi-assay-based structure activity relationship models: Improving structure activity relationship models by incorporating activity information from related targets. *Journal of Chemical Information and Modeling*, 49:2444–2456, 2009. doi: 10.1021/ci900182q.

CHAPTER 17

MODELING AND VALIDATION OF TRANSMEMBRANE PROTEIN STRUCTURES

MAYA SCHUSHAN and NIR BEN-TAL
Department of Biochemistry and Molecular Biology
The George S. Wise Faculty of Life Sciences
Tel Aviv University
Tel Aviv, Israel

17.1. INTRODUCTION

TM proteins comprise an estimated 15% to 30% of bacterial and eukaryotic genomes [1–3]. As gateways to the cell, TM proteins participate in a variety of processes, such as energy production, transport of metabolites, and cell–cell communication. Structural information is needed in order to uncover the components that contribute to these diverse functions and to the structure–function relationship. In addition, TM-protein structure might provide an interpretation at the molecular level for mutations and enable structure-based drug design [4]. However, despite the substantially growing numbers of reported TM-protein structures, owing to technical difficulties only some have been experimentally solved to date [5]. As a result, TM-protein structures currently account for less than 2% of the Protein Data Bank (PDB) [6]. Moreover, most of the available structures are of bacterial origin, whereas only a small minority are of eukaryotic TM proteins [7].

Polytopic TM domains exhibit one of two possible folds: an α-helix bundle or a β-barrel [8–10]. The α-helical proteins are widespread, whereas distribution of the β-barrels is limited to mitochondria, chloroplasts, and the outer membranes of gram-negative bacteria [9]. Because the two types display distinct characteristics, there is some variation in their three-dimensional (3D)

Introduction to Protein Structure Prediction: Methods and Algorithms,
Edited by Huzefa Rangwala and George Karypis

modeling. In this chapter we deal only with the α-helical type. Owing to the uniqueness of the membrane environment, many features of TM proteins are quite distinct from those of soluble proteins (e.g., References [11–14]). This has significant implications for the prediction of their structure.

17.1.1. Traits and Topology of Helical TM Proteins

TM proteins display an amino acid composition that is quite different from that of soluble proteins [11,15,16], for example, with regard to the proportion of hydrophobic residues [13,17]. Strongly polar residues are less abundant in TM proteins than in soluble ones [18], as their transfer from the aqueous phase to the hydrocarbon region of the membrane is associated with a high energetic penalty [10,19,20]. As might be expected, the extra-membrane regions in the TM proteins, which interact with the aqueous phase, are much more hydrophilic than the membrane-embedded helices. There are two other noteworthy characteristics of TM proteins: von Heijne's "positive-inside" rule [21] (discussed in Chapter 6) and the existence of "aromatic belts," that is, an abundance of Trp and Tyr residues near both ends of the TM helices [15,17]. Additionally, the structural context of proline residues in TM helices was explored in several studies [10,22,23], showing that in many cases proline residues induce distortions such as kinks or bends in central regions of TM helices, where they contribute to function, conformational changes, and folding (e.g., References [24–26]). In addition, Yohannan et al. showed that kinks often correspond to positions exhibiting an abundance of proline residues (>10%) in multiple sequence alignments (MSAs) [27]. Thus, even when a proline is not included in the sequence itself, identification of proline peaks in alignments might also offer some information about specific features of an α-helical TM protein.

The membrane topology of a query protein can often be identified on the basis of its amino acid sequence, and is addressed in detail in Chapter 6, with focus on TM prediction methods. Some 20 years ago, it was reported that algorithms for secondary structure prediction could not accurately predict the secondary structure of TM proteins [28]. A recent study, however, showed that modern algorithms, developed for soluble proteins, are almost as precise for TM proteins as they are for soluble proteins [29]. This suggested that TM and soluble α-helices are more similar than was previously assumed. Other studies have clearly demonstrated, however, that some of the secondary structure propensities of TM proteins are unique [14,30].

A recent study showed that five kinds of specific interactions are abundant in TM structures, and can even be employed to correctly reassemble the native helix packing, starting from the backbone of the individual helices [31]. The contacts consist of hydrogen bonds, salt bridges, aromatic interactions, and packing of small and of aliphatic residues. These findings support the hypothesis that the interactions constrain the helix backbone, and facilitate folding and stability in TM proteins.

17.1.2. Fold Space of Helical TM Structures

The contemporary view of the variety of folds presented by α-helical TM proteins differs significantly from the initial picture [7,10,16,19]. In the past, α-helical TM structures were thought to be composed of strictly canonical helices that span the entire membrane in an approximately vertical direction, in correspondence with the first TM-protein structures to be solved (e.g., bacteriorhodopsin [32], Fig. 17.1A). That view implied that the architecture of α-helical TM proteins was rather limited, and that their modeling might therefore be much simpler than that of water-soluble proteins, which manifest a variety of folds.

Once some additional structures were determined, however, it became clear that TM-protein structures can also possess non-canonical helices, half or discontinuous helices, and re-entrant loops. Furthermore, some helices are very short and do not span the entire membrane, while others are extremely long and are tilted relative to the membrane plane. These observations are exemplified by the structures of the bacterial Na^+/H^+ exchanger NhaA [33] and the glycerol channel GlpF [34] (Figs. 17.1B,C), as well as by numerous other TM structures. Overall, the fold space of the α-helical TM proteins is larger than initially estimated because their non-standard structural elements are broadly distributed. It is restricted, however, relative to the fold space of soluble proteins, owing to the membrane environment as well as the distinct secondary structural elements and composition [8–10,16]. This implies that the development of specific computational modeling tools for α-helical membrane

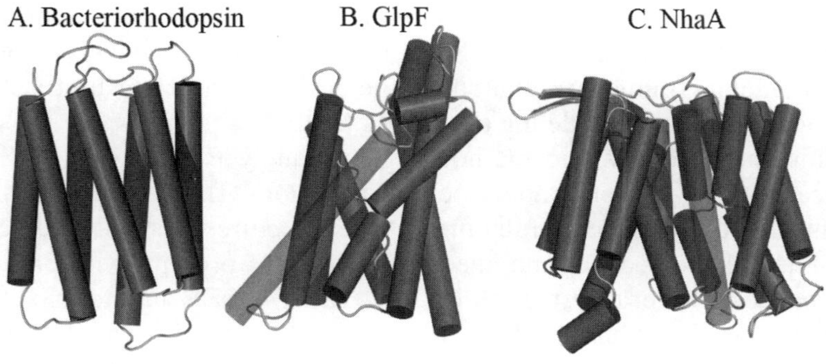

FIGURE 17.1 Simple versus compound helical TM structures. The cytoplasmic side is downward. In panels B and C, one of the TM helices is depicted as transparent for clarity. (A) The structure of bacteriorhodopsin [32] shows a "classical" architecture, composed of almost straight helices spanning the membrane. (B) In GlpF [34], many of the TM helices are tilted with respect to the membrane normal. The structure also features half-helices and intramembrane loops. (C) The structure of NhaA [33] encompasses an assembly in which two of the segments are discontinued helices located opposite to one another. The structure also contains bent helices.

structures is likely to be more complicated than originally thought, but is nevertheless still attainable.

17.1.3. General Computational Approaches to the Modeling of TM Proteins

Given the scarcity of experimentally derived structures of TM proteins, especially those of human or other eukaryotic origin [5,7,35], computational modeling techniques provide an appealing alternative. Depending on the availability of data, there are in general three different modeling approaches: (i) comparative (or homology) modeling, (ii) experimental data fitting, and (iii) template-free prediction. For comparative modeling, the query protein should be related to a similar protein with a solved high-resolution structure. Methods known as fold recognition (or threading) are highly effective when the query protein has template structures that are difficult to detect on the basis of sequence similarity (addressed in Chapters 9 and 10). Even in the absence of similarity to a known structure, experimental constraints combined with additional features such as the evolutionary profile might suffice to yield model-structures [7,36]. In the next sections, we will discuss the principles and application of these two approaches.

Should these two alternatives fail, it is possible to resort to template-free prediction, an approach that has yet to become generally applicable [9,37]. In principle, this approach utilizes only the rules of physics and chemistry to model the TM protein's structural features. However, "free modeling" techniques also include hybrid approaches. These methods incorporate the use of structural data in the form of libraries of the structures of short fragments, as well as "statistical potentials" that represent common proximities of amino acids or atoms in proteins [37,38]. Actually, the most advanced approaches, such as Rosetta and Threading ASSEmbly Refinement (TASSER), offer a unified modeling framework, combining the different modeling approaches to better address various modeling challenges [39].

Within this category, two methodologies, namely Rosetta and TASSER, have featured novel membrane-specific adaptations. The Rosetta methodology has been used to successfully predict the structures of small soluble proteins [40], and was recently modified for helical TM proteins with promising performance in several test cases [41–43]. The TASSER methodology (discussed in Chapter 12) has also been adapted for TM proteins, and was applied to predict the structures of hundreds of human G-protein-coupled receptors (GPCRs) [44]. Recently, the structure of a human GPCR protein, the β2-adrenergic receptor, was solved experimentally [45]. To assess the performance of the above two computational approaches, we compared their predicted β2-adrenergic receptor models with the native structure. Barth and Baker (unpublished results) predicted the structure of the β2-adrenergic receptor via the membrane-modified version of Rosetta. The starting point for the Rosetta model was a model obtained via homology modeling, with the

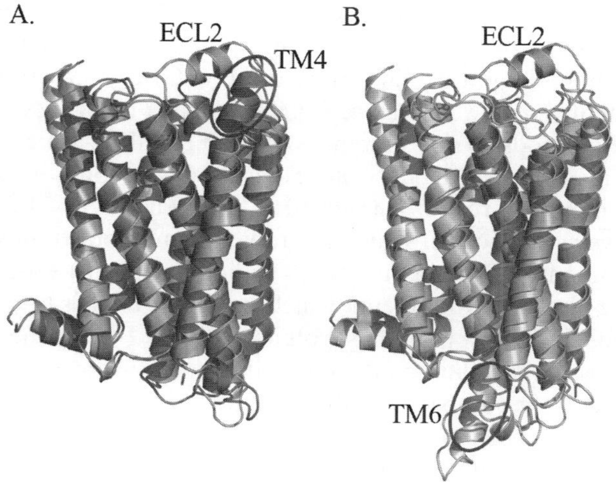

FIGURE 17.2 Performance of Rosetta and TASSER in predicting the structure of the human β2-adrenergic receptor. In both panels, the crystal structure of the β2-adrenergic receptor [45] is shown as pink ribbons; the cytoplasmic region is downward and the short helix in ECL2 is marked. The Rosetta model (Barth and Baker, unpublished results) (purple) and the best TASSER model [44] (green) are superimposed on the native structure in panels A and B, respectively. The prolonged segments in predicted TM4 and TM6 are marked, along with ECL2 of the native structure. (See color insert.)

structure of bovine rhodopsin serving as template. In the case of the TASSER algorithm this comparison was actually a blind test since the model was published, within the TASSER database of GPCRs [44], before the experimental structure came out. In both cases the TM region of the β2-adrenergic receptor model was reasonably accurate, with root-mean-square deviation (RMSD) values of 1.54 Å over 212 Cα atoms for the final Rosetta model (Fig. 17.2A) and 1.7 Å over 199 Cα atoms for the best TASSER model (Fig. 17.2B).

As might be expected, the extra-membrane regions were more difficult to predict than the TM regions. On examining the helical structural elements, we could see that the predicted TM4 in the Rosetta model was longer than the native TM4 in the X-ray structure (Fig. 17.2A), whereas in the TASSER-derived model a longer helical segment relative to the native structure was predicted for TM6 (Fig. 17.2B). Notably, one of the unique features of the β2-adrenergic receptor structure is an extra helix in the second extracellular loop (ECL2; Fig. 17.2). Both the TASSER and the Rosetta models failed to predict this helix (Fig. 17.2). However, the Rosetta model did predict a short helical segment in the region preceding the ECL2 helix (Fig. 17.2A). Overall, both produced reasonably good models.

GPCRs, the largest family of TM signal-transduction proteins, include about 1000 human isoforms [46,47]. Since they comprise approximately 50% of contemporary protein drug targets, there is particular interest in modeling their structures for purposes of drug discovery [48]. Up to now, efforts to model GPCRs have been based on a variety of computational approaches, ranging from homology modeling using the few GPCR structures available from experiments (reviewed, e.g., in Reference [49]) to specifically designed template-free methods (e.g., References [44,50,51]), all of which are tailored for GPCRs. This is a research field of its own, and is beyond the scope of this chapter. The interested reader is referred to References [52–54].

An objective assessment of structure prediction methods is provided by the Critical Assessment of Protein Structure Prediction (CASP) experiments. The object of these biennial experiments, which started in 1994, is to assess current abilities and inabilities in predicting protein structure. During the experiment, different groups submit blind modeling predictions of various proteins. These predictions are later compared with the native structures of the proteins, which have already been determined experimentally but are not yet known to the participating scientists during the experiment [55]. Unfortunately, TM proteins are not used as CASP targets; thus, there is currently no generally accepted way to assess, directly and without bias, the application of available modeling techniques for TM proteins.

17.2. COMPARATIVE MODELING

In comparative (or homology) modeling, currently the leading computational approach for generating protein models, a high-resolution, experimentally solved structure is used to produce a model-structure of a homologous protein for which structural data are not yet available [38,56–58]. The technique has been successfully applied to numerous soluble proteins and is considered to be the most accurate approach to structural modeling available today [56–58]. Recently, Forrest et al. showed that comparative modeling can also be applied to TM proteins to produce models of similar accuracy to those of soluble proteins [29]. To this end, the structures of 11 TM-protein families were examined, covering a range of folds and sequence similarities. In line with previous observations for soluble proteins [58], the analysis of Forrest et al. [29] showed that the accuracy of TM model-structures produced via comparative modeling depends, as anticipated, on the similarity between the query and the template sequences. The precision of the produced model-structures also decreases with decreasing similarity between the sequences, owing to two factors: alignment errors and inherent structural differences between the two proteins.

Using a previous division of TM proteins from 95 genomes into families [59], Granseth et al. examined the relationship between prokaryotic and eukaryotic TM proteins [35] with the object of assessing the extent to which comparative modeling could derive structural models of eukaryotic and even human TM proteins from available prokaryotic structures. Their analysis revealed that 13% of eukaryotic TM families include also members of the prokaryotic kingdom. Of these 256 families, solved structures exist for representatives of only 29 [35]. Although these data are not particularly encouraging, they nevertheless indicate that a significant number of eukaryotic TM models could be obtained by comparative modeling. In this respect it should also be noted that the sequence similarity between the eukaryotic query protein and the prokaryotic template is often low, further complicating the production of an accurate pairwise sequence alignment between them [29,57,58]. Since comparative modeling is largely dependent on pairwise alignment (discussed below), this presents a major obstacle in obtaining TM model-structures of high quality.

This difficulty is best illustrated by a description of some recent efforts to model TM human proteins based on their remote prokaryotic homologs. The serotonin transporter of the neurotransmitter:sodium symporter (NSS) family was modeled using the eubacterium *Aquifex aeolicus* leucine transporter (aaLeuT) as template [60]. An available alignment [61] was refined, because of low sequence identity between the prokaryotic and eukaryotic family members, by the use of various bioinformatics tools along with elaborate experimental data. Interestingly, the model-structure was utilized to identify a chloride ion-binding site in Cl^- dependent transporters, a prediction confirmed by experimental assays [60].

In another study, the extremely low sequence identity between the human Na^+/H^+ exchanger NHE1 and NhaA of *Escherichia coli* (<15%) prompted a composite modeling approach in which various state-of-the-art computational modeling tools were integrated to achieve correct alignment of the two proteins [62]. Supported by elaborate mutagenesis, the model revealed common properties of the inhibitor-binding sites of NHE1 and NhaA, as well as a putative ion-transport mechanism for NHE1 [62].

In yet another example, although the intriguing cystic fibrosis TM conductance regulator (CFTR) is a chloride channel, the structure of Sav1866, a multidrug transporter of the same superfamily, was used as template for its modeling [63]. To overcome the obstacle of low sequence identity, their divergent sequences were aligned through a technique of hydrophobic cluster analysis [64]. The model-structure, which demonstrated good correlation with experimental data, offers a molecular-level insight into the contacts that might be affected as a result of the deletion of Phe508, the most abundant cystic fibrosis-causing mutation [63].

17.2.1. Work Scheme

The scheme for predicting a structure by means of homology modeling is generally the same for soluble and TM proteins. It can be divided into four major steps: (i) template search and selection, (ii) pairwise sequence alignment of the query and the template sequences, (iii) model building, and (iv) evaluation and validation. It is noteworthy that depending on the outcome of the validation stage, it might be necessary to refine the model-structure by repeating the previous stages. This cycle can be repeated until a model of the best possible quality is produced. Because of the paucity of experimentally solved TM structures on the one hand and the exclusive features of these proteins on the other, it is necessary to develop unique methods for each step. Accurate prediction of the membrane topology (addressed in Chapter 6) can be helpful in the first two stages, which are the keys to proper modeling. However, because the structural data are limited, the computational aspects of membrane-specific homology modeling are still underdeveloped [37]. In the following sections we provide a description of the fine points of these work steps for homology modeling of α-helical TM proteins. The last step, model evaluation and validation, will be presented in Section 17.4, as it is identical for models generated via homology modeling (Section 17.2) and via experimental data fitting (Section 17.3).

17.2.2. Template Search and Selection

17.2.2.1. Simple and Advanced Search. When attempting to determine the structure of a TM protein, we first search for a potential template or templates. This is easy if the structure of a close homolog of the query protein has already been solved, but difficult if it has not. The number of available TM structures, however, is very small, which makes it hard to find suitable structural templates.

Because similar sequences adopt similar structures, the initial strategy in searching for a potential template is usually to employ sequence-search methods such as the Basic Local Alignment Search Tool (BLAST) [65]. Many TM proteins possess water-soluble domains in addition to the TM region, which may appear at the N- or C-termini or between consecutive TM helices. By excluding large extra-membrane regions and using only the sequence of the TM domain to be matched with PDB-derived sequences, it may be possible to produce more accurate results since they will be focused only on the region of interest. It is worth noting that the "low-complexity filter" included among the common search tools might remove hydrophobic segments, and should therefore not be applied [66]. In addition, Hedman et al. showed that the search for homologs (tested on a benchmark of GPCR sequences) can be improved by utilizing predictions of the location of the TM segments in the sequence [66].

When the simple search for a template fails, advanced sequence-based search could be employed. It is also possible to apply fold recognition or threading algorithms to detect putative structural templates. These algorithms perform two modeling steps: template identification and alignment of the query and the template sequences [38]. Chapters 9 and 10 describe these approaches, which are currently employed in the same manner for both soluble and TM proteins.

17.2.2.2. Template Selection. The next step is to select the most suitable template(s) from the detected hits. For TM proteins this is easy, simply because it is rare to find any templates at all. The challenge here is rather to estimate the suitability of a putative template whose sequence similarity to the query protein is often low. Because resemblance between the two sequences correlates with model accuracy [29], it is important—as in the case of soluble proteins—to assess their similarity and their evolutionary relationship [58]. The number of TM helices in the query protein and in the template proteins is likely to be the same. Thus, the known (or predicted) membrane topology of the query protein will probably aid in selection of the most suitable template. In view of the known relationship between structure and function, experimental evidence of functional similarity between the two proteins might be taken as an indication of the template's suitability. This was done, for example, in the modeling of NHE1 based on the structure of NhaA [62].

17.2.3. Aligning the Query and the Template Sequences

Aligning the query and the template sequences as accurately as possible is a crucial step in model building, as the alignment largely determines the 3D location of each of the residues of the query protein [29,67]. Interestingly, a recent study by Gao and Stern [67] showed that the accuracy of TM models is significantly improved when sequence identity between the query protein and the template exceeds 30%, and is substantially reduced at weaker sequence identities. Exactly the same threshold values for correct modeling also apply in the case of soluble proteins [68]. We next address two cases: alignment with high and with low sequence similarities (Fig. 17.3).

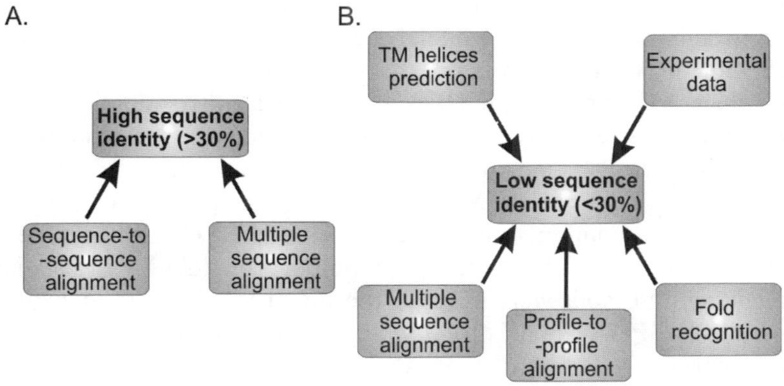

FIGURE 17.3 Computational approaches for the alignment of query and template proteins of high and low similarity. (A) If the query and the template sequences are close enough (>30% identity) it is possible to use simple sequence-to-sequence alignment, but it is often more accurate to extract the pairwise alignment from an MSA. (B) As sequence identity decreases, it becomes necessary to combine more sources of data in order to align the sequences correctly. These include fold recognition, profile-to-profile methods, TM prediction methods, MSAs, and available experimental data.

17.2.3.1. High Similarity. A rather simple sequence-to-sequence alignment might suffice when the query and the template proteins exhibit sequence identity of over 30%, covering all the TM segments of the sequence, whereas extraction of the pairwise alignment from an MSA might add essential evolutionary information and thus improve the alignment accuracy (Fig. 17.3A) [29]. To ensure MSA integrity and avoid sequence fragments, it might be useful to include in the MSA only those sequences that share all of the query protein's TM helices. Forrest et al., after examining the performance of different MSA algorithms in securing the pairwise alignments needed for TM-protein modeling [29], reported that advanced alignment methods, such as MUSCLE [69] or T-Coffee [70], are more effective than the traditional Clustal W [71].

Amino acid substitution matrices are essential for generating both pairwise and multiple sequence alignments. The widely used substitution matrices, such as PAM [72] and BLOSUM [73], were derived from datasets of homologous soluble proteins. Additional substitution matrices, for example, JTT TM [13], PHAT [74], and SLIM [75], were developed specifically for TM proteins. To the best of our knowledge, no study has yet compared all three matrices. However, when the abilities of both matrices in searching for homologs searches were compared using a dataset of GPCRs, SLIM was predicted to outperform PHAT [75]. When the same TM dataset was examined, both SLIM and PHAT were shown to be more accurate than the traditional BLOSUM. On the other hand, the PHAT matrix performed better than JTT TM [74]. In this case, the test set was composed of 100 sequences from 74 TM-protein families that were utilized as query proteins for database searching.

Theoretically, in bipartite alignments the membrane matrices should be utilized to align TM regions, while the extra-membrane regions should be aligned with the traditional matrices. Such an approach is implemented by STAM, which adds high gap penalties in predicted TM regions. However, the STAM method was evaluated only for GPCRs and was compared only with Clustal W [76]. Forrest et al., on examining the performance of bipartite alignments on a diverse dataset of TM proteins, found that performance was worse when they used a combination of the PHAT and BLOSUM matrices than BLOSUM alone [29]. Nevertheless, some improvements over the commonly used alignment methods were seen for the PRALINE™ method, which also incorporates the PHAT matrix for TM alignment [77]. This was attributed to a more accurate prediction of the TM segments. A systematic evaluation of the substitution matrices and their performance in bipartite alignments has yet to be carried out.

17.2.3.2. Low Similarity. When the query and the template proteins are distant homologs (sequence identity <30%), as is often the case when eukaryotic proteins are aligned to prokaryotic proteins (e.g., References [61–63]), the straightforward approach described above might not suffice [29]. In such cases it is rather difficult to produce a fully continuous alignment. However, the TM helices are typically more conserved than the extra-membrane regions and it is often possible to align them properly. Indeed, such fragmented alignments can be used for model building of the TM domain. So the problem becomes a matter of detecting the TM helices of the query protein and their subsequent alignment to the known helices of the template. The TM helices are not only strongly hydrophobic, but are also usually preserved within the protein family. Hence, in an MSA of the query protein and its homologs, these regions often appear as gap-less segments of a strongly hydrophobic nature; this observation simplifies their detection.

Fold recognition approaches, including profile-to-profile alignments, not only enable remote relationships between query and template proteins to be detected, but also produce a sensitive pairwise alignment. The HMAP method [78] produced more accurate alignments between query and template TM proteins than those produced via sequence-to-sequence and MSAs, especially in cases of low sequence similarity [29].

Overall, when attempting to properly align sequences of low identity, it is recommended to use a range of tools and all the available data (Fig. 17.3B). These include MSAs, results of fold recognition approaches, TM-helix predictions, and the available biochemical and biophysical experimental data (e.g., site-directed mutagenesis and accessibility measurements). In the optimal situation, data from all sources will overlap, thus consolidating the prediction. But even in less favorable situations, in which conflicting data might be obtained, often there is consensus at least with regard to the location of some of the TM helices. When data from various sources are in conflict concerning the location of a particular TM helix, and in the absence of a more compelling basis for resolution, decisions can be made based on the majority of (independent) data. Alternatively, 3D models can be built on the basis of more than one sequence alignment.

17.2.4. Building a 3D Model-Structure

The model-building process includes construction of the protein core based on an existing structural template, and modeling of the backbone and side chains of peripheral regions for which a template might not be available [57]. For α-helical TM proteins the core includes the TM helices, while the periphery contains the extra-membrane loops that tend to vary even between related TM proteins.

As in the case of soluble proteins, the building process is carried out via one of the many available applications, such as Modeller [79] and NEST [80]. The performance of model-building methods specifically for TM proteins was recently investigated in two studies. Reddy et al. [81] assessed the model-building performance of five methods (or combinations of methods): Modeller [58,79], the MOE [82] homology module of InsightII [83], Swiss-PdbViewer [84,85] and models produced via initial construction by InsightIIHomology followed by Modeller refinement. Although this analysis did not include state-of-the-art programs such as NEST [80], PLOP [86–88], or Rosetta's template-based module [89], it was an initial attempt to evaluate the performance of some algorithms for building models of TM proteins. The results indicated that for this particular dataset of TM structures, Modeller generally outperformed the other methods.

Gao and Stern [67] compared Modeller [58,79] with PLOP [86–88] while using a dataset of TM proteins that included both α-helix bundles and β-barrels. First they evaluated model-structures built via Modeller and PLOP on the basis of accurate pairwise alignments, constructed using structural alignments to eliminate alignment errors. PLOP outperformed Modeller in this case, probably as a result of its improved energy function. However, when those alignments were replaced with Clustal W-derived [71] pairwise alignments, which usually contain some alignment errors, Modeller and PLOP produced similar results, despite the fact that PLOP's energy function is considered to be superior to that of Modeller. The authors offered an explanation for this discrepancy by suggesting that the current sampling of conformational space by PLOP is not sufficient to detect the correct structural conformation. This suggestion was supported by their finding that in the refinement of the loop regions both methods performed poorly, again possibly owing to limited sampling of conformational space [67].

Since some traits of TM proteins differ from those of soluble proteins, future research should probably address specific adaptations of the above-mentioned packages for TM proteins. These might include, for example, the use of rotamers constructed from TM structures, as well as novel scoring functions. To the best of our knowledge, only the template-based modeling application of Rosetta [89] has so far been modified to include a membrane-specific force field [42]. This modified method was utilized, for example, in a study of voltage-gated potassium channels, where models derived via Rosetta's homology/*de novo* membrane mode were used to provide a mechanism for voltage-dependent gating [90].

Although many energy (or scoring) functions do not currently encompass membrane-specific adaptations, Gou and Stern examined the ability of several high-quality energy functions to distinguish native loops in TM structures from decoys of loop conformations, generated via molecular dynamics. This analysis included, *inter alia,* the energy functions implanted in Modeller [58,79], Rosetta [91,92], PLOP [86–88], and the DFIRE potential [93]. The results indicated that some of the examined energy functions can reliably discriminate the native loop from the decoy conformations. Moreover, all but one energy function successfully ranked the energy of the entire TM structure lower than those of decoy models produced by homology modeling. These findings raise hopes for the future implementation of energy functions in the refinement of TM models, possibly in parallel with the improvement of sampling in conformational space.

17.2.5. Useful Tips for TM Comparative Modeling

- It is useful to evaluate the topology of the TM protein. This prediction might come in handy for template identification and/or query-template alignment.
- Identification of a common origin can indicate a shared fold between two TM proteins. When sequences diverge, however, this functional relationship is not always easy to detect.
- Both simple and advanced similarity searches can be helpful in the identification and alignment of possible templates, especially metaservers that combine methods such as fold recognition and membrane topology predictions.
- When the similarity between query and template sequences is low, alignment accuracy might be improved by combining state-of-the-art bioinformatical tools with experimental data.
- It is sometimes useful to obtain and evaluate a number of models built using different alignments or templates until the best model or models are found.

17.3. EXPERIMENTAL DATA FITTING

In contrast to the section on comparative modeling, here we describe a computational approach that does not rely on the existence of a high-resolution structure of a similar protein. Instead, other types of available data can be exploited as constraints in order to produce a model-structure [7,36]. Essential data for this purpose might come from low-resolution structures (e.g. cryo-EM maps) and from mutagenesis studies. The former have been shown to produce more accurate models, and will be addressed here more thoroughly. It should be noted that only a few TM model-structures have been obtained by experimental data fitting (e.g., References [94–98,98a]). This is mainly because of a lack of the needed preliminary data, but might also result from a rather complicated modeling process, which requires manual intervention and specialized expertise. Nevertheless, the models obtained so far have raised considerable interest. As more experimental data emerge, especially cryo-EM maps of eukaryotic TM proteins, this approach is likely to become much more relevant and easier to use.

17.3.1. Starting from Electron-Density Maps at Intermediate Resolution

Cryo-EM maps occasionally provide an intermediate-resolution image of the protein structure, with an in-plane resolution of 5 Å–10 Å but a much lower resolution along the membrane normal. Owing to the low resolution, such maps

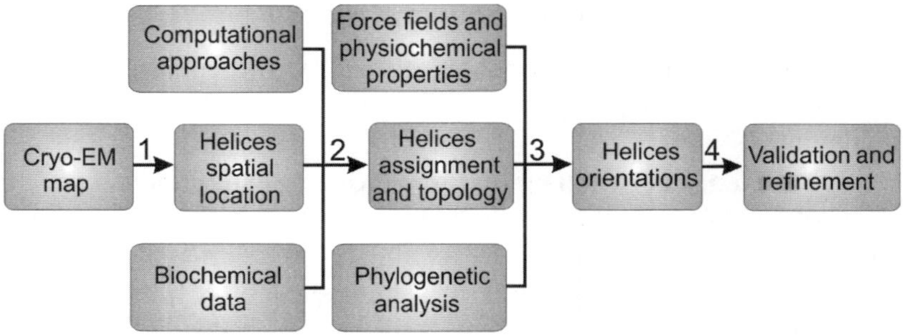

FIGURE 17.4 Predicting TM models from cryo-EM maps [7]. Step 1. Locations of TM helices, and specifically of their principal axes, are derived. Step 2. Using computational tools and biochemical data, the TM segments in the sequence are assigned to the helical rods in the density map. The topology of the TM protein in the map is also determined. Step 3. To correctly rotate each helix around its principal axis, additional data from phylogenetic analysis, physicochemical properties, and/or force fields can be exploited. A Cα-trace model is generated. Side chains can be added to obtain a full-atom model. Step 4. Data that were not employed for model building can be used for validation. The model can then be refined by reviewing the preceding modeling steps. In addition, the model-structure can be used to design mutagenesis experiments and undergo subsequent refinement on the basis of the results.

cannot reveal the exact features of TM-protein structures. In the case of α-helical TM proteins, they cannot even be used to assign the TM segments to the map helices, not to mention the coordinates of the amino acids of the TM helices in 3D space. Nevertheless, the maps provide important data concerning the number, tilt, and overall location of the TM helices in the structure.

For production of a molecular model from a cryo-EM map, additional data must be incorporated for the various modeling steps [94,99,100]. The overall modeling process is depicted by the flowchart in Figure 17.4 [7,36]. First, spatial locations of the helices are obtained from an available intermediate-resolution cryo-EM map. Using the map, the principal axes of the helices can be detected and extracted. Next, TM segments in the protein sequence are assigned to the helical density rods, usually by employing both biochemical data and computational approaches. TM helices, corresponding to the TM-sequence segments, are then constructed using the principal axes. During this step, their register along the axes must be determined. Additional sources of data, such as evolutionary conservation and physicochemical properties of the protein sequence, are subsequently exploited to orient the helices around their principal axes. The result is a Cα-trace model-structure, that is, the predicted location of the Cα atoms of the TM domain. The backbone atoms and the side chains of the residues can then be reconstructed in order to generate a full atom model of the TM protein. Finally, the model should be validated, typically via experimental data that were not used to build the model (described in Section 17.4). The results of the validation process can then be utilized for model refinement. Details of each step of this modeling process are addressed in the following sections.

17.3.1.1. Helix Assignment and Membrane Topology. For accurate modeling, the TM segments in the protein sequence must be identified and the topology of the protein in the membrane determined, as described above. These features are essential for TM-model building, but in most cases it is not easy to predict them with confidence, especially the precise helices boundaries. Thus, it is best to rely, as much as possible, on reliable experimental data.

For n helices, the number of possibilities for assigning the TM segments detected in the protein's sequence to the helices in the map is $n!$. Adding the two possible membrane orientations (the cytoplasmic- and the extracellular-facing sides of the cryo-EM map) in relation to the protein's topology, the number of possible models is $2 \times n!$. This implies that the crucial step of helix assignment and selection of the membrane orientation is extremely complicated even for TM proteins of moderate size; a TM protein with four helices, for example, will have 48 combinations.

An attempt was made to develop a graph-theory approach for assigning TM helices and predicting topology based on the lengths of the loops connecting the helices [101]. The method worked well for short loops of up to seven residues, but the accuracy of prediction depends strongly on exact determination of the boundaries of the TM helices, which is usually not available. Thus, there is no generally applicable way to determine the helix assignment and topology of a TM protein using a single automatic computational tool. The problem may occasionally be solved by combining manual analysis of biochemical data with use of the available computational tools. As in the modeling of the EmrE [95] and the hCTR1 transporters [98a], considerations from phylogenetic analysis, hydrophobicity, and experimental data might also be useful.

Generally speaking, the TM helices detected in the map can be divided into two groups: (i) core helices, surrounded by other TM helices in the bundle, and (ii) peripheral helices, which are in contact with the core, but also have at least one lipid-exposed face. Owing to dissimilar evolutionary pressures, both the hydrophobicity and the evolutionary conservation patterns of the two types of helices are quite distinct. Relying on these differences of TM helices, Adamian and Liang developed an automatic method to identify core TM helices, which are less accessible to the lipid membrane [102]. This method can help reduce the number of possibilities for helix assignment. Another useful observation for this modeling step is that interacting residues from neighboring TM helices might evolve cooperatively [96,103]. Hence, detection of coevolving positions by phylogenetic analysis (e.g., References [103,104]) can help guide the assignment of interacting TM-helix pairs. Experimental data such as distance constraints, site-directed mutagenesis, and accessibility assays can also be used.

The complexity of this step is best demonstrated by the example of the gap junction Cα-model, produced based on a cryo-EM map [96]. When a crystal structure (of a homologous protein) became available, it became clear that the helix assignment that was utilized for model-building was incorrect; only one of the four TM helices of each subunit in the homo-hexamer was assigned correctly [105]. The erroneous assignment was based on mutagenesis data that apparently was interpreted wrongly [106].

17.3.1.2. Helix Building and Rotation. An intermediate-resolution map for producing a model-structure of a TM protein was first employed by Baldwin et al. who constructed a Cα-trace model of vertebrate rhodopsin [97]. The model was generated from a structure at 7 Å resolution in the membrane plane using constraints derived from MSA and biochemical data. When the structure was later determined at high resolution by X-ray crystallography, the orientations of TM helices in the model-structure were found to be quite accurate (3.2 Å RMSD). Most of the variation was attributed to difficulties encountered in the precise modeling of two kinked helices [7].

Expanding on this pioneering approach, Fleishman et al. developed an automatic method for TM-model prediction based essentially on evolutionary conservation analysis [100]. For predicting the orientation of each helix, the algorithm included a scoring function that favored the burial of conserved (and charged) residues in the protein core, as well as the exposure of variable amino acids to the lipid membrane. This method, in which only the Cα atoms of the TM domain were constructed, was later applied to predict the structure of the gap-junction [96], the EmrE multidrug transporter [95], and recently the hCTR1copper transporter [98a]. The crystal structure of EmrE, determined a few months later, was remarkably similar to the model-structure, with an RMSD value of 1.4 Å for the core region [107] (and of 3.52 Å for the entire TM domain, Fig. 17.5).

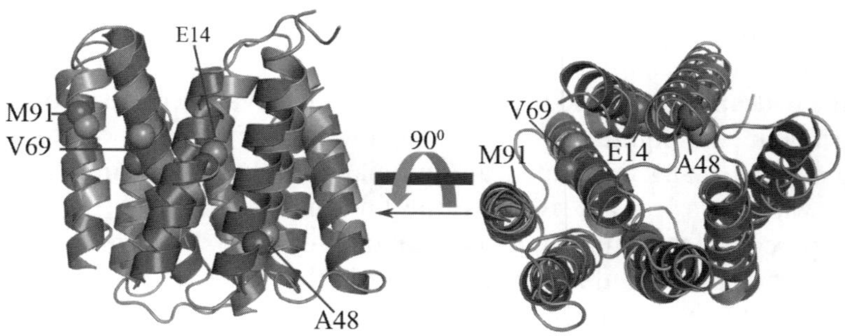

FIGURE 17.5 Model versus crystal structure of EmrE. A model of the EmrE homodimer (blue) was derived using a cryo-EM map and the computational approach of Fleishman et al. [95]. The crystal structure of EmrE (pink) was solved later [107]. The model and structure are aligned and viewed from the side (left) and top (right). The 3D location of specific Cα atoms (marked as spheres on one monomer of the model and structure) demonstrate the similarity between the model and native structure. (See color insert.)

Beuming and Weinstein proposed a similar method [94], in which helix orientations are selected by employing both evolutionary conservation and a knowledge-based scale of the propensities of the 20 amino acids to be exposed to lipids. In addition to the Cα atoms, the backbone and side chains of residues are also constructed, and this is followed by structure minimization and some manual adjustment. The result is a full-atom model. This method was used to predict a molecular model of the bacterial oxalate transporter OxlT, using its electron density map of 6.5 Å resolution in the membrane plane. The model was in agreement with cross-linking experiments and data concerning functional residues [94].

The most recent work in this field was done by Kovacs et al., who presented a new method [99] in which helix orientations are determined by minimizing an energy function that takes into account van der Waals interactions, electrostatics, hydrogen bonding, and torsional and density correlation terms. Side chains are also predicted. The best conformations are then energy-minimized by a complex procedure in which atoms of the helical backbone are restrained to fit the observed cryo-EM map densities. This minimization step also relies on a solvent-accessibility grid map of the density rods. It should be noted that in constructing this accessibility map the membrane boundaries must be selected within the cryo-EM map. Correct prediction of these boundaries is not a simple task, and deviations from the real (unknown) boundaries can affect the model-structure. Another limitation of the approach is that because of the complicated energy calculations required for an all-atom representation, it is feasible only for TM proteins of moderate size (up to four helices per symmetric subunit). The approach worked well in three test-cases, but has not yet been used for *de novo* predictions.

Overall, several different methods are available for modeling the TM domains of α-helical proteins employing restraints derived from electron-density maps of sufficient resolution. So far the starting point has always been a cryo-EM map, but maps from X-ray crystallography experiments at intermediate resolution can be used as well. Each of the above methods was developed and tested on only a small number of cases. Until the performance of all methods is examined by means of a large-scale assessment of known structures, their efficacies cannot be determined. In particular, it would be interesting to know whether the addition of side chains increases or decreases the accuracy. It may well be that each method is suitable only for certain specific cases. When constructing new structural models, therefore, it might be advisable to examine several potential models each produced by various different methods. The models can then be inspected on the basis of, for example, reliable experimental data along with other evaluation procedures (described in Section 17.4).

17.3.2. Modeling Based on Biochemical and Biophysical Data

Other computational approaches aimed at addressing cases for which there were no available structural data. These approaches have employed biochemical and biophysical data obtained, for example, from site-directed mutagenesis and chemical cross-linking, as the only constraints on the protein structure. Because these data are difficult to interpret in an unequivocal manner, this approach is inherently less reliable than modeling based on an intermediate resolution structure from cryo-EM maps, as described above. In particular, the results of mutagenesis often represent phenomena that are associated with more than one conformation of the TM protein, and the observed phenotypes of a mutation might be indicative of allosteric effects.

Sale et al. [108] developed an automatic method for TM-structure prediction based on distance restraints obtained from experimental assays such as chemical cross-linking, nuclear magnetic resonance (NMR), electron paramagnetic resonance (EPR) or fluorescence resonance energy transfer (FRET). A search of the conformational space for a TM model-structure that is compatible with the available distance restraints is followed by optimization using Monte-Carlo simulated annealing. The optimization samples models that correspond well both with the experimental restraints and with knowledge-based structural parameters derived from a dataset of known TM structures. Although this approach produced an accurate model-structure of the seven TM helices of bovine rhodoposin (RMSD of 3.2 Å) [108], it has yet to be applied for the prediction of novel TM-protein structures.

The bacterial lactose permease (also referred to as lacY), a galactosidase transporter, is perhaps the most extensively studied TM protein to date. This transporter was examined by means of various experimental approaches, including systematic site-directed mutagenesis, double-cysteine mutants, thiol cross-linking, engineered Mn(II) binding sites, *N*-ethylmaleimide (NEM) alkylation of single cysteine mutants, site-directed EPR, and discontinuous mAb epitope mapping [109,110]. The accumulated experimental data related to each position in the 12 TM helices of lacY, comprising 417 residues. By utilizing helical backbone restraints and 99 long-range restraints derived from thiol cross-linking and engineered Mn(II) binding sites, Sorgen et al. obtained a single cluster of models for lacY with small deviations from one another [98]. They achieved this using an algorithm based on torsion-angle dynamics-simulated annealing, which was initially developed and utilized in NMR structure determination [111].

The crystal structure of lacY was later determined [112], making it possible to evaluate the effectiveness of this modeling approach. Comparison between the model and the native structure showed that various local arrangements of functional residues, such as sugar- and proton-binding residues, were fairly accurate. However, the overall architecture of the model and its structure were not superimposable (Fig. 17.6) [7]. Given the crystal structure, it was possible to examine the experimentally measured distances that were used for modeling. While the distances on the periplasmic side of the crystal structure were in good agreement with the experimental data, many of the distances on the cytoplasmic side were underestimated. The distances obtained for the cytoplasmic side probably corresponded to the periplasmic-facing conformation or other conformational substates, and therefore did not agree with the crystal

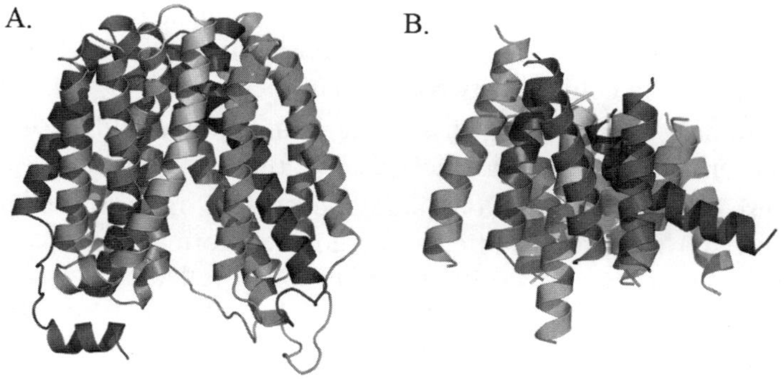

FIGURE 17.6 Comparison of the lacY model, produced via experimental constraints, and the solved crystal structure. In both panels the cytoplasmic side points downward. The lacY crystal structure (A) [112] and computational model (B) [98] are colored by rainbow. Although the overall fold and helix organization are quite distinct, there are regions of similarity, especially between the helices that contribute to the cytoplasmic-facing pore. (See color insert.)

structure, which was solved in an inward-facing conformation [113]. The fact that the experimental data probably reflect different conformations might account for the discrepancies between the model and the structure. Overall, this case study demonstrated the difficulty of producing an accurate model when the experimental data do not account for a single structural conformation. Obviously such "monochromatic" data are usually not available.

Prediction of the dimeric structure of the *E. coli* Na^+/H^+ antiporter NhaA is another example of constraint-based modeling of a TM protein. Although the protein in its physiological form is a dimer [114,115], its crystal structure depicts only the monomer; the physiological dimeric contacts are not exhibited [33]. To obtain the dimeric structure, two NhaA monomeric structures were considered as rigid bodies. Nine long-range EPR distance measurements were then used as constraints to build a dimer by docking the two monomers [116]. The model-dimer showed good agreement with the interfacial domain observed in cryo-EM 2D crystals, which exhibited the electron density of the NhaA dimer. The suggested dimeric interface was further supported by chemical cross-linking [115] and deletion assays [117]. Although this is not a classical example of the use of experimental constraints to model a helical TM protein, it shows how the membrane plane and intrinsic symmetry reduces the degrees of freedom of the modeling process. Thus, even a small number of distance constraints was sufficient for inferring the oligomeric conformation.

17.3.3. Tips for Modeling by Experimental Data Fitting

- A cryo-EM map is a good starting point. When the map's resolution is high enough to detect the TM helices, at least Cα-trace models of TM proteins can be produced.

- Because helix assignment is a crucial and complicated step, several data sources are usually needed in order to correctly assign the TM sequence segments to the helix density contours in the map.
- Evolutionary conservation, physicochemical features, and force fields are useful for rotating the TM helices around their principal axes.
- The use of empirical data to build models of TM proteins is complicated because: (i) the protein often undergoes conformational changes, and data relating to the effects of mutations might reflect a mixture of these conformations. (ii) Mutagenesis data might reflect both direct interactions and remote (allosteric) effects.

17.4. QUALITY ASSESSMENT

Computational methods of model evaluation are usually referred to as Model Quality Assessment Programs (MQAPs) [38,118]. As reviewed in Chapters 15 and 16, numerous computational methods have already been developed for local and global assessments of model-structures, indicating the importance of this step in protein modeling (e.g., References [119–124]). These are all general methods for model evaluation; to the best of our knowledge, specialized MQAPs for TM proteins are currently not available. Therefore, when these general evaluation tools are applied to TM model-structures, the results of the assessment should be viewed with caution. A good strategy for assessment of TM models predicted via homology modeling might be to apply the MQAPs to both model and template. The results of the template could be used thereafter as a reference point for the result of the model-structure. It is anticipated that in the future, existing approaches will be modified to better comply with the distinct traits of TM proteins. However, assessment of the performance of state-of-the-art MQAPs on TM structures might reveal that current methods are also adequate for evaluating this class of proteins.

The validity of the TM model-structure can be further assessed through an examination of its generic characteristics. These might comprise only external features, defined here as protein characteristics that were not accounted for during model building. A recent investigation of the local accuracy of TM models that were produced via homology modeling demonstrated that even when the membrane-embedded helices of the query and the template sequences are structurally similar, the extra-membrane regions that connect them might deviate in both sequence and length [29]. In another study it was demonstrated that refinement procedures cannot clearly improve the loop regions in TM proteins [67]. Thus, loops in TM model-structures should be considered a priori as regions of questionable accuracy, as in the modeling of soluble proteins.

17.4.1. Compatibility of the Model with General Characteristics of TM Proteins

In Section 17.1.1 we presented some of the general features that characterize α-helical TM proteins. These distinct traits were observed by analyzing TM structures that were solved experimentally. Those traits could therefore be utilized to assess the accuracy of TM model-structures, provided that they were not used in building the model. When model-structures are assessed in the future, it might be helpful to exploit the recent discovery that the structural determinants of TM helices appear to incorporate five specific types of inter-helical interactions [31]. Accordingly, the expectation would be that trustworthy models will feature these interactions, and that wrong models will not. When erroneous X-ray crystal structures were retracted from the PDB, they were indeed found to include only very few interactions of that sort, which would not suffice to keep the fold intact.

17.4.1.1. The "Positive-Inside" Rule and the "Aromatic-Belt." The "positive-inside" rule [21] can be used to examine the overall architecture of a TM model-structure, as the distribution of lysines and arginines in the extra-membrane regions of the model-structure can serve as an indication of whether the TM segments and extra-membrane regions are correctly approximated. Moreover, this rule can point out cases where the template selection or the query-template alignments are entirely erroneous. Clearly, however, this will not be of help in determining the exact TM boundaries, inter-helical structural arrangement, or packing. In addition, TM-protein structures frequently feature an "aromatic belt" near the borders of the hydrocarbon core region [17]. As with the "positive-inside" rule, this feature can also be evaluated to assess the overall topology of the TM model-structure, but will not help to validate its precise molecular details.

It should be mentioned that both of the above features are implemented in many of the advanced methods for predicting the membrane topology from the sequence. If such methods were used for building the model, it is fairly obvious that the model will inevitably be compatible with both thumb rules.

17.4.1.2. Hydrophobicity of Lipid-Facing Residues. Based on knowledge derived from available TM structures, most of the lipid-exposed residues of the TM model-structure are expected to be hydrophobic [17]. In some of the methods for predicting TM structure by the use of experimental constraints, this trait is exploited to produce the 3D model (e.g., References [94,100]). In comparative modeling, this trait can be addressed indirectly when integrating the results of TM-helix prediction for correct alignment of the query-template sequences.

Such examination is likely to be useful for validation provided that the nature of the lipid-exposed residues was not taken into consideration during model building. To ensure that this requirement is met, the lipid-exposed positions in the TM model-structure should be reviewed using a hydrophobicity scale (e.g., Reference [20]). Mapping of the hydrophobicity on the residues of the TM model-structure reveals its degree of correspondence with the structure's expected pattern. It should be noted that this evaluation process is useful for peripheral helices, in which residues are exposed to the membrane, but not for helices that are buried in the TM core. If, on examining a model generated by comparative modeling, a peripheral helix is found in which polar residues face the membrane while hydrophobic residues face the protein core, this might indicate that the pairwise alignment of the query-template in this region needs to be refined. During this evaluation process, moreover, the physiological oligomeric state should also be taken into account. Regions that appear to be lipid-exposed might actually participate in inter-protein interfaces within the oligomeric structure of the TM protein. These might include polar residues.

17.4.1.3. Prolines and Kinks. Cordes et al. have demonstrated that proline residues are more abundant at the ends of TM helices [22]. Proline residues interrupt helical segments and are also commonly found in irregular regions of TM helices [10,22,23]. Furthermore, inspection of hinge regions of TM proteins revealed that 60% of the prolines comprise the hinge itself or are located up to four residues (approximately one turn) before it [22]. Another study showed that positions in which the MSA exhibits a high content of prolines are likely to correspond to proline-induced kinked or otherwise disrupted regions [27]. Taking all of the above into consideration, it is interesting to examine the predicted location of proline residues and the positions in which the MSA shows an abundance of proline. Many of these positions, especially the conserved ones, would be expected to cluster at the ends of the helix or in the vicinity of helix irregularities in the model-structure. This can provide a rough validation for the TM model, especially with respect to the assignment of irregular TM helices.

17.4.2. Evolutionary Conservation Profile

Proteins are usually subjected to evolutionary pressure in areas of structural or functional importance. A number of studies have shown that α-helical TM proteins exhibit a distinct conservation pattern in which the protein core is conserved while the loops and lipid-exposed residues are rather variable [12,36,94,100,102,125–127]. Thus, peripheral helices frequently present distinct variable and conserved helical faces, with the variable faces exposed to the membrane. Because the core region contributes to structural stability and function, it is typically under stronger evolutionary pressure, and would accordingly generally exhibit a high level of conservation. This evolutionary conservation pattern has been demonstrated for various membrane proteins (for example, bacteriorhodopsin [7] and the sodium/proton transporter NhaA

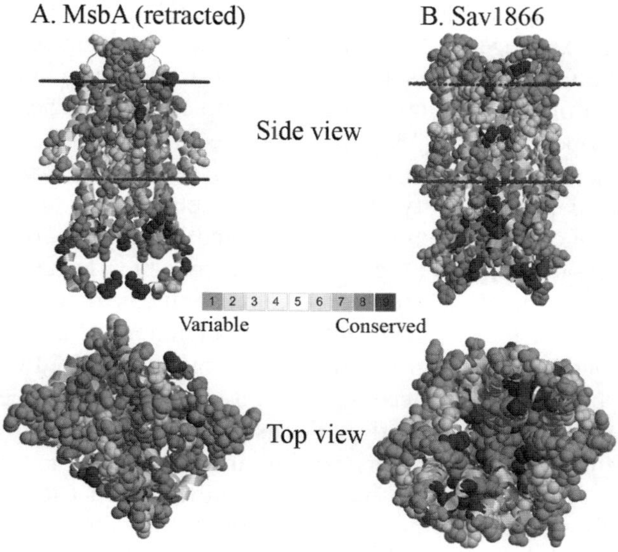

FIGURE 17.7 Conservation analysis of erroneous and correct structures of ABC transporters. The retracted structure of MsbA [131] (panel A) and the structure of sav1866 [133] (panel B) are colored according to conservation, using the ConSurf color scale [134]. Highly conserved residues, receiving grades of 8 or 9, along with the outermost variable (grades of 1 or 2) are shown as spheres. The two upper panels show a side view of the two proteins with their cytoplasmic sides facing down. Approximated membrane boundaries are shown in gray. The nucleotide-binding cytoplasmic domains of both MsbA and sav1866 were omitted for clarity. The two lower panels show a top (and closer) view of the same proteins. (See color insert.)

of *E. coli* [62,128]). By contrast, mapping of evolutionary conservation analysis on erroneous TM structures, such as two of the structures of the EmrE transporter [129,130], does not fit this paradigmatic pattern [36]. The empirical principle can also be demonstrated by a comparison of the conservation pattern of another retracted structure, the crystal structure of the ATP-binding cassette multidrug transporter MsbA [131,132], to that of the correct structure of homologous sav1866 [133]. The conserved residues of sav1866 are evidently buried in the protein core while variable residues face the lipids, as anticipated (Fig. 17.7B). However, the evolutionary profile of the retracted structure of MSbA shows a different pattern: some conserved residues face the lipids and some variable residues are buried in the core (Fig. 17.7A). In both proteins, however, the cytoplasmic ends are highly conserved and form contacts with the cytoplasmic domains.

Incorporating this notion, mapping of evolutionary conservation analysis on the model is highly effective in assessing putative structural models. The examination is applicable only if conservation was not taken into account during generation of the model. The approach was recently utilized, for example, to validate models produced for the SERT transporter [60] and the NHE1 Na^+/H^+ exchanger [62]. In both studies, conservation scores were calculated via the ConSurf web server (http://consurf.tau.ac.il [134]). This is the underlying logic of the ConQuass web server (http://bental.tau.ac.il/ConQuass/) that was developed recently [135]. Based on the correspondence of the model with the conservation pattern, this server assigns a score that can then be utilized to compare the quality of different models.

In the case of comparative modeling, examination of the evolutionary conservation analysis mapped on the model-structure can indicate if the pairwise alignment or the template selection procedures should be revised. For example, a common error such as a single shift in the alignment of a TM helix might result in the placement of conserved residues toward the lipid while variable residues are buried. Such inaccuracy is easily visible from the conservation analysis, but might be difficult to decipher using other evaluation tools. Overall, mapping of the evolutionary conservation analysis on TM model-structures can be considered a highly effective method of evaluation. Close examination of this analysis can allow large or small errors to be detected in the model-structure. This will help not only in the assessment of the TM model's local and global quality, but also in the refinement of problematic regions.

It is noteworthy that water-soluble proteins also exhibit similar evolutionary profiles. That is, their interior is more conserved than their exterior. Indeed, this property has been used to evaluate the quality of structural models [136–138].

17.4.3. Correspondence with Experimental and Clinical Data

As already mentioned, some types of experimental data provide constraints that are useful in predicting TM structures, and if not used for model building, these data can be utilized for validation. Generally speaking, residues in which mutations disrupt a protein's function would normally be found in the TM-protein core, and would typically be of structural or functional importance. They might, for example, contribute to stabilization through inter-helical interfaces or to a function such as direct binding of a substrate. By contrast, most of the positions that are less sensitive to mutations are typically exposed to the lipid, owing to the fact that membrane-facing positions are in general not directly involved in structural stabilization or in function. This paradigm was well exemplified by mapping of elaborate mutagenesis data on the model-structure of the Na^+/H^+ exchanger NHE1 [62].

The above general logic can be applied on examination of the model-predicted locations of polymorphisms and disease-causing mutations. The former are predicted to reside on peripheral regions of the TM protein, whereas the latter typically comprise the core. Besides site-directed mutagenesis and clinical data, other types of empirical data are also helpful in validating the structure of TM proteins. These include, for example, accessibility assays (employed, for instance, to evaluate the EmrE model-structure [95]) and distance assessments using chemical cross-linking (used, e.g., in assessing the model of CFTR [63]) or other measurements.

Nevertheless, it is worth reemphasizing that the experimental data should be treated with caution, especially with regard to intrinsic conformational changes. This was well illustrated in a recent study by Forrest et al. of the *Aquifex aeolicus* leucine transporter (aaLeuT), whose structure had been previously solved in its extra-cellular-facing conformation [139]. By exploiting the pseudo-symmetry observed in the crystal structure, they produced a model of the cytoplasmic-facing conformation of the aaLeuT transporter [140]. To assess their cytoplasmic-facing model-structure they used two inhibitors of the homologous SERT transporter, each of which stabilizes a distinct structural conformation (inward or outward). Accessibility measurements obtained for the inhibitor-stabilized inward state of the SERT transporter provided experimental support for the cytoplasmic-facing model of the aaLeuT transporter. These results demonstrated that when the available data correspond to a single conformational state of a TM protein, it is possible to obtain accurate validation of its model-structure.

17.4.4. Tips for Evaluation

- MQAPs should be used with caution as their performance on TM proteins has yet to be examined.
- It is helpful to inspect the predicted location of specific amino-acid types that exhibit special traits in TM structures.
- Membrane-exposed residues usually exhibit marked hydrophobicity. Thus, the presence of too many polar residues in lipid-facing regions, especially if the residues are charged, might be indicative of an inadequate model-structure.
- Evolutionary conservation analysis is a useful tool for TM-model assessment and refinement. It is important to bear in mind that such analysis is profoundly affected by the quality of the input MSA.
- TM-model validation via experimental data is extremely helpful. Data that are reliable and easy to interpret offer the best available external assessment of TM model-structures.

ACKNOWLEDGMENTS

This work was supported by Grant 611/07 from the Israel Science Foundation to N.B-T. M.S. was supported by the Edmond J. Safra Bioinformatics program at Tel Aviv University.

REFERENCES

1. J. Liu and B. Rost. Comparing function and structure between entire proteomes. *Protein Sci*, 10(10):1970–1979, 2001.
2. R.Y. Kahsay, G. Gao, and L. Liao. An improved hidden Markov model for transmembrane protein detection and topology prediction and its applications to complete genomes. *Bioinformatics*, 21(9):1853–1858, 2005.
3. S. Mitaku et al. Proportion of membrane proteins in proteomes of 15 single-cell organisms analyzed by the SOSUI prediction system. *Biophys Chem*, 82(2–3):165–171, 1999.
4. K. Lundstrom. Structural genomics and drug discovery. *J Cell Mol Med*, 11(2):224–238, 2007.
5. S.H. White. The progress of membrane protein structure determination. *Protein Sci*, 13(7):1948–1949, 2004.
6. H.M. Berman et al. The protein data bank. *Nucleic Acids Res*, 28(1):235–242, 2000.
7. S.J. Fleishman, V.M. Unger, and N. Ben-Tal. Transmembrane protein structures without X-rays. *Trends Biochem Sci*, 31(2):106–113, 2006.
8. N. Hurwitz, M. Pellegrini-Calace, and D.T. Jones. Towards genome-scale structure prediction for transmembrane proteins. *Philos Trans R Soc Lond B Biol Sci*, 361(1467):465–475, 2006.
9. M. Punta et al. Membrane protein prediction methods. *Methods*, 41(4):460–474, 2007.
10. J.U. Bowie. Solving the membrane protein folding problem. *Nature*, 438(7068):581–589, 2005.
11. Y. Liu, D.M. Engelman, and M. Gerstein. Genomic analysis of membrane protein families: Abundance and conserved motifs. *Genome Biol*, 3(10):research0054.0051–00.54.0012, 2002.
12. D. Donnelly et al. Modeling alpha-helical transmembrane domains: The calculation and use of substitution tables for lipid-facing residues. *Protein Sci*, 2(1):55–70, 1993.
13. D.T. Jones, W.R. Taylor, and J.M. Thornton. A mutation data matrix for transmembrane proteins. *FEBS Lett*, 339(3):269–275, 1994.
14. S.E. Blondelle et al. Secondary structure induction in aqueous vs membrane-like environments. *Biopolymers*, 42(4):489–498, 1997.
15. E. Wallin et al. Architecture of helix bundle membrane proteins: An analysis of cytochrome c oxidase from bovine mitochondria. *Protein Sci*, 6(4):808–815, 1997.
16. I. Ubarretxena-Belandia and D.M. Engelman. Helical membrane proteins: Diversity of functions in the context of simple architecture. *Curr Opin Struct Biol*, 11(3):370–376, 2001.
17. M.B. Ulmschneider, M.S. Sansom, and A. Di Nola. Properties of integral membrane protein structures: Derivation of an implicit membrane potential. *Proteins*, 59(2):252–265, 2005.
18. N.J. Tourasse and W.H. Li. Selective constraints, amino acid composition, and the rate of protein evolution. *Mol Biol Evol*, 17(4):656–664, 2000.

19. G. von Heijne. Membrane-protein topology. *Nat Rev Mol Cell Biol*, 7(12):909–918, 2006.

20. A. Kessel and N. Ben-Tal. Free energy determinants of peptide association with lipid bilayers. *Curr Top Membr*, 52:205–253, 2002.

21. G.V. Heijne. The distribution of positively charged residues in bacterial inner membrane proteins correlates with the trans-membrane topology. *EMBO J*, 5(11):3021–3027, 1986.

22. F.S. Cordes, J.N. Bright, and M.S. Sansom. Proline-induced distortions of transmembrane helices. *J Mol Biol*, 323(5):951–960, 2002.

23. D.J. Barlow and J.M. Thornton. Helix geometry in proteins. *J Mol Biol*, 201(3):601–619, 1988.

24. D.P. Tieleman et al. Proline-induced hinges in transmembrane helices: Possible roles in ion channel gating. *Proteins*, 44(2):63–72, 2001.

25. H. Lu, T. Marti, and P.J. Booth. Proline residues in transmembrane alpha helices affect the folding of bacteriorhodopsin. *J Mol Biol*, 308(2):437–446, 2001.

26. C.J. Brandl and C.M. Deber. Hypothesis about the function of membrane-buried proline residues in transport proteins. *Proc Natl Acad Sci U S A*, 83(4):917–921, 1986.

27. S. Yohannan et al. The evolution of transmembrane helix kinks and the structural diversity of G protein-coupled receptors. *Proc Natl Acad Sci U S A*, 101(4):959–963, 2004.

28. B.A. Wallace, M. Cascio, and D.L. Mielke. Evaluation of methods for the prediction of membrane protein secondary structures. *Proc Natl Acad Sci U S A*, 83(24):9423–9427, 1986.

29. L.R. Forrest, C.L. Tang, and B. Honig. On the accuracy of homology modeling and sequence alignment methods applied to membrane proteins. *Biophys J*, 91(2):508–517, 2006.

30. S.C. Li and C.M. Deber. A measure of helical propensity for amino acids in membrane environments. *Nat Struct Biol*, 1(6):368–373.1994.

31. S.E. Harrington and N. Ben-Tal. Structural determinants of transmembrane helical proteins. *Structure*, 17(8):1092–1103, 2009.

32. N. Grigorieff et al. Electron-crystallographic refinement of the structure of bacteriorhodopsin. *J Mol Biol*, 259(3):393–421, 1996.

33. C. Hunte et al. Structure of a Na+/H+ antiporter and insights into mechanism of action and regulation by pH. *Nature*, 435(7046):1197–1202, 2005.

34. D. Fu et al. Structure of a glycerol-conducting channel and the basis for its selectivity. *Science*, 290(5491):481–486, 2000.

35. E. Granseth et al. Membrane protein structural biology–how far can the bugs take us? *Mol Membr Biol*, 24(5–6):329–332, 2007.

36. S.J. Fleishman and N. Ben-Tal. Progress in structure prediction of alpha-helical membrane proteins. *Curr Opin Struct Biol*, 16(4):496–504, 2006.

37. A. Elofsson and G. von Heijne. Membrane protein structure: prediction versus reality. *Annu Rev Biochem*, 76:125–140, 2007.

38. Y. Zhang. Progress and challenges in protein structure prediction. *Curr Opin Struct Biol*, 18(3):342–348, 2008.

39. R. Das and D. Baker. Macromolecular Modeling with Rosetta. *Annu Rev Biochem*, 77(1):363–382, 2008.
40. P. Bradley, K.M. Misura, and D. Baker. Toward high-resolution de novo structure prediction for small proteins. *Science*, 309(5742):1868–1871, 2005.
41. V. Yarov-Yarovoy, J. Schonbrun, and D. Baker. Multipass membrane protein structure prediction using Rosetta. *Proteins*, 62(4):1010–1025, 2006.
42. P. Barth, J. Schonbrun, and D. Baker. Toward high-resolution prediction and design of transmembrane helical protein structures. *Proc Natl Acad Sci U S A*, 104(40):15682–15687, 2007.
43. P. Barth, B. Wallner, and D. Baker. Prediction of membrane protein structures with complex topologies using limited constraints. *Proc Natl Acad Sci U S A*, 106(5):1409–1414, 2009.
44. Y. Zhang, M.E. Devries, and J. Skolnick. Structure modeling of all identified G protein-coupled receptors in the human genome. *PLoS Comput Biol*, 2(2):e13, 2006.
45. V. Cherezov et al. High-resolution crystal structure of an engineered human beta2-adrenergic G protein-coupled receptor. *Science*, 318(5854):1258–1265, 2007.
46. S. Takeda et al. Identification of G protein-coupled receptor genes from the human genome sequence. *FEBS Lett*, 520(1–3):97–101, 2002.
47. K.L. Pierce, R.T. Premont, and R.J. Lefkowitz. Seven-transmembrane receptors. *Nat Rev Mol Cell Biol*, 3(9):639–650, 2002.
48. K. Lundstrom. Latest development in drug discovery on G protein-coupled receptors. *Curr Protein Pept Sci*, 7(5):465–470, 2006.
49. A. Patny, P.V. Desai, and M.A. Avery. Homology modeling of G-protein-coupled receptors and implications in drug design. *Curr Med Chem*, 13(14):1667–1691, 2006.
50. S. Shacham et al. PREDICT modeling and in-silico screening for G-protein coupled receptors. *Proteins*, 57(1):51–86, 2004.
51. R.J. Trabanino et al. First principles predictions of the structure and function of g-protein-coupled receptors: Validation for bovine rhodopsin. *Biophys J*, 86(4):1904–1921, 2004.
52. F. Fanelli and P.G. De Benedetti. Computational modeling approaches to structure-function analysis of G protein-coupled receptors. *Chem Rev*, 105(9):3297–3351, 2005.
53. L. Oliveira et al. Heavier-than-air flying machines are impossible. *FEBS Lett*, 564(3):269–273, 2004.
54. O.M. Becker et al. Modeling the 3D structure of GPCRs: Advances and application to drug discovery. *Curr Opin Drug Discov Devel*, 6(3):353–361, 2003.
55. J. Moult. A decade of CASP: Progress, bottlenecks and prognosis in protein structure prediction. *Curr Opin Struct Biol*, 15(3):285–289, 2005.
56. K. Ginalski. Comparative modeling for protein structure prediction. *Curr Opin Struct Biol*, 16(2):172–177, 2006.
57. D. Petrey and B. Honig. Protein structure prediction: Inroads to biology. *Mol Cell*, 20(6):811–819, 2005.

58. A. Fiser and A. Sali. Modeller: Generation and refinement of homology-based protein structure models. *Methods Enzymol*, 374:461–491, 2003.
59. A. Oberai et al. A limited universe of membrane protein families and folds. *Protein Sci*, 15(7):1723–1734, 2006.
60. L.R. Forrest et al. Identification of a chloride ion binding site in Na+/Cl-dependent transporters. *Proc Natl Acad Sci U S A*, 104(31):12761–12766, 2007.
61. T. Beuming et al. A comprehensive structure-based alignment of prokaryotic and eukaryotic neurotransmitter/Na+ symporters (NSS) aids in the use of the LeuT structure to probe NSS structure and function. *Mol Pharmacol*, 70(5):1630–1642, 2006.
62. M. Landau et al. Model structure of the Na+/H+ exchanger 1 (NHE1): Functional and clinical implications. *J Biol Chem*, 282(52):37854–37863, 2007.
63. J.P. Mornon, P. Lehn, and I. Callebaut. Atomic model of human cystic fibrosis transmembrane conductance regulator: Membrane-spanning domains and coupling interfaces. *Cell Mol Life Sci*, 65(16):2594–2612, 2008.
64. I. Callebaut et al. Deciphering protein sequence information through hydrophobic cluster analysis (HCA): Current status and perspectives. *Cell Mol Life Sci*, 53(8):621–645, 1997.
65. S.F. Altschul et al. Gapped BLAST and PSI-BLAST: A new generation of protein database search programs. *Nucleic Acids Res*, 25(17):3389–3402, 1997.
66. M. Hedman et al. Improved detection of homologous membrane proteins by inclusion of information from topology predictions. *Protein Sci*, 11(3):652–658, 2002.
67. C. Gao and H.A. Stern. Scoring function accuracy for membrane protein structure prediction. *Proteins*, 68(1):67–75, 2007.
68. D. Baker and A. Sali. Protein structure prediction and structural genomics. *Science*, 294(5540):93–96, 2001.
69. R.C. Edgar. MUSCLE: A multiple sequence alignment method with reduced time and space complexity. *BMC Bioinformatics*, 5:113, 2004.
70. C. Notredame, D.G. Higgins, and J. Heringa. T-Coffee: A novel method for fast and accurate multiple sequence alignment. *J Mol Biol*, 302(1):205–217, 2000.
71. J.D. Thompson, D.G. Higgins, and T.J. Gibson. CLUSTAL W: Improving the sensitivity of progressive multiple sequence alignment through sequence weighting, position-specific gap penalties and weight matrix choice. *Nucleic Acids Res*, 22(22):4673–4680, 1994.
72. M.O. Dayhoff, R.M. Schwartz, and B.C. Orcutt. A model of evolutionary change in proteins. In M.O. Dayhoff (Ed.), *Atlas of Protein Sequence and Structure*, Washington, DC: National Biomedical Research Foundation, 5(Suppl. 3):345–352, 1978.
73. S. Henikoff and J.G. Henikoff. Amino acid substitution matrices from protein blocks. *Proc Natl Acad Sci U S A*, 89(22):10915–10919, 1992.
74. P.C. Ng, J.G. Henikoff, and S. Henikoff. PHAT: A transmembrane-specific substitution matrix. Predicted hydrophobic and transmembrane. *Bioinformatics*, 16(9):760–766, 2000.
75. T. Muller, S. Rahmann, and M. Rehmsmeier. Non-symmetric score matrices and the detection of homologous transmembrane proteins. *Bioinformatics*, 17(1):S182–S189, 2001.

76. Y. Shafrir and H.R. Guy. STAM: Simple transmembrane alignment method. *Bioinformatics*, 20(5):758–769, 2004.
77. W. Pirovano, K.A. Feenstra, and J. Heringa. PRALINETM: A strategy for improved multiple alignment of transmembrane proteins. *Bioinformatics*, 24(4):492–497, 2008.
78. C.L. Tang et al. On the role of structural information in remote homology detection and sequence alignment: New methods using hybrid sequence profiles. *J Mol Biol*, 334(5):1043–1062, 2003.
79. A. Sali and T.L. Blundell. Comparative protein modelling by satisfaction of spatial restraints. *J Mol Biol*, 234(3):779–815, 1993.
80. D. Petrey et al. Using multiple structure alignments, fast model building, and energetic analysis in fold recognition and homology modeling. *Proteins*, 53(6):430–435, 2003.
81. C.S. Reddy et al. Homology modeling of membrane proteins: A critical assessment. *Comput Biol Chem*, 30(2):120–126, 2006.
82. K. Kelly. 3D bioinformatics and comparative protein modeling in MOE. *J Chem Comp Group*, available at http://www.chemcomp.com/journal/bio1999.htm, 1999.
83. H.E. Dayringer, A. Tramontano, and R.J. Fletterick. Interactive program for visualization and modelling of proteins, nucleic acids and small molecules. *J Mol Graph*, 4(2):82–87, 1986.
84. T. Schwede et al. SWISS-MODEL: An automated protein homology-modeling server. *Nucleic Acids Res*, 31(13):3381–3385, 2003.
85. N. Guex and M.C. Peitsch. SWISS-MODEL and the Swiss-PdbViewer: An environment for comparative protein modeling. *Electrophoresis*, 18(15):2714–2723, 1997.
86. M. Jacobson and A. Sali. Comparative protein structure modeling and its applications to drug discovery. *Annual Reports in Medicinal Chemistry*, 39:259–267, 2004.
87. M.P. Jacobson et al. A hierarchical approach to all-atom protein loop prediction. *Proteins*, 55(2):351–367, 2004.
88. M.P. Jacobson et al. Force field validation using protein side chain prediction. *J Phys Colloid Chem*, 106(44,):11673–11680, 2002.
89. C.A. Rohl et al. Modeling structurally variable regions in homologous proteins with Rosetta. *Proteins*, 55(3):656–677, 2004.
90. V. Yarov-Yarovoy, D. Baker, and W.A. Catterall. Voltage sensor conformations in the open and closed states in Rosetta structural models of K(+) channels. *Proc Natl Acad Sci U S A*, 103(19):7292–7297, 2006.
91. K.T. Simons et al. Assembly of protein tertiary structures from fragments with similar local sequences using simulated annealing and Bayesian scoring functions. *J Mol Biol*, 268(1):209–225, 1997.
92. K.T. Simons et al. Improved recognition of native-like protein structures using a combination of sequence-dependent and sequence-independent features of proteins. *Proteins*, 34(1):82–95, 1999.
93. H. Zhou and Y. Zhou. Distance-scaled, finite ideal-gas reference state improves structure-derived potentials of mean force for structure selection and stability prediction. *Protein Sci*, 11(11):2714–2726, 2002.

94. T. Beuming and H. Weinstein. Modeling membrane proteins based on low-resolution electron microscopy maps: A template for the TM domains of the oxalate transporter OxlT. *Protein Eng Des Sel*, 18(3):119–125, 2005.

95. S.J. Fleishman et al. Quasi-symmetry in the cryo-EM structure of EmrE provides the key to modeling its transmembrane domain. *J Mol Biol*, 364(1):54–67, 2006.

96. S.J. Fleishman et al. A Calpha model for the transmembrane alpha helices of gap junction intercellular channels. *Mol Cell*, 15(6):879–888, 2004.

97. J.M. Baldwin, G.F. Schertler, and V.M. Unger. An alpha-carbon template for the transmembrane helices in the rhodopsin family of G-protein-coupled receptors. *J Mol Biol*, 272(1):144–164, 1997.

98. P.L. Sorgen et al. An approach to membrane protein structure without crystals. *Proc Natl Acad Sci U S A*, 99(22):14037–14040, 2002.

98a. M. Schushan, Y. Barkan, T. Haliloglu, and N. Ben-Tal. C(alpha)-trace model of the transmembrane domain of human copper transporter 1, motion and functional implications. *PNAS*, 107(24):10908–10913, 2010.

99. J.A. Kovacs, M. Yeager, and R. Abagyan. Computational prediction of atomic structures of helical membrane proteins aided by EM maps. *Biophys J*, 93(6):1950–1959, 2007.

100. S.J. Fleishman et al. An automatic method for predicting transmembrane protein structures using cryo-EM and evolutionary data. *Biophys J*, 87(5):3448–3459, 2004.

101. A. Enosh et al. Assigning transmembrane segments to helices in intermediate-resolution structures. *Bioinformatics*, 20(1):i122–i129, 2004.

102. L. Adamian and J. Liang, Prediction of transmembrane helix orientation in polytopic membrane proteins. *BMC Struct Biol*, 6:13, 2006.

103. A. Fuchs et al. Co-evolving residues in membrane proteins. *Bioinformatics*, 23(24):3312–3319, 2007.

104. S.J. Fleishman, O. Yifrach, and N. Ben-Tal. An evolutionarily conserved network of amino acids mediates gating in voltage-dependent potassium channels. *J Mol Biol*, 340(2):307–318, 2004.

105. S. Maeda et al. Structure of the connexin 26 gap junction channel at 3.5[thinsp] A resolution. *Nature*, 458(7238):597–602, 2009.

106. I.M. Skerrett et al. Identification of amino acid residues lining the pore of a gap junction channel. *J. Cell Biol*, 159(2):349–360, 2002.

107. Y.J. Chen et al. X-ray structure of EmrE supports dual topology model. *Proc Natl Acad Sci U S A*, 104(48):18999–19004, 2007.

108. K. Sale et al. Optimal bundling of transmembrane helices using sparse distance constraints. *Protein Sci*, 13(10):2613–2627, 2004.

109. H.R. Kaback, M. Sahin-Toth, and A.B. Weinglass. The kamikaze approach to membrane transport. *Nat Rev Mol Cell Biol*, 2(8):610–620, 2001.

110. H.R. Kaback and J. Wu. From membrane to molecule to the third amino acid from the left with a membrane transport protein. *Q Rev Biophys*, 30(4):333–364. 1997.

111. E.G. Stein, L.M. Rice, and A.T. Brunger. Torsion-angle molecular dynamics as a new efficient tool for NMR structure calculation. *J Magn Reson*, 124(1):154–164, 1997.

112. J. Abramson et al. Structure and mechanism of the lactose permease of Escherichia coli. *Science*, 301(5633):610–615, 2003.
113. J. Abramson et al. The lactose permease of Escherichia coli: Overall structure, the sugar-binding site and the alternating access model for transport. *FEBS Lett*, 555(1):96–101, 2003.
114. K.A. Williams et al. Projection structure of NhaA, a secondary transporter from Escherichia coli, at 4.0 A resolution. *EMBO J*, 18(13):3558–3563, 1999.
115. Y. Gerchman et al. Oligomerization of NhaA, the Na+/H+ antiporter of Escherichia coli in the membrane and its functional and structural consequences. *Biochemistry*, 40(11):3403–3412, 2001.
116. D. Hilger et al. High-resolution structure of a Na+/H+ antiporter dimer obtained by pulsed electron paramagnetic resonance distance measurements. *Biophys J*, 93(10):3675–3683, 2007.
117. A. Rimon, T. Tzubery, and E. Padan. Monomers of the NhaA Na+/H+ antiporter of Escherichia coli are fully functional yet dimers are beneficial under extreme stress conditions at alkaline pH in the presence of Na+ or Li+. *J Biol Chem*, 282(37):26810–26821, 2007.
118. D. Fischer. Servers for protein structure prediction. *Curr Opin Struct Biol*, 16(2):178–182, 2006.
119. M. Fasnacht, J. Zhu, and B. Honig. Local quality assessment in homology models using statistical potentials and support vector machines. *Protein Sci*, 16(8):1557–1568, 2007.
120. D. Eisenberg, R. Luthy, and J.U. Bowie. VERIFY3D: Assessment of protein models with three-dimensional profiles. *Methods Enzymol*, 277:396–404, 1997.
121. M.J. Sippl. Recognition of errors in three-dimensional structures of proteins. *Proteins*, 17(4):355–362, 1993.
122. B. Wallner and A. Elofsson. Identification of correct regions in protein models using structural, alignment, and consensus information. *Protein Sci*, 15(4):900–913, 2006.
123. B. Wallner and A. Elofsson. Can correct protein models be identified? *Protein Sci*, 12(5):1073–1086, 2003.
124. S.C. Tosatto, The victor/FRST function for model quality estimation. *J Comput Biol*, 12(10):1316–1327, 2005.
125. J.A. Briggs, J. Torres, and I.T. Arkin. A new method to model membrane protein structure based on silent amino acid substitutions. *Proteins*, 44(3):370–375, 2001.
126. T.J. Stevens and I.T. Arkin. Substitution rates in alpha-helical transmembrane proteins. *Protein Sci*, 10(12):2507–2517, 2001.
127. D.T. Jones. Improving the accuracy of transmembrane protein topology prediction using evolutionary information. *Bioinformatics*, 23(5):538–544, 2007.
128. L. Kozachkov, K. Herz, and E. Padan. Functional and structural interactions of the transmembrane domain X of NhaA, Na+/H+ antiporter of Escherichia coli, at physiological pH. *Biochemistry*, 46(9):2419–2430, 2007.
129. O. Pornillos et al. X-ray structure of the EmrE multidrug transporter in complex with a substrate. *Science*, 310(5756):1950–1953, 2005.
130. C. Ma and G. Chang. Structure of the multidrug resistance efflux transporter EmrE from Escherichia coli. *Proc Natl Acad Sci U S A*, 101(9):2852–2857, 2004.

131. G. Chang et al. Retraction. *Science*, 314(5807):1875, 2006.
132. C.L. Reyes and G. Chang, Structure of the ABC transporter MsbA in complex with ADP.vanadate and lipopolysaccharide. *Science*, 308(5724):1028–1031, 2005.
133. R.J. Dawson and K.P. Locher. Structure of a bacterial multidrug ABC transporter. *Nature*, 443(7108):180–185, 2006.
134. M. Landau et al. ConSurf 2005: The projection of evolutionary conservation scores of residues on protein structures. *Nucleic Acids Res*, 33(Web Server issue):W299–W302, 2005.
135. M. Kalman and N. Ben-Tal. Quality assessment of protein model-structures using evolutionary conservation. *Bioinformatics*, 26(10):1299–1307, 2010.
136. O. Olmea, B. Rost, and A. Valencia. Effective use of sequence correlation and conservation in fold recognition. *J Mol Biol*, 293(5):1221–1239, 1999.
137. U.K. Muppirala and Z. Li. A simple approach for protein structure discrimination based on the network pattern of conserved hydrophobic residues. *Protein Eng Des Sel.*, 19(6):265–275, 2006.
138. I. Mihalek et al. Combining inference from evolution and geometric probability in protein structure evaluation. *J Mol Biol*, 331(1):263–279, 2003.
139. A. Yamashita et al. Crystal structure of a bacterial homologue of Na+/Cl– dependent neurotransmitter transporters. *Nature*, 437(7056):215–223, 2005.
140. L.R. Forrest et al. Mechanism for alternating access in neurotransmitter transporters. *Proc Natl Acad Sci U S A*, 105(30):10338–10343, 2008.

CHAPTER 18

STRUCTURE-BASED MACHINE LEARNING MODELS FOR COMPUTATIONAL MUTAGENESIS

MAJID MASSO and IOSIF I. VAISMAN
Laboratory for Structural Bioinformatics
Department of Bioinformatics and Computational Biology
George Mason University
Manassas, VA

18.1. INTRODUCTION

18.1.1. Structural and Functional Effects of Mutations

Proteins exhibit a wide range of functional consequences upon mutation. In this chapter, we will focus specifically on mutations that are the result of single or multiple amino acid substitutions. Experimentally well-studied functional effects of residue replacements in proteins include relative changes to protein activity or stability. The activities for a large number of single residue replacements in a particular protein, studied under identical experimental conditions and protocols, are typically reported quantitatively as percentages of the wild-type protein activity. More frequently, however, such mutants are qualitatively described as each belonging to one of a few categorical classes based on the degree of activity. Mutant stability changes can measured experimentally using a variety of quantitative measures: $\Delta\Delta G$ and $\Delta\Delta G^{H2O}$ represent the free-energy change of unfolding due to thermal and chemical denaturations, respectively, while ΔT_m refers to mutant thermal stability change. With zero as a cutoff value for these measures, mutant stability can also be described qualitatively as either increased or decreased relative to the wild-type protein. In the case of a protein that serves as a target for an inhibitor drug, a more broadly defined functional consequence refers to the relative change in

Introduction to Protein Structure Prediction: Methods and Algorithms,
Edited by Huzefa Rangwala and George Karypis

inhibitor binding and/or drug susceptibility upon mutation. Again, these can be measured either quantitatively (e.g., fold levels of resistance) or qualitatively (e.g., drug-sensitive or drug-resistant mutant).

Since protein structure dictates function, it follows that structural changes upon mutation drive these functional effects. For example, at buried positions in a protein, replacement of a hydrophobic residue with a polar or charged amino acid may significantly interrupt the network of hydrophobic core interactions. In this case, the mutation will likely have an adverse effect on structural stability, which in turn could damage function. On the other hand, two possibilities exist if the position is substituted with another hydrophobic residue. If the new residue is equally favorable, or even slightly more favorable, in the given structural environment (i.e., no adverse effect on sequence-structure compatibility), then structural stability and function will usually be maintained. Alternately, the new hydrophobic residue may negatively impact sequence-structure compatibility, which would likely be due to significant size differences with native amino acid. Therefore the protein might not fold properly, so that structure and possibly function could be affected.

A different set of circumstances emerges at surface positions. Since a catalytic residue on a protein surface interacts with an external molecule, it generally does not participate in interatomic non-covalent interactions with structurally nearby amino acids. The same holds true for many, but certainly not all, active and binding site surface positions. Hence at many of these functional (catalytic, active, binding) surface sites, substitutions generally lead to new residues that may interact more favorably with neighboring amino acids in the structure (i.e., increased sequence-structure compatibility) compared with the native residue. But this increase in the bonding network also typically damages function: catalytic and binding site residues are no longer as free to interact with external molecules and binding partners, while active site pockets lose flexibility and become too rigid to accommodate substrates. On the other hand, residue replacements at all other surface positions generally have a minimal effect on structural stability and function; however, certain substitutions may be damaging depending on their proximity and structural impact on functional sites. Additionally, some surface substitutions at both charged positions that participate in energetically favorable salt bridges and positions that are important for the solubility of the folded protein may have adverse consequences [1].

18.1.2. Computational Study of Mutations

A number of theoretical approaches have been developed and applied toward quantifying the functional consequences of protein mutations, each of which utilizes an energy function characterizing the physical forces involved in protein folding. Based on the degree of explicit structural detail described by the energy functions, the methods generally can be categorized into three groups: physical effective energy functions (PEEFs), statistical effective energy functions (SEEFs), and empirical effective energy functions (EEEFs) [2,3]. Effective energy refers to the free energy of a solvated protein system, and it is assumed that the free-energy minimum coincides with the protein native state.

PEEFs approximate the true energy and are characterized by molecular mechanics energy functions that provide a comprehensive account of forces between particles and incorporate the effects of solvation [2,4,5]. Model parameters are obtained experimentally based on physical measurements of simple systems. Examples include those force fields used by the CHARMM [6], GROMOS [7], and AMBER [8] packages, each of which also integrates the ability to perform molecular dynamics simulations [9–12]. Since these methods are computationally expensive, they are limited to the study of a small number of mutants. A novel method that combines Monte-Carlo methods with molecular dynamics (CMC/MD) allows for the possibility of *in silico* exhaustive mutagenesis based on a single free-energy calculation [13].

Alternatively, SEEFs and EEEFs utilize force fields that are based on pseudo-energy functions, which are much faster to compute and allow for the concomitant analysis of large datasets of mutants. SEEFs are derived from structural distribution data collected from known protein structures. They have an advantage in that known as well as unrecognized entropic effects and complex interactions, including the effect of solvation, are implicitly incorporated into these knowledge-based potentials by virtue of their statistical nature [14–23]. These methods typically utilize an empirical approximation for the denatured (reference) state. Although atom-based approaches have been developed, the majority of SEEFs are derived at the residue level, which requires proteins to be represented discretely using constituent amino acid Cα, Cβ, or side chain center of mass (CM) coordinates. Pairwise distance potentials, higher order interaction potentials, and torsion angle potentials, individually or in combination, characterize a majority of SEEF methods. Finally, SEEFs have been utilized in Monte-Carlo simulations [24] and for developing PEEF/SEEF hybrid models [25,26].

EEEFs are similar to SEEFs but utilize parameterized energy terms that are designed to optimally fit experimentally obtained data [3]. Two main challenges must be overcome in the successful development of an EEEF. First, the method requires a comprehensive accounting of all stabilizing forces in proteins using physically, statistically, or empirically based energy terms. Subsequently, scaling parameters in these terms must be approximated in order to generate an optimal fit with experimental thermodynamic data available for known mutants. Despite these difficulties, a number of EEEFs have been reported in the literature [27–29].

Recently, statistical machine learning techniques have been successfully applied toward large-scale prediction of the functional consequences associated with single-residue replacements in proteins [30–38]. These methods learn predictive models that are trained by using a diverse dataset of mutants for which the particular functional effect has already been experimentally determined. As will be more fully detailed in the section that follows, mutants are represented as identically ordered vectors of measurable sequence, structure, or evolutionary attributes (independent variables) that may explain the functional effect (dependent variable). The work described later in this chapter is based on a hybrid SEEF/machine learning combined approach.

18.2. METHODOLOGY

18.2.1. Computational Geometry Approach to Protein Structure

In a molecular system a computational geometry approach can be used for statistical analysis of the nearest neighbor atoms or groups of atoms, which are identified by irregular polyhedra obtained as a result of a specific tessellation in three-dimensional (3D) space. Voronoi tessellation partitions the space into convex polytopes called Voronoi polyhedra. In the case of a protein molecule the Voronoi polyhedron is a region of space around the atom (which may represent an entire amino acid residue), such that each point of this region is closer to this atom than to any other atom in the molecule. A group of four atoms, whose Voronoi polyhedra meet at a common vertex, forms another basic topological object called a Delaunay simplex (Fig. 18.1). The topological difference between these objects is that the Voronoi polyhedron defines the environment of individual atoms whereas the Delaunay simplex represents

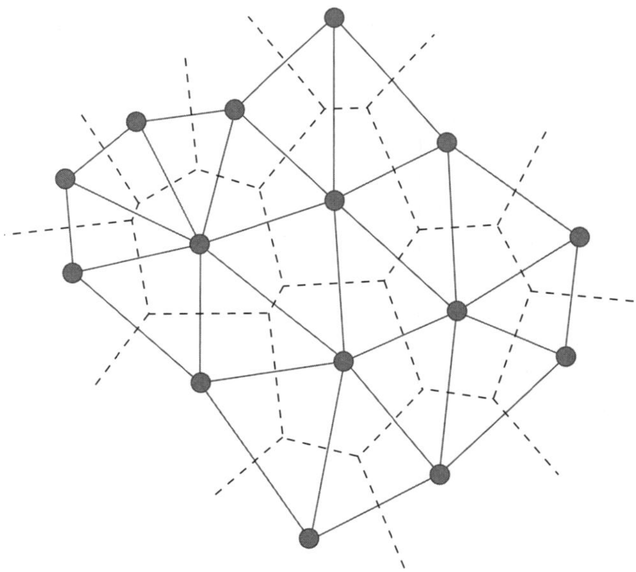

FIGURE 18.1 Voronoi/Delaunay tessellation in 2D space (Voronoi tessellation—dashed line, Delaunay tessellation—solid line).

the ensemble of neighboring atoms. The Voronoi polyhedra and the Delaunay simplices are topological duals and they are completely determined by each other. However the Voronoi polyhedra may have different numbers of faces and edges, while the Delaunay simplices are always tetrahedra in 3D space. These tetrahedra can be used to define objectively the nearest neighbor entities in molecular systems. The Delaunay tessellation of a set of points is a canonical tessellation of space based on nearest neighbors; it is equivalent to a convex hull of the set in one higher dimension [39,40]. The Delaunay tessellation can be performed using a number of different algorithms, including Quickhull, which is a variation of the randomized, incremental Clarkson and Shor algorithm [41]. Quickhull produces the Delaunay tessellation by computing the convex-hull of this set of points in four dimensions.

A computational geometry approach to study structure of molecular systems by partitioning space occupied by a molecule was originally developed for liquids and glasses [42,43]. Later it was extended to study packing and volume distributions in proteins using Voronoi tessellation [44–46]. A topological dual to Voronoi partitioning, the Delaunay tessellation has an additional utility as a method for non-arbitrary identification of neighboring points in the molecular systems represented by the extended sets of points in space. The Delaunay tessellation was successfully used to study various properties of liquids and solutions [47–50]. The first application of the Delaunay tessellation for proteins was designed for identification of nearest neighbor residues and derivation of a four-body statistical potential [51]. This potential has been successfully tested for inverse protein folding [52], fold recognition [53], decoy structure discrimination [54,55], protein design [56], protein folding on a lattice [57], thermodynamic stability of the protein core [58], computational mutagenesis [59–67], protein structure similarity comparison [68], secondary structure assignment [69], and protein structure classification [70]. Statistical compositional analysis of Delaunay simplices revealed highly nonrandom clustering of amino acid residues in folded proteins when all residues were treated separately as well as when they were grouped according to their chemical, structural, or genetic properties [71].

18.2.2. Multi-body Statistical Potential

A non-homologous training set of over 1400 high-resolution crystallographic protein structures with low primary sequence identity was selected from the Protein Data Bank (PDB) [72] for developing the knowledge-based potential. Each structure was represented as a discrete set of points in 3D space, corresponding to the Cα atomic coordinates or CM of the side chain of each of the constituent amino acid residues in the protein. A computational geometry construct known as Delaunay tessellation, applied to each discretized protein structure, generates an aggregate of non-overlapping, space-filling, irregular tetrahedral simplices by utilizing the points as vertices [51,71]. Hence, this approach objectively defines quadruplets of nearest neighbor amino acids in

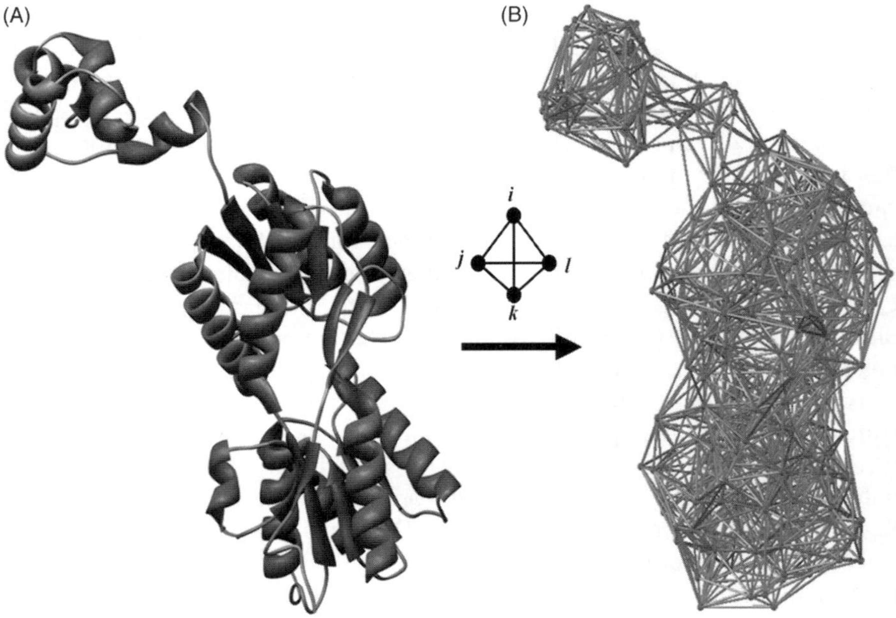

FIGURE 18.2 (A) Ribbon diagram of a single chain of the lac repressor homotetramer. (B) Delaunay tessellation of the same monomer of lac repressor, subject to a 12 Angstrom edge-length filter. PDB accession file: 1efa, chain B. (See color insert.)

a protein structure based on the residue identities represented by the vertices of the simplices formed by a protein tessellation (Fig. 18.2). As an added measure to ensure physically meaningful interactions, we only considered simplices in protein tessellations for which all six edge lengths were less than12 Angstroms.

Based on a 20-letter alphabet of amino acid building blocks for proteins, a total of 8855 distinct residue quadruplets can be generated if residue repeats are allowed in quadruplets and if all permutations of an already enumerated quadruplet are excluded [51,71]. For each quadruplet, we determined the observed proportion of simplices among all the protein tessellations whose vertices represented the four amino acids. We also computed a rate expected by chance for each quadruplet based on a multinomial reference distribution that utilized the frequency of each amino acid among the training set proteins. Modeled after the inverse Boltzmann law, an empirical potential of quadruplet interaction (log-likelihood score) was calculated as a logarithm of the ratio of observed to expected values. The four-body statistical potential is defined as the collection of 8855 quadruplet types along with each of their respective log-likelihood scores.

The four-body statistical potential is useful for empirically evaluating sequence-structure compatibility in any protein structure selected from the

PDB and subsequently tessellated. Every tetrahedral simplex in the Delaunay tessellation identifies a residue quadruplet at the vertices, and the global sum of these residue quadruplet log-likelihood scores is referred to as the *total potential* or *topological score* of the protein [59,61]. Two bordering tetrahedral simplices in the tessellation may share a triangular face (three vertex points, i.e., residues, in common), a linear edge (two residues in common), or a single vertex (residue). In general, each point is shared as a vertex by a number of tetrahedral simplices, so each residue in a protein is simultaneously a member of several quadruplets of nearest neighbors defined by these simplices. For each residue, the local sum of these quadruplet log-likelihood scores is referred to as the *individual residue potential* or the *residue environment score*, and a vector of such scores (where component numbers correspond to residue primary sequence positions) generates a 3D-1D potential profile [59,63,73].

18.2.3. Representation of Protein Mutants

In cases where structures have been resolved for both a native protein as well as the same protein with a single residue substitution (i.e., a single-point mutant), their Delaunay tessellations are frequently either highly similar or identical. This is due to both coarse-grained protein representations as residue point sets in 3D space, as well as robustness of Delaunay tessellation to small perturbations in the coordinates of these vertex points. Hence, since there are relatively few, if any, solved mutant structures for a given native protein with a solved structure, 3D-1D potential profiles for all single-point mutants of a protein are calculated by using the native (wild type) structure tessellation. For each single-point mutant, the amino acid identity is altered at the Cα or side chain CM coordinate representing the mutated position, and the log-likelihood scores of the simplicial quadruplets sharing the point are recomputed. In the wild-type 3D-1D potential profile vector, such a substitution only alters the residue environment scores of the mutated position itself, as well as all positions whose Cα/CM coordinates participate as vertices in simplices with the Cα/CM coordinate of the mutated position [59,61].

The *residual profile* of a mutant is defined as the difference between the mutant and wild-type protein 3D-1D potential profile vectors, and the value of each residual profile component is referred to as an *EC (environmental change) score* (Fig. 18.3). Specifically, the residual profile components with nonzero EC scores identify the mutated position and all of its structural nearest neighbors defined by the tessellation, and the values of these nonzero EC scores signify the degree of environmental perturbation at those positions caused by the specific type of residue replacement at the mutated position. Due to its significance as a scalar that empirically quantifies the relative change in mutant sequence-structure compatibility from wild type, the EC score at the mutated position component in a mutant residual profile is referred to as the *residual score* of the mutant. For a protein-specific dataset, consisting of a number of distinct single point mutants all selected from the same protein, the

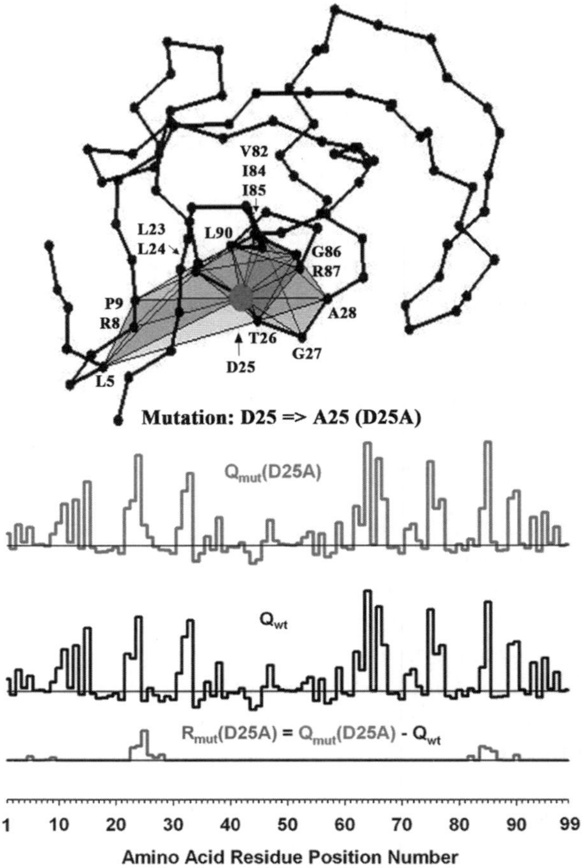

$EC_5 = 0.43$, $EC_8 = 0.01$, $EC_9 = 0.31$, $EC_{23} = 1.63$, $EC_{24} = 1.87$, $EC_{25} = 3.83$, $EC_{26} = 0.67$, $EC_{27} = 0.34$, $EC_{28} = 0.67$, $EC_{82} = 0.49$, $EC_{84} = 1.76$, $EC_{85} = 1.52$, $EC_{86} = 1.14$, $EC_{87} = 4.60 \times 10^{-5}$, $EC_{90} = 0.63$, $EC_i = 0$ for all other positions i

FIGURE 18.3 Representation of the D25A mutation in HIV-1 protease (PDB ID: 3phv). Top: A Cα trace of HIV-1 protease and a subset of its Delaunay tessellation highlighting only the tetrahedral simplices that all share the point representing residue D25 as a vertex (Cα enlarged and colored red). All nearest neighbor residues to D25, whose Cα coordinates participate as the additional vertices in these simplices, are labeled on the trace. Bottom: The 3D-1D potential profiles (Q) of the mutant and wild type protein are shown in red and black, respectively. Their difference is the mutant residual profile (R), shown in green, whose components are EC scores. $EC_{25} = 3.83$ is the residual score of the D25A HIV-1 protease mutant and provides an empirical scalar measure of the change in sequence-structure compatibility relative to the native protein. (See color insert.)

collective residual scores of these mutants have been used to successfully elucidate the protein structure-function relationship [61,65,67]. On the other hand, since the entire residual profile vector provides additional important sequence and structure information about each mutant, all the components are used as feature vector input attributes (along with native/new residues and mutated position number) to develop protein-specific mutagenesis models using machine learning tools.

Universal models, trained on datasets consisting of single-point mutants from a diverse collection of proteins, require an alternative approach due to the fact that residual profile vectors of mutants from distinct proteins differ in size [65]. Assuming that each Cα/CM coordinate in a protein structure tessellation participates in at least two simplices as the only shared vertex, every amino acid position is expected to have a minimum of six nearest neighbor residues defined by the tessellation. An attribute vector is generated for every single-point protein mutant under consideration, consisting of the residual score (i.e., EC score of the mutated position), followed by the EC scores of the six nearest neighbors, ordered by the 3D Euclidean distances of the neighboring Cα coordinates from that of the mutated position. Additional attributes that we evaluate include the identities of the native and replacement amino acids at the mutated position, the ordered identities of the amino acids at the six nearest neighbor positions and the ordered primary sequence distances between the nearest neighbors and the mutated position. Finally, in order to perform direct comparisons with published reports, we also include where appropriate the thermodynamic parameters of pH and temperature under which experimental mutagenesis and stability measurements were performed, in addition to relative solvent accessibility (RSA) and secondary structure of the mutated residue, as provided by those studies and described more fully in subsequent sections. For the protein structures under consideration, tessellations are performed only on single chains of multimeric proteins, and for nuclear magnetic resonance (NMR) structures, only a single model was tessellated unless a minimized average structure is available.

18.2.4. Machine Learning

Synthesis of missense mutations in proteins and experimental approaches for calculating changes to thermodynamic stability or enzymatic activity can be costly as well as time-and labor-intensive. Researchers utilize sequence or structure characteristics about a protein of interest or its homologs, including information previously obtained about other residue replacements, as a way to potentially identify a small subset of new mutations for further testing that may yield desired biophysical properties. Because undertaking a manual search for such data is itself time-consuming, automated models that try to predict the functional impact of protein mutations are frequently utilized for this purpose.

Such predictive models are usually based on the implementation of either supervised classification or regression algorithms from the field of statistical machine learning. Examples of supervised classification algorithms include neural networks (NNs), decision trees (DTs), support vector machines (SVMs), and random forests (RFs) for which the prediction output is a categorical attribute of relative functional change upon mutation (e.g., active/inactive or increased/decreased stability change from wild type). On the other hand, algorithms such as tree regression and support vector regression (SVR) learn models for which the output variable is real-valued (e.g., percentage of wild-type activity or numerical value for mutant stability change).

How are these models developed? All methods require the availability of a diverse representative collection of mutants, sampled from a given population (i.e., either protein-specific or universal datasets), for which the particular output attribute under consideration has already been experimentally determined. This collection is referred to as a *training set*, and the mutants are each represented using a common set of quantifiable protein sequence, structure, or evolutionary input attributes (i.e., predictors) that taken together are useful for explaining the mutant functional change given by the output variable. The ordered collection of mutant predictors is referred to as a *feature vector*. Although each algorithm is unique, they all share a common goal of learning a particular type of complex nonlinear function that best reflects the training set data, where predictors serve as the independent variables and the functional change is the dependent variable. A learned model represents a consistent set of relationships or rules between the input attributes of the mutants and the functional change output variable. Prediction of functional change for a previously unstudied mutant is subsequently obtained by providing its feature vector as input to the model. Many factors influence the reliability of predictions made by a trained model, including the choice of algorithm and of predictor input attributes, as well as the size and diversity of the training set mutants.

18.2.4.1. NNs. The NN architecture [74] consists of a layered feed forward topology of nodes. An input layer serves to introduce the data to the network and contains as many nodes as the number of components in the mutant feature vectors. A hidden layer of nodes act as neurons that are each connected to every input layer node, and neurons in the output layer are similarly each connected to every hidden layer node. All of these directed connections are weighted by values that are learned during training. A technique known as backpropagation is used for NN training, whereby gradient descent is performed to minimize an error function. A variety of early stopping conditions are available to avoid over-fitting the model to the training data, which may reduce the prediction performance on independent test sets of mutants. Once the feature vector for a test mutant is introduced at the input layer of a trained NN, a sigmoid transfer function is used to convert the sum of the weighted input values entering a node into a real number on the closed unit interval

[0,1]. Function values at the output nodes are then used in making a classification decision for the mutant.

18.2.4.2. DTs. DT learning [75] yields a classifier in the form of a rooted tree. A divide-and-conquer approach is employed during model training, whereby at each node starting from the root, an input attribute is selected that best separates the output classes. Learned trees are typically pruned to avoid overfitting. Starting from the root, a test mutant is sorted down the tree based on its value of the input attribute used to split on by the current node, and the appropriate branch is followed to the next node. The recursive process terminates once the mutant reaches a leaf node, where the mutant class is provided. The reduced-error pruning regression tree (REPTree) algorithm represents a modification of DT for the case where the output variable is real-valued rather than categorical.

18.2.4.3. SVMs. SVM classification [76] utilizes a kernel function to nonlinearly map feature vectors of the training set mutants into a higher dimensional space, where an optimal hyperplane between mutants belonging to two different output classes is more easily constructed. The hyperplane provides a maximal margin of separation between the classes and corresponds to a nonlinear decision boundary in the original space. A variety of kernel functions are available, including linear, polynomial, and radial basis function (RBF) kernels. Modifications to SVM for the case where the output variable is real-valued rather than categorical yield the SVR algorithm [77].

18.2.4.4. RFs. The RF algorithm [78] utilizes *bagging* (bootstrap aggregating) to generate multiple bootstrapped datasets, each of which trains a classification tree by random selection of a fixed-size subset of the available predictors for splitting at each node, and predictions are made by majority vote over all trees. Each bootstrapped dataset is the same size as the original training set and obtained by sampling *with replacement* from the training set. In each case, initially about one-third of the training set is randomly selected and left out of the sample for use in obtaining a running *out-of-bag (oob)* unbiased classification error estimate as the trees are added to the growing forest. Additionally, all trees are unpruned and grown to the largest extent possible.

18.2.4.5. Evaluation Methods. Cross-validation (CV) and two-subset random split are commonly used testing procedures for evaluating the performance of an algorithm on a training set [77]. These approaches are useful when a test set of already annotated mutants, independent of the training set, is not available for retrospective classification or regression. For example, implementation of tenfold CV begins with a random grouping of the training set mutants into ten equally sized subsets. With classification algorithms, stratification ensures that class proportions in the full training set are maintained in each of the subsets. Next, one of the subset is held out while the remaining nine subsets (90% of the original training set) are combined into one set that is used to train a model. The held-out subset (10% of the original training set) is then treated as a test set, and the trained model predicts each mutant in the subset. The procedure is repeated for a total of ten times, whereby for each iteration a different subset is held out as a test set, and the remaining nine subsets are combined to form a training set for learning a model that is used to predict the test set mutants. The iterative procedure yields a single (output class or numerical value) prediction for each of the mutants in the original training set, and algorithm performance is gauged by comparing the predictions to the actual classes or values of the output attribute. Alternative random initial grouping of the training set mutants into ten subsets introduces a degree of variability into the tenfold CV predictions; hence, the procedure is typically run multiple times, and the performance is averaged.

Unlike tenfold CV, the leave-one-out CV procedure (LOOCV, also known as the jackknife) is deterministic and yields identical results with each run. In this case, each mutant in the original training set initially forms a single-element subset (a singleton). Hence for a training set with N mutants, the method can be referred to as N-fold CV. With LOOCV, each mutant is predicted by a model trained using all N-1 remaining mutants from the original training set. The procedure is iterated N times, and as with tenfold CV, the end result is a single prediction for each of the N mutants in the original training set.

A typical two-subset random split consists of randomly selecting only 66% of the mutants in the original dataset to train a model, with the remaining 34% of the mutants serving as a separate test set for prediction. As with tenfold CV, the two subsets can be stratified in the case of classification. However, while the final tenfold CV and LOOCV performance results are based on predictions obtained for all mutants in the original dataset, the overall two-subset random split performance is based only on those predictions obtained for the subset of mutants in the test set.

Assuming that the mutants belong to a generic pair of classes, positive (P) and negative (N), classification algorithm predictions obtained by using the procedures described above can be enumerated in the form of the following *confusion matrix*:

		Predicted as	
		P	N
Actual	P	TP	FN
class	N	FP	TN

where TP (TN) represent the number of correctly predicted positive (negative) mutants, and FN (FP) are the number of misclassifications in the respective classes. These values are useful for obtaining the overall accuracy

$$Q = (TP + TN)/(TP + FN + TN + FP)$$

of the predictions. However, this measure may not be reliable in the case where class sizes are significantly unequal. For instance, if 80% of the dataset consists of P class mutants and the trained models are of the simplest form that always predict class P for a mutant, then Q = 0.80 despite the fact that all class N mutants are incorrectly predicted. For this reason, the following performance measures are introduced for the P class:

$$\text{sensitivity (or recall)} = TP/(TP + FN)$$

$$\text{specificity} = TN/(TN + FP)$$

$$\text{precision} = TP/(TP + FP),$$

and it is clear how they are analogously defined for the N class [77]. Sensitivity is also referred to as the true positive rate (TPR), and 1– specificity is equivalent to the false positive rate (FPR). The information represented by these terms, as defined for both classes, can be implicitly encapsulated into the following single measures of performance that are each relatively robust to unequal class distributions:

$$BER = 0.5 \times [FP/(FP + TN) + FN/(FN + TP)]$$

$$MCC = (TP \times TN - FP \times FN)/[(TP + FN)(TP + FP)(TN + FN)(TN + FP)]^{1/2},$$

where BER = balanced error rate and MCC = Matthew's correlation coefficient [79].

Finally, a similarly robust performance measure can be obtained from the area (AUC) under the receiver operating characteristic (ROC) curve, a plot of TPR versus FPR in the unit square [80]. With most classification algorithms, class predictions are based on probabilities assigned to the mutants by a decision function associated with each machine learning tool. The ROC curve for a particular class is obtained by ranking the mutants according to their predicted probabilities for belonging to the class, then plotting successive points based on the actual class memberships of mutants that lie above a steadily decreasing predicted probability threshold [77]. The AUC represents the probability that a classifier will rank a randomly chosen mutant from the reference class higher than one randomly selected from the alternative class, and it is equivalent to the non-parametric Wilcoxon-Mann Whitney test of ranks [81]. For a nearly linear ROC curve that rides along the diagonal connecting points (0,0) and (1,1) in the unit square, AUC is approximately 0.5 and suggests that a trained model is not likely to perform any better than random guessing. On the other hand, an AUC value of 1.0, corresponding to a piecewise linear ROC curve that joins (0,0) to (0,1) then (0,1) to (1,1), is indicative of a perfect classifier.

With regression algorithms, the performance of the testing procedures is measured as the correlation coefficient (r) of the predicted and experimental mutant quantities for the real-valued output variable. A measure of the standard error (SE) and an equation for the regression line are also reported.

18.3. PROTEIN-SPECIFIC MUTAGENESIS MODELS

18.3.1. Enzymatic Activity

Mutagenesis studies are one of the most widely used approaches for protein functional analysis. However, experimental mutagenesis techniques are expensive with regard to time, labor, and cost. Computational efforts to infer protein function have utilized a variety of statistical and evolutionary methods [82–87]. Similarly, by analyzing sequence and structure information, success has been achieved in understanding the functional effects of coding nonsynonymous single nucleotide polymorphisms (nsSNPs) [88–93]. An SNP is the result of a nucleotide variation at the same position in the genomic DNA of a given population, and a coding nsSNP leads to an amino acid substitution in the encoded protein sequence. Recent work has focused on using supervised learning techniques to predict the functionality of single site enzyme mutants resulting from coding nsSNPs [38,94]. Models developed with these latter methods are trained with a limited set of single site mutants, each belonging to one of a discrete set of activity classes based on experimental studies. Given that the models are expected to perform well at classifying mutants, they can subsequently be used to infer the activity classes to which the remaining uncharacterized enzyme mutants belong. Supervised classification algorithms require that the mutants of an enzyme be represented as vectors of the same dimension with each vector component describing a particular attribute of the mutants. Model performance is significantly influenced by the strength of the signals embedded in these vectors, coupled with the degree of disparity of signals associated with differing classes. The attributes explored in the cited literature include information readily available from sequence data (e.g., physicochemical classes of the native and new residues, hydrophobicity difference, and conservation score at the mutated residue position), and information directly predicted from protein structure (e.g., secondary structure, buried charge, and solvent accessibility). Additionally, those studies place the training set nsSNP mutant enzymes into two classes (activity is either unaffected or affected by the mutation), regardless of the number of classes originally used by the investigators in their experimental studies. Here we generalize the situation by including in our training and test sets all single-site mutants of an enzyme, rather than focusing exclusively on mutants generated by coding nsSNPs.

We developed models both by using a two-class (active/ inactive) labeling of the mutants, as well as by working with the larger number of mutant classes originally defined in the experimental studies. More importantly than these issues is the fact that each mutant attribute vector, derived using a four-body statistical potential function, consists of components that quantify the ECs from wild-type experienced at every residue position in the enzyme as a result of the specific amino acid substitution that generated the mutant. Hence, the dimensionality of every mutant attribute vector (i.e., mutant residual profile) is equivalent to the number of amino acids in the primary sequence of the protein under consideration. Both sequence and structure information is embedded in the residual profiles of the mutants, and both contribute to the overall signal strength and interclass signal disparity. In particular, the nonzero components of a mutant residual profile correspond to all amino acid positions that participate in nearest-neighbor topological contacts with the mutated residue position, as well as the mutated position itself, based on the Delaunay tessellation of the protein structure. Additionally, the values at these nonzero components are a unique reflection of the type of residue replacement occurring at the mutated position. The approach described above is applied to generating training and test sets for two enzyme systems of single site mutants: HIV-1 protease and bacteriophage T4 lysozyme. Since one monomer of the protease homodimer consists of 99 amino acids, there are $99 \times 19 = 1881$ possible single site mutants; similarly, the 164 amino acids forming the primary sequence of T4 lysozyme afford the possibility of $164 \times 19 = 3116$ mutants. Numerous mutagenesis experiments have been published analyzing both of these enzymes. However, the two most comprehensive studies experimentally measured activity levels for 536 mutants of HIV-1 protease (representing substitutions at all 99 positions), based on 336 published mutants [95] and 200 additional mutants courtesy of R. Swanstrom, as well as 2015 mutants of T4 lysozyme obtained by introducing the same 13 residues as replacements at positions 2–164 and resulting in 104 additional non-mutant controls [96]. Residual profiles of these mutants for which activity is known are used as a training set to build accurate inferential models for each enzyme, and these models are then used to predict the activity levels of the remaining 1345 protease and 1101 T4 lysozyme mutants in their respective test sets based on the signals embedded in their residual profiles. According to the experimental measurements, the protease mutants each belong to one of three activity classes (positive, intermediate, or negative), and the T4 lysozyme mutants each belong to one of four activity classes (high, medium, low, or negative). Viewed as a two-class system, the protease mutants are either active (positive and intermediate classes combined) or inactive (negative class). Similarly, T4 lysozyme mutants in the high and medium classes are considered active, while mutants in the low and negative classes are inactive [96]. The results of HIV-1 protease mutant classification are shown in Figure 18.4

18.3.2. Drug Resistance

Emergence of pathogens with mutations, which make them resistant to drugs, presents one of the main challenges in treatment of infectious and viral

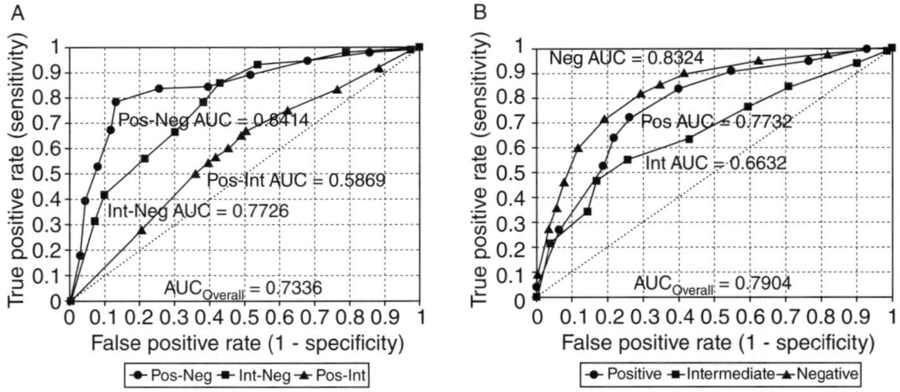

FIGURE 18.4 ROC curves generated with 536 HIV-1 protease mutants by applying (A) the 1-against-1 and (B) the 1-against-all methods for handling multiple classes with decision trees. In the legend for (A), Pos-Neg refers to the subset consisting of mutants experimentally determined to belong only to the positive or negative pair of activity classes (intermediate mutants removed from the full set of 536 mutants); the remaining legend entries are similarly defined. The class pair AUC values reflect intuitive ideas based on biological principles, whereby the signals embedded in the residual profiles of the positive and negative class mutants are the most disparate, complementing the great structural and functional differences between these mutants. On the other hand, positive and intermediate mutants are more or less active, and their overlapping signals pose a challenge for accurate discrimination between these classes. In the legend for (B), Positive refers to the full set of 536 mutants, where positive is the reference class, and intermediate and negative mutants are combined into a single "non-positive" class; the remaining legend entries are similarly defined. Note that the AUC value for the positive reference class ROC curve falls between the AUC values for the positive–negative and positive–intermediate class pair ROC curves; the same is true for the other reference classes.

diseases. One of the prominent examples is the evolution of HIV mutations, defying antiretroviral therapies that rely on inhibitors of HIV-1 protease. Mutations may affect the conformation of the active site, and thus the substrate binding affinity and catalytic activity. They also may improve viral fitness by active site structural rearrangements or by lowering the binding affinity of the inhibitor. From these alternative mutational patterns emerges a complex collection of drug-resistant and drug-sensitive mutants of HIV-1 protease relative to each inhibitor. Genotypic and phenotypic resistance assays are widely used to assess the degree of susceptibility of an HIV-1 protease mutant to a particular inhibitor [97]. Genotyping consists of sequencing the mutant protease in order to discern whether mutations previously known to be associated with resistance are present, while phenotyping entails a direct measurement of the susceptibility of a mutant to an inhibitor relative to a drug-sensitive control. Since phenotypic assays are costly and require a reporting time of approximately two weeks, various computational approaches have been developed to rapidly predict phenotype from genotype [98–102], including techniques that utilize models generated by supervised machine learning tools [103–106]. Structure-based predictions of resistance patterns have also been explored [98], using methods such as molecular modeling [107–111], fitness evolution [112], molecular dynamics simulation [113], and machine learning [105]. Incorporating both sequence and structure information in order to characterize each mutant of HIV-1 protease, through the use of a computational mutagenesis application of a four-body statistical contact potential, may improve the performance phenotype predictors.

With a focus on HIV-1 protease mutants for which phenotypic resistance testing has already been performed in response to a particular inhibitor, and by using appropriately chosen thresholds, each mutant can be classified as either sensitive or resistant to the drug. The residual profile vector representations and class labels of these mutants form the basis of a training set for supervised learning algorithms. We evaluate the performance of NN, DT, SVM, and RF learning schemes on the training set. The entire procedure is implemented repeatedly over seven distinct training sets of protease mutants, where each set consists of mutants that have been phenotypically assayed with respect to a specific FDA-approved protease inhibitor. For each training set, learned models can be used to make class predictions about protease mutants with unknown phenotypes

The datasets for the models were selected from Stanford HIV Drug Resistance Database, which consists of nearly 400 distinct mutational patterns for HIV-1 protease, isolated and sequenced multiple times from over 4000 patients enrolled in large-scale clinical trials [114]. For each of seven FDA-approved protease inhibitors (nelfinavir—NFV, saquinavir—SQV, indinavir—IDV, ritonavir—RTV, amprenavir—APV, lopinavir—LPV, atazanavir—ATV), phenotypic testing was performed on a subset of the observed mutants, ranging from a high of 152 for NFV to a low of 84 for ATV.

For each of the training sets, corresponding to the seven HIV-1 protease inhibitors under consideration, we evaluated the performance of four supervised classification algorithms by employing RF, SVM, DT, and NN. In each of the 28 cases, we derived a plot of the corresponding ROC curve, calculated the AUC, and evaluated the SE as illustrated in Figure 18.5. Overall, RF performed the best when considering the AUC values associated with all seven protease inhibitor training sets; in particular, RF performed better than SVM and DT for 5/7 inhibitors, and better than NN in all cases. Additionally, SVM performed better than DT for 4/7 inhibitors, and both SVM and DT performed better than NN in all cases. In general, impressive algorithm performance is demonstrated for all inhibitor training sets using RF, SVM, and DT. However, while NN performance is good on the NFV, SQV, IDV, and RTV training sets, it is surprisingly poor for APV, LPV, and ATV. Training set size (n) in Figure 18.5 refers to the number of distinct protease mutants for which phenotypic tests have been performed in response to the particular inhibitor. The relatively smaller sizes of the ATV and LPV training sets, and

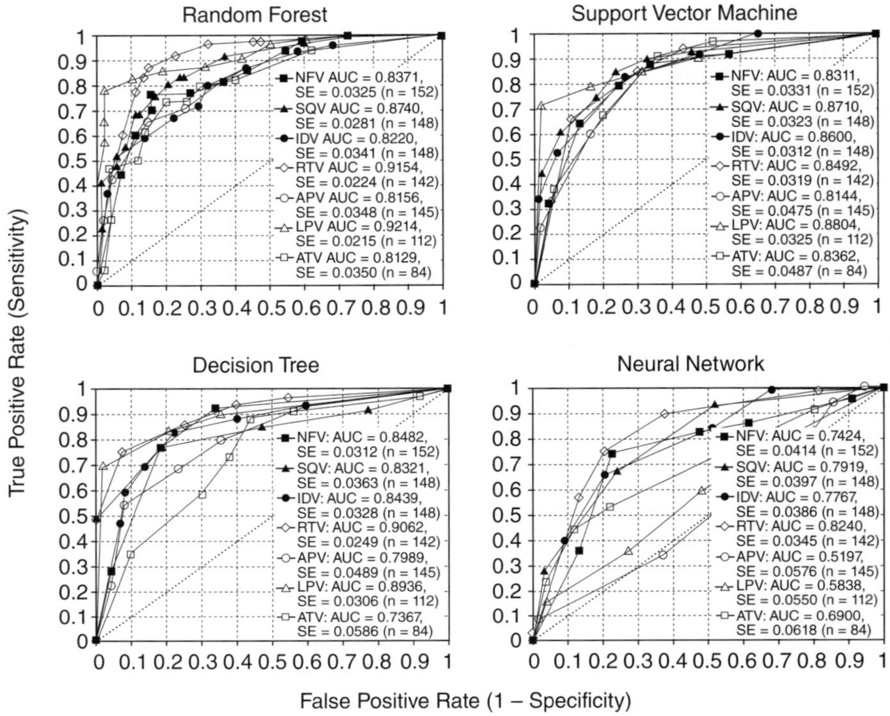

FIGURE 18.5 ROC curves for seven HIV-1 protease inhibitors using four supervised learning methods.

generally poorer performance on these sets, suggests that improvement may be possible once additional mutant phenotypic data becomes available for these inhibitors and is included in their training sets.

18.4. UNIVERSAL MODELS OF THERMODYNAMIC STABILITY

Experimental assessments of changes in protein stability, which result from amino acid residue substitutions, represent an important area of research in biochemistry and molecular biology. A complete understanding of the factors that influence protein folding can be developed through the analysis of this information, including the persistence or elimination of non-covalent contacts (hydrophobic and van der Waals interactions; hydrogen and ionic bonds) upon mutation as well as the secondary structure and solvent accessibility of each substituted position. Such studies are also useful for the elucidation of catalytic residues in enzymes and for developing a clearer understanding of the functional roles of other residue positions in proteins. Additionally, these data are keys for designing new proteins that possess desired attributes, such as specific levels of stability or enzymatic activity, avoidance of protein aggregation and enhancement or diminution of protein–protein interactions or DNA binding capability. Given the abundance and significance of applications, and due to the costly nature with respect to both time and expense of performing exhaustive wet-lab mutagenesis studies, accurate predictive models of protein stability changes upon single point substitutions are in great demand.

Predictions have been carried out by some groups through the application of force fields based on physical effective energy functions derived from molecular mechanics [2,4,5], which are often combined with molecular dynamics or Monte-Carlo simulations [9–13]. However, these approaches are computationally expensive, limiting their practical utility to small sets of protein mutants. Alternatively, methods described in the literature that utilize force fields based on pseudo-energy functions have been effectively applied to the stability analysis of large mutant datasets. These techniques are derived either from knowledge-based statistical potentials [14–23], or from physical descriptions of possible interactions combined with experimentally obtained empirical data [27–29].

Recently, supervised classification and regression machine learning techniques have been used successfully to predict the direction (increased or decreased) and value of mutant stability change, respectively [30–36]. These approaches are capable of utilizing local and nonlocal interactions that impact protein stability by learning complex nonlinear functions based on large training sets of protein mutants with experimentally measured stability changes. Independent variables (i.e., predictors) include information about the mutation as well as the protein sequence or structure, and each protein mutant is encoded as an ordered vector of these attributes.

The dataset used to build the model consists of 1925 experimentally studied single-point mutants selected from a structurally diverse set of 55 proteins. Each mutant was represented as an attribute vector consisting of the following features [65]:

1. Wild-type and replacement amino acid identities at the mutated position;
2. Residual score (i.e., EC score at the mutated position obtained from the residual profile vector);
3. For each of the six nearest neighbors of the mutated position, defined by the Delaunay tessellation of the structure and ordered by 3D Euclidean distance: EC scores obtained from the residual profile vector, amino acid identities at those positions and their signed primary sequence distances away from the mutated position;
4. Computed RSA and experimental temperature, pH and $\Delta\Delta G$ (sign for classification, actual value for regression).

The features in (1) and (4) are common to both structure-based attribute vectors [65] as well as those of the widely-known I-Mutant SVM model [33], while the features in (2) and (3) are specifically based on the four-body statistical potential. In order to make a direct comparison with the previous I-Mutant study, we trained an SVM classifier using an RBF kernel and a 20-fold CV testing procedure. The results using all the mutant attributes (Table 18.1) reveal an increase of 0.04 (5%) in accuracy over I-Mutant, assisted significantly by a 0.14 (25%) sensitivity increase in the "+" class and leading to a 0.08 (29%) drop in BER and a 0.10 (20%) increase in MCC. We also applied an RF classifier with 20-fold CV, where parameters chosen included growing 50 trees from bootstrapped datasets and selecting five random attributes (from among all the attributes described above) to split at every node of every tree. As shown in Table 18.1, accuracy improved by 0.06 (8%) over I-Mutant as a result of significant increases in all sensitivity and precision measures, resulting in a 0.10 (36%) drop in BER and a 0.15 (29%) increase in MCC. When only seven mutant attributes were provided, corresponding to the mutant residual score (i.e., EC score at the mutated position) and the ordered EC scores of the six nearest neighbors, the RF structure-based classifier (50 trees, four random attributes/node) again outperformed the I-Mutant SVM classifier [33] with respect to all reported measures (Table 18.1).

TABLE 18.1 Comparison of 20-fold CV Prediction Performance on 1925 Mutants Set

Method	Q	S(+)	P(+)	S(−)	P(−)	BER	MCC
RF (all attributes)	0.86	0.70	0.81	0.93	0.88	0.18	0.66
RF (only 7 EC scores)	0.82	0.61	0.75	0.91	0.84	0.24	0.55
SVM (all attributes)	0.84	0.70	0.75	0.90	0.87	0.20	0.61
I-Mutant (SVM)	0.80	0.56	0.73	0.91	0.83	0.28	0.51

S/P = sensitivity/precision; +/− = increased/decreased mutant stability change relative to native protein.

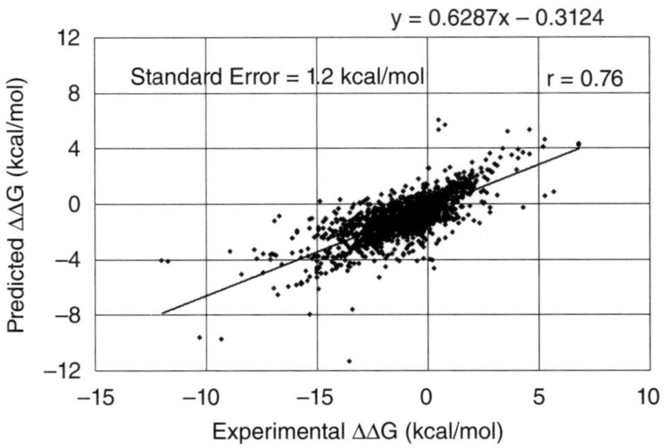

FIGURE 18.6 Correlation plot of the experimental and predicted values of ΔΔG based on the SVR method.

Finally, we replaced the increased/decreased mutant stability change class labels with the actual ΔΔG values as the output attribute for the mutant feature vectors, and we applied SVR to our dataset, using an RBF kernel and 20-fold CV for direct comparison (Fig. 18.6). Our results again outperformed those of I-Mutant, for which $r = 0.71$, SE = 1.3 kcal/mol, and the regression line had equation $y = 0.5223x - 0.4705$. When SVR was replaced with the REPTree algorithm in conjunction with our approach, we observed minor additional improvements with $r = 0.79$, SE = 1.1 kcal/mol, and the regression line had equation $y = 0.5357x - 0.4376$.

18.5. CONCLUSIONS

Energy-based methods learn functions for making predictions by fitting a linear combination of (pseudo-) energy data obtained from experimental or knowledge-based approaches. On the other hand, machine learning techniques learn complex nonlinear functions for making predictions that depend on a common set of measured attributes for all examples in a dataset. Currently, published reports on applications of machine learning to the prediction of activity or stability changes in proteins due to single residue substitutions have all utilized as attributes information about protein sequence or structure, or evolutionary information, without making use of the more strongly correlative data obtained from energy-based methods. In the series of works described here, for the first time, we have made explicit use of data obtained from a four-body, knowledge-based, statistical contact potential, by defining a computational mutagenesis procedure leading to mutant attributes that quantify the environmental perturbations occurring at the mutated position and its six closest neighbors. In some instances, by leveraging the power of machine learning on as few as these seven energy-based attributes, we have outperformed techniques that utilize a much larger number of predictors. In all cases, our results are at least comparable to those of previous studies when analogous sequence, structure, or experimental parameter attributes are included. Unlike other energy-based approaches, the simplicity with which the four-body potential and computational mutagenesis can be applied makes it ideally suited for use in conjunction with machine learning techniques as a way of developing improved models for predicting activity and stability changes in protein mutants.

Future developments of computational mutagenesis approaches will focus on improving model performance and predictive capability by utilizing other supervised classification algorithms, and by including additional components to the mutant residual profiles that incorporate sequence, structure, and evolutionary information. One of the key ingredients that will permit these methods for mutant functional inference to enjoy a practical utility among experimentalists is the development of a standardized way for identifying a minimal training set of protein mutants. Such a tool will provide researchers interested in a comprehensive exploration of the single point mutation space with an idea of the preliminary work required to reap the benefits of reliable predictions for the remaining mutants. The methodology for the analysis and prediction of mutant properties described here can be also used for protein engineering and protein design applications. Some of tools described in this chapter are available online at http://proteins.gmu.edu/automute [115].

REFERENCES

1. T.E. Creighton. *Proteins: Structures and Molecular Properties*. New York: W. H. Freeman and Company, 1984.
2. T. Lazaridis and M. Karplus. Effective energy functions for protein structure prediction. *Curr Opin Struct Biol*, 10(2):139–145, 2000.
3. J. Mendes, R. Guerois, and L. Serrano. Energy estimation in protein design. *Curr Opin Struct Biol*, 12(4):441–446, 2002.
4. J. Moult. Comparison of database potentials and molecular mechanics force fields. *Curr Opin Struct Biol*, 7(2):194–199, 1997.
5. Y. Wang et al. Position-dependent protein mutant profile based on mean force field calculation. *Protein Eng*, 9(6):479–484, 1996.
6. B.R. Brooks et al. CHARMM: A program for macromolecular energy, minimization, and dynamics calculations. *J Comput Chem*, 4:271–277, 1983.
7. M. Christen et al. The GROMOS software for biomolecular simulation: GROMOS05. *J Comput Chem*, 26(16):1719–1751, 2005.
8. J.W. Ponder and D.A. Case. Force fields for protein simulations. *Adv Protein Chem*, 66:27–85, 2003.
9. M. Prevost et al. Contribution of the hydrophobic effect to protein stability: Analysis based on simulations of the Ile-96—-Ala mutation in barnase. *Proc Natl Acad Sci U S A*, 88(23):10880–10884, 1991.
10. Y. Duan and P.A. Kollman. Pathways to a protein folding intermediate observed in a 1-microsecond simulation in aqueous solution. *Science*, 282(5389):740–744, 1998.
11. Y. Duan, L. Wang, and P.A. Kollman. The early stage of folding of villin headpiece subdomain observed in a 200-nanosecond fully solvated molecular dynamics simulation. *Proc Natl Acad Sci U S A*, 95(17):9897–9902, 1998.
12. P.A. Kollman et al. Calculating structures and free energies of complex molecules: Combining molecular mechanics and continuum models. *Acc Chem Res*, 33(12):889–897, 2000.
13. J.W. Pitera and P.A. Kollman. Exhaustive mutagenesis in silico: Multicoordinate free energy calculations on proteins and peptides. *Proteins*, 41(3):385–397, 2000.
14. D. Gilis and M. Rooman. Stability changes upon mutation of solvent-accessible residues in proteins evaluated by database-derived potentials. *J Mol Biol*, 257(5):1112–1126, 1996.
15. D. Gilis and M. Rooman. Predicting protein stability changes upon mutation using database-derived potentials: Solvent accessibility determines the importance of local versus non-local interactions along the sequence. *J Mol Biol*, 272(2):276–290, 1997.

16. C. Hoppe and D. Schomburg. Prediction of protein thermostability with a direction- and distance-dependent knowledge-based potential. *Protein Sci*, 14(10):2682–2692, 2005.

17. J.M. Kwasigroch et al. PoPMuSiC, rationally designing point mutations in protein structures. *Bioinformatics*, 18(12):1701–1702, 2002.

18. L. Meyerguz, J. Kleinberg, and R. Elber. The network of sequence flow between protein structures. *Proc Natl Acad Sci U S A*, 104(28):11627–11632, 2007.

19. M. Ota, S. Kanaya, and K. Nishikawa. Desk-top analysis of the structural stability of various point mutations introduced into ribonuclease H. *J Mol Biol*, 248(4):733–738, 1995.

20. V. Parthiban et al. Structural analysis and prediction of protein mutant stability using distance and torsion potentials: Role of secondary structure and solvent accessibility. *Proteins*, 66(1):41–52, 2007.

21. C.M. Topham, N. Srinivasan, and T.L. Blundell. Prediction of the stability of protein mutants based on structural environment-dependent amino acid substitution and propensity tables. *Protein Eng*, 10(1):7–21, 1997.

22. L. Wang et al. Can one predict protein stability? An attempt to do so for residue 133 of T4 lysozyme using a combination of free energy derivatives, PROFEC, and free energy perturbation methods. *Proteins*, 32(4):438–458, 1998.

23. H. Zhou and Y. Zhou. Distance-scaled, finite ideal-gas reference state improves structure-derived potentials of mean force for structure selection and stability prediction. *Protein Sci*, 11(11):2714–2726, 2002.

24. T. Haliloglu. Characterization of internal motions of Escherichia coli ribonuclease H by Monte Carlo simulation. *Proteins*, 34(4):533–539, 1999.

25. J. Lee et al. Calculation of protein conformation by global optimization of a potential energy function. *Proteins*, 37(S3):204–208, 1999.

26. J. Lee, A. Liwo, and H.A. Scheraga. Energy-based de novo protein folding by conformational space annealing and an off-lattice united-residue force field: Application to the 10-55 fragment of staphylococcal protein A and to apo calbindin D9K. *Proc Natl Acad Sci U S A*, 96(5):2025–2030, 1999.

27. A.J. Bordner and R.A. Abagyan. Large-scale prediction of protein geometry and stability changes for arbitrary single point mutations. *Proteins*, 57(2):400–413, 2004.

28. R. Guerois, J.E. Nielsen, and L. Serrano. Predicting changes in the stability of proteins and protein complexes: A study of more than 1000 mutations. *J Mol Biol*, 320(2):369–387, 2002.

29. K. Saraboji, M.M. Gromiha, and M.N. Ponnuswamy. Average assignment method for predicting the stability of protein mutants. *Biopolymers*, 82(1):80–92, 2006.

30. L.T. Huang et al. Knowledge acquisition and development of accurate rules for predicting protein stability changes. *Comput Biol Chem*, 30(6):408–415, 2006.

31. L.T. Huang et al. Prediction of protein mutant stability using classification and regression tool. *Biophys Chem*, 125:462–470, 2007.

32. E. Capriotti, P. Fariselli, and R. Casadio. A neural-network-based method for predicting protein stability changes upon single point mutations. *Bioinformatics*, 20(1):I63–I68, 2004.

33. E. Capriotti et al. Predicting protein stability changes from sequences using support vector machines. *Bioinformatics*, 21(2):ii54–ii58, 2005.
34. E. Capriotti, P. Fariselli, and R. Casadio. I-Mutant2.0: Predicting stability changes upon mutation from the protein sequence or structure. *Nucleic Acids Res*, 33(Web server issue):W306–W310, 2005.
35. J. Cheng, A. Randall, and P. Baldi. Prediction of protein stability changes for single-site mutations using support vector machines. *Proteins*, 62(4):1125–1132, 2006.
36. C.M. Frenz. Neural network-based prediction of mutation-induced protein stability changes in Staphylococcal nuclease at 20 residue positions. *Proteins*, 59(2):147–151, 2005.
37. R. Karchin et al. Functional impact of missense variants in BRCA1 predicted by supervised learning. *PLoS Comput Biol*, 3(2): e26, 2007.
38. V.G. Krishnan and D.R. Westhead. A comparative study of machine-learning methods to predict the effects of single nucleotide polymorphisms on protein function. *Bioinformatics*, 19(17):2199–2209, 2003.
39. F. Aurenhammer. Voronoi Diagrams - a Survey of a Fundamental Geometric Data Structure. *Computing Surveys*, 23(3):345–405, 1991.
40. K. Sugihara and H. Inagaki. Why Is the 3d Delaunay triangulation difficult to construct. *Inform Process Lett*, 54(5):275–280, 1995.
41. C.B. Barber, D.P. Dobkin, and H.T. Huhdanpaa. The quickhull algorithm for convex hulls. *ACM Trans Math Software*. 22:469–483, 1996.
42. J.D. Bernal. A geometrical approach to the structure of liquids. *Nature*, 183(4655):141–147, 1959.
43. K. Gotoh and J.L. Finney. Statistical geometrical approach to random packing density of equal spheres. *Nature*, 252(5480):202–205, 1974.
44. C. Chothia. Structural invariants in protein folding. *Nature*, 254(5498):304–308, 1975.
45. J.L. Finney. Volume occupation, environment and accessibility in proteins. The problem of the protein surface. *J Mol Biol*, 96(4):721–32, 1975.
46. F.M. Richards. The interpretation of protein structures: Total volume, group volume distributions and packing density. *J Mol Biol*, 82(1):1–14, 1974.
47. II. Vaisman and M.L. Berkowitz. Local structural order and molecular associations in water dmso mixtures - molecular-dynamics study. *J Am Chem Soc*, 114(20):7889–7896, 1992.
48. II. Vaisman, L. Perera, and M.L. Berkowitz. Mobility of stretched water. *J Chem Phys*, 98(12):9859–9862, 1993.
49. II. Vaisman, F.K. Brown, and A. Tropsha. Distance dependence of water-structure around model solutes. *J Physi Chem*, 98(21):5559–5564, 1994.
50. N.N. Medvedev, V.P. Voloshin, and Y.I. Naberukhin. Local environmental geometry of atoms in the Lennard-Jones systems. *Mater Chem Phys*, 14(6):533–548, 1986.
51. R.K. Singh, A. Tropsha, and I.I. Vaisman. Delaunay tessellation of proteins: Four body nearest-neighbor propensities of amino acid residues. *J Comput Biol*, 3(2):213–221, 1996.

52. A. Tropsha et al. Statistical geometry analysis of proteins: Implications for inverted structure prediction. *Pac Symp Biocomput*, 614–623, 1996.
53. W. Zheng et al. A new approach to protein fold recognition based on Delaunay tessellation of protein structure. *Pac Symp Biocomput*, 486–497, 1997.
54. P.J. Munson and R.K. Singh. Statistical significance of hierarchical multi-body potentials based on Delaunay tessellation and their application in sequence-structure alignment. *Protein Science*, 6(7):1467–1481, 1997.
55. B. Krishnamoorthy and A. Tropsha. Development of a four-body statistical pseudo-potential to discriminate native from non-native protein conformations. *Bioinformatics*, 19(12):1540–1548, 2003.
56. A. Babajide et al. Exploring protein sequence space using knowledge-based potentials. *J Theor Biol*, 212(1):35–46, 2001.
57. H.H. Gan, A. Tropsha, and T. Schlick. Lattice protein folding with two and four-body statistical potentials. *Proteins*, 43(2):161–174, 2001.
58. C.W. Carter et al. Four-body potentials reveal protein-specific correlations to stability changes caused by hydrophobic core mutations. *J Mol Biol*, 311(4):625–638, 2001.
59. M. Masso and I.I. Vaisman. Comprehensive mutagenesis of HIV-1 protease: A computational geometry approach. *Biochem Biophys Res Commun*, 305(2):322–326, 2003.
60. M. Barenboim, D.C. Jamison, and I.I. Vaisman, Statistical geometry approach to the study of functional effects of human nonsynonymous SNPs. *Hum Mutat*, 26(5):471–476, 2005.
61. M. Masso, Z. Lu, and I.I. Vaisman. Computational mutagenesis studies of protein structure-function correlations. *Proteins*, 64:234–245, 2006.
62. E. Mathe et al. Predicting the transactivation activity of p53 missense mutants using a four-body potential score derived from Delaunay tessellations. *Hum Mutat*, 27(2):163–172, 2006.
63. M. Masso and I.I. Vaisman. Accurate prediction of enzyme mutant activity based on a multibody statistical potential. *Bioinformatics*, 23(23):3155–3161, 2007.
64. M. Masso and I.I. Vaisman. A novel sequence-structure approach for accurate prediction of resistance to HIV-1 protease inhibitors. *Proc IEEE Bioinformatics and Bioengineering*, 2:952–958, 2007.
65. M. Masso and I.I. Vaisman. Accurate prediction of stability changes in protein mutants by combining machine learning with structure based computational mutagenesis. *Bioinformatics*, 24(18):2002–2009, 2008.
66. M. Barenboim et al. Statistical geometry based prediction of nonsynonymous SNP functional effects using random forest and neuro-fuzzy classifiers. *Proteins*, 71:1930–1939, 2008.
67. M. Masso et al. Modeling the functional consequences of single residue replacements in bacteriophage f1 gene V protein. *Protein Eng Des Sel*, 22(11):665–671, 2009.
68. D. Bostick and I.I. Vaisman. A new topological method to measure protein structure similarity. *Biochem Biophys Res Commun*, 304(2):320–325, 2003.
69. T. Taylor et al. New method for protein secondary structure assignment based on a simple topological descriptor. *Proteins Struct Funct Bioinf*, 60(3):513–524, 2005.

70. D.L. Bostick, M. Shen, and I.I. Vaisman. A simple topological representation of protein structure: Implications for new, fast, and robust structural classification. *Proteins*, 56(3):487–501, 2004.
71. I.I. Vaisman, A. Tropsha, and W. Zheng. Compositional preferences in quadruplets of nearest neighbor residues in protein structures: Statistical geometry analysis. *Proc IEEE Symp Intelligence and Systems*, 163–168, 1998.
72. H.M. Berman et al. The protein data bank. *Nucleic Acids Res*, 28(1):235–242, 2000.
73. J.U. Bowie, R. Luthy, and D. Eisenberg. A method to identify protein sequences that fold into a known three-dimensional structure. *Science*, 253(5016):164–170, 1991.
74. J.P. Bigus. *Data Mining with Neural Networks*. New York: McGraw Hill, 1996.
75. R. Quinlan. *C4.5: Programs for Machine Learning*. San Mateo, CA: Morgan Kaufman Publishers, 1993.
76. J. Platt. Fast training of support vector machines using sequential minimal optimization. In B. Scholkopf, C. Burges, and A. Smola (Eds.), *Advances in Kernel Methods - Support Vector Learning*. Cambridge, MA: MIT Press, 1998.
77. I.H. Witten and E. Frank. *Data Mining*. San Francisco, CA: Morgan Kaufmann, 2000.
78. L. Breiman. Random forests. *Machine Learning*, 45:5–32, 2001.
79. B.W. Matthews. Comparison of the predicted and observed secondary structure of T4 phage lysozyme. *Biochem. Biophys. Acta.*, 405(2):442–451, 1975.
80. J.A. Hanley and B.J. McNeil. The meaning and use of the area under a receiver operating characteristic (ROC) curve. *Radiology*, 143(1):29–36, 1982.
81. T. Fawcett. *ROC Graphs: Notes and Practical Considerations for Researchers*. Palo Alto: Hewlett-Packard Labs, 2003,
82. P.M. Bowers et al. Use of logic relationships to decipher protein network organization. *Science*, 306(5705):2246–2249, 2004.
83. P.D. Dobson and A.J. Doig. Predicting enzyme class from protein structure without alignments. *J Mol Biol*, 345(1):187–199, 2005.
84. L.Y. Han et al. Prediction of functional class of novel viral proteins by a statistical learning method irrespective of sequence similarity. *Virology*, 331(1):136–143, 2005.
85. T. Aita and Y. Husimi. Fitting protein-folding free energy landscape for a certain conformation to an NK fitness landscape. *J Theor Biol*, 253(1):151–161, 2008.
86. K. Sjolander. Phylogenomic inference of protein molecular function: Advances and challenges. *Bioinformatics*, 20(2):170–179, 2004.
87. W. Tian, A.K. Arakaki, and J. Skolnick, EFICAz: A comprehensive approach for accurate genome-scale enzyme function inference. *Nucleic Acids Res*, 32(21):6226–6239, 2004.
88. D. Chasman and R.M. Adams. Predicting the functional consequences of non-synonymous single nucleotide polymorphisms: Structure-based assessment of amino acid variation. *J Mol Biol*, 307(2):683–706, 2001.
89. P.C. Ng and S. Henikoff. Predicting deleterious amino acid substitutions. *Genome Res*, 11(5):863–874, 2001.

90. V. Ramensky, P. Bork, and S. Sunyaev. Human non-synonymous SNPs: Server and survey. *Nucleic Acids Res*, 30(17):3894–3900, 2002.
91. C.T. Saunders and D. Baker. Evaluation of structural and evolutionary contributions to deleterious mutation prediction. *J Mol Biol*, 322(4):891–901, 2002.
92. S. Sunyaev et al. Prediction of deleterious human alleles. *Hum Mol Genet*, 10(6):591–597, 2001.
93. Z. Wang and J. Moult. SNPs, protein structure, and disease. *Hum Mutat*, 17(4):263–270, 2001.
94. R. Karchin et al. LS-SNP: Large-scale annotation of coding non-synonymous SNPs based on multiple information sources. *Bioinformatics*, 21(12):2814–2820, 2005.
95. D.D. Loeb et al. Complete mutagenesis of the HIV-1 protease. *Nature*, 340(6232):397–400, 1989.
96. D. Rennell et al. Systematic mutation of bacteriophage T4 lysozyme. *J Mol Biol*, 222(1):67–88, 1991.
97. J. Sebastian and H. Faruki. Update on HIV resistance and resistance testing. *Med Res Rev*, 24(1):115–125, 2004.
98. Z.W. Cao et al. Computer prediction of drug resistance mutations in proteins. *Drug Discov Today*, 10(7):521–529, 2005.
99. A.G. DiRienzo et al. Non-parametric methods to predict HIV drug susceptibility phenotype from genotype. *Stat Med*, 22(17):2785–2798, 2003.
100. E. Puchhammer-Stockl et al. Comparison of virtual phenotype and HIV-SEQ program (Stanford) interpretation for predicting drug resistance of HIV strains. *HIV Med*, 3(3):200–206, 2002.
101. B. Schmidt et al. Simple algorithm derived from a geno-/phenotypic database to predict HIV-1 protease inhibitor resistance. *AIDS*, 14(12):1731–1738, 2000.
102. M. Zazzi et al. Comparative evaluation of three computerized algorithms for prediction of antiretroviral susceptibility from HIV type 1 genotype. *J Antimicrob Chemother*, 53(2):356–360, 2004.
103. N. Beerenwinkel et al. Diversity and complexity of HIV-1 drug resistance: A bioinformatics approach to predicting phenotype from genotype. *Proc Natl Acad Sci U S A*, 99(12):8271–8276, 2002.
104. N. Beerenwinkel et al. Geno2pheno: Estimating phenotypic drug resistance from HIV-1 genotypes. *Nucleic Acids Res*, 31(13):3850–3855, 2003.
105. S. Draghici and R.B. Potter. Predicting HIV drug resistance with neural networks. *Bioinformatics*, 19(1):98–107, 2003.
106. D. Wang and B. Larder. Enhanced prediction of lopinavir resistance from genotype by use of artificial neural networks. *J Infect Dis*, 188(5):653–660, 2003.
107. Y.Z. Chen, X.L. Gu, and Z.W. Cao. Can an optimization/scoring procedure in ligand-protein docking be employed to probe drug-resistant mutations in proteins? *J Mol Graph Model*, 19(6):560–570, 2001.
108. A.C. Nair et al. Computational studies of the resistance patterns of mutant HIV-1 aspartic proteases towards ABT-538 (ritonavir) and design of new derivatives. *J Mol Graph Model*, 21(3):171–179, 2002.
109. M.D. Shenderovich et al. Structure-based phenotyping predicts HIV-1 protease inhibitor resistance. *Protein Sci*, 12(8):1706–1718, 2003.

110. W. Wang and P.A. Kollman. Computational study of protein specificity: The molecular basis of HIV-1 protease drug resistance. *Proc Natl Acad Sci U S A*, 98(26):14937–14942, 2001.

111. I.T. Weber and R.W. Harrison. Molecular mechanics analysis of drug-resistant mutants of HIV protease. *Protein Eng*, 12(6):469–474, 1999.

112. D. Stoffler et al. Evolutionary analysis of HIV-1 protease inhibitors: Methods for design of inhibitors that evade resistance. *Proteins*, 48(1):63–74, 2002.

113. X. Chen, I.T. Weber, and R.W. Harrison. Molecular dynamics simulations of 14 HIV protease mutants in complexes with indinavir. *J Mol Model*, 10(5–6):373–381, 2004.

114. S.Y. Rhee et al. Distribution of human immunodeficiency virus type 1 protease and reverse transcriptase mutation patterns in 4,183 persons undergoing genotypic resistance testing. *Antimicrob Agents Chemother*, 48(8):3122–3126, 2004.

115. M. Masso and I.I. Vaisman. AUTO-MUTE: Web-based tools for predicting stability changes in proteins due to single amino acid replacements. *Protein Eng*. 23(8):683–687, 2010.

CHAPTER 19

CONFORMATIONAL SEARCH FOR THE PROTEIN NATIVE STATE

AMARDA SHEHU
Department of Computer Science
George Mason University
Fairfax, VA

19.1. THE QUEST FOR THE PROTEIN NATIVE STATE

From the first formulation of the protein folding problem by Wu in 1931 to the experiments of Mirsky and Pauling in 1936, chemical and physical properties of protein molecules were attributed to the amino acid composition and structural arrangement of the protein chain [1,2]. Mirsky and Pauling hypothesized that denaturing conditions like heating abolished chemical and physical properties of a protein by "melting away" the protein structure. The relationship between sequence, structure, and function in protein molecules was under significant debate until the revolutionizing 1960s experiments by Christian Anfinsen and his coworkers at the National Institute of Health [3].

The Anfinsen experiments showed that ribonuclease would spontaneously reassume its structure and enzymatic activity after denaturation. This unique ability to regain both structure and function was confirmed on thousands of proteins. After a decade of experiments, Anfinsen concluded that the amino acid sequence governed the folding of a protein chain into a "biologically active conformation" under a "normal physiological milieu" [4]. Anfinsen used the terms "conformation" and "structure" interchangeably to describe a three-dimensional (3D) arrangement of the chain connecting amino acids in a protein. To this day, no distinction is drawn between structure and conformation.

The Anfinsen experiments posited that, if one were to understand how the amino acid sequence determines the biologically active conformation, one

Introduction to Protein Structure Prediction: Methods and Algorithms,
Edited by Huzefa Rangwala and George Karypis

could find this conformation in silico. A computer algorithm would follow simple rules or instructions to compute the biologically active conformation. Computing this conformation from the knowledge of amino acid sequence alone remains a grand challenge of computational biology [5,6]. Nonetheless, significant computational progress has been made. Many methods now provide useful mechanistic insight into the biological function of a protein through a structural description of the functionally relevant (native) state.

19.1.1. Native Structure versus Native Conformational Ensemble

The survey in this chapter focuses on methods that search the space of different conformations of a protein chain to find those associated with the native state. Significant computational research on describing the protein native state in conformational detail focuses on computing a single representative conformation, often referred to as the native structure [7–14]. Such research, the focus of CASP [15], is justified in many proteins.

The single structure view of the native state does not fully take into account the inherent flexibility of a protein chain. Long gone are the days when protein molecules were considered rigid and solid-like [16]. In the words of Richard Feynmann [17], "everything that living things do can be understood in terms of the jiggling and wiggling of atom."

When flexibility under native conditions consists of mostly local insignificant deviations around an average structure, the single structure view is well warranted. For many proteins, however, an accurate description of flexibility may involve large-scale conformational rearrangements. This is often the case in proteins involved in biochemical processes like molecular recognition, enzymatic catalysis, and signal transduction [18–20]. Evidence of functionally relevant flexibility, not necessarily around a unique structure, advocates a more general description of the native state through an ensemble of conformations, also referred to as the native state ensemble [21–23].

19.1.2. Thermodynamic versus Kinetic Hypothesis

The current understanding of what drives the folding of a protein chain directly impacts the assumptions and strategies employed by computational methods to find the native state. Historically, two hypotheses have competed to explain the process of folding. The thermodynamic hypothesis states that the native state of a protein minimizes free energy, whereas the kinetic hypothesis attributes the native state to that which is kinetically accessible.

Anfinsen made the case that his experiments and those of other researchers established the generality of the thermodynamic hypothesis:

> This hypothesis states that the three-dimensional structure of a native protein in its normal physiological milieu (solvent, pH, ionic strength, presence of other components such as metal ions or prosthetic groups, temperature, etc.) is the

one in which the Gibbs free energy of the whole system is lowest; that is, that the native conformation is determined by the totality of interatomic interactions and hence by the amino acid sequence, in a given environment. (Nobel Lecture, December 11, 1972)

The thermodynamic hypothesis suggests that a naive computer algorithm can be written to systematically enumerate the distinct structures or conformations assumed by a protein chain. The algorithm can sum over the interatomic interactions to evaluate the energy of each computed conformation. Assuming the algorithm can properly identify the computed conformation(s) where the free energy reaches its global minimum value, the biologically active state will have been captured in structural detail by simple enumeration.

Enumeration is possible when (1) the number of parameters employed to represent a protein conformation is small, and when (2) these parameters take values from a finite set. In other words, the space of possible conformations, referred to as the conformational space, has to be low dimensional and discretizable for enumeration to be a feasible strategy. Early computational research on finding native conformations of a protein assumed a low-dimensional and discretizable space that was amenable to enumeration [24–26]. Such simplification allowed the application of exhaustive search to the discovery of native conformations, albeit at the cost of a systematic inability to capture subtle structural details of the native state [27,28].

Back-of-the-envelope calculations led Cyrus Levinthal to a supposed paradox. Even assuming a small number of configurations of the peptide bond connecting two consecutive amino acids (e.g., 3) in a short protein chain of 101 amino acids, the number of ensuing conformations is 3^{100}. Assuming a rate of 10^{13} conformations per second, it would still take 10^{27} years for a protein to sample all these conformations. This dramatic example showed that a protein could not possibly find its native structure by searching at random within a vast and high-dimensional conformational space [29].

Levinthal was concerned with the time that it would take a protein to find its lowest free energy state, that is, the actual kinetics of folding. Given that many proteins refold in a few microseconds after denaturation, random sampling of the conformational space does not explain the process of folding. Levinthal's paradox illustrates that (1) diffusion cannot be the only guiding force behind folding, and (2) random searches are infeasible strategies for sampling the native conformation(s) of a protein chain.

Levinthal's calculations cast the process of finding the native structure as searching for a needle in a haystack. Early research showed in simulation that the problem of computing the lowest free energy state was indeed hard [30]. On the one hand, theoretical research in computer science proved that the problem, even when employing simple lattice models to represent conformations of a protein chain, is NP-hard [31,32]. On the other hand, simulation studies showed that proteins could get trapped in structures that were energetically similar but topologically different from the native structure obtained in the wet lab [33]. An alternative hypothesis was offered to explain the discrepancy through the possibility of kinetic traps. The kinetic hypothesis suggested that proteins folded into structures that were kinetically accessible.

Despite the complexity associated with searching the protein conformational space and the seemingly competing views of protein folding, efficient algorithms exist today. These algorithms operate in a high-dimensional and continuous search space. This is made possible by a better understanding of protein physics and the process of folding accompanied by a steady increase in the number of calculations that can be performed in one CPU cycle. The predictive power of these algorithms and their ability to reproduce observations in the wet lab with high accuracy has significantly improved in the last decade. Significant progress in our understanding of protein folding came with the introduction of the energy landscape view, which unified the thermodynamic and kinetic hypotheses. This view is the focus of the next section.

19.1.3. The Energy Landscape View of Protein Folding

Levinthal's paradox ignored the energetic bias against unfavorable protein conformations. Seminal work that interpreted evidence emerging from folding experiments through the theory of statistical mechanics presented an energy landscape view of protein folding [34,35] that reconciled the thermodynamic and kinetic hypotheses. The "new view" [34] offered a statistical description

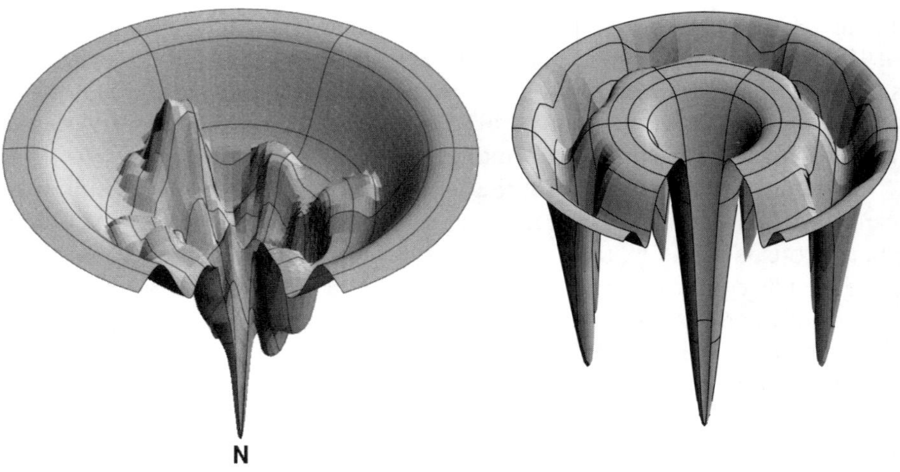

FIGURE 19.1 The schematic diagram illustrates different folding scenarios. The vertical axis plots the internal free energy of a protein. The conformational space is projected on two coordinates (horizontal axes). The landscape on the left illustrates the classic scenario, where the native state labeled N is associated with the global energy minimum. The landscape on the right illustrates how a protein can be trapped in multiple deep minima. Reprinted from Reference [34] with permission of Ken Dill.

of the complex energy surface of a protein through an energy landscape. Despite the high dimensionality of the conformational space and the intricate number of interatomic interactions in a protein, the energy surface associated with the conformational space can be projected into a few coordinates to obtain an energy landscape view of how proteins fold.

Under this view, three main classes of energy surfaces emerge. They range from surfaces with a single global free energy minimum (Fig. 19.1, left), corresponding to proteins with a very strong stability point, to surfaces with a few minima (Fig. 19.1, right) and surfaces with a shallow native basin [36]. Actual energy surfaces of proteins may combine these three main cases.

The energy landscape view employs statistical mechanics to organize the multitude of protein conformations in terms of a minimal number of collective parameters [35,37,38]. This statistical formulation allows capturing essential features of the free energy surface of a protein with only a limited set of parameters. Various computational methods exploit this formulation to focus the search for native or near-native conformations of a protein chain to minima in the energy landscape that emerge when projecting the energy surface over the employed parameters. However, the general existence of underlying collective parameters that guide a protein to quickly locate its lowest free energy state remains open to debate [38].

The energy landscape view provides a theoretical framework to explain how a protein may assume different low-energy conformations, for instance, upon binding [39]. The free energy minimum in the energy landscape could be populated by different low-energy conformations of a protein chain, which map to the same region in the space of the underlying collective parameters. Since understanding protein function requires obtaining a comprehensive view of the conformational space associated with the free energy minimum (or minima) in the energy landscape [22], many computational methods describe the protein native state as an ensemble of conformations [40–42].

19.1.4. Computational Issues in the Search for Native Conformations

Figure 19.1 allows visualizing the thermodynamic versus the kinetic hypothesis [43,44]. The energy landscape view brings into focus fundamental questions and computational issues that need to be addressed by algorithms designed to search a protein's conformational space. These issues and potential strategies to address them are summarized below. Methods that implement these strategies are then detailed in the rest of the chapter.

19.1.4.1. Does Search for the Native State Need the How? If one wants to design an algorithm to compute conformations of a protein chain under native conditions, should the algorithm consider the actual process of folding? Should physical timescales be associated with computed conformations? Considering how a protein chain tumbles down the energy landscape may help capture possible kinetic traps and find the actual native state. Many computational methods follow the physical process of folding to let a protein system "sample" the native state in simulation. Consideration of "the how" can result in very long simulations. Currently, methods that employ mainly the thermodynamic hypothesis exhibit faster sampling efficiency. Recent evidence emerging from the most successful methods suggests that incorporating features of the physical process of folding can actually improve both efficiency and accuracy [8,45,46].

19.1.4.2. Realms of Discretization. The energy landscape view advocates that, if one were to know "the true" collective parameters (often referred to as reaction coordinates) that guide the folding reaction, the energy landscape obtained by projecting the protein energy surface over these parameters is not complex. The search for native conformations can be conducted over a discretization of the projection of the conformational space over the coordinates. At the very least, the discretization of the projected space can be employed to keep track and guide the search in the high-dimensional conformational space. Finding reaction coordinates, however, is challenging. A rich body of research beyond the scope of this survey pursues finding collective parameters that can serve as general reaction coordinates for protein systems [47].

19.1.4.3. Sampling over Enumeration. A fundamental problem in obtaining a structural description of the protein native state is that of efficiently computing conformations associated with the global minimum (or deepest minima) in a rugged energy surface constellated with local minima. Multiscale modeling, which combines coarse-grained and fine-grained detail when modeling a protein conformation, and a naturally inspired discretization of the process through which conformations are assembled are currently the most successful strategies [9,10,38,42]. The paradigm in some of the most successful methods is away from a systematic search and toward a probabilistic walk or probabilistic sampling of the conformational space [11,48–51].

19.1.4.4. Guiding the Search and Narrowing the Search Space. The vast high-dimensional conformational space available to a protein chain raises practical computability problems. Many computational methods (detailed below) resort to employing additional information to narrow the conformational space relevant for their search. This information, either in the form of thermodynamic averages over conformations of the native state or in the form of an average native structure captured in experiment, allows constraining the search for native conformations by what is observed in experiment.

19.1.4.5. Enhancing Sampling of a Vast High-Dimensional Space. Ab initio methods that employ only knowledge of amino acid sequence for a protein at hand have to enhance sampling of the vast conformational space. Enhanced sampling methods include simulated annealing, importance and umbrella sampling, replica exchange (also known as parallel tempering), local elevation, activation relaxation, local energy flattening, jump walking, multicanonical ensemble, conformational flooding, Markov state models, discrete time step molecular dynamics (MD), and many more. Since a complete survey of these methods is not possible, the following summary focuses on a few successful representative methods.

19.1.4.6. Combining the Discrete and the Continuous. This survey of conformational search methods for the protein native state concludes with a discussion of potential benefits to future research by a combination of discrete and continuous exploration. The discussion focuses on combining search in a discretized energy landscape with search on a continuous conformational space. The chapter concludes with an outlook on how knowledge of collective parameters that can serve as reaction coordinates can be employed to guide conformational search toward relevant energy minima in the underlying energy landscape.

19.2. EXHAUSTIVE SEARCH: DISCRETIZATION OF CONFORMATIONAL SPACE

Early simulations of protein chains showed that important physical properties could be obtained with considerably less than atomic detail. Coarse-grained modeling of protein conformations opened up the possibility of exhaustively searching a simplified conformational space through explicit enumeration of possible conformations of a protein chain.

Some of the first coarse-grained models were based on lattices, explicitly modeling a representative (often C_α) atom of each amino acid in a protein chain and restricting atoms to lie on a lattice [52]. Lattice modeling not only allowed computing native conformations of very long protein chains, but incidentally exposed an interesting complexity result: Finding the lowest-energy conformation on a 3D cubic lattice is NP-hard [31]. Despite this complexity result, lattice models offered both analytical and computational simplicity [53]. In addition to very fast integer math evaluations of conformational energies on a lattice, lattice modeling made it feasible for exhaustive searches to explicitly enumerate conformations [24–26].

Despite the simplicity, exhaustive search methods that employ lattice modeling can reproduce the backbone with accuracy no greater than half the lattice spacing [28]. These methods cannot capture subtle structural details and may bias toward specific secondary structures [27]. Research on improving accuracy and getting the full computational benefits of searching in a discretized conformational space remains active. Indeed, some of the most successfully enhanced sampling methods implicitly employ discretization by assembling conformations of a protein chain with naturally occurring structures of short fragments defined over the chain [9,10,12,42,45,46].

Coarse-grained models that capture realistic protein structures are predominantly off-lattice [38,54]. Rather than discretize the conformational space, these models simplify and lower the dimensionality of this space. Sophisticated force fields designed for these models associate potential energies with computed conformations [54–58]. For instance, backbone-resolution models, where only heavy backbone atoms are explicitly modeled, allow obtaining highly accurate native conformations (some of these models include C_β atoms to represent side chains) [10–12,42,59].

Recent methods combine coarse- and fine-grained modeling to enhance sampling of the conformational space [60–62]. For instance, methods that predict native conformations from the amino acid sequence conduct most of their search in a coarse-grained space, adding atomic detail (as in Reference [63]) only when it is imperative to refine conformations or to determine which low-energy minima are relevant for the native state [9,42,64]. Multiscaling is one of the many computational strategies to enhance the exploration of the high-dimensional protein conformational space.

19.3. SYSTEMATIC SEARCH: MD

Computational methods that follow the physics of folding to let a protein "sample" its native state implement the MD approach [65]. MD-based methods systematically search the conformational space by numerically solving Newton's equations of motion. The solution accuracy dictates a small time step in the order of femtoseconds between consecutive conformations in an MD trajectory. As a result, MD-based simulations may demand long trajectories before attaining native conformations. Moreover, when no knowledge of the global energy minimum is available, multiple trajectories may need to be computed in order to determine that no significantly lower energies can be obtained. The issue of convergence has practical implications for time demands. Most MD studies circumvent this issue by employing similarity between computed conformations and experimentally available native structures of tested proteins as a termination criterion.

Since MD-based methods follow the process of folding, they offer more than just a set of the conformations relevant for the native state. They also reveal kinetic information, that is, how the unfolded protein chain reaches its native state and in what timescales. This added information increases the computational requirements of MD-based methods. These requirements are often alleviated by distributing the MD search of the conformational space on supercomputers [66] and grids of desktops [67]. Specific architectures like the IBM BLUE Gene and Anton and distributed MD implementations like Desmond are devoted to surpassing computational milestones and achieving high-resolution native conformations through MD simulations [68,69].

19.4. BIASED RANDOM WALK: METROPOLIS MONTE CARLO (MC)

Rather than solving Newton's equations of motions, random search techniques such as MC conduct biased random (probabilistic) walks in conformational space to obtain a sequence of conformations [65,70]. The random walk ensures through the Metropolis criterion [71] that a conformation is obtained with frequency proportional to its Boltzmann probability. While exhibiting higher sampling efficiency than MD simulations, MC simulations also obtain conformations sequentially. Like MD, they also spend considerable time sampling rare events such as crossing maxima in the energy landscape.

The tendency of MD- and MC-based methods to converge to local minima in the energy surface that underlies the protein conformational space underscores the fact that MD and MC are local optimization techniques. This tendency is usually circumvented in two ways: (1) by narrowing the search to specific regions in the energy surface or conformational space through experimentally available information, and (2) by enhancing the local optimization in the MD systematic search or the MC sampling through an array of enhanced sampling strategies beyond multiscaling. These two (not mutually exclusive) groups of methods are discussed next.

19.5. GUIDED SEARCH OF CONFORMATIONAL SPACE

A special class of conformational search methods employs experimental data to guide MD or MC trajectories to the relevant search space. The data help to focus computational resources to regions of the energy surface or the conformational space that are relevant for the native state and to quickly guide the exploration toward native conformations. These data come in the form of thermodynamic averages (over the underlying native state ensemble) obtained from nuclear magnetic resonance (NMR) experiments, density maps obtained from X-ray crystallography or cryo-electron microscopy (cryoEM), or an average structure obtained from X-ray or NMR.

19.5.1. Guiding the Search with Thermodynamic Averages

Methods that employ NMR thermodynamic averages such as nuclear Overhauser effect (NOE) distance constraints, S^2 order parameters, three-bond scalar couplings (^3J), residual dipolar couplings (RDCs), chemical shifts, ϕ or ψ values, or protection factors often incorporate these averages in an additional term in the potential energy function [41,50,72–74]. The resulting pseudo-energy function biases trajectories launched in conformational space away from conformational ensembles that, while low in energy, do not reproduce the NMR averages.

The NMR data are averages over an ensemble of molecules over time. While structures obtained in NMR is refined to agree with these averages, the

FIGURE 19.2 One hundred forty-four conformations computed in Reference [74] are superimposed in transparent over the first one (in opaque) of ubiquitin. The ensemble is obtained from the Protein Data Bank (PDB) under id 2nr2. Reproduced with permission of Michele Vendruscolo. (See color insert.)

refinement cannot capture the possibly nonlocal conformational heterogeneity present in solution. For this reason, methods that conduct a guided exploration of the conformational space are able to obtain a broader picture of the native state through an ensemble of conformations whose statistical averages reproduce the NMR observables better than a single structure. Figure 19.2 shows one such ensemble obtained for ubiquitin that reproduces the NMR data better than a single native structure.

The strategy of incorporating experimental data in the energy function is often described as a way to overcome possible nonphysical biases in the current generation of (semiempirical) molecular mechanics force fields [75]. The pseudo-energy function distorts the energy surface by deepening those low-energy regions that reproduce the experimental data. In this way, local optimization techniques have a better chance of converging to the funnel of the true energy surface of a protein.

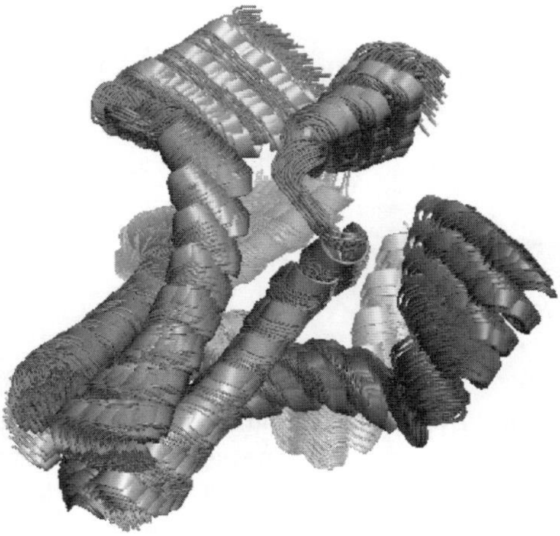

FIGURE 19.3 One hundred eighty-four phospholamban conformations (under id 2hyn in the PDB) computed in Reference [77] are shown superimposed over one another. The five monomers of the complex are shown in different colors. Courtesy of Chris Bailey-Kellogg. (See color insert.)

Rather than modify the underlying energy surface by enforcing agreement with experimental data, recent methods use the experimental data to either build probabilistic models of relevant regions of the conformational space or explicitly disqualify regions from further exploration [76,77]. In particular, the work in Reference [77] presents a complete method that subdivides the search space into regions worthy of further exploration and regions corresponding to structures in direct violation of NMR NOE distance constraints. A branch-and-bound search computes native structures of cyclic complexes such as the phospholamban protein shown in Figure 19.3.

19.5.2. Narrowing the Search with a Template Structure

Other methods elucidate structural details of the native state by searching with geometric or rigidity constraints. These constraints are often extracted from an average structure obtained in experiment [51,78–80]. By constraining their search around an experimental structure, these methods capture the conformational heterogeneity in proteins where flexibility under native conditions consists of local fluctuations around a representative structure. The representative structure is essentially employed as a semirigid template.

Work in References [79,80] is inspired from the constraint theory in the context of mechanical engineering considerations in bar and joint frameworks. Rigidity analysis over the template structure reveals under-constrained degrees of freedom (angles) at room temperature. These angles define a search space that is explored to obtain conformations that obey the rigidity constraints and exhibit as much internal mobility as allowed by the template [80].

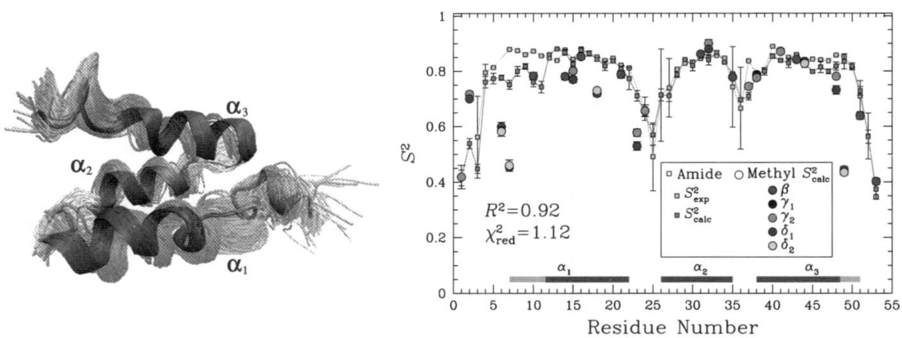

FIGURE 19.4 Left: The lowest-energy conformations computed with the method described in Reference [81] are drawn in transparent over the opaque X-ray structure of ALB8-GA. Right: Amide S^2_{calc} data (orange squares) calculated over the ensemble are compared with the available NMR S^2_{exp} data (yellow squares). Methyl S^2_{calc} data are shown in colored circles (no NMR data are available for comparison). Horizontal bars on the x-axis show the position of the three α-helices (also annotated over the ensemble). The parts of the bars in lighter colors indicate amino acids found in unfolded configurations. Reprinted from Reference [81] with permission. (See color insert.)

Other works, inspired by the treatment of inverse kinematics in robotics, conduct a geometrically constrained probabilistic sampling of the conformational space around the template structure [51,78,81]. Local fluctuations are obtained around the representative structure by computing geometrically constrained conformations of consecutive overlapping fixed-length fragments defined over the protein chain. Figure 19.4 shows fragment conformations superimposed over the reference structure the ALB8-GA protein. Thermodynamic averages calculated over the conformations reproduce data well obtained in experiment.

19.6. ENHANCED SAMPLING OF CONFORMATIONAL SPACE

Stochastic search is one of the strategies employed to enhance the sampling of the high-dimensional protein conformational space. Stochastic search (or stochastic optimization) is a powerful strategy to solve global optimization problems on surfaces marked by abundance of local minima [82]. When the energy surface is complex and decisions are made locally about which conformations map to minima in the energy surface, stochastic search becomes a viable strategy to explore the protein conformational space. For instance, work in References [51,64] employs a robotics-inspired probabilistic sampling to compute geometrically constrained conformations.

Other strategies enhance sampling in the context of a trajectory-based exploration by replicating trajectories, varying temperature (such as simulated annealing), and exchanging conformations from which trajectories are launched in conformational space. An incomplete list of successfully enhanced sampling strategies applied to searching the protein conformational space include importance sampling, simulated annealing, umbrella sampling, genetic algorithms, replica exchange (also known as parallel tempering), local elevation, activation relaxation, local energy flattening, jump walking, multicanonical ensemble, conformational flooding, Markov state models, discrete time step MD, fragment-based assembly (FA), and many more (cf. Reference [65]).

19.6.1. Principles of Self-Organization in Protein Chains

Analysis of conformational search methods identifies two main ingredients as essential for success: (1) a powerful sampling strategy to obtain a broad coverage of the conformational space and (2) an accurate energy function that allows a near-native conformation to converge to the nearby native basin. The design of accurate energy functions remains an active area of research and is pursued vigorously by many computational groups [9,10,55,57]. Energy functions provide a search method with a local view of the energy surface. This local view biases the search toward low-energy regions of the emerging energy surface. The energy function should not significantly distort the energy surface of an amino acid sequence under consideration.

Research shows that proteins have been designed by evolution to fold in spite of errors [83]. These findings advocate that a sampling strategy should not be highly sensitive to minor distortions of the energy surface and should be able to succeed as long as the energy function maps the explored conformational space on an energy surface that is funneled toward the native state. Significant computational efforts target the design of powerful sampling strategies to rapidly cover the conformational space [5,9]. Work in Reference [84] highlights that "the ultimate speed limit in protein folding is conformational search."

A good coverage of the conformational space should yield diverse conformations that are near energy minima relevant for the native state. Local optimization can then push near-native conformations to the native basin(s). It is worth mentioning that the notion of coverage is well defined and employed in the AI and robotics community [85]. The complexity and high dimensionality of the conformational space makes it very costly to estimate coverage [78]. Recent conformational search methods inspired in robotics are employing simple estimates of coverage to guide the search for the native state [86].

A powerful sampling strategy that does not incapacitate the predictive power of a method searching for the native state allows testing different hypotheses on how self-organization emerges in protein chains. Some of these hypotheses focus on determining the amount of detail that is necessary to capture the native state. For instance, work in Reference [87] suggests a backbone-based theory of protein folding. Results emerging from multiscaling studies of the protein native state advocate employment of different scales [38].

Growing evidence points toward a hierarchical organization of the native structure, where local interactions bias the local structure that emerges in protein chains. This in turn limits the number of ways low-energy conformations can be put together. This realization is not new and continues to emerge from various studies [8–10,12,13,54,88]. FA methods, the focus of the next section, employ this realization to efficiently compute conformations by assembling them from smaller local structures.

19.6.2. Local Structure Limits Global Arrangements

FA methods are emerging as successful ab initio methods in predicting the native state from knowledge of the amino acid sequence [46]. The basic process in FA methods is to assemble conformations of a protein chain with local structures of fragments of the chain. The assembly can be implemented either in the context of an MC-based [89] or MD-based search [8–10,12,42,86]. The sequence of the protein under consideration is divided into short fragments. Rarely, additional information is associated with the fragments from discretized Ramachandran maps of the backbone angles [10,12].

The key feature is that conformations of a protein chain are assembled from local structures of short fragments. Candidate local structures for the fragments are compiled from a nonredundant database of protein structures, often extracted from the Protein Data Bank (PDB). These local structures constitute a limited move set considered in an MC or MD framework to put together global tertiary structures (conformations) of a protein chain. The extent to which the limited move set reflects the naturally occurring biases that the fragment sequence has on local structure depends on the length of the fragment and the richness of the PDB. Employed fragment lengths range from 3 to 9. Deciding on a suitable fragment length depends on the richness of the PDB to provide a comprehensive picture of the extent to which the fragment sequence determines the structures in which the fragment can be found in nature.

By considering a limited set of structures for a protein fragment, FA methods discretize the relevant conformational space to be explored. Yet, the success of these methods does not lie in this superficial observation. FA methods implement the experimental observation that the local sequence (implemented in the definition of a fragment) biases but does not uniquely decide the local structure of the fragment (implemented in the sampling of structures of a fragment from a database). These local structural biases limit the number of ways low-energy conformations can be assembled together.

Recent work in Reference [45] improves on the classic FA framework by iteratively fixing the secondary structure assignments of amino acids during the generation of conformations in an MC simulated annealing search. The available search space of local fragment structures is progressively narrowed by "locking in" predominant secondary structure assignments that emerge during the search. Besides improving the efficiency of the search and the

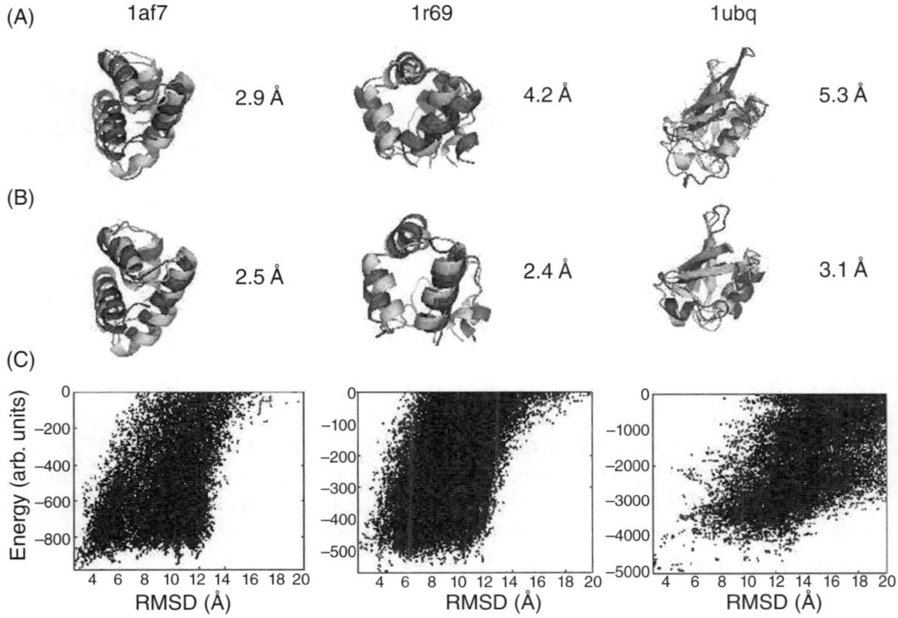

FIGURE 19.5 Alignments of the predictions generated in Reference [45] (in red) have (A) the lowest energy and (B) the lowest least root-mean-squared deviation (lRMSD) to the native structure (blue) for three proteins (PDB codes at the top, lRMSD values indicated). Images are created using the PyMol visualization software. (C) Scatter plots of lRMSD versus energy. Courtesy of Tobin Sosnick. (See color insert.)

accuracy of the resulting lowest-energy conformations (see Fig. 19.5), the method outperforms homology-based secondary structure prediction methods while using only a coarse-grained modeling with no explicit side chains. The success of this method may shed insight into the actual process of folding not just as a hierarchical process but a process that employs information on which secondary structures dominate a robust and efficient folding pathway.

Some studies suggest that biases on local structure, even when combined with nonspecific compaction forces (which promote compact conformations), are sufficient to result in a rapid sampling of native-like conformations of small proteins [46]. The extent to which the energy function determines the success of FA methods is under some debate [89]. Application to larger proteins with multiple competing conformational ensembles under native conditions suggests a larger role for both a sophisticated energy function and an enhanced sampling strategy in order to sufficiently populate possibly multiple energy minima relevant for the native state [42,89]. Figure 19.6 shows the energy landscape and competing conformational ensembles computed for calmodulin at room temperature with an FA-based MC simulated annealing [42].

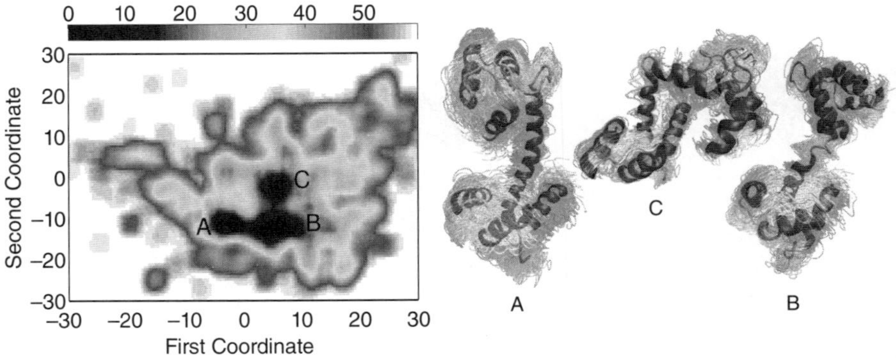

FIGURE 19.6 Left: The 2D energy pseudo-free energy landscape obtained for the calmodulin sequence with the method described in Reference [42] is shown in a red-to-blue color scheme that denotes high-to-low energy values. The deepest minima are labeled A, B, and C. Right: Computed conformational ensembles that correspond to the minima are shown and labeled accordingly. Conformations are superimposed in transparent over lowest-energy ones drawn in opaque. Reproduced from Reference [42] with permission. (See color insert.)

19.7. DISCUSSION OF FUTURE RESEARCH DIRECTIONS

Enhanced sampling on a simplified search space and sophisticated energy functions are allowing FA methods to achieve high prediction accuracy of the native state and, in the process, to shed light on the physical process of folding. While the prediction accuracy has improved both in proteins with high and low homology [9,45] and even in proteins with multiple functional states [42], research on improving the efficiency of FA methods is active. Time demands remain a point to address through further research.

One way to bring the computation of native-like conformations to a few hours on a single CPU is to discourage sampling of similar low-energy conformations. Currently, it seems difficult to ensure that computed conformations are geometrically distinct and not representative of few regions of the conformational space [42]. Part of the difficulty lies in the inability to find a few meaningful parameters on which to define distance measures. The classic measures like least root-mean-squared deviation (lRMSD) and radius of gyration (Rg) often mask away differences among conformations [42]. The robotics-inspired method in Reference [86] proposes a way to address this problem.

The tree-based method in Reference [86] advocates that conformational search be guided through low-dimensional projections of the conformational space and the energy surface. The projections afford a discretized view of the explored conformational space and its corresponding energy surface and allow defining a two-layered probability distribution by which to guide the search toward conformations that are both low energy and geometrically distinct.

While the projection coordinates employed in Reference [86] are not proposed as general reaction coordinates, the two-layered search may be an interesting framework through which to maximize the sampling of low-energy conformations that populate a desired conformational subspace.

The survey in this chapter has highlighted the complexity of computing conformations that populate the native state from minimal information such as amino acid sequence. Given that the prediction of native conformations is a stringent test of the ability of computers to fold protein sequences, research on effective and accurate conformational search for the protein native state will be active. Considering the interdisciplinary challenges that arise in the context of this problem, contributions will likely emerge from collaborations that reach across exact and life science communities of researchers.

The future holds promises for both communities: computer (and computational) scientists will be challenged and will learn how to mimic in silico the efficient steps that proteins apparently employ to fold within a few microseconds; physicists, biologists, and chemists will complement their experimental and theoretical understanding of the process of folding by testing diverse hypotheses in silico. Continued scientific progress is expected to result from discoveries of efficient search algorithms in computer science and further improvements in our understanding of protein physics.

REFERENCES

1. H. Wu. Studies on denaturation of proteins XIII. A theory of denaturation. *Chin J Physiol*, 5(4):321–344, 1931.
2. A.E. Mirsky and L. Pauling. On the structure of native, denatured and coagulated proteins. *Proc Natl Acad Sci U S A*, 22:439–447, 1936.
3. C.B. Anfinsen, E. Haber, M. Sela, and F.H. Jr. White. The kinetics of formation of native ribonuclease during oxidation of the reduced polypeptide chain. *Proc Natl Acad Sci U S A*, 47(9):1309–1314, 1961.
4. C.B. Anfinsen. Principles that govern the folding of protein chains. *Science*, 181(4096):223–230, 1973.
5. K.A. Dill, B. Ozkan, M.S. Shell, and T.R. Weikl. The protein folding problem. *Annu Rev Biophys*, 37:289–316, 2008.
6. Y. Zhang. Progress and challenges in protein structure prediction. *Curr Opin Struct Biol*, 18(3):342–348, 2008.
7. B. Kuhlman, G. Dantas, G.C. Ireton, G. Varani, B.L. Stoddard, and D. Baker. Design of a novel globular protein fold with atomic-level accuracy. *Science*, 302(5649):1364–1368, 2003.
8. G. Chikenji, Y. Fujitsuka, and S. Takada. A reversible fragment assembly method for de novo protein structure prediction. *J Chem Phys*, 119(13):6895–6903, 2003.
9. P. Bradley, K.M.S. Misura, and D. Baker. Toward high-resolution de novo structure prediction for small proteins. *Science*, 309(5742):1868–1871, 2005.

10. A. Colubri, A.K. Jha, M.-Y. Sheri, A. Sali, R.S. Berry, T.R. Sosnick, and K.F. Freed. Minimalist representations and the importance of nearest neighbor effects in protein folding simulations. *J Mol Biol*, 363(4):835–857, 2006.
11. S. Oldziej, C. Czaplewski, A. Liwo, M. Chinchio, M. Nanias, J.A. Vila, M. Khalili, Y.A. Arnautova, A. Jagielska, M. Makowski, H.D. Schafroth, R. Kazmierkiewicz, D.R. Ripoll, J. Pillardy, J.A. Saunders, Y.K. Kang, K.D. Gibson, and H.A. Scheraga. Physics-based protein-structure prediction using a hierarchical protocol based on the UNRES force field: Assessment in two blind tests. *Proc Natl Acad Sci U S A*, 102(21):7547–7552, 2005.
12. H. Gong, P.J. Fleming, and G.D. Rose. Building native protein conformations from highly approximate backbone torsion angles. *Proc Natl Acad Sci U S A*, 102(45):16227–16232, 2005.
13. S.B. Ozkan, G.H.A. Wu, J.D. Chodera, and K.A. Dill. Protein folding by zipping and assembly. *Proc Natl Acad Sci U S A*, 104(29):11987–11992, 2007.
14. Y. Zhang. Template-based modeling and free modeling by I-TASSER in CASP7. *Proteins*, 8(S1):108–117, 2007.
15. J. Moult, K. Fidelis, A. Kryshtafovych, B. Rost, T. Hubbard, and A. Tramontano. Critical assessment of methods of protein structure prediction (GASP) round VII. *Proteins*, 69(S8):3–9, 2007.
16. E. Schroedinger. *What is Life?* Cambridge, UK: Cambridge University Press, 1944.
17. R.P. Feynman, R.B. Leighton, and M. Sands. *The Feynman Lectures on Physics: Volume I*. Boston: Addison Wesley Longman, 1964.
18. A.J. Wand. Dynamic activation of protein function: A view emerging from NMR spetroscopy. *Nat Struct Biol*, 8(11):926–931, 2001.
19. G.A. III Palmer. NMR characterization of the dynamics of biomacromolecules. *Annu Rev Biophys Biomol Struct*, 104(8):3623–3640, 2004.
20. E.Z. Eisenmesser, O. Millet, W. Labeikovsky, D.M. Korzhnev, M. Wolf-Watz, D.A. Bosco, J.J. Skalicky, L.E. Kay, and D. Kern. Intrinsic dynamics of an enzyme underlies catalysis. *Nature*, 438(7064):117–121, 2005.
21. Y.P.J. Huang and G.T. Montellione. Structural biology: Proteins flex to function. *Nature*, 438(7064):36–37, 2005.
22. M. Karplus and J. Kuriyan. Molecular dynamics and protein function. *Proc Natl Acad Sci U S A*, 102(19):6679–6685, 2005.
23. V.J. Hilser, B. Garcia-Moreno, G.T. Oas, G. Kapp, and S.T. Whitten. A statistical thermodynamic model of the protein ensemble. *Chem Rev*, 106(5):1545–1558, 2006.
24. D.A. Hinds and M. Levitt. Exploring conformational space with a simple lattice model for protein structure. *J Mol Biol*, 243(4):668–682, 1994.
25. A. Kolinski and J. Skolnick. Monte Carlo simulations of protein folding. I. Lattice model and interaction scheme. *Proteins*, 18(4):338–352, 1994.
26. K. Ishikawa, K. Yue, and K.A. Dill. Predicting the structures of 18 peptides using Geocore. *Protein Sci*, 8(4):716–721, 1999.
27. B.H. Park and M. Levitt. The complexity and accuracy of discrete state models of protein structure. *J Mol Biol*, 249(2):493–507, 1995.
28. B.A. Reva, A.V. Finkelstein, M.F. Sanner, and A.J. Olson. Adjusting potential energy functions for lattice models of chain molecules. *Proteins*, 25(3):379–388, 1996.

29. C. Levinthal. Are there pathways for protein folding? *J Chem Phys*, 65(1):44–45, 1968.

30. K.F. Lau and A.K. Dill. A lattice statistical mechanics model of the conformational and sequence spaces of proteins. *Macromolecules*, 22(10):3986–3997, 1989.

31. R. Unger and J. Moult. Finding lowest free energy conformation of a protein is an NP-hard problem: Proof and implications. *Bull Math Biol*, 55(6):1183–1198, 1993.

32. W.E. Hart and S. Istrail. Robust proofs of NP-hardness for protein folding: General lattices and energy potentials. *J Comp Biol*, 4(1):1–22, 1997.

33. T. Lazaridis and M. Karplus. New view of protein folding reconciled with the old through multiple unfolding simulations. *Science*, 278(5345):1928–1931, 1997.

34. K.A. Dill and H.S. Chan. From Levinthal to pathways to funnels. *Nat Struct Biol*, 4(1):10–19, 1997.

35. J.N. Onuchic, Z. Luthey-Schulten, and P.G. Wolynes. Theory of protein folding: The energy landscape perspective. *Annu Rev Phys Chem*, 48:545–600, 1997.

36. O.M. Becker and M. Karplus. The topology of multidimensional potential energy surfaces: Theory and application to peptide structure and kinetics. *J Chem Phys*, 106(4):1495–1517, 1997.

37. M. Gruebele. Protein folding: The free energy surface. *Curr Opin Struct Biol*, 12(2):161–168, 2002.

38. C. Clementi. Coarse-grained models of protein folding: Toy-models or predictive tools? *Curr Opin Struct Biol*, 18(1):10–15, 2008.

39. G.R. Smith, M.J.E. Sternberg, and P.A. Bates. The relationship between the flexibility of proteins and their conformational states on forming protein-protein complexes with an application to protein-protein docking. *J Mol Biol*, 347(5):1077–1101, 2005.

40. V.J. Hilser, T. Oas, D. Dowdy, and E. Freire. The structural distribution of cooperative interactions in proteins: Analysis of the native state ensemble. *Proc Natl Acad Sci U S A*, 95(17):9903–9908, 1998.

41. K. Lindorff-Larsen, R.B. Best, M.A. DePristo, C.M. Dobson, and M. Vendruscolo. Simultaneous determination of protein structure and dynamics. *Nature*, 433(7022):128–132, 2005.

42. A. Shehu, L.E. Kavraki, and C. Clementi. Multiscale characterization of protein conformational ensembles. *Proteins*, 76(4):837–851, 2009.

43. D. Baker and D.A. Agard. Kinetics versus thermodynamics in protein folding. *Biochemistry*, 33(24):7505–7509, 1994.

44. S. Govindarajan and R.A. Goldstein. On the thermodynamic hypothesis of protein folding. *Proc Natl Acad Sci U S A*, 95(10):5545–5549, 1997.

45. J. DeBartolo, A. Colubri, A. Jha, J.E. Fitzgerald, K.F. Freed, and T.R. Sosnick. Mimicking the folding pathway to improve homology-free protein structure prediction. *Proc Natl Acad Sci U S A*, 106(10):3734–3739, 2009.

46. G. Chikenji, Y. Fujitsuka, and S. Takada. Shaping up the protein folding funnel by local interaction: lesson from a structure prediction study. *Proc Natl Acad Sci U S A*, 103(9):3141–3146, 2006.

47. P. Das, M. Moll, H. Stamati, L.E. Kavraki, and C. Clementi. Low-dimensional free energy landscapes of protein folding reactions by nonlinear dimensionality reduction. *Proc Natl Acad Sci U S A*, 103(26):9885–9890, 2006.

48. Z. Li and H.A. Scheraga. Monte Carlo-minimization approach to the multiple-minima problem in protein folding. *Proc Natl Acad Sci U S A*, 84(19):6611–6615, 1987.
49. R. Abagyan and M. Totrov. Biased probability Monte Carlo conformational searches and electrostatic calculations for peptides and proteins. *J Mol Biol*, 235(3):983–1002, 1994.
50. R.B. Best and M. Vendruscolo. Determination of ensembles of structures consistent with NMR order parameters. *J Am Chem Soc*, 126(26):8090–8091, 2004.
51. A. Shehu, C. Clementi, and L.E. Kavraki. Modeling protein conformational ensembles: From missing loops to equilibrium fluctuations. *Proteins*, 65(1):164–179, 2006.
52. H. Taketomi, Y. Ueda, and N. Go. Studies on protein folding, unfolding and fluctuations by computer simulation: The effect of specific amino acid sequence represented by specific inter-unit interactions. *Int J Pept Protein Res*, 7(6):445–459, 1975.
53. K. Yue, K.M. Fiebig, P.D. Thomas, H.S. Chan, E.I. Shakhnovich, and K.A. Dill. A test of lattice protein folding algorithms. *Proc Natl Acad Sci U S A*, 92(1):325–329, 1995.
54. T.H. Hoang, A. Trovato, F. Seno, J.R. Banavar, and A. Maritan. Geometry and symmetry presculpt the free-energy landscape of proteins. *Proc Natl Acad Sci U S A*, 101(21):7960–7964, 2007.
55. G.A. Papoian, J. Ulander, M.P. Eastwood, Z. Luthey-Schulten, and P.G. Wolynes. Water in protein structure prediction. *Proc Natl Acad Sci U S A*, 101(10):3352–3357, 2004.
56. S. Matysiak and C. Clementi. Optimal combination of theory and experiment for the characterization of the protein folding landscape of S6: How far can a minimalist model go? *J Mol Biol*, 343(8):235–248, 2004.
57. P. Das, S. Matysiak, and C. Clementi. Balancing energy and entropy: A minimalist model for the characterization of protein folding landscapes. *Proc Natl Acad Sci U S A*, 102(29):10141–10146, 2005.
58. S. Matysiak and C. Clementi. Minimalist protein model as a diagnostic tool for misfolding and aggregation. *J Mol Biol*, 363(1):297–308, 2006.
59. C. Hardin, Z. Luthey-Schulten, and P.G. Wolynes. Backbone dynamics, fast folding, and secondary structure formation in helical proteins and peptides. *Proteins*, 34(3):281–294, 1999.
60. W. Kwak and U.H. Hansmann. Efficient sampling of protein structures by model hopping. *Phys Rev Lett*, 95(13):138102, 2005.
61. E. Lyman, F.M. Ytreberg, and D.M. Zuckermann. Resolution exchange simulations. *Phys Rev Lett*, 96(2):028105, 2006.
62. S. Christakos, C. Gabrielides, and W.B. Rhoten. Multigraining: An algorithm for simultaneous fine-grained and coarse-grained simulation of molecular systems. *J Chem Phys*, 125(15):154106, 2006.
63. A.P. Heath, L.F. Kavraki, and C. Clementi. From coarse-grain to all-atom: Towards multiscale analysis of protein landscapes. *Proteins*, 68(3):646–661, 2007.
64. A. Shehu, L.E. Kavraki, and C. Clementi. Unfolding the fold of cyclic cysteine-rich peptides. *Protein Sci*, 17(3):482–493, 2008.

65. W.F. van Gunsteren, D. Bakowies, R. Baron, I. Chandrasekhar, M. Christen, X. Daura, P. Gee, D.P. Geerke, A. Glättli, P.H. Hünenberger, M.A. Kastenholz, C. Oostenbrink, M. Schenk, D. Trzesniak, N.F. van der Vegt, and H.B. Yu. Biomolecular modeling: Goals, problems, perspectives. *Angew Chem Int Ed Engl*, 45(25):4064–4092, 2006.

66. Y. Duan and Y.A. Kollman. Pathways to a protein folding intermediate observed in a 1-μs simulation in aqueous solution. *Science*, 282(5389):740–744, 1998.

67. M. Shirts and V.J. Pande. Computing: Screen savers of the world unite! *Science*, 290(5498):1903–1904, 2000.

68. J.W. Pitera and W. Swope. Understanding folding and design: Replica-exchange simulations of Trp-cage miniproteins. *Proc Natl Acad Sci U S A*, 100(13):7587–7592, 2003.

69. D.E. Shaw et al. Anton, a special-purpose machine for molecular dynamics simulation. *Commun ACM*, 51(7):91–97, 2008.

70. D.R. Ripoll, J.A. Vila, and H.A. Scheraga. Folding of the villin headpiece subdomain from random structures. Analysis of the charge distribution as a function of pH. *J Mol Biol*, 339(4):915–925, 2004.

71. N. Metropolis, A.W. Rosenbluth, M.N. Rosenbluth, A.H. Teller, and E. Teller. Equation of state calculations by fast computing machines. *J Chem Phys*, 21(6):1087–1092, 1953.

72. G.M. Clore and C.D. Schwieters. How much backbone motion in ubiquitin is required to account for dipolar coupling data measured in multiple alignment media as assessed by independent cross-validation? *J Am Chem Soc*, 126(9):2923–2938, 2004.

73. G.M. Clore and C.D. Schwieters. Concordance of residual dipolar couplings, backbone order parameters and crystallographic B-factors for a small α/β protein: A unified picture of high probability, fast atomic motions in proteins. *J Mol Biol*, 355(5):879–886, 2006.

74. B. Richter, J. Gsponer, P. Várnai, X. Salvatella, and M. Vendruscolo. The MUMO (minimal under-restraining minimal over-restraining) method for the determination of native state ensembles of proteins. *J Biomol NMR*, 37(2):117–135, 2007.

75. V. Hornak, R. Abel, A. Okur, B. Strockbine, A. Roitberg, and C. Simmerling. Comparison of multiple amber force fields and development of improved protein backbone parameters. *Proteins*, 65(3):712–725, 2006.

76. F. DiMao, D. Kondrashov, E. Bitto, A. Soni, G. Bingman, G. Phillips, and J. Shavlik. Creating protein models from electron-density maps using particle-filtering methods. *Bioinformatics*, 23(21):2851–2858, 2007.

77. S. Potluri, A.K. Yarr, J.J. Chou, B.R. Donald, and C. Bailey-Kellogg. Structure determination of symmetric homo-oligomers by a complete search of symmetry configuration space, using NMR restraints and van der Waals packing. *Proteins*, 65(1):203–219, 2006.

78. A. Shehu, C. Clementi, and L.E. Kavraki. Sampling conformation space to model equilibrium fluctuations in proteins. *Algorithmica*, 48(4):303–327, 2007.

79. S. Wells, S. Menor, B. Hespenheide, and M.F. Thorpe. Constrained geometric simulation of diffusive motion in proteins. *J Phys Biol*, 2(4):127–136, 2005.

80. M. Chubunsky, B. Hespenheide, D.J. Jacobs, L.A. Kuhn, M. Lei, S. Menor, A.J. Rader, M.F. Thorpe, W. Whiteley, and M.I. Zadovsky. Constraint theory applied to proteins. *Nanotechnol Res J*, 2(1):61–72, 2008.

81. A. Shehu, L.E. Kavraki, and C. Clementi. On the characterization of protein native state ensembles. *Biophys J*, 92(5):1503–1511, 2007.

82. K.A. De Jong. *Evolutionary Computation: A Unified Approach*. Cambridge, MA: MIT Press, 2006.

83. M.C. Prentiss, C. Hardin, M.P. Eastwood, C. Zong, and P.G. Wolynes. Protein structure prediction: The next generation. *J Chem Theory Comput*, 2(3):705–716, 2006.

84. K. Ghosh, S.B. Ozkan, and K.A. Dill. The ultimate speed limit to protein folding is conformational searching. *J Am Chem Soc*, 129(39):11920–11927, 2007.

85. H. Choset, K.M. Lynch, S. Hutchinson, G. Kantor, W. Burgard, L.E. Kavraki, and S. Thrun. *Principles of Robot Motion: Theory, Algorithms, and Implementations*, 1st edition. Cambridge, MA: MIT Press, 2005.

86. A. Shehu. An Ab-initio tree-based exploration to enhance sampling of low-energy protein conformations. In *Proceedings of Robotics: Science and Systems*, J. Trinkle, Y. Matsuoka, and J.A. Castellanos (Eds.), pp. 31–39. Seattle, WA, June, 2009.

87. G.D. Rose, P.J. Fleming, J.R. Banavar, and A. Maritan. A backbone-based theory of protein folding. *Proc Natl Acad Sci U S A*, 103(45):16623–16633, 2006.

88. N. Haspel, C.J. Tsai, H. Wolfson, and R. Nussinov. Hierarchical protein folding pathways: A computational study of protein fragments. *Proteins*, 51(2):203–215, 2003.

89. J. Hegler, J. Laetzer, A. Shehu, C. Clementi, and P.G. Wolynes. Restriction vs. guidance: Fragment assembly and associative memory Hamiltonians for protein structure prediction. *Proc Natl Acad Sci U S A*, 106(36):15302–15307, 2009.

CHAPTER 20

MODELING MUTATIONS IN PROTEINS USING MEDUSA AND DISCRETE MOLECULE DYNAMICS

SHUANGYE YIN, FENG DING, and NIKOLAY V. DOKHOLYAN
Department of Biochemistry and Biophysics
University of North Carolina
Chapel Hill, NC

20.1. INTRODUCTION

The advance of high-throughput sequencing techniques [1] and genomic-wide screening [2] has accelerated the rate of discovering mutations associated with human diseases. In search of medical cure for these diseases, the understanding of the functional and structural roles of the mutant protein is of vital importance, especially for structure-based drug design [3]. In addition, mutations are often introduced to proteins during site-directed mutagenesis experiments designed to probe the mechanism of enzymes, protein–protein, and protein–ligand interactions [4,5]. Structural model of the engineered proteins can significantly help design these experiments. Despite recent developments of structural determination techniques and structural genomic efforts, the number of solved protein structures is still much smaller than the number of known gene sequences. For example, there are only a few G-protein-coupled receptor (GPCR) structures solved to date, even though they represent about half of the targets of modern clinical drugs [6,7]. Therefore, there is a crucial need for novel methods to determine structures of mutant proteins.

The deficiency of experimental determination of mutant proteins can be partly compensated by computational modeling. Although *ab initio* prediction of protein structure [8–10] is still a formidable task, the tertiary structure can

Introduction to Protein Structure Prediction: Methods and Algorithms,
Edited by Huzefa Rangwala and George Karypis

now be predicted with high fidelity if the protein has a homolog with solved tertiary structure [8]. The foundation of this comparative modeling (or homology modeling) approach [11,12] is the observation that homologous proteins often adapt a similar tertiary fold. Therefore, by sequence alignment with its homologous proteins, the approximate tertiary structural model of a protein can be built. Further optimization is subsequently applied to the model to determine the conformation of unaligned amino acids and to optimize the overall conformation of the protein. Comparative modeling methods are especially useful if the sequence similarity is high (>30%) between the template structure and the target protein. In modeling these mutant proteins, the mutated residues (including substitution mutation, addition, and deletion) are sparse and likely do not change the overall structure except for the residues in the vicinity of mutation sites. The root-mean-square deviation (RMSD) of the backbone atoms from the experimental structure is often less than 1 Å if the sequence identity between the protein and the template is more than 50% [8]. For modeling these highly homologous proteins, the main challenge is to generate models that are accurate in atomic resolution, which is essential for structure-based drug design [3,13,14] or protein design [15–19].

We model the substitution mutations using our recently developed Medusa protein design suite [20]. For substitution mutations, we assume that the protein backbones only vary slightly from the template. The major task is to determine the side chain conformation of the mutated residue and nearby residues. In Medusa, the side chain conformations are efficiently sampled by Monte Carlo (MC) simulations using backbone-dependent rotamers [21]. The utilization of rotamer libraries not only provides accurate description of intra-residue energies but also allows fast sampling of protein side chain conformation. The quality of the side chain packing is evaluated with the Medusa force field [20]. The Medusa force field is an atomistic force field that combines physical, empirical, and statistical potentials. It describes the crucial interactions within proteins, including van der Waals, solvation, hydrogen bonding, and intra-residue rotamer energies. We perform multiple simulated annealing calculations in Medusa and use the lowest energy side chain packing conformations as the putative mutant structure.

For mutations involving insertion and deletion of amino acids, a larger backbone perturbation is expected. In this case, we first create the mutation without considering actual physical interactions and then use discrete molecular dynamics (DMD) simulations [22–24] to relax the structure. In contract to traditional molecular dynamics, we do not rely on fine-tuned physical force field parameters in DMD simulations. Instead, we mimic natural interactions by assigning a simplified stepwise interaction potential between atoms. For a stepwise potential, the derivatives of the potential are either zero (acceleration is zero), or infinity (collision occurs with instantaneous velocity change). Unlike molecular mechanics simulations that are driven by physical forces, DMD simulations are driven by collision events due to the discrete nature of interaction potential. Thus, instead of solving the equation of motion with small, discretized time steps (usually femtoseconds in molecular dynamics [MD] simulations), DMD drives the system by collision events where two atoms collide with an instantaneous potential energy change. As a result, the DMD algorithm gains efficiency over traditional MD in two ways: (1) on average a larger time step and (2) a faster searching/updating algorithm since only the colliding atoms needs to be updated at each collision. Although the actual speed up of DMD compared with traditional MD varies from system to system, it can reach 3–10 orders of magnitude.

Another advantage of DMD is its versatility. In DMD, constraints can be assigned between any atom pairs and the interaction potential can be arbitrary functions of pairwise distance. For example, protein insertion and deletion often features drastic backbone changes near the mutation site. Such backbone changes cannot be efficiently modeled using Medusa. Using DMD, with its high computational efficiency and flexibility, insertion and deletion within each protein can be modeled. To make sure that the structural relaxation is localized, we also constrain the structure far from the site of interest. We found that DMD simulation allows us to find the relaxed structure with proper backbone peptide geometries through rapid optimization in simulated annealing simulations.

A closely related question for modeling protein mutations is how much the protein's stability will change upon mutations. A protein's thermodynamic stability, measured as the free energy difference between the folded and unfolded states ($\Delta G = G^{\text{fold}} - G^{\text{unfold}}$), determines the fraction of the protein in folded states, and thus, the biological activity of the protein. It is known that most proteins are marginally stable [9] so that introduction of a single mutation can drastically alter the stability and affect the activities of the protein. There are attempts to use physical force field and extensive MD simulations in explicit solvents to calculate such stability changes [25,26]. Although $\Delta\Delta G$ values have been reproduced with 1 kcal/mol accuracy, such methods are, in general, too computationally expensive to be applied to large numbers of mutations or large proteins. Other methods have therefore been developed that feature empirical and implicit solvent models to estimate $\Delta\Delta G$ and do not require MD simulations [27–32]. There are also approaches that use machine learning techniques [28], taking advantage of a vast amount of protein stability data generated from protein engineering experiments [33]. However, many heuristic methods often do not explicitly model the atomic structure of each amino acid, so they are inapplicable in structural modeling of mutant proteins.

We estimate the $\Delta\Delta G$ of mutations using a modified Eris protocol [32,34]. In Eris, $\Delta\Delta G$ values are estimated by comparing the Medusa energies of the mutation and native structures. Due to the stochastic nature of the MC algorithm, we generate 20 structures for both native and mutant proteins, and the Medusa energies are averaged over the 20 structures. We can also use an optional prerelaxation step to optimize the backbone conformations of the input structure. Such prerelaxation improves $\Delta\Delta G$ prediction when high-resolution structures are not available. We have demonstrated the performance of our $\Delta\Delta G$ estimation method by applying our protocols to a large set of 595 protein mutations belonging to five different proteins families. The estimated $\Delta\Delta G$ values of those mutations agree well with experiments. We also find that incorporating backbone flexibility significantly improves the prediction accuracy if clashes are introduced by mutations, for example, by mutating a buried small residue into a larger residue. The Eris $\Delta\Delta G$ estimation method is available online at http://eris.dokhlab.org.

20.2. METHODS

20.2.1. Protocol Overview

We model 31 mutants of the T4 lysozyme protein. The wild-type protein contains 162 residues, and the number of amino acid substitutions, deletions, and insertions in these mutants varies from 1 to 10. We use the following protocols to model the mutant proteins:

- For mutants that involve insertion or deletion of amino acids, we first perform DMD simulations to reconstruct the backbone conformations near the mutation sites. Then, we use Medusa to optimize the side chain packing.
- For mutants that have only amino acid substitution, we keep the backbone fixed and directly use Medusa to repack the side chains near the mutation sites.

For side chain repacking, we use our in-house Medusa method rather than other side chain repacking algorithms [35–37] since we have shown that Medusa is able to accurately recapitulate the naturally occurring amino acid sequence and side chain packing [20]. Most importantly, using Medusa for side chain repacking after DMD backbone reconstruction ensures the best compatibility of protein models after DMD sampling and Medusa side chain repacking since the interaction potential in all-atom DMD protein models is essentially a discretized version of the Medusa force field [38].

20.2.2. DMD Algorithm

DMD force field is based on pairwise spherically symmetrical potentials that are discontinuous functions of the interatomic distance r. Each atom has a specific type—A, B, C, ...—that determines its interaction with other atoms. Each type is characterized by its mass m. The interaction potential between atoms A and B is a step function of their distance r, characterized by distances $0 < r_{\min}^{AB} < r_1^{AB} \ldots < r_{\max}^{AB}$. If the distance r between two atoms A and B satisfies the inequality $r_i^{AB} < r < r_{i+1}^{AB}$, the pair potential has a value of U_i^{AB}. If $r < r_{\min}^{AB}$, $U^{AB} = \infty$ and $r_{\min}^{AB}$ is the hardcore collision distance. If $r > r_{\max}^{AB}$, $U^{AB} = 0$ and $r_{\max}^{AB}$ is the maximal range of the A–B interaction. If atoms A and B are linked by a covalent bond, they interact according to a different potential characterized by values $\tilde{r}_{\max}^{AB}$ and $\tilde{U}_{\max}^{AB}$. In this case, if $r > \tilde{r}_{\max}^{AB}$, $\tilde{U}_{\max}^{AB} = \infty$, which indicates that the bond is permanent and cannot be broken under any conditions.

In DMD, all atoms move with a constant velocity unless their distance becomes equal to r_i^{AB}. At this moment of time, their velocities change instantaneously. This change satisfies the laws of energy, momentum, and angular momentum conservation. When the kinetic energy of the particles is not sufficient to overcome the potential change $e_i^{AB} = \pm\left(U_{i-1}^{AB} - U_i^{AB}\right)$, the atoms undergo a hardcore reflection with no potential energy change. The ± sign of the potential energy change depends on the relative velocity between the two particles. The main challenge of this method is the effective storage, sorting, and update of the collision times. However, it is possible to make the speed of the algorithm inversely proportional to $N \ln N$, where N is the total number of atoms. For a sufficiently large number of steps in the energy potentials, the method becomes equivalent to a regular MD based on Newtonian dynamics. In order to effectively simulate the collisions, the system is divided into cells. The dimension of the cell is assigned to be the largest interaction range of all the atom pairs. Thus, all possible interacted atoms of specific atoms are within the neighbor cells $n_{\text{cell}} = 3^d$ where d is the dimensionality of the simulation system. In addition, migration of different atoms from one cell into another cell has to be included as additional collision events. In order to determine the soonest collision time t_i for atom i, the calculation only needs to be done between this atom and the atoms in the neighbor 3^d cells. Then, the smallest t_i will be the soonest collision of the system.

Since each atom moves with constant velocity in between collisions, it is the state—the position of previous collision and the velocity, as well as the time of previous collision—that has to be kept track of. The DMD simulation maintains a set of all possible collisions, namely a collision table, and determines the soonest collision. Once the soonest collision is determined between q and p after time Δt, the states of atoms p and q will be updated accordingly by satisfying the conservation of energy and momentum. If an event is cell crossing, the old collisions between this atom and atoms in the non-neighboring cells are removed from the table, and collisions with respect to atoms in the newly emerged cells are incorporated into the table. The system time is proceeded by Δt. Then, all the outdated collision events related to p and q will be updated with respect to the new position and velocity. Then, a search in the partially updated collision table is performed to find the next soonest collision. Therefore, during each collision event, only the evolved atom pairs need to be updated to keep track of their new state, and the rest of the system need not be updated. To facilitate the search for the soonest collision, a priority queue is used.

20.2.3. All-Atom DMD

We have developed a series of protein models for DMD simulations ranging from one-bead per amino acid to all-atom models. The coarse-grained protein models have the advantage of efficient sampling of conformational spaces with the price of reduced protein resolution. In order to study the atomistic conformations changes of proteins upon deletion and insertions, the all-atom protein model is required. We have developed the all-atom protein model for DMD simulations [38]. The all-atom DMD method employs a united atom protein model, where heavy atoms and polar hydrogen atoms are explicitly modeled. We include van der Waals, solvation, and environment-dependent hydrogen bond interactions. We adopt the Lazaridis–Karplus solvation model [39] and use the fully solvated conformation as the reference state. Due to the strong screening effect of solvent, distant charges have weak polar interactions. For salt bridges, we expect the hydrogen bonds to partially account for their polar interactions. Similar neutralization of charged residues was also employed in the implicit solvent model of the effective energy function of CHARMM19 [39,40]. DMD has been shown to have a higher sampling efficiency than traditional MD and has been used to study protein folding thermodynamics and protein aggregation [41,42]. All-atom DMD features a transferable force field and has been successfully used to fold several small proteins *ab initio* (<60 amino acids [38,43]).

20.2.4. Insertion and Deletion in DMD

To model the deletion in a protein, we remove deleted residues and assign constraints between the two flanking residues to ensure that planar peptide bond constraints are not violated. Similarly, we model the insertion by carefully assigning constraint potentials between inserted regions and the corresponding residues in the original protein. The inserted residues are initially set in an extended conformation. For the secondary structures adjacent to the inserted/deleted residues, we allow full conformational flexibility during DMD simulations. For the rest of the protein, we assign harmonic constraints to minimize the deviation from their original conformations while allowing them to relax.

20.2.5. Medusa Force Field

We use the Medusa force field to describe the physical interactions within proteins [20]. The Medusa force field uses a combination of van der Waals, solvation, hydrogen bonding, and rotameric statistical energies to approximate the interactions within a protein. The protein in the unfolded state is modeled as a fully solvated chain so that the free energy of the unfolded state can be approximated as a linear sum of reference energies that are specific for each type of amino acid. More specifically, the free energy of the protein is expressed as,

$$\Delta G = W_{vdw\_attr}E_{vdw\_attr} + W_{vdw\_rep}E_{vdw\_rep} + W_{solv}E_{solv} + W_{bb\_hbond}E_{bb\_hbond} + W_{sc\_hbond}E_{sc\_hbond} + W_{bb\_sc\_hbond}E_{bb\_sc\_hbond} + W_{aa|\phi,\psi}E_{aa|\phi,\psi} + W_{rot|\phi,\psi,aa}E_{rot|\phi,\psi,aa} - E_{ref}. \quad (20.1)$$

Here E_{vdw_attr} and E_{vdw_rep} are the attractive and repulsive parts of the van der Waals energy, respectively; E_{solv} is the solvation free energy; E_{bb_hbond}, E_{sc_hbond}, and $E_{bb\_sc\_hbond}$ are the hydrogen bonding energies between backbone atoms, between side chain atoms, and between backbone and side chain atoms; $E_{aa|\phi,\psi}$ is the statistical energy of the amino acid for any given backbone dihedral angles ϕ and ψ; $E_{rot|\phi,\psi,aa}$ is the statistical energy for the side chain rotamer for any given amino acid and backbone dihedral angles; E_{ref} is the sum of reference energies for all amino acids; W_{vdw_attr}, W_{vdw_rep}, W_{solv}, W_{bb_hbond}, W_{sc_hbond}, $W_{bb\_sc\_hbond}$, $W_{aa|\phi,\psi}$, and $W_{rot|\phi,\psi,aa}$ are factors to weigh the different energy terms.

The parameters for the energy terms are taken from different sources. The CHARMM19 force field parameters [39,40] are used for the van der Waals interaction. The EEF1 implicit solvent model developed by Lazaridis and Karplus [39] is adapted as solvation energy. The hydrogen bonding interaction is calculated using a statistical model developed by Kortemme et al. [44]. The rotameric statistical energies are derived from the Dunbrack backbone-dependent rotamer library [21]. All these energy terms are summed with a weighing factor to balance the contributions. We obtain those weighing factors as well as the 20 reference energies by training over a dataset of 34 high-resolution protein structures so that the native amino acid sequences tend to have the lowest energies.

20.2.6. MC Side Chain Repacking

Side chain conformations are optimized using an MC simulated annealing algorithm. In the algorithm, the rotameric states of the side chains are first randomized at high temperature. Next, a series of MC simulations are performed to randomly perturb the side chain conformations and accept or reject the perturbation based on the energy difference and the simulation temperature (Metropolis criterion [45]). To minimize the energy, we gradually reduced the simulation temperatures in 20 stages, starting at 10 kcal/R and ending at 0.1 kcal/R, where R is the ideal gas constant. As the temperature gradually decreases, the optimal packing of the side chains is obtained at the end of the simulation. The whole MC simulated annealing is repeated 20 times to increase the sampling space, and the output structure with the lowest Medusa total energy is selected as the final predicted model.

To reduce the conformational space for sampling, we only allow the most favorable rotamers of the amino acid (probability >0.01 for the given backbone conformation) for each site. The native rotamer is used if no rotamer is available within the probability cutoff. We also enable off-rotamer sampling by allowing small deviations from the average dihedral angles during each MC step.

Similar to DMD, grid techniques are used for pairwise energy evaluation so that only atoms located in cells within the interaction range are searched. To further improve the calculation speed, the total energy is expressed as inter-residue interactions so that only one residue needs energy reevaluation after each trial step.

20.2.7. Flexible Backbone Minimization

During the side chain repacking simulation, we can also incorporate backbone movement when steric clashes are detected. We identify the clashes by monitoring the acceptance rate during the MC simulation. If the acceptance is less than 0.05 for an MC stage, it indicates that there are a large number of unresolved atomic clashes in the current backbone conformation. In this case, the backbone dihedral angles are optimized using a conjugated gradient minimization algorithm at the end of the MC stage. The next MC simulation stage will start from the backbone-relaxed structure. The application of this algorithm on a benchmark set demonstrates that the flexible backbone minimization algorithm can further optimize the total Medusa energy for the structure. Compared with a fixed-backbone algorithm, the total energy is lowered by approximately 10–20 kcal/mol. This minimization is especially useful if the backbone conformation is of low resolution (e.g., from a structure determined using nuclear magnetic resonance [NMR] spectroscopy) [34].

20.2.8. Protein Stability Change

The $\Delta\Delta G$ is calculated by the energy difference between the wild-type and mutant proteins using $\Delta\Delta G = \Delta G^{mut} - \Delta G^{wt}$, where ΔG^{mut} and ΔG^{wt} are the Medusa energy calculated using Equation 20.1 on mutant and wild-type proteins, respectively. Since MC simulated annealing is a nondeterministic algorithm, it gives slightly different results with different initial conditions. To reduce the uncertainty from the MC simulations, we perform the minimization 20 times and average the energies over the 20 output structures. The structural minimizations are performed for both wild-type and mutant proteins at the same condition (temperatures and constraints) so that only the changes related to mutations are captured in the energy differences.

20.2.9. Prerelaxation

High-resolution input structures are not always available, so we include a prerelaxation step before the $\Delta\Delta G$ evaluation. The main purpose of this step is to optimize the backbone conformation of the wild-type protein before $\Delta\Delta G$ calculation. In the prerelaxation step, the flexible backbone minimization is applied 20 times to the input wild-type protein structure, and the output structure with the lowest energy is selected as the input structure for further $\Delta\Delta G$ calculation. The structure prerelaxation is especially useful if the only available wild-type structure is from the NMR spectroscopy or from computational modeling. Our previous benchmark results show that applying the prerelaxation on the NMR structure of chicken alpha-spectrin protein significantly improves the $\Delta\Delta G$ prediction accuracy [34].

20.2.10. Cutoff Distance

Repacking the side chain conformation can be challenging for large proteins because the size of the conformational space grows exponentially with the number of residues. Even with multiple MC simulations, the global minimum energy state may not be reached during sampling, resulting in errors for energy evaluation. To avoid insufficient sampling and to accelerate the simulation speed, we restrain the side chain movements only to the neighboring residues of the mutation sites. We assume that only residues in the proximity to the mutation sites may undergo conformational changes upon mutation. Such assumption is consistent with the current knowledge of the structural effects of protein mutations [46,47]. For the T4 lysozyme protein mutations, we allow every residue that is within 5 Å from the mutation site to be freely movable (all rotamer conformations are allowed) and allow residues that are within 10 Å from the mutation site to be slightly movable near the native rotamer state (only off-rotamer conformations within the native rotamer are allowed). Here, the distance between two residues is measured as the smallest distance between any two pairs of heavy atoms from each residue.

20.2.11. Mutation Set

We apply the above protocols on the set of mutations from the CASPM experiments (Critical Assessment of Techniques for Protein Structure Prediction—Continuous CASP Experiment on Mutants Modeling; see http://predictioncenter.gc.ucdavis.edu/caspm1). There are 31 mutations in the dataset, all for T4 lysozyme. The dataset includes 22 substitution mutations and 9 insertion/deletion mutations. All the tertiary structure and $\Delta\Delta G$ of the mutants have been experimentally determined. Comparing the mutant and wild-type protein structures reveals that there are 11 mutations with small structure rearrangement compared with wild-type proteins, 14 with medium changes, and 6 with relatively large changes. Therefore, the dataset is ideal for assessing the current methods' capability of modeling various types of mutations. To avoid possible complication resulting from disulfide bonding, a pseudo wild-type cysteine-less construct containing C54T and C97A mutations (PDB ID 1L63) is used as the reference structure during the structure modeling.

20.3. RESULTS AND DISCUSSION

We illustrate our modeling methods by applying our protocols to the 31 CASPM mutants and compare the predicted structure with the crystal

TABLE 20.1 List of the CASPM Targets and the Overall Accuracy of Our Models

Target	Mutation	PDBID	RMSD-Cα	RMSD-ALL	RMSD-Cα Selected[a]	RMSD-ALL Selected[a]
TM001	M6L	230L	0.17	0.71	0.20	1.03
TM002	I29A	241L	0.23	0.69	0.31	1.35
TM003	L32T/T34K/K35V/ S36D/P37G/ S38N/L39S	176L	0.71	1.32	1.04	1.65
TM004	N40-[A]	102L	0.23	0.34	0.40	0.70
TM005	A42S	206L	0.11	0.29	0.14	0.36
TM006	A42V	1QTB	0.25	0.61	0.46	0.87
TM007	S44-[AAA]	205L	0.45	0.60	0.37	0.62
TM008	L46A	1L67	0.10	0.21	0.14	0.44
TM009	K48-[HP]	201L	0.81	1.28	1.29	2.17
TM010	I58A	243L	0.13	0.43	0.19	0.68
TM011	I58T	1G1V	0.13	0.42	0.15	0.68
TM012	D72P	1L76	0.46	0.66	0.34	0.58
TM013	A73Δ	210L	0.48	0.84	0.41	1.40
TM014	I78V	1P2R	0.24	0.63	0.20	1.08
TM015	I78V/V87M/ M120Y/L133F/ V149I/T152V	1PQI	0.32	0.78	0.34	0.90
TM016	L84M/V87M/ L91M/L99M/ V111M/L118M/ L121M/L133M	1LWG	0.77	1.25	0.79	1.50
TM017	V87I/I100V/ M102L/V103I/ M106I/V111A/ M120Y/L133F/ V149I/T152V	1PQD	0.47	0.95	0.52	1.11
TM018	A98M	1QTH	1.08	1.46	0.43	0.79
TM019	L99A	1L90	0.10	0.29	0.10	0.26
TM020	L99A/E108V	1QUO	0.15	0.64	0.16	0.51
TM021	L99G	1QUD	0.64	0.76	0.77	0.92
TM022	L99G/E108V	1QUH	0.22	0.64	0.28	0.66
TM023	I100A	244L	0.16	0.54	0.19	0.77
TM024	E108-[A]	211L	0.22	0.38	0.48	0.87
TM025	T115-[A]	215L	0.56	0.68	1.31	1.59
TM026	S117F	1TLA	0.22	0.46	0.22	0.90
TM027	R119-[A]	214L	2.33	3.05	3.82	5.71
TM028	L121A/A129M	199L	0.20	0.64	0.30	1.09
TM029	A129L	195L	0.21	0.74	0.29	1.76
TM030	V131-[A]	218L	2.25	2.94	2.64	4.23
TM032	V149A	237L	0.11	0.65	0.11	0.48

The data are extracted from the CASPM Web site (http://predictioncenter.gc.ucdavis.edu/caspm1), where result for other predictors can also be found.

[a]For selected atoms near the mutation sites.

structure for both backbone conformation and side chain rotamers. The list of the mutants and the overall accuracy of our models are shown in Table 20.1. In general, we find significant agreement with the experiments for substitution mutations. But for the insertion/deletion mutations, the deviation is relatively large. In the following section, we select several representative examples from the CASPM targets and compare the model with the known crystal structure, from which we infer the advantage and limitations of our methods.

20.3.1. Substitution Mutations

One successful example of prediction for mutation I58T is shown in Figure 20.1. The predicted side chain of the mutated threonine (Thr58) adapts the same rotamer as in the crystal structure. In addition, all the neighboring side chains have been faithfully reproduced except Lys16 and Val57. The deviation for Lys16 is not surprising because this residue is exposed to solvent, and the energy difference between the predicted and crystal conformation is expected to be small. Such subtle energetic difference is probably not captured in our

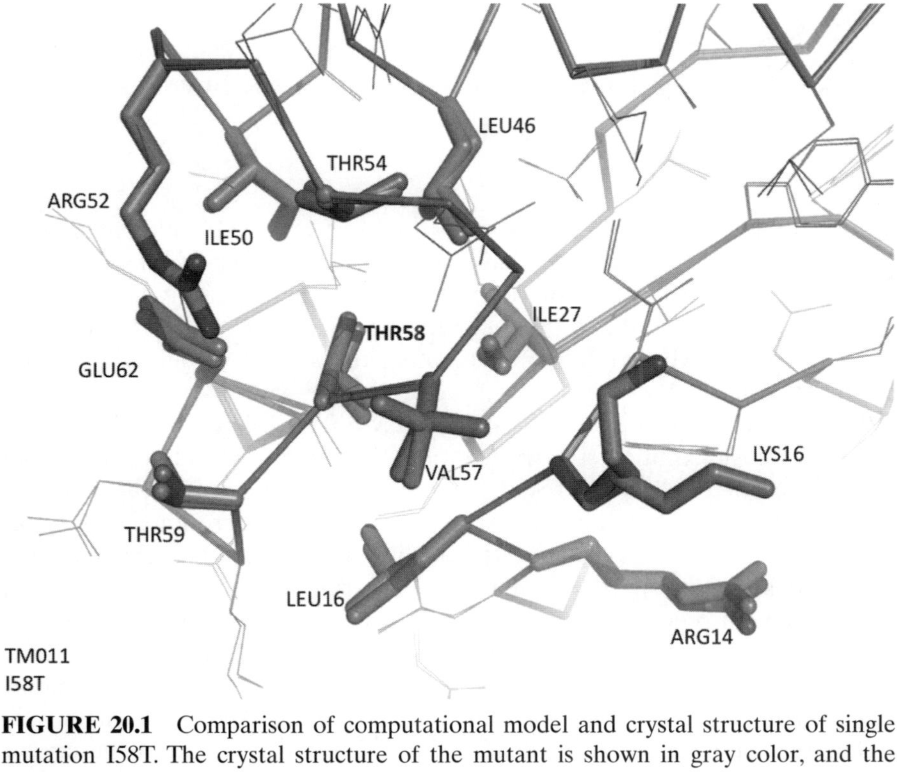

FIGURE 20.1 Comparison of computational model and crystal structure of single mutation I58T. The crystal structure of the mutant is shown in gray color, and the model in magenta. (See color insert.)

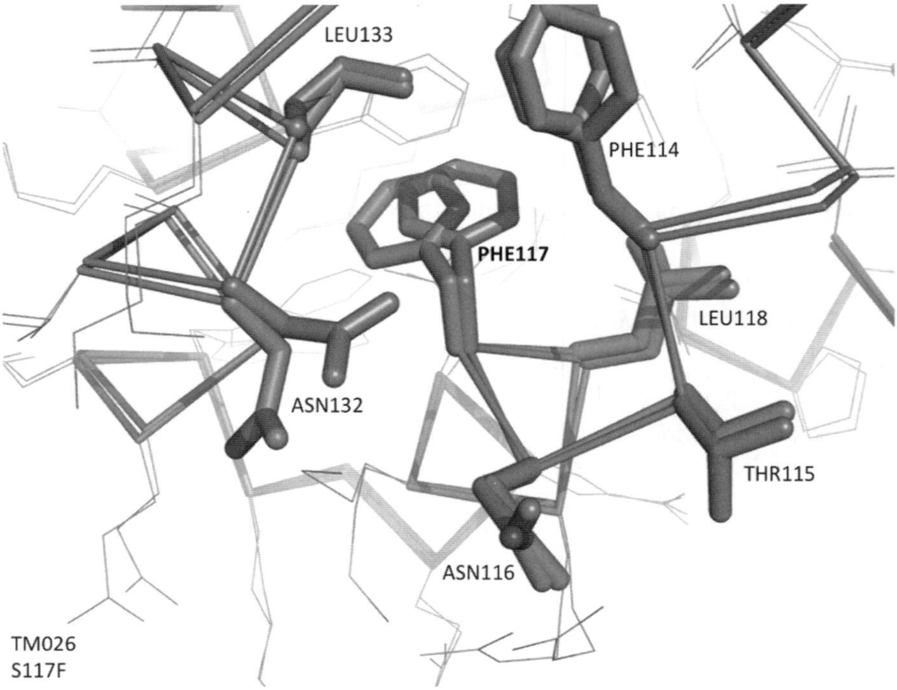

FIGURE 20.2 Comparison of computational model and crystal structure of single mutation S117F. (See color insert.)

force field. In fact, our prediction tends to move the Lys16 side chain away from other atoms, thereby minimizing the solvation energy of the amino group. The incorrect prediction of the Lys16 side chain possibly leads also to the error of assigning the Val57 rotamer, since the Val57 will be too close to the misplaced Lys16 side chain if adapting the same rotamer as from the crystal structure.

For the multiple-site mutant I78V V87M M120Y L133F V149I T152V (Fig. 20.2), the six substitution sites are distributed through one side of the whole protein. In our protocol, 36 residues are sampled within the native rotamer conformations, and 76 residues are sampled for all possible rotamer conformations. Therefore, it represents a considerably difficult sampling challenge for side chain sequence space sampling. By comparing with the crystal structure, we find that our prediction agrees well with experiments. The predicted rotamers are correct for mutation residues Tyr120, Phe133, Val152, and Val78. Deviation is observed for χ_2 dihedral angle of Ile149, although the χ_1 dihedral angle is still correctly recapitulated. For Met87, the predicted side chain conformation points to solvent instead of being buried, as found in the crystal structure. We note that the rotamers of neighboring residues of Met87 are all correctly predicted in the model. However, in the crystal structure, the helix

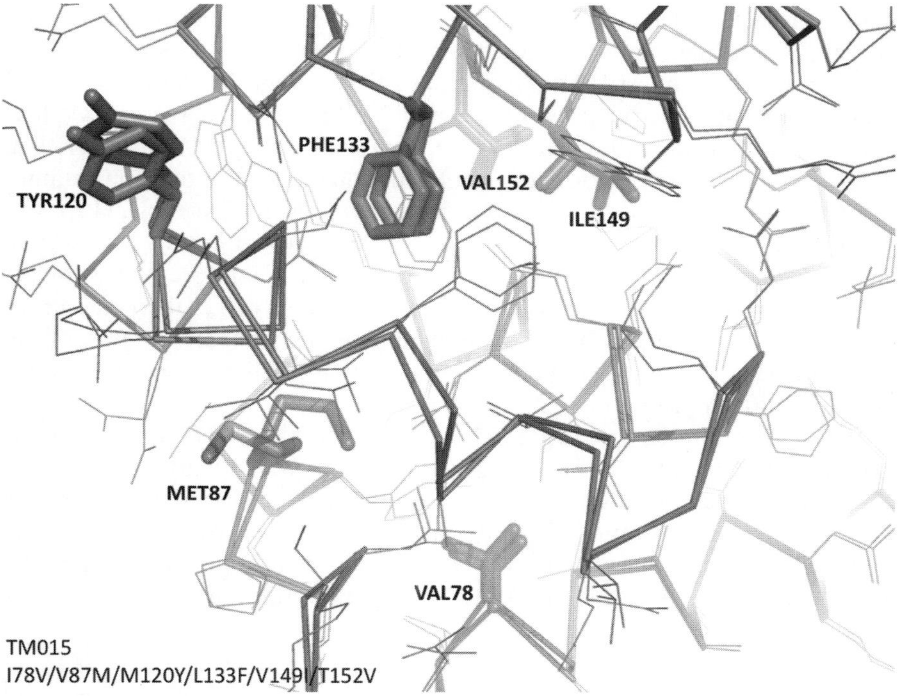

FIGURE 20.3 Comparison of computational model and crystal structure of multiple mutation I78V V87M M120Y L133F V149I T152V. (See color insert.)

above Met87 bends outward slightly to allow Met87 to make contact with Val118. This backbone movement is not modeled in our protocol, and the Met87 side chain will clash with Val118 if adapting the same rotamer as the crystal structure. This example emphasizes the need for accurate modeling of backbone conformation in order to achieve high fidelity in side chain packing.

Another interesting example is mutation S117F (Fig. 20.3), where the small and buried side chain of serine is replaced with a much larger side chain of phenylalanine. Analyzing the crystal structure reveals that there exists a small cavity in the wild-type structure, and hence, the protein can accommodate the mutated phenyl head without undergoing a large backbone adjustment. Instead, the structural rearrangement is mainly realized by flipping the side chain of Leu133 away from the cavity so that the phenyl ring of Phe117 can fit in. Our protocol correctly predicts this structural response for the Leu133 side chain and places the phenyl ring correctly in the cavity. However, the predicted side chain of a neighboring residue, Asn132, has the χ_1 angle shift by about 60 degrees from the conformation in the crystal structure. We examined the predicted structure for all the 20 MC runs (See Methods) and found that half of the 20 predicted structures actually have the correct χ_1 conformation and the other half predictions have the incorrect χ_1 angles as the final model. (Recall that the final model is the structure with the lowest Medusa energy.) We further separate the 20 structures into two groups according to the rotamer state of Asn132 and compare the average Medusa energies of these two groups of structures. Surprisingly, we find that the average Medusa energy of the group with correct χ_1 conformation is 1.2 kcal/mol lower than that of the group with incorrect χ_1 conformation, indicating that the rotamer of Asn132 in the crystal structure is indeed energetically favorable. However, the final model with the wrong χ_1 is still chosen because it has the lowest total energy even with one slightly unfavorable rotamer. Therefore, the incorrect prediction of Asn 132 is a result of insufficient sampling rather than inaccuracy of the force field per se. This example highlights the stochastic nature of the MC method in searching for global minima so that prediction errors can arise due to insufficient sampling.

We next examine a structure prediction that has relatively large deviations from the experimental structure. Figure 20.4 shows the structure comparison of the model and crystal structure of the mutation A129L. In this case, similar to S117F, a small and buried alanine residue is replaced by a larger leucine residue. We correctly predicted the conformation of several neighboring residues, for example, Arg125 and Met120. However, there are significant dis-

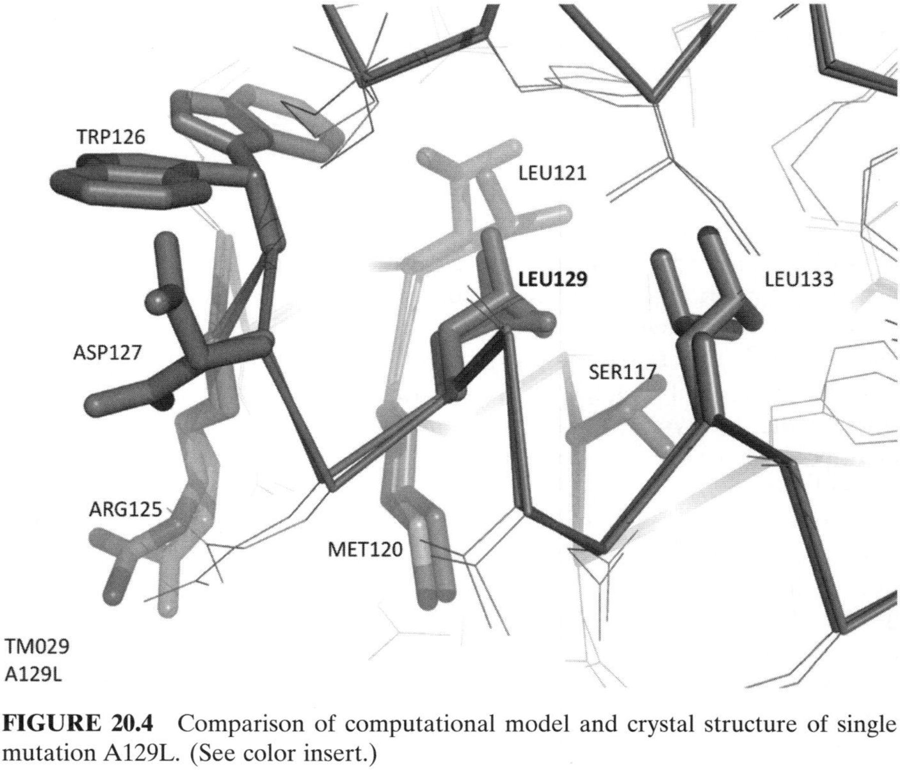

FIGURE 20.4 Comparison of computational model and crystal structure of single mutation A129L. (See color insert.)

crepancies between the predicted model and the crystal structure for other residues. In the model, residues Trp126 and Asp127 have their side chains pointing to the opposite directions; Leu121, Leu129, and Leu133 have opposite χ_2 dihedral angles, although their χ_1 dihedral angles are only shifted by 88, 11, and 87 degrees, respectively, and Ser117 also has a χ_1 dihedral angle shifted by about 101 degrees. We checked all the 20 predicted structures and found that there is one prediction that has the correct side chain rotamers for Leu129, Leu133, and Ser117, although the side chains of Met120, Leu121, Trp126, and Asp127 are still wrong. This particular prediction is ranked seventh among the 20 predictions in terms of Medusa energy, and its energy is about 2.4 kcal/mol higher than the lowest energy model. We postulate that the incorrect prediction is due to the small movement of the mutant backbone conformation so that the position and orientation of the backbone of Leu121 and Leu129 can allow the rotamers observed in the crystal structure. On the other hand, adapting such rotamers on the wild-type backbone for these residues would cause atomic clashes. To avoid the clash, all the predicted structures are forced to take alternative rotamers that stay slightly away from Leu129. Unfortunately, the new rotamer for Leu121 stays right in the way of the Trp126. Instead of forming tight packing with Leu121 and Leu129 as observed in the crystal structure, Trp126 in the predicted structure has to point toward the solvent. Consequently, the misplaced Trp126 side chain pushes Asp127 further away from the correct rotamer state This "chain reaction" explains the rather large range of discrepancy of side chain conformations between the predicted and crystal structure for mutant A129L.

20.3.2. Insertion/Deletion in Proteins

For insertion mutation N40-[A], the additional alanine is inserted right after the asparagine located at the N-terminus of a short helix. By comparing the crystal structure of the mutant with the wild type, we find that the N-terminus of the helix unwinds slightly to accommodate this additional alanine (Fig. 20.5). Our predicted model captures these structural responses correctly. In the model, only the backbone conformation of two residues, Asn40 and Ala40A (the inserted alanine after Asn40), undergoes significant movement from that in the wild type, and other residues are essentially not altered, agreeing with the experiments. However, the predicted backbone conformation still deviates slightly from the crystal structure, which may be due to insufficient backbone relaxation during DMD simulations. As a result of the imperfect backbone, the predicted side chain conformation also slightly deviates from the crystal structure. The RMSD for the Cα atom of Ala40A and Asn40 is only 1.68 and 1.63 Å from the crystal structure, respectively. For other residues, our predicted model correctly recapitulates the backbone as well as the side chain conformation.

Mutation E108-[A] is similar to N40-[A], where an alanine is inserted at the N-terminus of a helix. In this case, the crystal structure shows that the

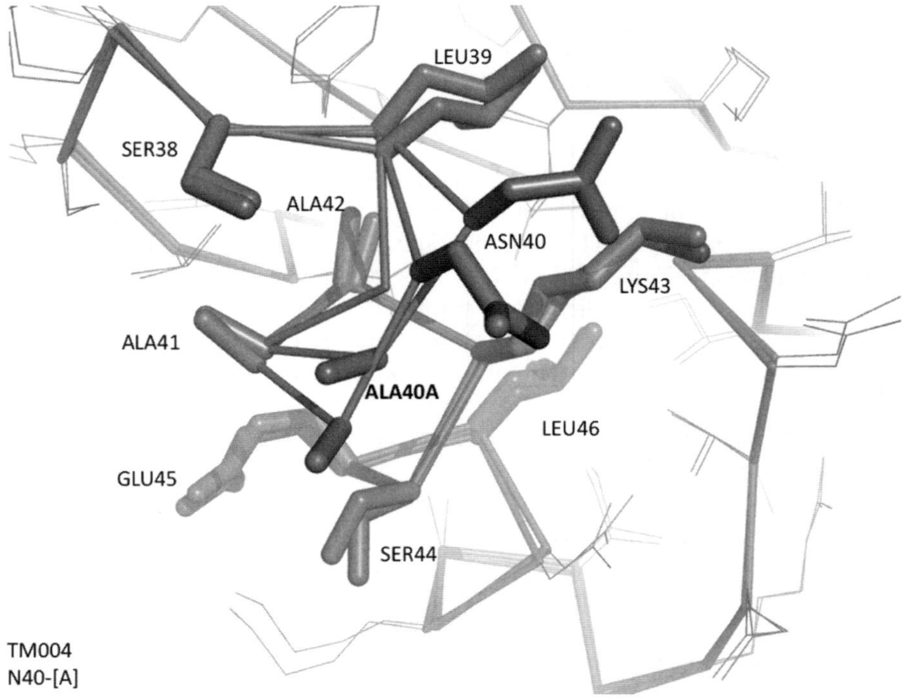

FIGURE 20.5 Comparison of computational model and crystal structure of single insertion mutation N40-[A]. The backbone of the wild-type protein is shown in green. (See color insert.)

inserted alanine adapts the position of Glu108 at the end of the helix and leaves the helix unaltered (Fig. 20.6). To accommodate this Glu108, additional backbone adjustment is created in the helix-turn region mainly by changing the backbone conformation of Gly107. Our prediction successfully recapitulates this backbone adjustment, and the Cα atoms of Gly107 and Glu108 residues only slightly deviate from the crystal structure by 0.82 and 1.07 Å, respectively. Although the backbone deviation is small, the rotamer states of Glu108 and the neighboring Thr109 are still not correctly predicted. This observation is consistent with what we have demonstrated in the earlier substitution mutation examples, that a small change in backbone conformation is sufficient to affect the side chain prediction accuracy.

For mutation T115-[A], an additional alanine is inserted again at the N-terminus of a helix (Fig. 20.7). However, different from N40-[A] and E108-[A], the backbone conformation for this mutant alters in a larger range, spanning from Thr109 to Thr115. The inserted alanine is placed in between a backbone kink created between site Thr115 and Asn116. The predicted model does not place the inserted alanine correctly in between Thr115 and Asn116. Instead, the alanine is positioned near the Thr115 sites. Subsequently, Thr115 is incorrectly aligned with Phe114 site, and a kink is created between residue

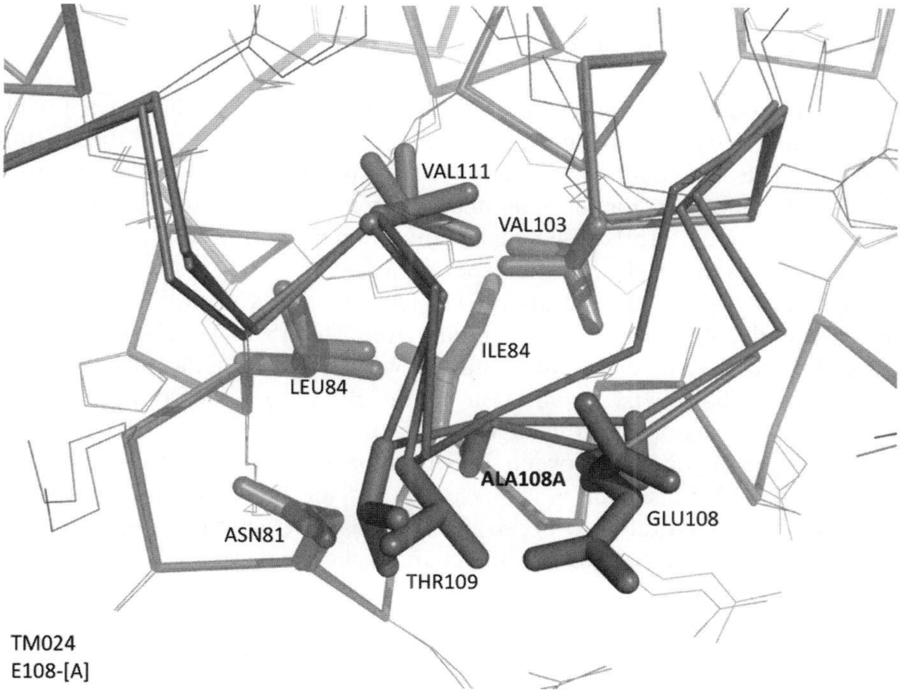

FIGURE 20.6 Comparison of computational model and crystal structure of mutation E108-[A]. (See color insert.)

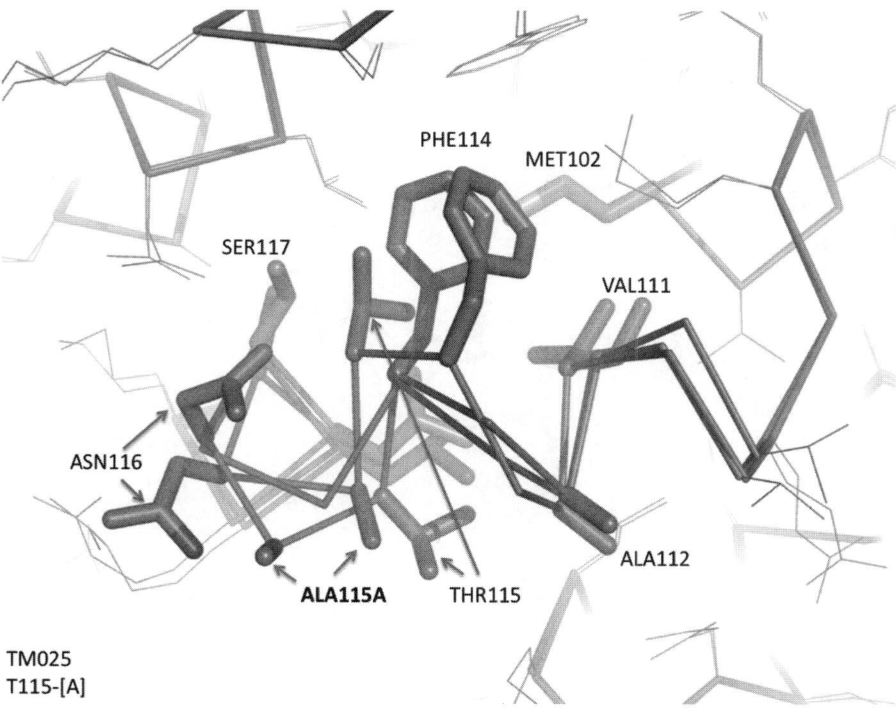

FIGURE 20.7 Comparison of computational model and crystal structure of mutation E115-[A]. (See color insert.)

site of 113 and 114 to hold the shifted Phe114. For other residues, our model predicts minor backbone conformation change, in agreement with the crystal structure. The side chain conformations are also correctly predicted for those residues. In summary, for this example, our prediction features wrong trends for the backbone perturbation near the mutation sites. As a result, the prediction does not agree well with the experiments.

In the last example, the mutation R119-[A] introduces an additional alanine in the middle of a short helix spanning from Thr115 to Lys124 (Fig. 20.8). From the crystal structure of this mutation, we observe a significant rearrangement of the whole helix. In addition, the neighboring helix from Arg125 to Arg137 is also significantly changed. In fact, the average Cα RMSD from residue range of 106–141 between the wild-type and mutant protein is 2.64 Å. We note that this insertion is in the middle of the helix, and it is generally energetically unfavorable to break the secondary structure. However, if we preserve the secondary structure, the charged chemical groups in Arg119 will be pointing inside the protein, which is also unfavorable in terms of energy. To prevent breaking the secondary structure or burying Arg119, both of the

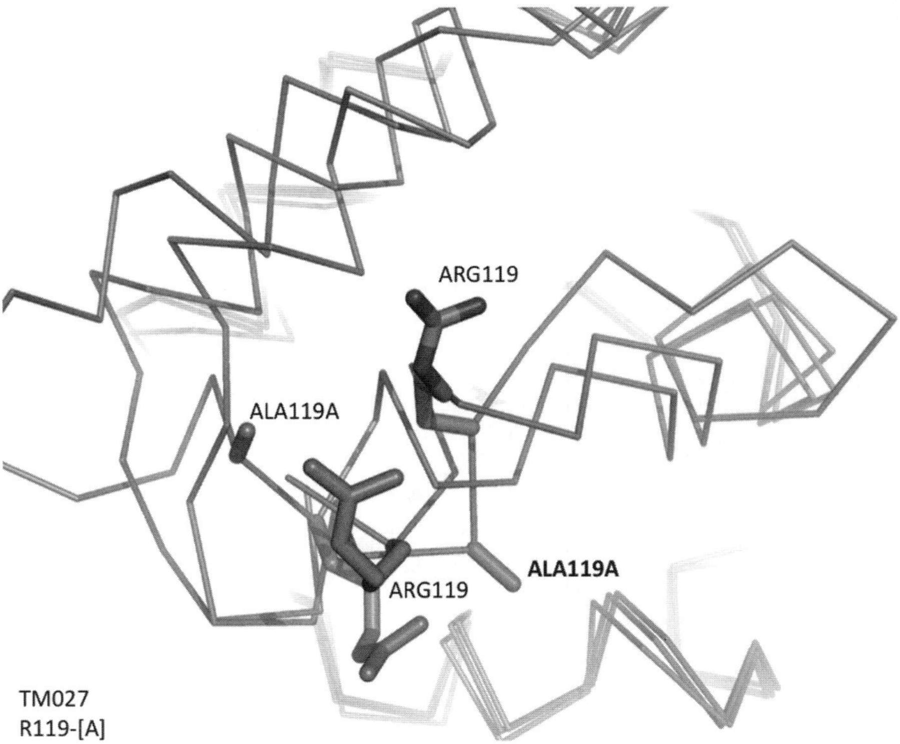

FIGURE 20.8 Comparison of computational model and crystal structure of single insertion mutation R119-[A]. (See color insert.)

helices are reoriented in the crystal structure, resulting in large conformational changes compared with the wild-type structure. Our predicted model correctly put the charged Arg119 head group toward the solvent. However, the predicted conformations of the two helices are not in agreement with the crystal structure. In fact, the short helix from Thr115 to Lys124 becomes slightly unfolded after DMD simulation. We find that the sampling requirement to capture such drastic structure arrangement is beyond our current prediction protocol.

20.3.3. Stability Prediction

We apply the Eris protocol on substitution mutations to predict the stability change and compare with the experimental measurements. The predicted $\Delta\Delta G$ values in general agree with experimental data, except for some outliers. These outliers all have exceptionally large $\Delta\Delta G$ estimations in the range of 10–20 kcal/mol (Fig. 20.9). In contrast, the experimental $\Delta\Delta G$ values of these mutations lay in the reasonable range of −1 to 5 kcal/mol. To find the reason of the overestimation of $\Delta\Delta G$ for these outliers, we examined the different energy terms (Equation 20.1) contributing to the total $\Delta\Delta G$ prediction and

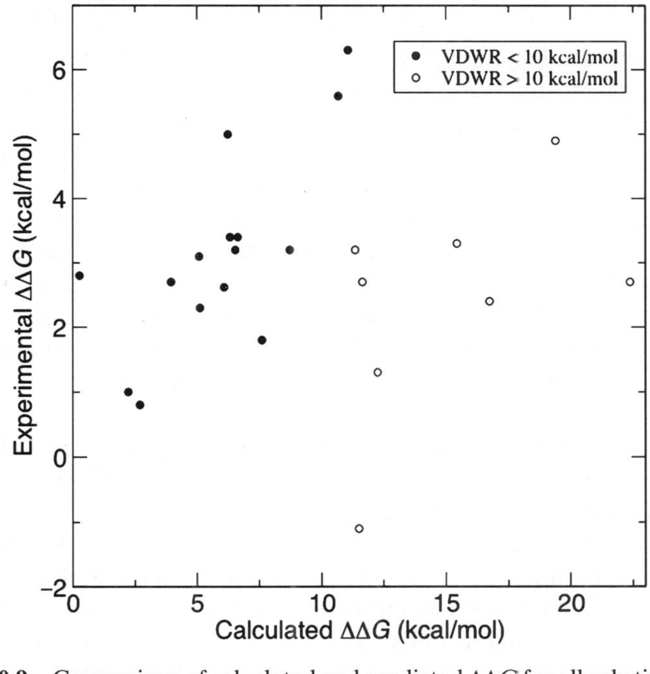

FIGURE 20.9 Comparison of calculated and predicted $\Delta\Delta G$ for all substitution mutations. The predicted $\Delta\Delta G$ values in general correlate with the experiments except for several outliers that have exceptionally large van der Waals repulsion terms.

found that these outliers feature exceptionally high van der Waals repulsion (VDWR) energies. After removing those mutations with high VDWR terms, the $\Delta\Delta G$ predictions correlate well with the experiments. For the mutations with VDWR contribution less than 10 kcal/mol, the correlation coefficient is 0.72 between the prediction and experimental measurements.

This observation is likely due to the fact that VDWR terms are very sensitive to the interatomic distance so that a small variation in the conformation can result in rather large energy differences. We have demonstrated in an early structure analysis that for some mutations, insufficient sampling can result in incorrect side chain prediction. Enhanced sampling of backbone will help to remove those clashes and improve structure prediction. In principle, those VDWR terms will be reduced if enough sampling is performed. A similar phenomenon has been observed for protein–ligand interactions [48,49]. In a previous study [48], we demonstrated that removing the VDWR term also helps to remedy the oversensitivity of the VDWR energy. Similarly, in the case of T4 lysozyme mutations, we also found that removing the VDWR terms slightly improves the $\Delta\Delta G$ prediction accuracy without performing further structure refinement.

20.4. CONCLUSION AND FUTURE DIRECTIONS

We model a series of T4 lysozyme mutations using Medusa and DMD methods. Analysis of the results demonstrates that most of the predictions are in agreement with experimental results. However, some prediction errors still exist in a few mutants, especially in cases of insertion and deletion of amino acids. Comparison of the models with the corresponding crystal structures suggests that the prediction inaccuracy is mainly due to backbone conformational change induced by mutations. Accurate modeling of the backbone movements still represents significant challenges for current mutation modeling methods.

Although the method is tested only on relatively small proteins, it is in principle applicable to larger proteins. The computational sampling efficiency is not a limiting factor since the side chain and backbone sampling is only restricted to residues near the mutation sites while the rest of protein is constrained.

Our method is limited to mutations where the conformational changes are localized near the mutation sites and do not involve global backbone adjustments, often resulting from multiple amino acid changes. This limitation is not due to the difficulty in sampling the expanded side chain conformational space with an increased number of mutations. In fact, even for multiple-site mutations, our method still features consistent performance as long as the backbone is not changed significantly. Based on the results from this study and earlier experience, we find that the major obstacle for reliable modeling of a large number of mutations is related to an extensive protein backbone change. As the number of mutations increases, the global backbone conformational change is more likely to happen. Although the altered backbone is still structurally similar to the wild-type protein with typical RMSD less than 1 Å, the exact conformations are difficult to predict by the current methods, which consequently leads to inaccurate side chain repacking.

It is clear that there is an urgent need for the future development of efficient algorithms for the accurate prediction of protein backbone conformations. Several interesting developments have emerged, where backbone template [50] or MD simulation [51] are used to further refine models built from homology modeling or from NMR experiments. For the particular goal of modeling protein mutations, it may be more efficient to perform sampling that couples side chain rotamer change and the local backbone rearrangements, such as the backrub motion [52].

ACKNOWLEDGMENTS

The authors are grateful to the CASPM organizers for conducting the critical assessment experiments, which are beneficial for the protein modeling community. The authors also want to thank Bryan Der for proofreading the manuscript. The Eris $\Delta\Delta G$ estimation method is available online at http://eris.dokhlab.org. The Eris stand-alone software package is distributed by Molecules in Actions, LLC.

REFERENCES

1. N. Hall. Advanced sequencing technologies and their wider impact in microbiology. *J Exp Biol*, 210(9):1518–1525, 2007.
2. H. Davies, G.R. Bignell, C. Cox, P. Stephens, S. Edkins, S. Clegg, J. Teague, H. Woffendin, M.J. Garnett, W. Bottomley, N. Davis, N. Dicks, R. Ewing, Y. Floyd, K. Gray, S. Hall, R. Hawes, J. Hughes, V. Kosmidou, A. Menzies, C. Mould, A. Parker, C. Stevens, S. Watt, S. Hooper, R. Wilson, H. Jayatilake, B.A. Gusterson, C. Cooper, J. Shipley, D. Hargrave, K. Pritchard-Jones, N. Maitland, G. Chenevix-Trench, G.J. Riggins, D.D. Bigner, G. Palmieri, A. Cossu, A. Flanagan, A. Nicholson, J.W.C. Ho, S.Y. Leung, S.T. Yuen, B.L. Weber, H.F. Siegler, T.L. Darrow, H. Paterson, R. Marais, C.J. Marshall, R. Wooster, M.R. Stratton, and P.A. Futreal. Mutations of the BRAF gene in human cancer. *Nature*, 417(6892):949–954, 2002.
3. I.D. Kuntz. Structure-based strategies for drug design and discovery. *Science*, 257(5073):1078–1082, 1992.
4. A.A. Bogan and K.S. Thorn. Anatomy of hot spots in protein interfaces. *J Mol Biol*, 280(1):1–9, 1998.
5. W.E. Stites. Protein-protein interactions: Interface structure, binding thermodynamics, and mutational analysis. *Chem Rev*, 97(5):1233–1250, 1997.
6. A.L. Hopkins and C.R. Groom. The druggable genome. *Nat Rev Drug Discov*, 1(9):727–730, 2002.

7. S.R. George, B.F. O'Dowd, and S.R. Lee. G-protein-coupled receptor oligomerization and its potential for drug discovery. *Nat Rev Drug Discov*, 1(10):808–820, 2002.
8. D. Baker and A. Sali. Protein structure prediction and structural genomics. *Science*, 294(5540):93–96, 2001.
9. K.A. Dill, S.B. Ozkan, M.S. Shell, and T.R. Weikl. The protein folding problem. *Ann Rev Biophys*, 37:289–316, 2008.
10. C. Hardin, T.V. Pogorelov, and Z. Luthey-Schulten. Ab initio protein structure prediction. *Curr Opin Struct Biol*, 12(2):176–181, 2002.
11. A. Sali and T.L. Blundell. Comparative protein modeling by satisfaction of spatial restraints. *J Mol Biol*, 234(3):779–815, 1993.
12. N. Guex and M.C. Peitsch. SWISS-MODEL and the Swiss-PdbViewer: An environment for comparative protein modeling. *Electrophoresis*, 18(15):2714–2723, 1997.
13. I. Halperin, B.Y. Ma, H. Wolfson, and R. Nussinov. Principles of docking: An overview of search algorithms and a guide to scoring functions. *Proteins*, 47(4):409–443, 2002.
14. A. Wlodawer and J.W. Erickson. Structure-based inhibitors of HIV-1 protease. *Annu Rev Biochem*, 62:543–585, 1993.
15. B.I. Dahiyat and S.L. Mayo. De novo protein design: Fully automated sequence selection. *Science*, 278(5335):82–87, 1997.
16. J.R. Desjarlais and T.M. Handel. De-novo design of the hydrophobic cores of proteins. *Protein Sci*, 4(10):2006–2018, 1995.
17. L. Jiang, E.A. Althoff, F.R. Clemente, L. Doyle, D. Rothlisberger, A. Zanghellini, J.L. Gallaher, J.L. Betker, F. Tanaka, C.F. Barbas, D. Hilvert, K.N. Houk, B.L. Stoddard, and D. Baker. De novo computational design of retro-aldol enzymes. *Science*, 319(5868):1387–1391, 2008.
18. B. Kuhlman, G. Dantas, G.C. Ireton, G. Varani, B.L. Stoddard, and D. Baker. Design of a novel globular protein fold with atomic-level accuracy. *Science*, 302(5649):1364–1368, 2003.
19. D. Rothlisberger, O. Khersonsky, A.M. Wollacott, L. Jiang, J. DeChancie, J. Betker, J.L. Gallaher, E.A. Althoff, A. Zanghellini, O. Dym, S. Albeck, K.N. Houk, D.S. Tawfik, and D. Baker. Kemp elimination catalysts by computational enzyme design. *Nature*, 453(7192):190–194, 2008.
20. F. Ding and N.V. Dokholyan. Emergence of protein fold families through rational design. *PLoS Comput Biol*, 2(7):725–733, 2006.
21. R.L. Dunbrack and F.E. Cohen. Bayesian statistical analysis of protein side-chain rotamer preferences. *Protein Sci*, 6(8):1661–1681, 1997.
22. N.V. Dokholyan, S.V. Buldyrev, H.E. Stanley, and E.I. Shakhnovich. Discrete molecular dynamics studies of the folding of a protein-like model. *Fold Des*, 3(6):577–587, 1998.
23. Y.Q. Zhou, C.K. Hall, and M. Karplus. First-order disorder-to-order transition in an isolated homopolymer model. *Phys Rev Lett*, 77(13):2822–2825, 1996.
24. Y.Q. Zhou and M. Karplus. Interpreting the folding kinetics of helical proteins. *Nature*, 401(6751):400–403, 1999.

25. P.A. Bash, U.C. Singh, R. Langridge, and P.A. Kollman. Free-energy calculations by computer-simulation. *Science*, 236(4801):564–568, 1987.

26. P. Kollman. Free-energy calculations—Applications to chemical and biochemical phenomena. *Chem Rev*, 93(7):2395–2417, 1993.

27. A.J. Bordner and R.A. Abagyan. Large-scale prediction of protein geometry and stability changes for arbitrary single point mutations. *Proteins*, 57(2):400–413, 2004.

28. E. Capriotti, P. Fariselli, R. Calabrese, and R. Casadio. Predicting protein stability changes from sequences using support vector machines. *Bioinformatics*, 21:ii54–58, 2005.

29. D. Gilis and M. Rooman. Predicting protein stability changes upon mutation using database-derived potentials: Solvent accessibility determines the importance of local versus non-local interactions along the sequence. *J Mol Biol*, 272(2):276–290, 1997.

30. R. Guerois, J.E. Nielsen, and L. Serrano. Predicting changes in the stability of proteins and protein complexes: A study of more than 1000 mutations. *J Mol Biol*, 320(2):369–387, 2002.

31. K. Saraboji, M.M. Gromiha, and M.N. Ponnuswamy. Average assignment method for predicting the stability of protein mutants. *Biopolymers*, 82(1):80–92, 2006.

32. S.Y. Yin, F. Ding, and N.V. Dokholyan. Eris: An automated estimator of protein stability. *Nat Methods*, 4(6):466–467, 2007.

33. M.M. Gromiha, J. An, H. Kono, M. Oobatake, H. Uedaira, and A. Sarai. ProTherm: Thermodynamic database for proteins and mutants. *Nucleic Acids Res*, 27(1):286–288, 1999.

34. S. Yin, F. Ding, and N.V. Dokholyan. Modeling backbone flexibility improves protein stability estimation. *Structure*, 15(12):1567–1576, 2007.

35. A.A. Canutescu, A.A. Shelenkov, and R.L. Dunbrack. A graph-theory algorithm for rapid protein side-chain prediction. *Protein Sci*, 12(9):2001–2014, 2003.

36. A.P. Heath, L.E. Kavraki, and C. Clementi. From coarse-grain to all-atom: Toward multiscale analysis of protein landscapes. *Proteins*, 68(3):646–661, 2007.

37. J.B. Xu. Rapid protein side-chain packing via tree decomposition. In *Research in Computational Molecular Biology*, S. Istrail, P. Pevzner, and M. Waterman (Eds.), pp. 423–439. Berlin: Springer-Verlag, 2005.

38. F. Ding and N.V. Dokholyan. Dynamical roles of metal ions and the disulfide bond in Cu, Zn superoxide dismutase folding and aggregation. *Proc Natl Acad Sci U S A*, 105(50):19696–19701, 2008.

39. T. Lazaridis and M. Karplus. Effective energy function for proteins in solution. *Proteins*, 35(2):133–152, 1999.

40. B.R. Brooks, R.E. Bruccoleri, B.D. Olafson, D.J. States, S. Swaminathan, and M. Karplus. CHARMM—A program for macromolecular energy, minimization, and dynamics calculations. *J Comput Chem*, 4(2):187–217, 1983.

41. S.D. Khare, F. Ding, K.N. Gwanmesia, and N.V. Dokholyan. Molecular origin of polyglutamine aggregation in neurodegenerative diseases. *PLoS Comput Biol*, 1(3):230–235, 2005.

42. S. Peng, F. Ding, B. Urbanc, S.V. Buldyrev, L. Cruz, H.E. Stanley, and N.V. Dokholyan. Discrete molecular dynamics simulations of peptide aggregation. *Phys Rev E*, 69(4):014908, 2004.

43. T. Hegedus, A.W.R. Serohijos, N.V. Dokholyan, L.H. He, and J.R. Riordan. Computational studies reveal phosphorylation-dependent changes in the unstructured R domain of CFTR. *J Mol Biol*, 378(5):1052–1063, 2008.
44. T. Kortemme, A.V. Morozov, and D. Baker. An orientation-dependent hydrogen bonding potential improves prediction of specificity and structure for proteins and protein-protein complexes. *J Mol Biol*, 326(4):1239–1259, 2003.
45. N. Metropolis, A.W. Rosenbluth, M.N. Rosenbluth, A.H. Teller, and E. Teller. Equation of state calculations by fast computing machines. *J Chem Phys*, 21(6):1087, 1953.
46. V.G.H. Eijsink, A. Bjork, S. Gaseidnes, R. Sirevag, B. Synstad, B. van den Burg, and G. Vriend. Rational engineering of enzyme stability. *Journal of Biotechnology*, 113:105–120, 2004.
47. M. Thunnissen, P.A. Franken, G.H. Dehaas, J. Drenth, K.H. Kalk, H.M. Verheij, and B.W. Dijkstra. Crystal-structure of a porcine pancreatic phospholipase-A2 mutant—A large conformational change caused by the F63V point mutation. *J Mol Biol*, 232(3):839–855, 1993.
48. S. Yin, L. Biedermannova, J. Vondrasek, and N.V. Dokholyan. MedusaScore: An accurate force field-based scoring function for virtual drug screening. *J Chem Inf Model*, 48(8):1656–1662, 2008.
49. J. Meiler and D. Baker. ROSETTALIGAND: Protein-small molecule docking with full side-chain flexibility. *Proteins*, 65(3):538–548, 2006.
50. B. Qian, S. Raman, R. Das, P. Bradley, A.J. McCoy, R.J. Read, and D. Baker. High-resolution structure prediction and the crystallographic phase problem. *Nature*, 450(7167):259–257, 2007.
51. J.H. Chen and C.L. Brooks. Can molecular dynamics simulations provide high-resolution refinement of protein structure? *Proteins*, 67(4), 922–930, 2007.
52. I.W. Davis, W.B. Arendall, D.C. Richardson, and J.S. Richardson. The backrub motion: How protein backbone shrugs when a sidechain dances. *Structure*, 14(2):265–274, 2006.

INDEX

Introduction to Protein Structure Prediction: Methods and Algorithms,
Edited by Huzefa Rangwala and George Karypis